Modern Switching Theory
and
Digital Design

Modern Switching Theory

and

Digital Design

Samuel C. Lee

Professor of Electrical Engineering and Computing Sciences
University of Oklahoma

Prentice-Hall, Inc., Englewood Cliffs, New Jersey   07632

Library of Congress Cataloging in Publication Data

LEE, SAMUEL C. (date)
  Modern switching theory and digital design.

  Includes index.
  1. Switching theory. 2. Digital electronics.
I. Title.
[QA268.5.L43]   621.3815′37   77-22096
ISBN 0-13-598680-X

10 9 8 7 6 5 4 3 2 1

Printed in the United States of America

PRENTICE-HALL INTERNATIONAL, INC., *London*
PRENTICE-HALL OF AUSTRALIA PTY. LIMITED, *Sydney*
PRENTICE-HALL OF CANADA, LTD., *Toronto*
PRENTICE-HALL OF INDIA PRIVATE LIMITED, *New Delhi*
PRENTICE-HALL OF JAPAN, INC., *Tokyo*
PRENTICE-HALL OF SOUTHEAST ASIA PTE. LTD., *Singapore*
WHITEHALL BOOKS LIMITED, *Wellington, New Zealand*

To my wife Elisa

Contents

Preface   *x*

Chapter 1   **Boolean Algebra and Boolean Function**   *1*

    1.1   *Sets, Ordered Sets, and Algebras*   1
    1.2   *Lattice and Its Basic Properties*   11
    1.3   *Boolean Algebra and Its Properties*   24
    1.4   *Boolean Function and Its Canonical Forms*   35

Chapter 2   **Vector Switching Algebra and Vector Switching Function**   *48*

    2.1   *Switching Algebra and Switching Function*   48
    2.2   *The Generalized Complement and the Rotation Operator*   56
    2.3   *Properties of Vector Switching Algebra*
        and Vector Switching Function   61
    2.4   *Canonical Forms*   62

Chapter 3   **Boolean Differential Calculus**   *73*

    3.1   *The Partial Derivative*   74
    3.2   *Series Expansions of a Boolean Function*   79
    3.3   *The Total Differential and the Total Variation*   89
    3.4   *Graphical Method and Computer Algorithm*
        for Computing Boolean Derivatives and Differentials   97

Chapter 4   **Special Switching Functions**   *104*

    4.1   *Monotonic Functions*   104
    4.2   *Threshold Functions*   115
    4.3   *Symmetric Functions*   123
    4.4   *Functionally Complete Functions*   137

Chapter 5 **Multivalued Logic**   *147*

    5.1 m-Valued Logic Functions 147
    5.2 Fuzzy Logic and m-Class Logic 157
    5.3 Analysis of m-Class Logic Functions 162
    5.4 Realization of m-Class Logic Functions 166

Chapter 6 **Fault Detection in Combinational Circuits**   *182*

    6.1 Test Generation by Boolean Derivative (Difference) 182
    6.2 Derivation of Minimal Fault-Detection Experiments
          for Monotonic Two-Level Circuits 184
    6.3 Analysis of Multiple Faults in Combinational Circuits 192
    6.4 The Literal Proposition Method 202
    6.5 D-Calculus 210
    6.6 The D-Algorithm 222
    6.7 DALG-II and TEST-DETECT 226

Chapter 7 **Sequential Machines**   *236*

    7.1 Basic Models 237
    7.2 The S. P. Partition and the Lattice L_M 241
    7.3 State Minimization Using the O.C.S.P. Partition 251
    7.4 State Assignment Using S. P. Partitions 262
    7.5 Serial Decomposition of Sequential Machine 283
    7.6 Parallel Decomposition of Sequential Machine 288

Chapter 8 **Regular Expressions**   *298*

    8.1 Regular Sets and Regular Expressions 299
    8.2 Nondeterministic Sequential Machines:
          Equivalence and Minimization 313
    8.3 Construction of Transition Diagrams
          of $R_1 R_2$, $R_1 \cup R_2$, $R_1 \cap R_2$, $R_1 \triangle R_2$, etc. 328

Chapter 9 **Sequential Machine Realization**   *339*

    9.1 Delay Flip-Flop Sequential Circuit Realization
          of Binary Machines 340
    9.2 Analysis of Clocked Flip-Flop Circuits 344
    9.3 A General Synthesis Procedure 353
    9.4 Sequential Machine Realization
          Using Pulse-Mode Sequential Circuits 360
    9.5 Sequential Machine Realization
          Using Fundamental-Mode (Pulseless) Sequential Circuits 365

Chapter 10 **Sequential Machine Fault Detection**   *386*

    10.1 Homing Sequence, Synchronizing Sequence,
          and Experiment Initialization 386
    10.2 Distinguishing Sequence and State Identification 391
    10.3 Design of Complete Fault-Detection Experiment 397

Chapter 11 **Digital Design Using Integrated Circuits**   *405*

11.1   *Digital Integrated Circuits*   406
11.2   *Digital Design Using TTL/SSI Integrated Circuits*   415
11.3   *Digital Design Using TTL/MSI Integrated Circuits*   424
11.4   *Digital Design Using MOS/LSI Integrated Circuits*   445

Chapter 12 **Digital Design Using Microprocessors Via Simulation**   *453*

12.1   *A Typical 8080 Microcomputer System*   453
12.2   *8080 Instruction Set and Assembly Language Programming*   463
12.3   *Microcomputer Simulation of Digital Design*   480

Index   *495*

Preface

Switching theory is recognized as the foundation for computer science and digital design. It is, therefore, no surprise that almost every major university in the nation offers a course, or a sequence of two courses, in the departments of computer science, electrical engineering, and/or applied mathematics. This course is usually a requirement for all graduate students who are pursuing an advanced degree in computer science, digital systems, or combinatory mathematics. This book is a compilation of classnotes of a course on switching theory taught by the author for the past eight years. Senior and graduate students taking this course were familiar with the following subjects:

1. Truth tables.
2. Logic gates.
3. Karnaugh maps.
4. Switching function minimization methods.
5. Flip-flops.
6. Basic digital devices, such as, counters, registers, basic binary adders and subtractors, etc.

This was, in fact, the only background required for this course. In selecting a textbook for this course, the author was a little surprised to find no textbooks on switching theory at this level which were published since 1970. However, there has been a tremendous amount of research in the area since 1970. It seemed apparent to this author that a textbook including these recent research results was needed. This book was born out of this need.

This book contains twelve chapters which comprise four parts:

1. Boolean Algebra and Boolean Differential Calculus (Chapters 1, 2, and 3).
2. Combinational Logic (Chapters 4, 5, and 6).
3. Sequential Logic (Chapters 7, 8, 9, and 10).
4. Digital Design (Chapters 11, 12).

A brief description by chapter is as follows.

Chapter 1 describes Boolean algebra and its properties. It provides the mathematical foundation of switching theory and digital design. A generalization of two-valued Boolean algebra or switching algebra is introduced in Chapter 2. In this generalized algebra, every element is presented by a binary vector and two new operations; the rotation operation and the generalized complement operation are defined. It is shown that DeMorgan's theorem, Shannon's theorem, and the expansion theorem are generalized into more general forms which include their corresponding ordinary versions as special cases. Chapter 3 introduces the partial derivative, the partial differential, the total differential, and the total variation of a Boolean function. Many properties of these differential and variational operators of Boolean functions are presented. Based on the partial derivative of a Boolean function, the MacLaurin expansion and the Taylor expansion of a Boolean function are derived; the expansions are analogous to those of real functions of real variables. In addition, two convenient ways for computing Boolean derivatives and differentials of a switching function are included.

Special switching functions are useful in switching circuit design. Four special functions, monotonic, threshold, symmetric, and functionally complete functions are presented in Chapter 4. Many properties of these functions and algorithms for determining them are included. Chapter 5 presents multivalued switching functions, particularly their analysis and realization. As the number of components in an integrated circuit (IC) chip increases, the need for circuit testing becomes a necessary part of the manufacturing process. Several methods for deriving fault detection experiments for single and multiple logic faults in combinational circuits are presented in Chapter 6. Two computer-oriented algorithms, D-ALGorithm version II (DALG-II) and TEST-DETECT, are included which are applicable to the fault detection test generation for large combinational circuits.

In Chapter 7, three important problems, the state minimization, the state assignment, and the machine decomposition of sequential machines, are solved by the use of the substitution property (S. P.) partition of the states. A systematic representation of the Rabin-Scott machine, known as the regular expression, is presented in Chapter 8. This includes both the deterministic and the nondeterministic sequential machines. Two types of equivalence, indistinguishability equivalence and tape equivalence, are discussed. It is shown that for deterministic sequential machines, the two imply each other. A systematic procedure for obtaining a regular expression from a transition diagram and a procedure for constructing a transition diagram from a regular expression are presented. In Chapter 9, it is shown that any sequential machine can be physically realized by a clocked sequential circuit and obtained by it. The realizations of sequential machines using the pulse-mode and fundamental-mode circuits are also studied in detail. The analysis and design of pulse-mode sequential circuits are similar to those of clocked sequential circuits. But the analysis and design of fundamental-mode sequential circuits are different from those of clocked sequential circuits. The two undesirable transient phenomena of fundamental circuits, race and hazards and their elimination, are discussed. Chapter 10 introduces a method for designing a fault-detection experiment for a sequential machine. It is shown that the problem of designing a fault-detection experiment is actually a restricted problem of machine identification. The construction of a fault-detection experiment consists of three phases: the initialization phase, the state identification phase, and the transition verification phase. They are described in detail and are illustrated by examples.

The last two chapters are devoted to the discussion of modern digital (circuits and systems) design. Chapter 11 describes digital design using digital integrated circuits. In particular, digital design using various types of MSI and LSI integrated circuits is presented. Chapter 12 presents the digital design using microprocessors—the state-of-art. The basic design procedure is outlined and the hardwares and softwares of commonly used microprocessors are presented. Several digital design microcomputer systems are used to illustrate this new digital design method.

Many of the materials included in this book are recent research results which have never been included in any textbooks.

It is the author's experience that students always welcome good examples, particularly in illustrating difficult concepts and theory. A special feature of this book is that it includes many such examples throughout. In order to make sure that the student not only understands the theory but also knows how to apply it, a large number of exercises are given at the end of almost every section.

A picture is worth a thousand words. Figures, tables, and flow-charts are given throughout the book to help the reader to "see through" the theory.

I would like to thank Dr. M. E. Van Valkenburg, Dr. K. S. Fu, Dr. H. S. Hayre, and Dr. M. S. Ghausi for their advice and friendship. Special thanks are due to Dr. W. R. Upthegrove and Dr. C. R. Haden for their encouragement and support. I am also thankful to Mrs. Mary-Allen Kanak for preparing the drawings of Chapters 11 and 12, and to Mr. Mike Weible for proofreading the manuscript.

SAMUEL C. LEE

Norman, Oklahoma

Modern Switching Theory
and
Digital Design

1

Boolean Algebra and Boolean Function

Switching theory deals primarily with the analysis (characterization, minimization, etc.) and synthesis (realization) of a special type of function, defined on a special type of algebra known as *switching algebra*. Switching algebra is, in turn, a special type of *Boolean algebra;* and the special type of function, known as the *switching function*, is a mapping defined on switching algebra. Switching algebra that contains two elements, 0 and 1, is the two-element Boolean algebra (the simplest nondegenerate Boolean algebra). To understand how switching algebra is derived, one must first learn Boolean algebra, its mathematical foundation. In fact, Boolean algebra is the mathematical foundation of the entire field of switching theory.

The algebraic structure of Boolean algebra is derived from the ordered set. We begin the chapter by introducing ordered sets and the one-to-one relationship between elements in set theory and elements in algebra. Before introducing Boolean algebra, we first define *lattice*, which is a special subclass of the class of ordered sets. Boolean algebra is a special class of a subclass of lattices known as the *complemented distributive lattice*, or the *Boolean lattice*. Important properties of Boolean algebra are discussed in detail. Finally, the formal definition of Boolean function and its canonical forms are presented. The existence of the canonical forms for every Boolean function provides us with a convenient means of determining the equivalence between two Boolean functions and with a basis for deriving switching-function minimization methods, which will be discussed in Chapter 2.

1.1 Sets, Ordered Sets, and Algebras

Set theory is often referred to as the "root" of mathematics. We can consider every branch of mathematics to be a study of sets of objects of one kind or another. For instance, roughly speaking, geometry is a study of sets of points. Algebra is concerned with sets of numbers and operations on those sets. Analysis deals mainly with sets of functions. The study of sets and their use in the foundations of mathematics was begun in the latter part of the nineteenth century by the German mathematician Georg

Cantor (1845–1918). Since then, set theory has unified many seemingly disconnected ideas and has, in an elegant and systematic way, helped to reduce many mathematical concepts to their logical foundations.

The objectives of this section are threefold. The first is to serve as a review of some of the relevant materials in set theory. The second is to study three types of ordering: partial ordering, total ordering (a special case of partial ordering), and well-ordering (a special case of total ordering), and their corresponding types of sets. The third is to show the analogous quantities among sets, ordered sets, and algebras.

A *set* is a collection of objects in which nothing special is assumed about the nature of the individual objects. The individual objects in the collection are called *elements* or *members* of the set, and they are said *to belong to* (or *to be contained in*) the set. A group of people, a bunch of flowers, and a sequence of numbers are examples of sets. Here, people, flowers, and numbers are elements or members of these sets. It is important to know that a set itself may also be an element of some other set. For example, a line is a set of points; the set of all lines in the plane is a set of sets of points. In fact, a set can be a set of sets of sets, and so on. Let A be a set and x and y be elements of A. Define the relation "$x \leq y$" as "y includes x" and the relation "$x < y$" as "y strictly includes x."

DEFINITION 1.1.1

A relation $\leq$ on a set A is said to be a *partial ordering* on A if it satisfies the following axioms.
(01) Reflexive: For all $x \in A$, $x \leq x$.
(02) Antisymmetric: If $x,y \in A$, $x \leq y$, and $y \leq x$, then $x = y$.
(03) Transitive: If $x,y,z \in A$, $x \leq y$, and $y \leq z$, then $x \leq z$.
A set P over which a relation $\leq$ of partial ordering is defined is called a *partially ordered set*, or a *poset*.

In the definition, the reason for qualifying "partial" is that some questions about order may be left unanswered.

DEFINITION 1.1.2

A relation R on A is said to be *connected* whenever $x,y \in A$ implies that $x \leq y$ or $y \leq x$.

From Definitions 1.1.1 and 1.1.2, we define:

DEFINITION 1.1.3

A relation $\leq$ on a set A is said to be a *total ordering* of A if (a) it is a partial ordering of A, and (b) in addition, it satisfies the following axiom:
(04) Connected: Whenever $x,y \in A$ implies that $x \leq y$ or $y \leq x$.
A set C over which a relation of total ordering is defined is called a *totally ordered set*, or a *simply ordered set*, or a *chain*.

As mentioned before, the elements of a set may themselves be sets. A special class of such sets is the power set.

DEFINITION 1.1.4

Let A be a given set. The *power set* of A, denoted by $P(A)$, is a family (set) of sets such that when $X \subseteq A$, then $X \in P(A)$. Symbolically, $P(A) = \{X \mid X \subseteq A\}$.

Example 1.1.1

The power set of the empty set $\varnothing$ is a singleton $\{\varnothing\}$.

Example 1.1.2

Let $A = \{a, b, c\}$. The power set of A is

$$P(A) = \{\varnothing, \{a\}, \{b\}, \{c\}, \{a, b\}, \{b, c\}, \{c, a\}, \{a, b, c\}\}$$

THEOREM 1.1.1

Prove that if a set A has exactly n elements, then $P(A)$ will have exactly 2^n elements.

Proof: One way of proving this theorem is to tabulate the number of possible subsets of A as follows:

TABLE 1.1.1 Number of Possible Subsets of a Set A
with n Elements

Number of elements contained in a subset of A	Number of subsets
0	C_0^n
1	C_1^n
.	.
.	.
.	.
n	C_n^n

$$\left(C_i^n = \frac{n!}{i!(n-i)!} \right)$$

Thus, the total number of subsets of A is $C_0^n + C_1^n + \ldots + C_n^n$. From the binomial theorem,

$$(1 + x)^n = C_0^n + C_1^n x + C_2^n x^2 + \ldots + C_n^n x^n$$

where x is a real number and n is a positive integer. When we let $x = 1$ in the expression above we find that $C_0^n + C_1^n + C_2^n + \ldots + C_n^n = 2^n$. Hence, the theorem is proved.

Another, more intuitive proof may be given as follows: Each element of A is either in or is not in some subset. Thus, there are n independent binary choices, or 2^n ways to choose a subset. ∎†

†∎ shows the end of a proof.

We sometimes use the symbol $(A, \leq)$ to denote a poset (chain), where A is a set and $\leq$ is a partial (total) ordering relation in A. Before we proceed further, let us see some simple examples of posets and chains.

Example 1.1.3

Let A be a set. The set-theoretic inclusion relation $\subseteq$ is a partial ordering in the power set $P(A)$. It is a total ordering if A is an empty set or a singleton.

Example 1.1.4

Another interesting example of a relation of partial ordering is arithmetic divisibility. Let A be a set of divisors of 100: $A = \{1, 2, 4, 5, 10, 20, 25, 50, 100\}$. Define a relation "$x \leq y$" as "$x$ is a divisor of y." We can indicate this relation among the elements of A by using an *inclusion diagram* (Fig. 1.1.1). It is obvious that the relation "x is a divisor of y" is a partial ordering relation in A, not a total ordering relation, because

$$2 \quad \text{and} \quad 5$$
$$4, 10, \quad \text{and} \quad 25$$
$$20 \quad \text{and} \quad 50$$

are not related by this relation.

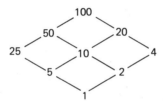

Fig. 1.1.1 The inclusion diagram of $(A, \leq)$ of Example 1.1.4.

Example 1.1.5

Now, if we consider the same relation "$x \leq y$" for "x is a divisor of y," but the set A is the set of divisors of 8, that is, $A = \{1, 2, 4, 8\}$, then the relation $\leq$ is a total-ordering relation in A, as shown in Fig. 1.1.2.

8

|

4

|

2

|

1

Fig. 1.1.2 The inclusion diagram of $(A, \leq)$ of Example 1.1.5.

In a poset A, if B is a nonempty subset of A and $b_0 \in B$, we call b_0 a *least element* of B if $b_0 \leq b$, for all $b \in B$; and we call b_0 a *minimal element* of B if there is no $b \in B$ such that $b < b_0$. A least element is necessarily minimal; but a minimal element need not be a least element, because a partial-ordering relation does not imply connected.

A *greatest element* and *maximal element* are defined in the corresponding way. By axiom 02, B can have at most one least and one greatest element, whereas it can have many minimal and maximal elements. The least and greatest elements of a poset are denoted by 0 and 1, respectively, whenever they exist.

From the definition of partial ordering, the following consequence is immediate.

THEOREM 1.1.2

Any finite subset X of a poset P has minimal and maximal elements.

Proof: Suppose that X is a singleton $X = \{x_1\}$. By the first condition of partial ordering, $x_1 \leq x_1$, x_1 may be considered as both a minimal element and a maximal element in X. If X contains two elements, $X = \{x_1, x_2\}$. There are two possible cases. One is that x_1, x_2 are related (i.e., either $x_1 \leq x_2$ or $x_2 \leq x_1$); hence, one is a minimal element and the other is a maximal element. Another case is that x_1, x_2 are unrelated. In this case, x_1, x_2 may be considered as both minimal and maximal elements in X. The induction of this argument for X containing n finite number of elements should be clear. ∎

From the definition of total ordering, it follows immediately that

THEOREM 1.1.3

For a chain, the notions of minimal and least (maximal and greatest) are equivalent. Hence, any finite chain has a least (first) and a greatest (last) element.

Proof: Since a least (greatest) element is necessarily minimal (maximal), we need only show that in a chain, a minimal (maximal) element is also a least (greatest) element of the chain. Let C be a chain. By Theorem 1.1.2, C has minimal and maximal elements. Let a be a minimal element of C (i.e., there is no x in C such that $x < a$). By axiom 04, $x \geq a$ for every x in A. Hence, a is also a least element of C. The proof that a maximal element of A is also a greatest element of A follows similarly. ∎

Example 1.1.6

In Example 1.1.4, consider a subset B of A, $B = \{4, 5, 10, 20, 25, 50\}$, as shown in Fig. 1.1.3. It is clear that B is a poset and has minimal elements 4 and 5 and maximal elements 20 and 50.

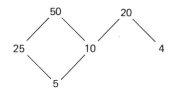

Fig. 1.1.3 The inclusion diagram of $(B, \leq)$ of Example 1.1.6.

Note that in Example 1.1.5 the element 1 is the least (first) element and 8 is the greatest (last) element.

DEFINITION 1.1.5

Let S and T be two sets (algebraic systems). If there exists a one-to-one correspondence between S and T, the correspondence is called an *isomorphism*, S and T are said to be isomorphic, or one set (system) is said to be *isomorphic* to the other. If the correspondence is not one-to-one, but many-to-one from S to T, the correspondence is called a *homomorphism*, or S is said to be *homomorphic* to T.

From Theorem 1.1.3 it follows that

THEOREM 1.1.4

Every finite chain of k elements is isomorphic to the ordered set $N_k = \{1, 2, \ldots, k\}$, where k is a positive integer. In other words, there always exists a mapping f from a chain C of k elements to N_k.

Proof: By Theorem 1.1.3 in C there exists a least and a greatest element. Let f map the least element of C to 1 and the least of the remaining elements into 2, and so on. Since both the chain C and N_k have the same number of elements, the greatest element of C, in this way, will be mapped to the greatest element k in N_k. Hence, the theorem is proved. ∎

The algebraic structure of a poset may be extended to a set of ordered pairs.

THEOREM 1.1.5

Let P be the Cartesian product of two posets A and B. Define an inclusion relation as

$$(a_1, b_1) \leq (a_2, b_2) \qquad \text{iff } a_1 \leq a_2 \text{ in } A \text{ and } b_1 \leq b_2 \text{ in } B$$

The set P with the product-inclusion relation is a poset. More generally, if P is the Cartesian product $A_1 \times A_2 \times \ldots \times A_n$ with an ordering relation defined as $(a_1, a_2, \ldots, a_n) \leq (a_1', a_2', \ldots, a_n')$ iff $a_1 \leq a_1'$ in A_1, $a_2 \leq a_2'$ in A_2, $\ldots$, $a_n \leq a_n'$ in A_n, then the set P with the product-inclusion relation is a poset.

Proof: The proof is evident.

Example 1.1.7

Let $A = \{1, 2, 3\}$ and $B = \{4, 5, 10, 20, 25, 50\}$ be two posets with ordering relation defined by arithmetic divisibility. The inclusion diagram for A is shown in Fig. 1.1.4; the inclusion diagram of B was shown in Fig. 1.1.3. The inclusion diagram of the Cartesian product $A \times B$ with the product-inclusion relation defined above is given in Fig. 1.1.5(a), and

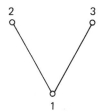

Fig. 1.1.4 The inclusion diagram of $(A, \leq)$ of Example 1.1.7.

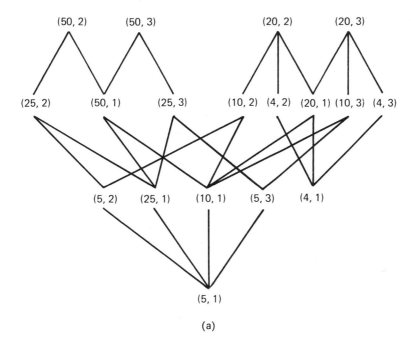

(a)

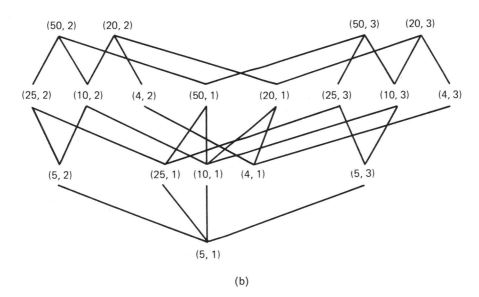

(b)

Fig. 1.1.5 (a) The inclusion diagram of the Cartesian product $A \times B$ of Example 1.1.7. (b) A more systematic way of generating the inclusion diagram of the Cartesian product $A \times B$ of Example 1.1.7.

a more systematic way of generating this inclusion diagram is shown in Fig. 1.1.5(b). From these diagrams, it is seen that the set $A \times B$ is a poset.

Now we introduce the third type of ordering.

DEFINITION 1.1.6

A relation $\leq$ on a set A is said to be a *well-ordering* of A if (a) it is a total ordering of A, and (b) it is such that every nonempty subset of A has a least element.

Here are several simple examples of well-ordered sets. In the first example, we demonstrate that a totally ordered set need not be well ordered.

Example 1.1.8

Let Q, Z, Z_e, and Z_o be the sets of rational numbers, integers, even integers, and odd integers, respectively. The set R^1 of all real numbers is totally ordered by the arithmetical relation $\leq$, but neither in R^1 nor in its subsets Q, Z, Z_e, Z_o, etc., is there any least element.

Example 1.1.9

The set N of natural numbers with the arithmetical relation $\leq$ is a well-ordered set.

Example 1.1.10

The set $P = \{\{A_1\}, \{A_1, A_2\}, \{A_1, A_2, A_3\}, \ldots\}$ with the set-theoretic inclusion relation $\subseteq$ is a well-ordered set.

Now we want to show that ordered sets provide a link between sets and algebras. We begin this discussion with the following definition.

DEFINITION 1.1.7

Let P be a poset and x and y be two elements of A. An element b in P is said to be a *lower bound* of x and y if $b \leq x$ and $b \leq y$. An element m in P is said to be a *greatest lower bound* (g.l.b.) of x and y if $m \leq x$ and $m \leq y$, $b \leq m$ for all b such that $b \leq x$ and $b \leq y$. Dually, we define an element u in P to be an upper bound of x and y if $x \leq u$ and $y \leq u$, and an element l in P to be a *least upper bound* (l.u.b.) of x and y if $x \leq l$ and $y \leq l$, $l \leq u$ for all u such that $x \leq u$ and $y \leq u$.

Now we define:

DEFINITION 1.1.8

An element m in P is a *meet* of x and y if it is a g.l.b. of x and y. An element l in P is a *join* of x and y if it is a l.u.b. of x and y. We shall denote the meet and join of x and y by $m = x \cap y$ and $l = x \cup y$, respectively.†

†The symbols "∩" and "∪" are called "cap" and "cup" by some authors.

It is worth noting that

1. For given x and y, the meet and join are *unique* if they exist.
2. Meet and join are *order duals* of each other. By order dual, we mean that the definition of meet may be obtained by the definition of join simply by replacing the relation $x \leq y$ by its converse, and vice versa.

The second property is obvious from Definitions 1.1.7 and 1.1.8. The proof of the first property is as follows. Suppose that m and m' are both meets of x and y. Definition 1.1.7 implies that $m \leq m'$ and $m' \leq m$. By the antisymmetry axiom of partial ordering, m' should be equal to m. Hence, a meet, when it exists, is unique. By a similar argument, a join of x and y is also unique when it exists. We shall use the symbols O and I to denote the (unique) least and greatest elements of a partially ordered set whenever they exist.

We can show, among other properties, that the meet and join operations satisfy the absorption property:

$$x \cap (x \cup y) = x$$
$$x \cup (x \cap y) = x$$

The proofs of these identities will be given in the next section. Here, we just use them as examples to show the analogy between sets and algebras. The role that ordered sets play in linking sets and algebras is shown in Table 1.1.2.

TABLE 1.1.2   From Sets to Algebras

		Algebras	
Sets	Ordered sets	Numbers or symbols in general	Integers or rational numbers
A	x	a	a
B	y	b	b
$\cap$	$\cap$	$\cdot$	$\cdot$
$\cup$	$\cup$	$+$	$+$
$A \cap (A \cup B) = A$	$x \cap (x \cup y) = x$	$a \cdot (a + b) = $ g.l.b.$[a,$ l.u.b.$(a, b)] = a$	
			$a \cdot (a + b) = $ min $[a,$ max $(a, b)] = a$
$A \cup (A \cap B) = A$	$x \cup (x \cap y) = x$	$a + a \cdot b = $ l.u.b.$[a,$ g.l.b.$(a, b)] = a$	
			$a + a \cdot b = $ max $[a,$ min $(a, b)] = a$

The analogous quantities among sets, ordered sets, and algebras are tabulated in Table 1.1.3.

**TABLE 1.1.3 Analogous Quantities Among Sets,
Ordered Sets, and Algebras**

Quantity	Sets	Ordered sets	Algebras
Relation	$\subseteq$: containment $X \subseteq Y$ means that Y contains X	$\leq$: inclusion relation $x \leq y$ means that y includes x	$\leq$: inequality $a \leq b$ means that b is equal to or greater than a
Operation	$\cup$: union $X \cup Y$ means the set whose elements are in X, or in Y, or in both	$\cup$: join $x \cup y$ means the l.u.b. of x and y	$+$: l.u.b. $a + b$ means the l.u.b. of a and b
	$\cap$: intersection $X \cap Y$ means the set whose elements are in both X and Y	$\cap$: meet $x \cap y$ means the g.l.b. of x and y	$\cdot$: g.l.b. $a \cdot b$ means the g.l.b. of a and b
Element	A The universal set	I The greatest element	1 The largest number
	$\emptyset$ The empty set	O The least element	0 The smallest number

Exercise 1.1

1. Show that any subset of a poset is itself a poset, relative to the same inclusion relation.

2. Let R be a binary relation on a set A. The *converse* of R, denoted by $\bar{R}$, is defined as follows: If $x,y \in A$, $x \, R \, y$ iff $y \, R \, x$. Show that the converse of any partial ordering is itself a partial ordering.

3. Prove that if a set A has n elements, then $\underbrace{P(P \ldots (P(A)) \ldots)}_{m}$ will have exactly $2^{2^{\cdot^{\cdot^{2^n}}}}$

    elements.

4. (a) Let A and B be two sets. Prove that

    $$P(A) \cap P(B) = P(A \cap B)$$

    (b) Is the following relation true?

    $$P(A) \cup P(B) = P(A \cup B)$$

5. Prove that any subset of a chain is a chain.

6. Prove that the converse of any total ordering is itself a total ordering.

7. Let $A = \{a_1, a_2, \ldots, a_n\}$ be a poset. Show that $a_1 \leq a_2 \leq \ldots \leq a_n \leq a_1$ implies that $a_1 = a_2 = \ldots = a_n$.

8. Draw an inclusion diagram for each of the following posets.
    (a) The set of positive integral factors of 16 under the partial ordering defined by divisibility.
    (b) The positive integral divisors of 24 under the partial ordering defined by divisibility.
    (c) The set of positive even integers up to 12 with divisibility again the ordering relation.

9. Let (x_1, y_1) and (x_2, y_2) be two points in the plane. What is the least upper bound of the set of these two points under the ordering $(x_3, y_3) \geq (x_4, y_4)$ iff $x_3 \geq x_4$ and $y_3 \geq y_4$?

10. Let $S = \{1, 2, \ldots, 10\}$. Let p be a binary relation on S defined by $a \, p \, b$ if $2a > b$. Is (S, p) a partially order set? Why or why not? What about (X, p) where $X = \{1, 2\}$?

11. Let $(A, \leq)$ be a poset. The poset $(A, \geq)$, where $\geq$ is the converse relation of $\leq$, is called the *dual* of the poset $(A, \leq)$. Show that if two posets are self-dual, so is their Cartesian product poset.

12. Show that the Cartesian product of two chains, each of which contains two or more elements, cannot be a chain.

13. Show that every countably infinite chain is isomorphic to the set of natural numbers.

14. Show, by an example, that not every infinite chain is self-dual.

15. Show that the poset of the positive integral divisors of 12 and the poset of the positive integral divisors of 45 are isomorphic.

16. Let R be a relation in A. R *is an equivalence relation in A* if the following conditions are satisfied:
    (1) $x \, R \, x$ for all $x \in A$ (R is reflexive).
    (2) If $x \, R \, y$, then $y \, R \, x$, for all $x, y \in A$ (R is symmetric).
    (3) If $x \, R \, y$ and $y \, R \, z$, then $x \, R \, z$ for all $x, y, z \in A$ (R is transitive).
    Prove that an isomorphism between posets is an equivalence relation.

17. If $(A, \leq)$ is a well-ordered set and $B \subseteq A$, then $(B, \leq)$ is a well-ordered set.

18. Let A_1 and A_2 be two nonempty sets and R_1 and R_2 be two well-ordering relations on A_1 and A_2, respectively. (A_1, R_1) is said to be *similar* to (A_2, R_2) if there exists a function f from A_1 to A_2 satisfying the following conditions:
    (1) A and B are isomorphic under f.
    (2) If $x \, R_1 \, y$, then $f(x) \, R_2 \, f(y)$.
    Consider two similar systems (A_1, R_1) and (A_2, R_2) under f. Prove the following:
    (a) If a is a least element in A_1, then $f(a)$ is a least element in A_2.
    (b) If b is a greatest element in A_1, then $f(b)$ is a greatest element in A_2.
    (c) If (A_1, R_1) is a well-ordered set, then (A_2, R_2) is also a well-ordered set.

1.2 Lattice and Its Basic Properties

Since in this section we are interested primarily in the algebraic aspects of ordered sets, we shall use $\cdot$ and $+$ to represent the meet $(\cap)$ and join $(\cup)$ operations. We shall use x, y, and z to denote generic elements in an ordered set and a, b, and c to denote specific elements.

From the definitions of meet and join introduced in the previous section (Definition 1.1.8), we have

LEMMA 1.2.1

In any algebraic poset A, the meet and join of two elements of A, if they exist, have the following property:

$$x \leq y \qquad \text{iff } x \cdot y = x \quad \text{and} \quad x + y = y$$

This property is often referred to as the *consistency property*.

Proof: The proof is obvious and can be omitted.

The following theorem states the properties of the meet and join in a poset.

THEOREM 1.2.1

In any poset A, the meet and join operations of two elements of A, if they exist, satisfy the idempotent, commutative, associative, and absorptive properties; that is, for all x, y, and z in A,

(a)   Idempotent, L1.: $x \cdot x = x$; $\quad$ L1$_+$: $x + x = x$

(b)   Commutative, L2.: $x \cdot y = y \cdot x$; $\quad$ L2$_+$: $x + y = y + x$

(c)   Associative, L3.: $x \cdot (y \cdot z) = (x \cdot y) \cdot z$;   L3$_+$: $x + (y + z) = (x + y) + z$

(d)   Absorptive, L4.: $x \cdot (x + y) = x$; $\quad$ L4$_+$: $x + (x \cdot y) = x$

Proof: The properties L1 and L2 follow directly from the definitions of meet and join. The properties L3 are evident since $x \cdot (y \cdot z)$ and $(x \cdot y) \cdot z$ are both equal to the g.l.b. of x, y, and z, and $x + (y + z)$ and $(x + y) + z$ are both equal to the l.u.b. of x, y, and z.

To prove L4, consider the following two cases:

(1)   If $x \leq y$, then

$$x \cdot (x + y) = x \cdot y \qquad \text{by Lemma 1.2.1}$$
$$ = x \qquad \text{by Lemma 1.2.1}$$

and

$$x + (x \cdot y) = x + x \qquad \text{by Lemma 1.2.1}$$
$$ = x \qquad \text{by L1}_+$$

(2)   If $y \leq x$, then

$$x \cdot (x + y) = x \cdot (y + x) \qquad \text{by L2}_+$$
$$ = x \cdot x \qquad \text{by Lemma 1.2.1}$$
$$ = x \qquad \text{by L1.}$$

and

$$x + (x \cdot y) = x + (y \cdot x) \qquad \text{by L2.}$$
$$ = x + y \qquad \text{by Lemma 1.2.1}$$
$$ = x \qquad \text{by Lemma 1.2.1}$$

Hence, the meet and join operations also satisfy the absorption property.   ■

A lattice is defined by

DEFINITION 1.2.1

A *lattice* is a poset L in which any two elements x and y have both a meet and a join.

A lattice has the following properties.

THEOREM 1.2.2

In any lattice,
(a) All the elements satisfy L1–L4 of Theorem 1.2.1.
(b) All elements satisfy the *isotone* property; that is, if $x \le y$, then $x \cdot z \le y \cdot z$ and $x + z \le y + z$.
(c) All elements satisfy the *modular inequality*, which is, if $x \le z$, then $x + (y \cdot z) \le (x + y) \cdot z$.
(d) The *distributive inequalities* are satisfied:

$$x \cdot (y + z) \ge (x \cdot y) + (x \cdot z)$$
$$x + (y \cdot z) \le (x + y) \cdot (x + z)$$

Proof: Property (a) is evident from the definition of lattice.

Properties (b), (c), and (d) may be proved by the following algebra.

Proof of property (b): If $x \le y$, then

$$x \cdot z = (x \cdot y) \cdot (z \cdot z) \qquad \text{by Lemma 1.2.1 and L1.}$$
$$= (x \cdot z) \cdot (y \cdot z) \qquad \text{by L2. and L3.}$$

which, by Lemma 1.2.1, implies that $(x \cdot z) \le (y \cdot z)$. The second inequality may be proved in a similar way.

Proof of Property (c): Since $x \le z$ and $x \le x + y$,

$$x \le (x + y) \cdot z$$

and since $y \cdot z \le z$ and $y \cdot z \le y \le x + y$,

$$y \cdot z \le (x + y) \cdot z$$

Combining these results and in view of the definition of $+$, we obtain

$$x + (y \cdot z) \le (x + y) \cdot z$$

Proof of Property (d): The proof of this property is similar to that of Property (c). Since $x \cdot y \le x$ and $x \cdot y \le y \le y + z$,

$$x \cdot y \le x \cdot (y + z)$$

From the relations $x \cdot z \leq x$ and $x \cdot z \leq z \leq y + z$,

$$x \cdot z \leq x \cdot (y + z)$$

Hence,

$$x \cdot (y + z) \geq (x \cdot y) + (x \cdot z)$$

Again, the second inequality may be proved in a similar way. ∎

We have shown in Theorem 1.1.2 that every finite chain has a least and a greatest element. Now we want to show that every finite lattice also has this property.

THEOREM 1.2.3

Every finite lattice has a least element and a greatest element.

Proof: Let the elements of a finite lattice L be $x_1, x_2, \ldots, x_n$. The least element of L is the element $x_1 \cdot x_2 \cdot \ldots \cdot x_n$, and the greatest element of L is the element $x_1 + x_2 + \ldots + x_n$. ∎

Example 1.2.1

Consider the set L of the positive integral divisors of the natural number 216, $L = \{1, 2, 3, 4, 6, 8, 9, 12, 18, 24, 27, 36, 54, 72, 108, 216\}$. The inclusion diagram of L is as shown in Fig. 1.2.1. The multiplication (g.l.b.) and addition (l.u.b.) of any two elements of L may be found directly from the diagram. For instance, the meet and join of 12 and 18 are 6 and 36. The meet and join of 24 and 9 are 3 and 72. The meet and join of 4 and 108 are 4 and 108. It is seen that any two elements have a meet and a join. The inclusion diagram provides a convenient means to find them. Hence, L is a lattice. One may easily verify that all the properties of Theorems 1.2.1 and 1.2.2 are satisfied.

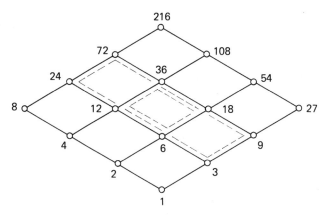

Fig. 1.2.1   The inclusion diagram of L of Example 1.2.1.

Now we want to prove that any nonempty set satisfying axioms L1–L4 is a lattice. First we define:

DEFINITION 1.2.2

A nonempty set with a single binary operation which satisfies L1–L3 is called a *semilattice*.

LEMMA 1.2.2

If P is a poset in which any two elements have a meet (join), P is a semilattice and is called a *meet (join) lattice*.

Conversely, we have

LEMMA 1.2.3

Let S be a semilattice under a binary operation $\circ$, and x and y be two elements of S.

(a) The definition

$$x \leq y \qquad \text{iff } x \circ y = x$$

makes S a poset in which $x \circ y = \text{g.l.b. } \{x, y\}$; that is, $\circ$ is a meet operation on S.

(b) The definition

$$x \leq y \qquad \text{iff } x \circ y = y$$

makes S a poset in which $x \circ y = \text{l.u.b. } \{x, y\}$; that is, $\circ$ is a join operation on S.

Proof: The proof of (a) contains two parts: First, we want to prove that the relation $\circ$ is a partial ordering in A, and then to show that $m \leq x$ and $m \leq y$, and $b \leq m$ for all b such that $b \leq x$ and $b \leq y$.

To show that the relation $\circ$ is a partial ordering in A is to show that the relation $\circ$ satisfies the reflexive law, antisymmetric law, and transitive law. By the definition of semilattice, the relation $\circ$ is idempotent, commutative, and associative. The idempotent law $x \circ x = x$ implies that the reflexive law $x \leq x$. If $x \leq y$ (iff $x \circ y = x$) and $y \leq x$ (iff $y \circ x = y$), by the commutative law, then $x \circ y = y \circ x$ and $x = x \circ y = y \circ x = y$. This proves that the relation $\circ$ satisfies the antisymmetric law. By the associative law, $x \leq y$ (iff $x \circ y = x$) and $y \leq z$ (iff $y \circ z = y$) implies that $x = x \circ y = x \circ (y \circ z) = (x \circ y) \circ z = x \circ z$, that is, $x \leq z$. This shows that the transitive law is also satisfied by the relation $\circ$. Hence, $\circ$ is a partial-ordering relation in S.

Next, we want to show that $\circ$ is a meet operation in S.

$$
\begin{aligned}
(x \circ y) \circ x &= x \circ (x \circ y) && \text{by L2} \\
&= (x \circ x) \circ y && \text{by L3} \\
&= x \circ y && \text{by L1} \\
(x \circ y) \circ x &= x \circ y \Longleftrightarrow x \circ y \leq x
\end{aligned}
$$

Similarly, we can show that $x \circ y \leq y$. Now if $b \leq x$ and $b \leq y$, then $b \circ (x \circ y) = (b \circ x) \circ y = b \circ y = b$, which implies that $b \leq x \circ y$. This proves that $x \circ y = \text{g.l.b. } \{x, y\}$. The proof of (b) may be obtained similarly. Hence, the lemma is proved. ∎

LEMMA 1.2.4

Let S be a nonempty set with the multiplication and addition operations defined on it. If the multiplication and addition operations satisfy the absorption properties L4. and L4$_+$, then $x \leq y \Longleftrightarrow x \cdot y = x$ and $x \leq y \Longleftrightarrow x + y = y$.

Proof: Let x and y be two elements of S. Since the multiplication and addition operations satisfy L4$_+$, $x \cdot y = x$ implies that $x + y = (x \cdot y) + y = y$ and $x + y = y$ implies that $x \cdot y = x \cdot (x + y) = x$. Conversely, if $x \leq y$,

$$x = x \cdot (x + y) = x \cdot y \qquad \text{by L4.}$$
$$y = x + (x \cdot y) = x + y \qquad \text{by L4}_+$$

Thus, $x \leq y \Rightarrow x \cdot y = x$ and $x \leq y \Rightarrow x + y = y$. Hence, the lemma is proved.   ∎

From Lemmas 1.2.3 and 1.2.4, we immediately have the following theorem.

THEOREM 1.2.4

Any nonempty set L with two binary operations which satisfy L1–L4 is a lattice.

Proof: The proof is evident from Lemmas 1.2.3 and 1.2.4.

As with the poset, a lattice may be a Cartesian product of two sets.

THEOREM 1.2.5

The Cartesian product of two lattices is a lattice. More generally, the Cartesian product $A_1 \times A_2 \times \ldots \times A_n$, where $A_1, A_2, \ldots, A_n$ are lattices, is a lattice.

Proof: Again, the proof is evident.

Let us examine an example of a Cartesian product lattice known as the *factorization lattice*.

Example 1.2.2

Let L be the set of *all* the factors of some natural number $x = p_1^{m_1} \times p_2^{m_2} \times \ldots \times p_n^{m_n}$, where the p_i are distinct primes and $p_i > 1$. From elementary number theory it can be shown that for any integer $x > 1$, the representation of x as a product of a prime is unique up to the order of the factors. Since every member of L is of the form $p_1^\alpha \times p_2^\beta \times \ldots \times p_n^\nu$, where $\alpha = 0, 1, \ldots, m_1$, $\beta = 0, 1, \ldots, m_2, \ldots$, and $\nu = 0, 1, \ldots, m_n$, we see that the number of factors is given by the product of the exponents, each having been increased by 1; for instance, $113{,}400 = 2^3 \cdot 3^4 \cdot 5^2 \cdot 7$ has $(3 + 1)(4 + 1)(2 + 1)(1 + 1) = 120$ factors. Now, let us examine some simple examples of this class of lattices. For instance,

$$L_1: \quad x = 4 = 2^2 \qquad\qquad (n = 1)$$
$$L_2: \quad x = 12 = 2^2 \cdot 3 \qquad\qquad (n = 2)$$
$$L_3: \quad x = 60 = 2^2 \cdot 3 \cdot 5 \qquad\qquad (n = 3)$$
$$L_4: \quad x = 2940 = 2^2 \cdot 3 \cdot 5 \cdot 7^2 \qquad (n = 4)$$

The four lattices are shown in Fig. 1.2.2. In Fig. 1.2.2(a), $n = 1$, and the factors of 4 form a chain. When $n = 2$, for instance, $x = 12 = 2^2 \cdot 3$, the two-dimensional lattice (L_2) can be constructed as shown in Fig. 1.2.2(b). As is shown in Fig. 1.2.2(c), the factorization expression of each element of the lattice can be by a 3-tuple, and the values of the coordinates are the powers of primes of the factorization expression. The lattice for $x = 2940$ (there are 36 factors), which is an $n = 4$ case, is shown in Fig. 1.2.2(d). It should be clear that we can construct a k-dimensional lattice for $n = k$, that is, the set of all factors of a finite natural number, by the method indicated.

There are three special classes of lattices of interest to us. They are modular lattices, distributive lattices, and Boolean lattices (complemented distributive lattices). First we introduce the modular lattice.

DEFINITION 1.2.3

A lattice L is *modular* if it satisfies the following modular identity: For all x, y, and z in L,

$$\text{L5:}\quad \text{If } x \leq z, \text{ then } x + (y \cdot z) = (x + y) \cdot z$$

Some of the properties of modular lattices are given in problem 7. One is presented below because it is needed to prove other theorems.

THEOREM 1.2.6

A lattice is modular iff the three relations

$$x \leq z \qquad x \cdot y = z \cdot y \qquad x + y = z + y$$

jointly imply that $x = z$.

Proof: (1) *Proof of the necessity:* Suppose that a lattice L is modular and that the three relations hold. Let x, y, and z be elements of L.

$$
\begin{aligned}
x &= x \cdot (x + y) & &\text{by L4.} \\
  &= x \cdot (z + y) & &\text{by hypothesis} \\
  &= z + (y \cdot x) & &\text{by L5} \\
  &= z + (y \cdot z) & &\text{by hypothesis} \\
  &= z & &\text{by L4}_+
\end{aligned}
$$

(2) *Proof of the sufficiency:* If $x \leq z$ and $x \cdot y = z \cdot y$ and $x + y = z + y$ imply that $x = z$, then

$$x + (y \cdot z) = x + (x \cdot y) = x = z = z \cdot (x + z) = z \cdot (x + y)$$

Hence, the lattice is modular. ∎

Another important special class of lattices is the distributive lattice.

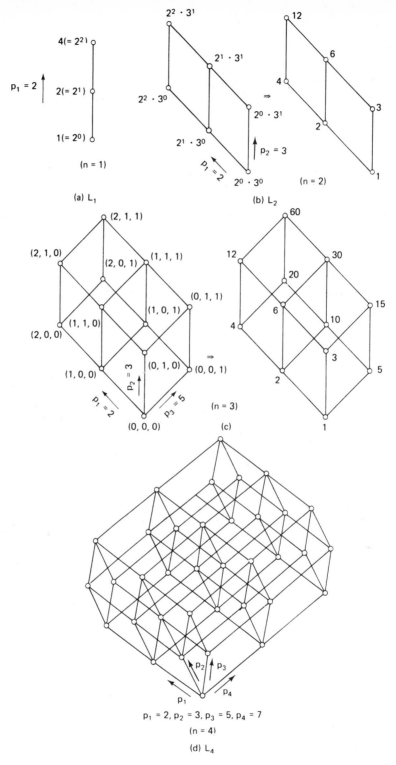

Fig. 1.2.2 Factorization lattices of Example 1.2.2.

DEFINITION 1.2.4

A lattice L is called *distributive* if it satisfies the distributive equalities

$$\text{L6a:} \quad x \cdot (y + z) = (x \cdot y) + (x \cdot z)$$
$$\text{L6b:} \quad x + (y \cdot z) = (x + y) \cdot (x + z)$$

for all x, y, and z in L.

It can be shown that a lattice which satisfies one of the two distributive inequalities will automatically satisfy the other. The proof of this property is left to the reader (problem 13).

Similar to Theorem 1.2.6, for distributive lattices we have

THEOREM 1.2.7

A lattice L is distributive iff the two conditions

$$x \cdot y = z \cdot y \quad \text{and} \quad x + y = z + y$$

jointly imply that $x = z$ for all x, y, and z in L.

Proof: (1) *Proof of the necessity:* Suppose that a lattice is distributive and that the two equations hold for all x, y, z in L. Then

$$
\begin{aligned}
x &= x \cdot (x + y)   && \text{by L4.} \\
  &= x \cdot (z + y)   && \text{by hypothesis} \\
  &= x \cdot z + x \cdot y && \text{by L6a} \\
  &= x \cdot z + z \cdot y && \text{by hypothesis} \\
  &= z \cdot (x + y)   && \text{by L6a} \\
  &= z \cdot (z + y)   && \text{by hypothesis} \\
  &= z   && \text{by L4.}
\end{aligned}
$$

(2) *Proof of the sufficiency:* If $x \cdot y = z \cdot y$ and $x + y = z + y$ imply that $x = z$, then

$$x \cdot (y + z) = x \cdot (x + y) = x = z = z \cdot y + z = z \cdot y + z \cdot z = x \cdot y + x \cdot z$$

Hence, the lattice is distributive. ∎

Some of the properties of distributive lattices are given in problem 18.

A special class of distributive lattice is the *complemented distributive lattice*, or the *Boolean lattice*.

DEFINITION 1.2.5

Let L be a lattice having both a zero element (least element) 0 and a unit element (greatest element) 1. Let x and y be elements of L. If $x \cdot y = 0$ and $x + y = 1$, then y is a *complement* of x and x is a *complement* of y.

DEFINITION 1.2.6

A lattice L is called *complemented* if all its elements have complements.

In general, an element of a lattice may have no complement, such as element 12 of the lattice in Fig. 1.2.1; may have several complements, such as element a of the lattice M_5 in Fig. 1.2.3(b), which has two complements (b and c), or element b of the pentagonal lattice in Fig. 1.2.3(a), which has two complements (a and c); or may have a unique complement, such as element a (or c) of the lattice of Fig. 1.2.3(a), which has a unique complement b.

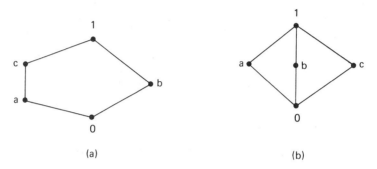

(a)   (b)

Fig. 1.2.3   (a) The pentagon lattice and (b) the lattice M_5.

From Theorem 1.2.3 and the definitions of 0 and 1, 0 is the *unique* complement of 1 and 1 is the unique complement of 0.

Example 1.2.3

Consider the power set $P(I)$ of all subsets S of a set I. The set $(P(I), \subseteq)$, with the meet and join operations defined by the set-theoretic intersection and set-theoretic union, is a lattice. Since the usual set-theoretic complement $\bar{S}$ of S in I does satisfy the relations $\bar{S} \cap S = \varnothing$ (zero element) and $\bar{S} \cup S = I$ (unit element), the algebraic system $(P(I), \subseteq, \cap, \cup)$ is a Boolean lattice.

Example 1.2.4

As another example, consider two lattices, $A = \{1, 2, 4, 8\}$ and $B = \{1, 2, 3, 6\}$, both relative to the positive integral divisibility relation. The Cartesian product $A \times B$ is shown in Fig. 1.2.4. Obviously, it is a Boolean lattice.

In the examples above we have seen that the complement of every element of $P(I)$ and $A \times B$ is unique. In fact, every Boolean lattice has such a property. This is stated in the following theorem.

THEOREM 1.2.8

In a Boolean lattice, each element has one and only one complement.

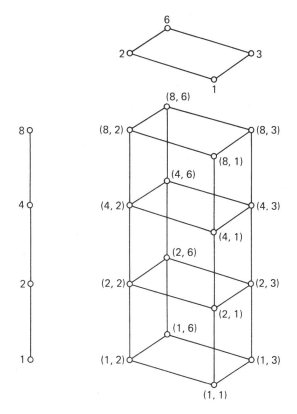

Fig. 1.2.4 The Cartesian product of two lattices of Example 1.2.4.

Proof: Let x be an element of a Boolean lattice. Suppose that x has two complements, y and z. Then

$$x \cdot y = x \cdot z = 0$$
$$x + y = x + z = 1$$

By Theorem 1.2.7, $y = z$. ∎

Before leaving this section, we would like to summarize the inclusion relations among the ordered sets that have been discussed in this and the previous sections. The ordered set is a subset of the set of sets. A lattice is an ordered set in which any two elements, x and y, have both a meet and a join. Thus, the set of lattices is a subset of the set of ordered sets. A modular lattice is a lattice that satisfies the modular identity (L5); so the set of modular lattices is a subset of the set of lattices. It can be shown (the proof is left to the reader) that every distributive lattice [i.e., lattice that satisfies the distributive inequalities (L6a and L6b)] is a modular lattice (problem 18). Hence, the set of distributive lattices is a subset of the set of modular lattices. Finally, it is clear that the set of Boolean lattices (complemented distributive lattices) is a subset of distributive lattices. These inclusion relations are depicted in Fig. 1.2.5.

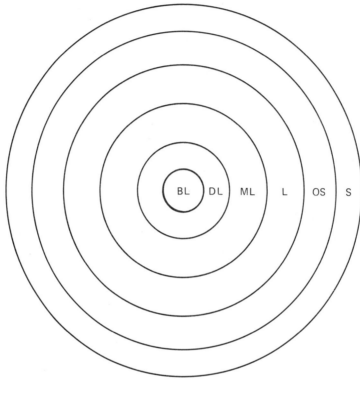

S = Set
OS = Ordered set
L = Lattice in general
ML = Modular lattice
DL = Distributive lattice
BL = Boolean lattice

Fig. 1.2.5 Inclusion relations among lattices.

Exercise 1.2

1. Show that not every lattice is a chain but that every chain is a lattice.

2. A *sublattice* M of a lattice L is a subset of L that satisfies the following condition: $x, y \in M \Rightarrow x \wedge y \in M$ and $x \vee y \in M$. Show that a lattice L is a chain iff every subset of L is a sublattice.

3. A lattice L is called *complete* if every subset of L has a l.u.b. and a g.l.b. Prove that a lattice is not complete unless it has a subset that forms an infinite chain.

4. Let L_1 and L_2 be two lattices. Show that the product lattice $L_1 \times L_2$ is isomorphic to the product lattice $L_2 \times L_1$. Also, show that the projection $L_1 \times L_2 \longrightarrow L_1$ is homomorphic to the projection $L_1 \times L_2 \longrightarrow L_2$.

5. Prove that if in a lattice for any x, y, and z,

$$\text{L5}': \quad (x+y)\cdot(y+z)\cdot(z+x) = x\cdot y + y\cdot z + z\cdot x$$

the lattice is modular.

6. Show that the pentagon lattice is not modular.

7. Prove the following properties of modular lattices.
   (a) Every sublattice of a modular lattice is modular.
   (b) A lattice is modular iff it does not contain a sublattice isomorphic to the pentagonal lattice [Fig. 1.2.3(a)].
   (c) For any pair of comparable elements a and b of a lattice L where $a \le b$, the subset X of L consisting of all elements x, $a \le x \le b$, is called the *interval* $[a, b]$. If x and y are any two elements of a modular lattice L, the two intervals $[a\cdot b, b]$ and $[a, a+b]$ are isomorphic sublattices of L.
   (d) Let α and β be elements of a poset P. We say that β *covers* α if $\alpha < \beta$ and there is no element $x \in P$ such that $\alpha < x < \beta$. Let L be a modular lattice and let a and b be elements of L. Then, $a+b$ covers a iff b covers $a\cdot b$.

8. Show that the following lattices are modular.
   (a) A lattice that contains less than four elements.
   (b) The lattice M_5.
   (c) A chain.

9. Show that the following lattices are nonmodular.
   (a) A polygonal lattice that is greater than a pentagon.
   (b) A lattice that has two or more noncomparable pairs.

10. Consider the following sublattices of the lattice of Fig. 1.2.1:
   (a) $L_1 = \{1, 2, 3, 6\}$
   (b) $L_2 = \{1, 2, 3, 4, 6, 9, 12, 18, 36\}$
   (c) $L_3 = \{6, 12, 18, 24, 36, 72, 108, 216\}$
   (d) $L_4 = \{6, 12, 18, 24, 54, 72, 108, 216\}$
   (e) $L_5 = \{1, 2, 3, 4, 6, 8, 54, 72, 216\}$
   Which of these are nonmodular lattices? Give your reason.

11. Prove that the homomorphic image of a modular lattice is modular.

12. Show that the Cartesian product $L \times M$ of two lattices with the inclusion relation defined in Theorem 1.1.5 is modular iff L and M are modular.

13. Prove that if any lattice satisfies one of the two distributive equalities, it will automatically satisfy the other.

14. Prove that a lattice L is distributive if it satisfies either

$$\text{L6a}': \quad x\cdot(y+z) \le (x\cdot y) + (x\cdot z) \qquad \text{for all } x, y, z \text{ in } L$$

or

$$\text{L6b}': \quad x + (y\cdot z) \ge (x+y)\cdot(x+z) \qquad \text{for all } x, y, z \text{ in } L$$

15. Prove that a lattice L is distributive iff it satisfies the condition

$$\text{L6}'': \quad (x + y) \cdot (y + z) \cdot (z + x) = xy + yz + zx$$

for all x, y, and z in L.

16. Show that the relations $x \cdot y = z \cdot y$ and $x + y = z + y$ cannot be simultaneously held for $x > z$.

17. Prove that the pentagon lattice and the lattice M_5 of Fig. 1.2.3 are nondistributive.

18. Prove the following properties of distributive lattices.
   (a) Every sublattice of a distributive lattice is distributive.
   (b) A lattice is distributive iff it does not contain a sublattice isomorphic to either the pentagon lattice or the lattice M_5 (Fig. 1.2.3).
   (c) Every distributive lattice is a modular lattice, but not the converse.

19. Show that the following are distributive: (a) any lattice containing no more than four elements, and (b) a chain.

20. Prove that the set of positive integers partially ordered by divisibility with the meet and join defined by the greatest common divisor (meet) and least common multiple (join) operations is a distributive lattice.

21. For any lattice L, prove the following inequality:

$$x \cdot y + y \cdot z + z \cdot x \le (x + y) \cdot (y + z) \cdot (z + x)$$

holds for all x, y, and z in L.

22. Prove that a lattice L is distributive iff for every set of elements x, y, and z in L, the inequalities

$$x \cdot z \le y \le x + z$$

imply that

$$x \cdot y + y \cdot z = y = (x + y) \cdot (y + z)$$

23. Draw all the lattices with five elements or less and indicate which of them are modular and which are distributive.

1.3 Boolean Algebra and Its Basic Properties

A Boolean lattice considered as an algebra closed with respect to the three operations, complementation, meet, and join is called a *Boolean algebra*. Conventionally, in Boolean algebras, we call the operations $\cdot$ and $+$ *multiplication* and *addition*, respectively, instead of meet and join. The element $x \cdot y$ is called the *product* of x and y; and the element $x + y$ is called the *sum* of x and y.

An alternative definition of Boolean algebra may be given as follows:

DEFINITION 1.3.1

A *Boolean algebra* is an algebraic system $(B; +, \cdot, '; 0, 1)$ consisting of a set B, binary operations $+$ and $\cdot$, and a unary operation $'$, satisfying axioms L1–L4, L6, and the following postulate:

L7   For every element x in B, there exists a complement† x' such that $x \cdot x' = 0$ and $x + x' = 1$. A Boolean algebra is called *degenerate* if its multiplicative identity is equal to its additive identity; that is, $0 = 1$.

A simple example of such an algebra is the one that is isomorphic to the power set $P(\varnothing)$ of the empty set, $P(\varnothing) = \{\varnothing\}$. In the future we shall consider *only* nondegenerate algebras (i.e., Boolean algebras in which 0 and 1 are distinct). As a consequence of Theorem 1.2.8, we have

THEOREM 1.3.1

Any (nondegenerate) Boolean algebra must contain an even number of elements.

Example 1.3.1

Consider the (simplest nondegenerate) two-element Boolean algebra $B_2 = \{0, 1\}$; 0 and 1 are complements of each other. From the definitions of meet (product) and join (sum), we can easily construct the following AND, OR, and NOT tables.

$\cdot$	0	1
0	0	0
1	0	1

$+$	0	1
0	0	1
1	1	1

$'$	
0	1
1	0

Next, consider the power set $P(S)$ of a set S which is a singleton. Clearly, $P(S)$ contains only two elements, $\varnothing$ and S. Three tables similar to those shown above for the set-theoretic intersection, union, and complementation of $\varnothing$ and S may be constructed as follows:

$\cap$	$\varnothing$	S
$\varnothing$	$\varnothing$	$\varnothing$
S	$\varnothing$	S

$\cup$	$\varnothing$	S
$\varnothing$	$\varnothing$	S
S	S	S

$\sim$	
$\varnothing$	S
S	$\varnothing$

The above two sets of tables of operations show that these two Boolean algebras are isomorphic to each other.

†In this book, a prime $(')$ and a bar $(^-)$ will be used interchangeably to denote the complementation of a Boolean variable or a Boolean expression.

Example 1.3.2

Consider the four-element Boolean algebra $B_4 = \{0, a, b, 1\}$. The algebra is described by the following tables:

·	0	a	b	1
0	0	0	0	0
a	0	a	0	a
b	0	0	b	b
1	0	a	b	1

+	0	a	b	1
0	0	a	b	1
a	a	a	1	1
b	b	1	b	1
1	1	1	1	1

′	
0	1
a	b
b	a
1	0

Now consider the power set $P(I)$, where $I = \{a, b\}$. The subsets of I are $\{a\}$, $\{b\}$, and the empty set. Denote them by S_a, S_b, and $\varnothing$, respectively. As shown in Example 1.2.3, $P(I)$ is a Boolean algebra with the three operations shown in the following tables.

∩	$\varnothing$	S_a	S_b	I
$\varnothing$	$\varnothing$	$\varnothing$	$\varnothing$	$\varnothing$
S_a	$\varnothing$	S_a	$\varnothing$	S_a
S_b	$\varnothing$	$\varnothing$	S_b	S_b
I	$\varnothing$	S_a	S_b	I

∪	$\varnothing$	S_a	S_b	I
$\varnothing$	$\varnothing$	S_a	S_b	I
S_a	S_a	S_a	I	I
S_b	S_b	I	S_b	I
I	I	I	I	I

~	
$\varnothing$	I
S_a	S_b
S_b	S_a
I	$\varnothing$

Again, it is seen that B_4 is isomorphic to $P(I)$, where $I = \{a, b\}$.

From the two examples above, one may ask: Is every Boolean algebra isomorphic to a power-set algebra? The answer is yes. Before proceeding with the proof, we need to introduce the following definition and lemmas.

DEFINITION 1.3.2

In a lattice with zero element 0, any element covering 0 is called an *atom*.

Let B be a Boolean algebra and x be an element of B. If we define $A(x)$ to be the set of all atoms a such that $a \leq x$, it is clear that $A(0) = \varnothing$ and $A(1) = A$. Since B is a lattice, for any x other than 0, $A(x)$ is nonempty.

LEMMA 1.3.1

Define A to be the set of all atoms of a Boolean algebra B. Then
(a) $A(x) \cap A(x') = \varnothing$.
(b) $A(x) \cup A(x') = A$.
(c) $A(x) = A(y)$ iff $x = y$ for all x, y in B.
(d) $A(x) \subseteq A(y)$ iff $x \leq y$.

Proof: (a) Suppose that $A(x) \cap A(x') \neq \varnothing$. Then there must exist an atom, say z, in the intersection. $z \in A(x)$ implies that $z < x$; and $z \in A(x')$ implies that $z \leq x'$. These jointly imply that $z \cdot z = z \leq x \cdot x' = 0$ or $z = 0$. This is a contradiction, since z is an atom, $z \neq 0$. Hence, $A(x) \cap A(x') = \emptyset$.

(b) Let a be an element in A. Then $a \cdot x \leq a$. Since a is an atom, $a \cdot x$ is either a or 0. If $a \cdot x = a$, then $a \in A(x)$. If $a \cdot x = 0$, then $a \cdot x' = a$, since $a = a \cdot (x + x') = a \cdot x + a \cdot x' = a \cdot x'$; hence, $a \in A(x')$. Therefore, $A(x) \cup A(x') = A$.

(c) If $x = y$, it is clear that $A(x) = A(y)$. Conversely, assume that $A(x) = A(y)$ and let $x \neq y$. Then either $x \nleq y$, or $y \nleq x$, or both. If $x \nleq y$, then $x \cdot y \neq x$, which implies that $x \cdot y' \neq 0$. This is because if $x \cdot y' = 0$, then $x = x \cdot y + xy' = x \cdot y$, which contradicts the fact that $x \cdot y \neq x$. If $x \cdot y' \neq 0$, then there must exist at least an atom a such that $a \leq x \cdot y'$, which in turn implies that $a \leq x$ and $a \leq y'$. This means that $a \in A(x)$ and $a \in A(y')$. Since $a \in A(y')$, by (a) and (b), $a \notin A(y)$; then $A(x) \neq A(y)$, which contradicts our assumption. Thus, $x \nleq y$ cannot be the case. By symmetry, $y \nleq x$ also implies that $A(x) \neq A(y)$. Hence, x must be equal to y.

(d) If $x \leq y$, then for any atom a in $A(x)$, $a \leq x \leq y$; hence, $A(x) \subseteq A(y)$. Conversely, assume that $A(x) \subseteq A(y)$, but $x \geq y$. It follows that if a is an atom in $A(y)$, then $a \leq y \leq x$, which means that a must also be in $A(x)$. From this we conclude that $A(x) \supseteq A(y)$, which is a contradiction. Hence, $x \leq y$. ∎

LEMMA 1.3.2

For any x and y in a Boolean algebra,
(a) $A(x \cdot y) = A(x) \cap A(y)$.
(b) $A(x + y) = A(x) \cup A(y)$.
(c) $A(x') = A - A(x)$.

Proof: From the definition of $\cdot$, we have that $a \leq x \cdot y$ iff $a \leq x$ and $a \leq y$. If a is an atom in $A(x \cdot y)$, this relation implies that $a \in A(x)$ and $a \in A(y)$, or $A(x) \cap A(y)$; and also if a is an element of $A(x) \cap A(y)$, the relation above implies that $a \in A(x \cdot y)$. Equation (b) may be shown using the relation $a \leq x + y$ iff $a \leq x$ or $a \leq y$. Equation (c) follows directly from Lemma 1.3.1(b). ∎

LEMMA 1.3.3

Let $a_1, a_2, \ldots, a_n$ be n *distinct* atoms of a Boolean algebra. Then
(a) $A(a_1 \cdot a_2 \cdot \ldots \cdot a_n) = A(a_1) \cap A(a_2) \cap \ldots \cap A(a_n) = \emptyset$.
(b) $A(a_1 + a_2 + \ldots + a_n) = A(a_1) \cup A(a_2) \cup \ldots \cup A(a_n) = \{a_1, a_2, \ldots, a_n\}$.

Proof: Since the operations $\cdot$ and $+$ satisfy the associative laws, by repeatedly applying Lemma 1.3.3(a) and (b), we obtain

$$A(a_1 \cdot a_2 \cdot \ldots \cdot a_n) = A(a_1) \cap A(a_2) \cap \ldots \cap A(a_n)$$

$$A(a_1 + a_2 + \ldots + a_n) = A(a_1) \cup A(a_2) \cup \ldots \cup A(a_n)$$

Since a_i are atoms, $A(a_i) = \{a_i\}$. This is because if in $A(a_i)$ there is another atom a_j, then $a_j \leq a_i$. Since we assume that $a_j \neq a_i$, a_j must be less than a_i. In other words, $a_j = 0$. This is a contradiction. Hence, $A(a_1 \cdot a_2 \cdot \ldots \cdot a_n) = \emptyset$ and $A(a_1 + a_2 + \ldots + a_n) = \{a_1, a_2, \ldots, a_n\}$. ∎

From the above lemmas, we have the following important theorems.

THEOREM 1.3.2

Let $B = \{x_1, x_2, \ldots, x_n\}$ be a Boolean algebra. Define F to be the family of sets of $A(x_i)$; that is, $F = \{A(x_1), A(x_2), \ldots, A(x_n)\}$. Then $(B, \cdot, +, ')$ and $(F, \cap, \cup, \sim)$ are isomorphic.

Proof: Since all x_i are distinct, from Lemma 1.3.1(c), the transformation $\varphi = x_i \longrightarrow A(x_i)$ from B to F is one to one and onto. Also, the transformation φ carries the Boolean operations $\cdot$, $+$, and $'$ into the set-theoretic operations $\cap$, $\cup$, and $\sim$, respectively, from B to F. Hence, the theorem is proved. ∎

THEOREM 1.3.3

Let A be the set of all atoms $a_1, a_2, \ldots, a_m$ of a Boolean algebra B. Then $F = P(A)$.

Proof: Obviously, every element of F is an element of $P(A)$. For the converse, consider an element in $P(A)$, say $A_k = \{a_1, a_2, \ldots, a_k\}$, where a_i are assumed to be distinct, and $k \leq m$.

$$A_k = A(a_1) \cup A(a_2) \cup \ldots \cup A(a_k) = A(a_1 + a_2 + \ldots + a_k) = A(x_k)$$

which is in F, where $x_k = a_1 + a_2 + \ldots + a_k$. Hence, $F = P(A)$. ∎

Combining Theorems 1.3.2 and 1.3.3, we have

THEOREM 1.3.4

Let A be the set of all atoms of a Boolean algebra B. Then B is isomorphic to $P(A)$.

COROLLARY 1.3.1

The number of elements of a Boolean algebra is a power of 2. In other words, for a given Boolean algebra B, there exists a positive integer n such that the number of elements of B is 2^n.

COROLLARY 1.3.2

If B and C are two Boolean algebras with the same number of elements, then B and C are isomorphic.

COROLLARY 1.3.3

A 2^n-element Boolean algebra B has exactly n atoms.

Proof: By Theorem 1.3.4, B is isomorphic to $P(A)$, and it has been shown in Theorem 1.1.1 that if A has n distinct elements, then $P(A)$ has 2^n distinct elements. ∎

THEOREM 1.3.5

The Cartesian product of two sets B and C is a Boolean algebra iff B and C are Boolean algebras.

The proof is routine and is left to the reader.

COROLLARY 1.3.4

Let B and C be two Boolean algebras with finite m and n atoms, respectively. Then the Cartesian product algebra $B \times C$ has $2^{m \times n}$ elements.

COROLLARY 1.3.5

Any Boolean algebra B is isomorphic to a Cartesian product of n two-element Boolean algebras $B_2 = \{0, 1\}$, namely, $B_2^n = \underbrace{B_2 \times B_2 \times \ldots \times B_2}_{n}$ for some integer n.

For example, the four-element Boolean algebra B_4 in Example 1.3.2 is isomorphic to $B_2^2 = B_2 \times B_2 = \{(0, 0), (0, 1), (1, 0), (1, 1)\}$ as shown in Fig. 1.3.1. The Boolean algebras B_2^n, $n = 3$ and 4, are shown in Fig. 1.3.2.

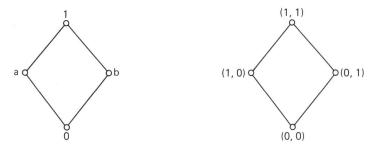

Fig. 1.3.1 Two isomorphic lattices.

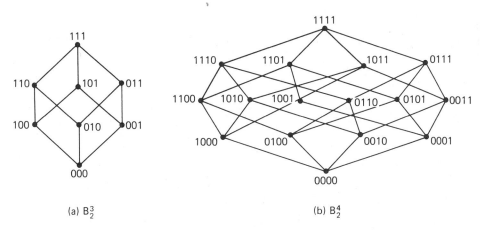

(a) B_2^3 (b) B_2^4

Fig. 1.3.2 Two Cartesian product algebras.

Example 1.3.3

Consider the eight-element Boolean algebra (B_8) shown in Fig. 1.3.3(a). It is easy to find the products and sums of all the pairs of elements and the complement of each element of this

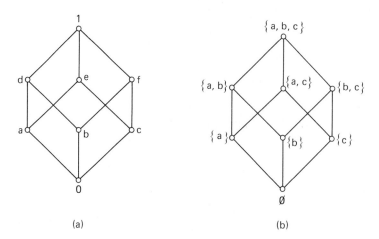

Fig. 1.3.3　Two eight-element Boolean algebras.

algebra. The tables for the three operations are as follows:

·	0	a	b	c	d	e	f	1		+	0	a	b	c	d	e	f	1		,	
0	0	0	0	0	0	0	0	0		0	0	a	b	c	d	e	f	1		0	1
a	0	a	0	0	a	a	0	a		a	a	a	d	e	d	e	1	1		a	f
b	0	0	b	0	b	0	b	b		b	b	d	b	f	d	1	f	1		b	e
c	0	0	0	c	0	c	c	c		c	c	e	f	c	1	e	f	1		c	d
d	0	a	b	0	d	a	b	d		d	d	d	d	1	d	1	1	1		d	c
e	0	a	0	c	a	e	c	e		e	e	e	1	e	1	e	1	1		e	b
f	0	0	b	c	b	c	f	f		f	f	1	f	f	1	1	f	1		f	a
1	0	a	b	c	d	e	f	1		1	1	1	1	1	1	1	1	1		1	0

The atoms are a, b, and c, and the sets $A(x)$ are

$$A(a) = \{a\} \qquad A(d) = \{a, b\} \qquad A(0) = \{\varnothing\}$$
$$A(b) = \{b\} \qquad A(e) = \{a, c\} \qquad A(1) = \{a, b, c\}$$
$$A(c) = \{c\} \qquad A(f) = \{b, c\}$$

The algebra is isomorphic to the power set $P(A)$ [where $A = \{a, b, c\}$, as shown in Fig. 1.3.3(b)], as seen from the diagram and the relation between x and $A(x)$ shown above.

　　The smallest infinite power-set algebra is the power set whose cardinal number is equal to the cardinal number of the set of natural numbers α_0. The Boolean algebras isomorphic to this power set contain $2^{\alpha_0} = c$ elements, each of which is an infinite sequence of 0's and 1's, where c is the cardinality of the real numbers. Thus, there are no power-set algebras of cardinality α_0. However, we may obtain Boolean algebras that are subalgebras of direct products of B_2 and have cardinality α_0.

We now want to turn to some basic properties of Boolean algebras.

THEOREM 1.3.6 De Morgan's Laws or De Morgan's Theorem

In a Boolean algebra,

$$(x \cdot y)' = x' + y'$$
$$(x + y)' = x' \cdot y'$$

Proof: Since

$$
\begin{aligned}
(x \cdot y) + (x' + y') &= [x + (x' + y')] \cdot [y + (x' + y')] && \text{by L6b} \\
&= [(x + x') + y'] \cdot [(y + y') + x'] = 1 && \text{by L3}_+ \text{ and L7}
\end{aligned}
$$

and

$$
(x \cdot y) \cdot (x' + y') = (x \cdot y \cdot x') + (x \cdot y \cdot y') = 0 \qquad \text{by L6a and L7}
$$

Therefore, $x \cdot y$ and $x' + y'$ are complementary to each other. By Theorem 1.2.8, $x \cdot y$ is the unique complement of $x' + y'$ and vice versa. Hence,

$$(x \cdot y)' = x' + y'$$

Dually, $x' \cdot y'$ is the complement of $x + y$. ∎

THEOREM 1.3.7

In a Boolean algebra,

$$x \leq y \Longleftrightarrow x \cdot y' = 0 \Longleftrightarrow x' + y = 1$$

Proof: By Theorem 1.2.2(b),

$$x \leq y \Longleftrightarrow x \cdot y' \leq y \cdot y' = 0 \Longleftrightarrow x \cdot y' = 0$$

Conversely, if $x \cdot y' = 0$, then

$$
\begin{aligned}
x + y &= (x + y) \cdot (y' + y) && \text{by definition of} \\
& && \text{meet and L7} \\
&= (x \cdot y') + y && \text{by modular equality,} \\
& && \text{L6a, and L7} \\
&= 0 + y && \text{by hypothesis} \\
&= y && \text{by definition of join}
\end{aligned}
$$

Thus, $x \leq y$. So far, we have shown that $x \leq y \Longleftrightarrow x \cdot y' = 0$.

Next, we want to show $x \cdot y' = 0 \Longleftrightarrow x' + y = 1$. Suppose that $x \cdot y' = 0$. Then,

$$
\begin{aligned}
x' + y &= (x \cdot y')' && \text{by Theorem 1.3.6} \\
&= 0' = 1
\end{aligned}
$$

Conversely, if $x' + y = 1$, then, by Theorem 1.3.6, $x \cdot y' = (x' + y)' = 1' = 0$. Hence, the theorem is proved.   ∎

THEOREM 1.3.8

(a) The dual of a Boolean algebra is a Boolean algebra.
(b) Every interval of a Boolean algebra is a Boolean algebra.
(c) Any homomorphic image of a Boolean algebra is a Boolean algebra.

Proof: The proofs of (a) and (b) are obvious.
The proof of (c) is as follows: Let B be a Boolean algebra and x, y, and z be any elements of the image H of B under a homomorphic h. Then there exists at least one set of elements a, b, and c in B such that $h(a) = x$, $h(b) = y$, and $h(c) = z$. It follows that

$$x \cdot (y + z) = h(a) \cdot [h(b) + h(c)]$$
$$= h(a) \cdot h(b + c)$$
$$= h[a \cdot (b + c)]$$
$$= h[a \cdot b + a \cdot c] \qquad \text{by L6a}$$
$$= h(a \cdot b) + h(a \cdot c)$$
$$= h(a) \cdot h(b) + h(a) \cdot h(c)$$
$$= x \cdot y + x \cdot z$$

Thus, H is distributive. Now define x' be the image of a' under h. Then

$$x \cdot x' = h(a) \cdot h(a')$$
$$= h(a \cdot a')$$
$$= h(0)$$
$$= 0$$

Similarly, we can show that

$$x + x' = 1$$

Hence, H is a Boolean algebra.   ∎

Exercise 1.3

1. Let x, y, and z be any elements of a Boolean algebra B. Prove the following properties of a Boolean algebra.
    (a) $x + x = x$
    (b) $x \cdot x = x$
    (c) $x + 1 = 1$
    (d) $x \cdot 0 = 0$
    (e) $x + xy = x$
    (f) $x(x + y) = x$

(g) $x(x' + y) = xy$
(h) $x + x'y = x + y$
(i) $(x + y)(x' + z) = xz + x'y$

2. Let a and b be two fixed elements of a Boolean algebra B. If $ax = bx$ and $ax' = bx'$, then $a = b$.

3. Is a sublattice of a Boolean algebra always a Boolean algebra? Justify your answer.

4. Show that the Cartesian product $P(A) \times P(B)$ of two power sets with finite m and n atoms is isomorphic to a power set with $m + n$ atoms.

5. For any element x and y in a Boolean algebra, show that $(x + y) \cdot (x' + z) = x \cdot z + x' \cdot y$

6. In any Boolean algebra B, prove the following:
(a) $(x')' = x$
(b) If x is the complement of y, then y must be the complement of x.
(c) $0' = 1$ and $1' = 0$
(d) If $x \le y$, then $y' \le x'$.
(e) If $x' \le y'$, then $y \le x$.

7. Either give a Boolean algebra with three elements or show that none exists.

8. If x is an element of the interval $[a, b]$ of a lattice L and $y \in L$ such that $x \cdot y = a$ and $x + y = b$, y is called a *complement of x relative to the interval* $[a, b]$ (for a definition of interval $[a, b]$, see problem 7(c) of Exercise 1.2). Show that a Boolean algebra is relatively complemented.

9. A lattice for which every subset has both a g.l.b. and a l.u.b. is said to be *complete*. Prove that the infinite distributive laws

$$x \cdot (y_1 + y_2 + \ldots) = x \cdot y_1 + x \cdot y_2 + \ldots$$
$$x + (y_1 \cdot y_2 \cdot \ldots) = (x + y_1) \cdot (x + y_2) \cdot \ldots$$

hold in a complete Boolean algebra.

10. Consider the eight-element Boolean algebra $B_8 = \{0, a, b, c, d, e, f, 1\}$. The multiplication, addition, and complementation tables are given below.

·	0	a	b	c	d	e	f	1
0	0	0	0	0	0	0	0	0
a	0	a	a	a	0	0	0	a
b	0	a	b	a	0	f	f	b
c	0	a	a	c	d	d	0	c
d	0	0	0	d	d	d	0	d
e	0	0	f	d	d	e	f	e
f	0	0	f	0	0	f	f	f
1	0	a	b	c	d	e	f	1

+	0	a	b	c	d	e	f	1
0	0	a	b	c	d	e	f	1
a	a	a	b	c	c	1	b	1
b	b	b	b	1	1	1	b	1
c	c	c	1	c	c	1	1	1
d	d	c	1	c	d	e	e	1
e	e	1	1	1	e	e	e	1
f	f	b	b	1	e	e	f	1
1	1	1	1	1	1	1	1	1

′	
0	1
a	e
b	d
c	f
d	b
e	a
f	c
1	0

(a) Find all the atoms of this algebra.
(b) Find the sets $A(0), A(a), \ldots, A(1)$.

(c) Draw the inclusion diagram of this algebra and show that it is isomorphic to a power-set algebra.

11. Draw a picture of a lattice that is not a Boolean algebra. Prove that, by showing that it is not distributive or not complemented. For example, the following lattice *is* a Boolean algebra.

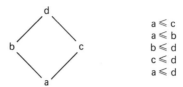

$$a \leqslant c$$
$$a \leqslant b$$
$$b \leqslant d$$
$$c \leqslant d$$
$$a \leqslant d$$

12. Let φ be a transformation mapping from B_8 of Example 1.3.3 into itself which is defined by

$$\varphi(0) = 0$$
$$\varphi(a) = \varphi(b) = \varphi(c) = a$$
$$\varphi(d) = \varphi(e) = \varphi(f) = c$$
$$\varphi(1) = e$$

Show that φ is homomorphic; hence, the image of φ is a Boolean algebra.

13. A Boolean algebra B consists of eight elements $a, b, \ldots, h$. Its multiplication table is

·	a	b	c	d	e	f	g	h
a	a	g	a	a	g	g	g	a
b	g	b	b	b	g	b	g	g
c	a	b	c	c	g	b	g	a
d	a	b	c	d	e	f	g	h
e	g	g	g	e	e	e	g	e
f	g	b	b	f	e	f	g	e
g	g	g	g	g	g	g	g	g
h	a	g	a	h	e	e	g	h

The complementation table is

$$a' = f, \quad b' = h, \quad c' = e, \quad d' = g$$

(a) Find the identities for addition and multiplication.
(b) Find the atoms of the algebra and express each element as a sum of atoms.
(c) Construct the set analog of the algebra and illustrate it on a Venn diagram.

14. Let $S = \{1, 2, \ldots, n\}$, where n is even, and let $\subseteq$ be the usual partial relation defined by $A \subseteq B$ iff A is contained in B.
(a) Let $O(S)$ be the collection of subsets of S which contain an odd number of elements, e.g., $\{1, 2, 4\}$. Is $O(S)$ a Boolean algebra under the above partial ordering? Why or why not?

(b) Let $R(S)$ be the collection of subsets of S all of whose elements are odd numbers, for example, $\{1, 5\}$. Is $R(S)$ a Boolean algebra under the partial ordering above? Why or why not?

(c) Let A be a proper subset of S; that is, $A \subset S$ but $A \neq S$. Let $T(S)$ be the set of all subsets of S that contain A. Is A a Boolean algebra under the partial ordering above? Why or why not?

(d) How many elements do $O(S)$, $R(S)$, and $T(S)$ contain?

1.4 Boolean Function and Its Canonical Forms

In this section we shall discuss Boolean functions (or Boolean expressions, or Boolean formulas, or Boolean forms) and Boolean mappings (or functions) from a Boolean algebra into itself. We shall present two canonical forms for these functions and their various representations. The type of definition that we shall use is that known as a "recursive" definition. In it, two simple types of functions, the *constant function* and the *projection function*, are defined, and rules are given for constructing all other Boolean functions from these.

DEFINITION 1.4.1

An element of a Boolean algebra B is called a *constant on B*.

DEFINITION 1.4.2

A symbol that may represent any one of the elements of B is called a (*Boolean*) *variable on B*.

A Boolean function of a Boolean algebra is defined as:

DEFINITION 1.4.3

Let $x_1, x_2, \ldots, x_n$ be variables on a Boolean algebra B. A mapping f of B into itself is a *Boolean function* of n variables, denoted by $f(x_1, \ldots, x_n)$, if it can be constructed according to the following rules.

1. Let a denote a constant on B. $f(x_1, \ldots, x_n) = a$ and $f(x_1, \ldots, x_n) = x_i$ are Boolean functions. The former is called the *constant function* and the latter is called the *projection function*.

2. If $f(x_1, \ldots, x_n)$ is a Boolean function, then $(f(x_1, \ldots, x_n))'$ is a Boolean function.

3. If $f_1(x_1, \ldots, x_n)$ and $f_2(x_1, \ldots, x_n)$ are Boolean functions, then $f_1(x_1, \ldots, x_n) + f_2(x_1, \ldots, x_n)$ and $f_1(x_1, \ldots, x_n) \cdot f_2(x_1, \ldots, x_n)$ are Boolean functions.

4. Any function that can be constructed by a finite number of applications of the rules above, and only such a function, is a Boolean function.

Thus, a Boolean function is any function that can be constructed from the constant function and projection functions by finitely many uses of the operations of $'$, $+$, and $\cdot$. For a function of one variable, the projection function is the *identity* function $f(x) = x$.

In writing a Boolean function we often drop the "·" and the unnecessary parentheses as we do in ordinary algebra; for instance, for a function such as

$$f_1(x_1, x_2) = [x_1 + (x_1' \cdot x_2)] \cdot x_2'$$

we shall simply write

$$f_1(x_1, x_2) = (x_1 + x_1'x_2)x_2' \qquad (1.4.1)$$

This function is equivalent to

$$f_2(x_1, x_2) = x_1x_2' + x_1'x_2x_2' \qquad (1.4.2)$$

or

$$f_3(x_1, x_2) = x_1x_2' \qquad (1.4.3)$$

A complete list of the values of a Boolean function is called a *truth table*. The truth table, for example, of the function of Eq. (1.4.1) is given in Table 1.4.1.

TABLE 1.4.1

x_1	x_2	$f(x_1, x_2)$ of Eq. (1.4.1)
0	0	0
0	1	0
1	0	1
1	1	0

DEFINITION 1.4.4

Two Boolean functions are *equivalent* if they have identical truth tables.

For example, the functions of Eqs. (1.4.1)–(1.4.3) have the same truth table, thus are equivalent.

Now we would like to ask the question: How many different† $f(x_1, \ldots, x_n)$ Boolean functions on B_2 are there? Before answering this question, let us examine how many different mappings there are which map from the Cartesian product set $\{0, 1\}^n$ into $\{0, 1\}$? This is answered in the following theorem.

THEOREM 1.4.1

There are 2^{2^n} different mappings which map from the Cartesian product set $\{0, 1\}^n$ into $\{0, 1\}$.

†"Different" here means "nonequivalent."

Proof: In the domain, it is clear that there are $\underbrace{2 \times 2 \times \ldots \times 2}_{n} = 2^n$ elements. Since each of these 2^n elements may take either of two values 0 or 1 independently, there are

$$\underbrace{2 \times 2 \times \ldots \times 2}_{2^n} = 2^{2^n} \tag{1.4.4}$$

different mappings. ∎

According to Theorem 1.4.1 for $n = 1$ (one variable), there are $2^{2^1} = 2^2 = 4$ different mappings, and for $n = 2$ (two variables), there are $2^{2^2} = 2^4 = 16$ different mappings. The four different one-variable mappings from $\{0, 1\}$ into $\{0, 1\}$ are shown in Fig. 1.4.1(a). It is seen that all four functions $f_1(x) = 0$, $f_2(x) = x$, $f_3(x) = x'$, and $f_4(x) = 1$ can be constructed according to the rules given in Definition 1.4.3 and thus are Boolean functions. Furthermore, if we denote $\Phi_1 = \{f_1(x), f_2(x), f_3(x)$, and $f_4(x)\}$, then $(\Phi_1; +, \cdot, '; f_1(x), f_4(x))$ form a Boolean algebra as shown in Fig. 1.4.1(b). Note that the 0 and 1 in Fig. 1.4.1(b) are the 0-function and the 1-function, respectively. This Boolean algebra is isomorphic to the Cartesian product algebra B_2^2.

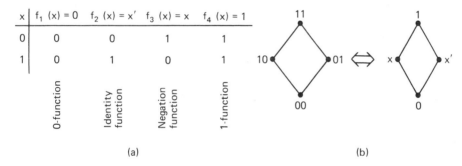

x	$f_1(x) = 0$	$f_2(x) = x'$	$f_3(x) = x$	$f_4(x) = 1$
0	0	0	1	1
1	0	1	0	1
	0-function	Identity function	Negation function	1-function

(a)

(b)

Fig. 1.4.1 (a) The 4 one-variable functions. (b) The Boolean algebra (lattice) formed by the 4 one-variable functions.

Now let us examine the 16 two-variable mappings which are tabulated in Fig. 1.4.2(a). Among the 16 mappings, we readily see that $f_1, f_2, f_4, f_6, f_8, f_{11}, f_{13}$, and f_{16} are Boolean functions. The remaining eight, $f_3, f_5, f_7, f_9, f_{10}, f_{12}, f_{14}$, and f_{15}, need to be examined.

THEOREM 1.4.2

The eight mappings $f_3, f_5, f_7, f_9, f_{10}, f_{12}, f_{14}$, and f_{15}, can be expressed in terms of the three functions $+$, $\cdot$, and $'$ as follows:

(a) $f_3(x_1, x_2) = x_1 \not\supset x_2 = x_1 x_2'$ \hfill (1.4.5)
(b) $f_5(x_1, x_2) = x_1 \not\subset x_2 = x_1' x_2$ \hfill (1.4.6)
(c) $f_7(x_1, x_2) = x_1 \oplus x_2 = x_1 x_2' + x_1' x_2$ \hfill (1.4.7)†

†The symbol "≡" is used by logicians for $\oplus$.

x_1 x_2	$f_1(x_1,x_2) = 0$	$f_2 = x_1 \cdot x_2$	$f_3 = x_1 \not\supset x_2$	$f_4 = x_1$	$f_5 = x_1 \not\subset x_2$	$f_6 = x_2$	$f_7 = x_1 \oplus x_2$	$f_8 = x_1 + x_2$
0 0	0	0	0	0	0	0	0	0
0 1	0	0	0	0	1	1	1	1
1 0	0	0	1	1	0	0	1	1
1 1	0	1	0	1	0	1	0	1
	0-function	AND function (conjunction function)	nonconditional implication function	x_1-projection function	nonconverse implication function	x_2-projection function	EXCLUSIVE-OR function (nonequivalence function)	OR function (disjunction function)

x_1 x_2	$f_9 = x_1 \downarrow x_2$	$f_{10} = x_1 \not\equiv x_2$	$f_{11} = x'_2$	$f_{12} = x_1 \subset x_2$	$f_{13} = x'_1$	$f_{14} = x_1 \supset x_2$	$f_{15} = x_1 \mid x_2$	$f_{16} = 1$
0 0	1	1	1	1	1	1	1	1
0 1	0	0	0	0	1	1	1	1
1 0	0	0	1	1	0	0	1	1
1 1	0	1	0	1	0	1	0	1
	NOR function (Pierce arrow)	equivalence function	NOT x_2-projection function	converse implication function	NOT x_1-projection function	conditional implication function	NAND function (Sheffer stroke)	1-function

Fig. 1.4.2(a)   The 16 two-variable functions.

(d) $f_9(x_1, x_2) = x_1 \downarrow x_2 = x'_1 x'_2$ (1.4.8)

(e) $f_{10}(x_1, x_2) = x_1 \not\equiv x_2 = x_1 x_2 + x'_1 x'_2$ (1.4.9)

(f) $f_{12}(x_1, x_2) = x_1 \subset x_2 = x_1 + x'_2$ (1.4.10)

(g) $f_{14}(x_1, x_2) = x_1 \supset x_2 = x'_1 + x_2$ (1.4.11)

(h) $f_{15}(x_1, x_2) = x_1 \mid x_2 = x'_1 + x'_2$ (1.4.12)

Proof: The proof of these eight expressions can be obtained by comparing their truth tables. Take (a), for instance; the truth tables of $x_1 \not\supset x_2$ and $x_1 x'_2$ are as follows:

x_1	x_2	$x_1 \not\supset x_2$	$x_1 x'_2$
0	0	0	0
0	1	0	0
1	0	1	1
1	1	0	0

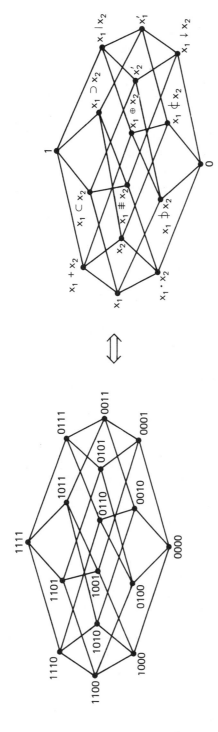

Fig. 1.4.2(b)   The Boolean algebra (lattice) formed by the 16 two-variable functions.

It is seen that they are identical; hence, $x_1 \,\oplus\, x_2 = x_1 x_2'$. The rest may be verified similarly.

Theorem 1.4.2 shows that all the 16 mappings from $\{0, 1\}^2$ into $\{0, 1\}$ shown in Fig. 1.4.2(a) are Boolean functions. More interestingly, the system $(\Phi_2; +, \cdot, '; 0, 1)$, where $\Phi_2 = \{f_1(x_1, x_2), \ldots, f_{16}(x_1, x_2)\}$, is also a Boolean algebra [see Fig. 1.4.2(b)], which is isomorphic to the Cartesian product algebra B_2^4 in Fig. 1.3.2(b). The proof is left to the reader as an exercise (problem 1).

The isomorphism relation between Φ_2 and B_2^4 makes the finding of the expressions of $f_3, f_5, f_7, f_9, f_{10}, f_{12}, f_{14}$, and f_{15} in terms of the functions $+$, $\cdot$, and $'$ easy. Take $f_7(x_1, x_2) = x_1 \oplus x_2$, for instance. Since the mapping $x_1 \oplus x_2$ is

x_1	x_2	$x_1 \oplus x_2$
0	0	0
0	1	1
1	0	1
1	1	0

we place the function $x_1 \oplus x_2$ at the 0110 position of the 16-element lattice. Then we see that $x_1 \oplus x_2$ is the meet of $(x_1 + x_2)$ and $x_1 \,|\, x_2$, and the latter is the join of x_1' and x_2'. Thus,

$$x_1 \oplus x_2 = (x_1 + x_2)(x_1' + x_2') = x_1 x_2' + x_1' x_2$$

The other seven expressions of Theorem 1.4.2 can all be obtained in this way.

From the above discussion, we have the following theorem.

THEOREM 1.4.3

Let Φ_n be the set of 2^{2^n} different mappings which map from $\{0, 1\}^n$ into $\{0, 1\}$. Then

(a)   Every one of the 2^{2^n} mappings is a Boolean function on B_2. So there are 2^{2^n} distinct Boolean functions.

(b) $(\Phi_n; +, \cdot, '; 0, 1)$ is a Boolean algebra and is isomorphic to $B_2^{(2^n)}$.

From the above, it is found that f_1, f_2, and f_3 of Eqs. (1.4.1)–(1.4.3) are equivalent because they have the same truth table. One may ask: Other than checking all combinations of argument values, is there an analytical procedure for determining whether two Boolean expressions represent the same function? The answer is yes. The required criterion is obtained by constructing a *canonical form* for each Boolean function and by showing that two expressions have the same canonical form iff they are equivalent. This means that the canonical form of a Boolean expression is *unique*.

DEFINITION 1.4.5

A *literal* is defined to be a constant, variable, or complemented variable.

There are two major canonical forms for Boolean expressions: the *canonical sum-of-products form* or *disjunctive normal form* and the *canonical product-of-sums form* or *conjunctive normal form*.

The canonical sum-of-products form:

$$f(x_1, \ldots, x_n) = \sum f(e_1, \ldots, e_n) x_1^{e_1} x_2^{e_2} \ldots x_n^{e_n} \qquad (1.4.13)$$

where

$$x_i^{e_i} = \begin{cases} 1 & \text{if } x_i = e_i \\ 0 & \text{otherwise} \end{cases}$$

The canonical product-of-sums form:

$$f(x_1, \ldots, x_n) = \prod (f(e_1, \ldots, e_n) + x_1^{e_1{}'} + x_2^{e_2{}'} + \ldots + x_n^{e_n{}'}) \qquad (1.4.14)$$

where

$$x_i^{e_i{}'} = \begin{cases} 0 & \text{if } x_i = e_i \\ 1 & \text{otherwise} \end{cases}$$

The proofs of these two canonical forms are evident from their truth tables and are left to the reader as an exercise (problem 5).

Example 1.4.1

Construct the canonical sum-of-products form of the function defined on $B_4 = \{0, a, b, 1\}$ whose truth table is

x_1	x_2	$f(x_1, x_2)$
0	0	a
0	b	1
1	a	a
a	b	b
otherwise		0

The canonical sum-of-products form of this function is

$$f(x_1, x_2) = a x_1^0 x_2^0 + x_1^0 x_2^b + a x_1^1 x_2^a + b x_1^a x_2^b \qquad (1.4.15)$$

The canonical form of a Boolean function guarantees that different functions have different canonical forms. Thus, in determining whether two logical formulas are logically equivalent, we *need not* construct their truth tables; we need only find their canonical forms. If they are equivalent, they should have the same form; otherwise, their canonical forms will be different.

It is interesting to note that the only constants in a Boolean function on B_2 are 0 and 1. Any Boolean function multiplied by 0 is 0; and any Boolean function multiplied by 1 is unchanged. Thus, in expressing the canonical sum-of-products form of a Boolean function on B_2, the terms are omitted from the expression if their multiplicative constants are 0, and are retained if they are 1.

Example 1.4.2

Determine whether the following two functions are equivalent.

$$f_1(x_1, x_2) = x_1 + x_2 + (x_1 \downarrow x_2) \tag{1.4.16}$$

and

$$f_2(x_1, x_2) = [(x_1 x_2 + x_1 + x_2) \supset (x_1' + (x_1 | x_2'))] \not\equiv [x_1' \supset ((x_2 \subset x_1) \downarrow x_2)] \tag{1.4.17}$$

Boolean functions such as the above two, involving operations other than the $+$, $\cdot$, and $'$ operations, need to be converted into a function that includes exclusively the three basic operations $+$, $\cdot$, and $'$, if the second procedure described above is to be applied. For example, by using the formulas given in Theorem 1.4.2, Eq. (1.4.16) and (1.4.17) can be expressed exclusively in the three basic operations.

$$f_1(x_1, x_2) = x_1 + x_2 + x_1' x_2' \qquad \text{by Eq. (1.4.8)}$$
$$f_2(x_1, x_2) = \{[(x_1 x_2 + x_1 + x_2)' + x_1' + (x_1' + x_2')]' + [(x_1')' + (x_2 + x_1')'(x_2')']'\}$$
$$= \{[(x_1 x_2 + x_1 + x_2)' + (x_1' + (x_1' + x_2))] + [(x_1')' + (x_2 + x_1')'(x_2')']\}$$
$$\text{by Eqs. (1.4.7) and (1.4.10)-(1.4.12)}$$

The canonical sum-of-products forms of these functions are

$$f_1(x_1, x_2) = x_1' x_2' + x_1' x_2 + x_1 x_2' + x_1 x_2$$

and

$$f_2(x_1, x_2) = x_1' x_2' + x_1' x_2 + x_1 x_2'$$

Hence, they are not equivalent.

Example 1.4.3

Consider

$$f_3(x_1, x_2) = (x_1 \supset x_2) \not\equiv (x_2' \supset x_1')$$
$$= (x_1' + x_2)(x_2 + x_1') + (x_1' + x_2)'(x_2 + x_1')' \tag{1.4.18}$$

and

$$f_4(x_1, x_2) = (x_1 + x_2)' \not\equiv x_1' x_2'$$
$$= (x_1 + x_2)'(x_1' x_2') + (x_1 + x_2)(x_1' x_2')' \tag{1.4.19}$$

It is easy to verify that the canonical sum-of-products forms of the two functions are the same, which is $x_1'x_2' + x_1'x_2 + x_1x_2' + x_1x_2$, and thus they are equivalent. As a matter of fact, the functions f_1, f_3, and f_4 are equivalent, since they all have the same canonical form.

As a final remark, the canonical forms of a Boolean function provide us not only with a convenient means of determining whether two given functions are equivalent, but also with a way of constructing a function from its truth table. For instance, a truth table such as

x_1	x_2	x_3	$f(x_1, x_2, x_3)$		Terms in canonical sum-of-products form of f
0	0	0	0		
0	0	1	0		
0	1	0	1	$\longrightarrow$	$x_1'x_2x_3'$
0	1	1	1	$\longrightarrow$	$x_1'x_2x_3$
1	0	0	1	$\longrightarrow$	$x_1x_2'x_3'$
1	0	1	0		
1	1	0	0		
1	1	1	1	$\longrightarrow$	$x_1x_2x_3$

can be immediately represented by the canonical form

$$f(x_1, x_2, x_3) = x_1'x_2x_3' + x_1'x_2x_3 + x_1x_2'x_3' + x_1x_2x_3 \qquad (1.4.20)$$

Exercise 1.4

1. Show, in detailed steps, that Φ_2 is isomorphic to B_2^4.

2. Prove that the following Boolean functions on B_2 are equivalent to the 1-function.
   (a) $(x_1x_2) \supset x_1$
   (b) $[x_1(x_1 \supset x_2)] \supset x_2$
   (c) $(x_1 \supset x_2) \not\equiv (x_1' + x_2)$
   (d) $[(x_1 \supset x_2)(x_3 \supset x_4)] \supset [(x_1 + x_3) \supset (x_2 + x_4)]$

3. Prove that the following Boolean functions on B_2 are equivalent to the 0-function.
   (a) $(x_1 + x_2)x_1'x_2'$
   (b) $(x_1 \supset (x_1' + x_2'))'$
   (c) $(x_1 \supset x_2) \not\equiv (x_1x_2')$
   (d) $(x_1 \supset (x_1 + x_2))'$

4. Prove that on B_2,

$$f_1(x_1, x_2) = (x_1 + x_2)' + (x_1 \downarrow x_2)'$$
$$f_2(x_1, x_2) = (x_1 \supset x_2)' \not\equiv (x_2' \supset x_1')'$$

and

$$f_3(x_1, x_2) = (x_1 + x_2) \not\equiv (x_1'x_2')'$$

are equivalent.

5. Prove that every Boolean function can be expressed in the canonical sum-of-products and product-of-sums forms of Eqs. (1.4.13) and (1.4.14), respectively.

6. Let x_1, x_2, x_3, and x_4 be variables on a Boolean algebra.
   (a) Find the canonical sum-of-products form for the Boolean function

   $$f(x_1, x_2, x_3) = (ax_1x_2 + bx_2x_3)(a'x_1x_3 + b'x_1x_2)'$$

   where f is defined on $B_4 = \{0, a, b, 1\}$.
   (b) Repeat (a) for $f(x_1, x_2, x_3, x_4) = x_1 + x_2'x_3 + x_1'x_3x_4$ being defined on B_2.

7. Prove that

   $$x_1x_4' + x_1x_3x_4 + x_1'x_3' = x_3'x_4' + x_1x_3 + x_1'x_3'x_4$$

8. (a) Find the canonical sum-of-products form for the function described by the following truth table:

x_1	x_2	x_3	$f(x_1, x_2, x_3)$
0	0	0	0
0	0	1	1
0	1	0	1
0	1	1	0
1	0	0	0
1	0	1	0
1	1	0	1
1	1	1	0

   (b) Find the canonical sum-of-products form for the complement of the function described in (a).

9. Show that the Boolean functions

   $$f_1(x_1, x_2, x_3) = x_1x_2 + x_2x_3 + x_3x_1'$$

   and

   $$f_2(x_1, x_2, x_3) = x_1x_2 + x_3x_1'$$

   which are defined on a Boolean algebra, have the same canonical sum-of-products form, hence are equivalent.

10. (a) Derive a procedure similar to that given in this section for obtaining the canonical product-of-sums form.
    (b) Give the canonical product-of-sums form of each of the following Boolean functions on B_4.
    (1) $f(x_1, x_2, x_3) = x_1x_2' + x_1x_3 + x_1'x_2$
    (2) $f(x_1, x_2, x_3) = (x_1 + x_2 + x_3)(x_1x_2 + x_1'x_3)$
    (3) $f(x_1, x_2, x_3) = x_1 + x_2 + x_3 + x_1'x_2' + x_2'x_3' + x_1'x_3'$

11. Prove that the two Boolean functions of problem 9 are equivalent using the canonical product-of-sums form.

12. Let

$$g(x_1, x_2, x_3) = x_1'x_2'x_3' + x_1'x_2x_3' + x_1'x_2x_3 + x_1x_2x_3'$$
$$h(x_1, x_2, x_3) = x_1'x_2'x_3' + x_1'x_2'x_3 + x_1x_2x_3$$

(a) Show that the canonical product-of-sums form of $g(x_1, x_2, x_3)$ is

$$g(x_1, x_2, x_3) = (x_1 + x_2 + x_3')(x_1' + x_2 + x_3)(x_1' + x_2 + x_3')(x_1' + x_2' + x_3')$$

(b) Write the canonical sum-of-products form for $g + h$.
(c) Write $(g(x_1, x_2, x_3))'$ in canonical sum-of-products form and also in canonical product-of-sums form.

13. A Boolean function $f(x_1, x_2, x_3)$ is 0 whenever two or more of the arguments are 1, and $f = 1$ otherwise. Obtain the following Boolean formulas for f:
(a) The canonical sum-of-products form.
(b) The canonical product-of-sums form.
(c) A form simpler than either (a) or (b).

Bibliographical Remarks

The material in this chapter has been carefully selected to meet the objective of being concise and providing the necessary mathematical background for this book. All the references listed have extensive discussions on their respective subject. References 5, 17, 26, and 35 are particularly recommended.

References

Set Theory

1. ALLENDOERFER, C. B., and C. O. OAKLEY, *Principles of Mathematics*, McGraw-Hill, New York, 1955.
2. BREUER, J., *Introduction of the Theory of Sets*, Prentice-Hall, Englewood Cliffs, N.J., 1958.
3. FRAENKEL, A. A., *Abstract Set Theory*, North-Holland, Amsterdam, 1953.
4. HALMOS, P. R., *Naive Set Theory*, Van Nostrand Reinhold, New York, 1960.
5. KEMENY, J. G., H. MIRKIL, J. L. SNELL, and G. L. THOMPSON, *Finite Mathematical Structure*, Prentice-Hall, Englewood Cliffs, N.J., 1958.
6. KLEENE, S. C., *Introduction to Metamathematics*, Van Nostrand Reinhold, New York, 1952.
7. Mathematical Association of America, Committee on the Undergraduate Program, *Elementary Mathematics of Sets*, Ann Arbor, Mich., 1958.
8. QUINE, W. V. O., *Set Theory and Its Logic*, Harvard University Press, Cambridge, Mass., 1963.
9. ROTMAN, B., and G. T. KNEEBONE, *The Theory of Sets and Transfinite Numbers*, Spottiswoode, Ballantyne, London, 1966.

10. RUBIN, J. E., *Set Theory for the Mathematician*, Holden-Day, San Francisco, 1967.

11. RUDIN, W., *Principles of Mathematical Analysis*, McGraw-Hill, New York, 1953.

12. STROLL, R. R., *Set Theory and Logic*, W. H. Freeman, San Francisco, 1963.

13. TAYLOR, A. E., *Introduction to Functional Analysis*, Wiley, New York, 1958.

14. ZEHNA, P. W., and R. L. JOHNSON, *Elements of Set Theory*, Allyn and Bacon, Boston, 1962.

Lattice Theory

15. ABBOTT, J. C., *Sets, Lattices, and Boolean Algebras*, Allyn and Bacon, Boston, 1969.

16. ABBOTT, J. C., *Trends in Lattice Theory*, Van Nostrand Reinhold, New York, 1969.

17. BIRKHOFF, G., *Lattice Theory*, 3rd ed., American Mathematical Society, Providence, R.I., 1967.

18. DONNELLAN, T., *Lattice Theory*, Pergamon Press, Oxford, 1968.

19. DUBISCH, R., *Lattices to Logic*, Blaisdell, New York, 1963.

20. FUCHS, L., *Partially Ordered Algebraic Systems*, Pergamon Press, Oxford, 1963.

21. *Lattice Theory*, Proc. Symp. Pure Math., Vol. 2, American Mathematical Society, Providence, R.I., 1961.

22. LIEBER, L. R., *Lattice Theory*, Galois Institute of Mathematics and Art, Brooklyn, N.Y., 1959.

23. RUTHREFORD, D. E., *Introduction to Lattice Theory*, Hafner, New York, 1965.

24. SKORNJAKOV, L. A., *Complemented Modular Lattices and Regular Rings*, Gos. Izd.-Mat. Lit., Moscow, 1961.

25. SZASZ, G., *Introduction to Lattice Theory*, Academic Press, New York, 1963.

Boolean Algebra

26. ABBOTT, J. C., *Sets, Lattices, and Boolean Algebras*, Allyn and Bacon, Boston, 1969.

27. ADELFIO, S. A., JR., and C. F. NOLAN, *Principles and Applications of Boolean Algebra*, Hayden, New York, 1964.

28. ARNOLD, B. H., *Logic and Boolean Algebra*, Prentice-Hall, Englewood Cliffs, N.J., 1962.

29. BOOLE, G., *An Investigation into the Laws of Thought*, Open Court, Chicago, 1854/1940.

30. BOWRAN, A. P., *A Boolean Algebra*, Macmillan, London, 1965.

31. FLEGG, H. G., *Boolean Algebra and Its Applications*, Wiley, New York, 1964.

32. GOODSTEIN, R. L., *Boolean Algebra*, Pergamon Press, Oxford, 1966.

33. HALMOS, P. R., *Lectures on Boolean Algebras*, Van Nostrand Reinhold, New York, 1963.

34. HOHN, F. E., *Applied Boolean Algebra*, Macmillan, New York, 1960.

35. MACLANE, S., and G. BIRKHOFF, *Algebra*, Macmillan, New York, 1967.

36. SAMPATH KUMARACHAR, E., *Some Studies in Boolean Algebra*, Karnatak University, Dharwar, 1967.

37. SIKORSKI, R., *Boolean Algebras*, Springer-Verlag, New York, 1964.

38. WHITESITT, J. E., *Boolean Albegra and Its Applications*, Addison-Wesley, Reading, Mass., 1961.

2

Vector Switching Algebra and Vector Switching Functions

In Chapter 1, we discussed properties of Boolean algebra in general. In this chapter, we shall study the two-element Boolean algebra (B_2), or *switching algebra*. Boolean functions defined on B_2 are called *switching functions*. General theorems of switching functions are discussed.

The main objective of this chapter, however, is to introduce *binary-vector switching algebra*, or simply *vector switching algebra*, which is a generalization of (scalar) switching algebra. In this generalized algebra, every element is represented by a binary vector; and in addition to ordinary AND, OR, and NOT operations, a new operation, called the *rotation operation*, which rotates (rightward or leftward) the components of a binary vector, is introduced. Moreover, the NOT or COMPLEMENTATION operation is extended to a more general operation called the *generalized complement*, which includes the total complement (ordinary complement), the null complement (no complement), and newly introduced partial complements. Because of this generalization, all axioms and theorems of ordinary switching algebra are generalized. In particular, it is shown that De Morgan's theorem, Shannon's theorem, and the expansion theorem are generalized into more general forms, which include their corresponding ordinary versions as special cases.

It is shown that any 2^n-valued logic truth table can be represented by a vector switching function. Compact canonical (sum-of-products and product-of-sums) forms of this function are presented. These forms may be used in describing digital systems with multibused inputs and outputs.

2.1 Switching Algebra and Switching Function

Among all the Boolean algebras, the two-element Boolean algebra (B_2), known as *switching algebra*, is the most useful one. It is the mathematical foundation of the analysis and design of switching circuits that make up digital systems. Switching algebra contains the two extreme elements, the largest number represented by 1, and the smallest number represented by 0.

First, let us review axioms L1–L4, L6, and L7 for switching algebra. Let x, y, and z be binary variables in B_2; that is, they can take on two possible values, 0 and 1.

L1. Idempotent: $\qquad x \cdot x = x$ $\qquad\qquad\qquad\qquad$ $x + x = x$

L2. Commutative: $\quad x \cdot y = y \cdot x$ $\qquad\qquad\qquad$ $x + y = y + x$

L3. Associative: $\quad\; x \cdot (y \cdot z) = (x \cdot y) \cdot z$ $\qquad$ $x + (y + z) = (x + y) + z$

L4. Absorptive: $\quad\;\; x \cdot (x + y) = x$ $\qquad\qquad$ $x + (x \cdot y) = x$

L6. Distributive: $\quad\; x \cdot (y + z) = (x \cdot y) + (x \cdot z)$ $\quad$ $x + (y \cdot z) = (x + y) \cdot (x + z)$

L7. Complement: $\quad x \cdot x' = 0$ $\qquad\qquad\qquad$ $x + x' = 1$

Additionally, since 0 and 1 are the smallest and largest numbers in B_2,

$$x \cdot 0 = 0 \qquad x + 1 = 1$$

Note that the operations $x \cdot y$ and $x + y$ are *not* ordinary algebraic operations; they are the least upper bound (l.u.b.) and greatest lower bound (g.l.b.) operations of x and y, respectively. In switching algebra, we call the former the *AND* (*Boolean product*) *operation* and the latter the *OR* (*Boolean sum*) *operation*. The complement operation is often referred to as the *NOT operation* or the *negation operation*.

In the following, we shall present several general theorems of switching functions. A very important and interesting one is De Morgan's theorem, which was presented in Theorem 1.3.6 for two variables. A more general version of it is presented below.

THEOREM 2.1.1 Generalized De Morgan's Theorem

(a) $\quad (x_1 + x_2 + \ldots + x_n)' = x_1' \cdot x_2' \cdot \ldots \cdot x_n'$

(b) $\quad (x_1 \cdot x_2 \cdot \ldots \cdot x_n)' = x_1' + x_2' + \ldots + x_n'$

Proof: We shall prove Theorem 2.1.1(a) by mathematical induction. For $n = 1$, the equality holds. Now assume that the equality holds for $n = k$.

$$(x_1 + x_2 + \ldots + x_k)' = x_1' \cdot x_2' \cdot \ldots \cdot x_k'$$

we want to prove that it holds for $n = k + 1$.

$$
\begin{aligned}
(x_1 + x_2 + \ldots + x_k + x_{k+1})' &= [(x_1 + x_2 + \ldots + x_k) + x_{k+1}]' \\
&= (x_1 + x_2 + \ldots + x_k)' \cdot x_{k+1}' \qquad \text{by Theorem 1.3.6} \\
&= (x_1' \cdot x_2' \cdot \ldots \cdot x_k') \cdot x'_{k+1} \qquad \text{by assumption}
\end{aligned}
$$

Theorem 2.1.1(a) is thus proved. Theorem 2.1.1(b) can be proved similarly. ∎

De Morgan's theorem does not convey completely the relation between complementary functions. Shannon has suggested a generalization of the theorem which is described below.

THEOREM 2.1.2 Shannon's Theorem

$$(f(x_1, x_2, \ldots, x_n, +, \cdot))' = f(x_1', x_2', \ldots, x_n', \cdot, +)$$

In words, this theorem says that the complement of any function is obtained by replacing each variable by its complement and, at the same time, by interchanging the AND and OR operations.

Proof: Since any $(f(x_1, \ldots, x_n, +, \cdot))'$ may be expressed by

$$(f(x_1, \ldots, x_n, +, \cdot))' = (f_1(x_1, \ldots, x_n, +, \cdot) + f_2(x_1, \ldots, x_n, +, \cdot))'$$
$$= (f_1(x_1, \ldots, x_n, +, \cdot))' \cdot (f_2(x_1, \ldots, x_n, +, \cdot))'$$

and/or

$$(f(x_1, \ldots, x_n, +, \cdot))' = ((f_1(x_1, \ldots, x_n, +, \cdot) \cdot (f_2(x_1, \ldots, x_n, +, \cdot))'$$
$$= (f_1(x_1, \ldots, x_n, +, \cdot))' + (f_2(x_1, \ldots, x_n, +, \cdot))'$$

where f_1 and f_2 represent two partial functions of f. Repeat the same argument for f_1 and f_2, and so forth, until De Morgan's theorem is applied to every variable. Since each time De Morgan's theorem is applied to f (its partial function), the complement of f (its partial function) is obtained by replacing each partial function (subpartial function) by its complement and, at the same time, by interchanging the AND and OR operations. After De Morgan's theorem is applied to every variable, the final result should be $f(x_1', x_2', \ldots, x_n', \cdot, +)$. ∎

The following example illustrates the application of this theorem.

Example 2.1.1

If $f(x_1, x_2, x_3) = (x_1 x_2)' x_3 + x_1(x_2 + x_3')$, then the complement of this function is

$$f'(x_1, x_2, x_3) = ((x_1' + x_2')' + x_3') \cdot (x_1' + x_2' x_3)$$

Another interesting theorem of switching algebra is the expansion theorem.

THEOREM 2.1.3   Expansion Theorem

(a) $f(x_1, x_2, \ldots, x_n) = x_1 \cdot f(1, x_2, \ldots, x_n) + x_1' \cdot f(0, x_2, \ldots, x_n) = x_1 \cdot f(1, x_2, \ldots, x_n) \oplus x_1' \cdot f(0, x_2, \ldots, x_n)$

(b) $f(x_1, x_2, \ldots, x_n) = [x_1 + f(0, x_2, \ldots, x_n)][x' + f(1, x_2, \ldots, x_n)]$

The function in this form is said to be *expanded about* x_1. Similarly, expressions can be written to represent expansions about any of the variables.

Proof: If we first substitute $x_1 = 1$ and $x_1' = 0$ and then substitute $x_1 = 0$ and $x_1' = 1$ in each equation, they reduce to identities. ∎

Any switching function in n variables may be developed into a series by using the expansion theorem.

Example 2.1.2

The function $f(x_1, x_2, x_3) = (x_1 x_2)' x_3 + x_1(x_2 + x_3')$ can be expressed as

$$f(x_1, x_2, x_3) = x_1 \cdot (x_2' x_3 + x_2 + x_3') + x_1' \cdot (x_3)$$

or

$$f(x_1, x_2, x_3) = [x_1 + x_3][x_1' + (x_2'x_3 + x_2 + x_3')]$$

From the expansion theorem, it follows that

THEOREM 2.1.4

(a) $\begin{cases} (1) & x_1 \cdot f(x_1, x_2, \ldots, x_n) = x_1 \cdot f(1, x_2, \ldots, x_n) \\ (2) & x_1 + f(x_1, x_2, \ldots, x_n) = x_1 + f(0, x_2, \ldots, x_n) \end{cases}$

(b) $\begin{cases} (1) & x_1' \cdot f(x_1, x_2, \ldots, x_n) = x_1' \cdot f(0, x_2, \ldots, x_n) \\ (2) & x_1' + f(x_1, x_2, \ldots, x_n) = x_1' + f(1, x_2, \ldots, x_n) \end{cases}$

Using the axioms of switching algebra and the above general theorems (Theorems 2.1.1–2.1.4), we can simplify all the possible 2^{2^n} switching functions of n variables in canonical sum-of-products form. For $n = 2$, this is demonstrated in Table 2.1.1.

TABLE 2.1.1 Simplified Forms of All Possible Switching Functions of Two Variables in Canonical Form

Number of terms in canonical sum-of-products form	$x_1'x_2'$	$x_1'x_2$	x_1x_2'	x_1x_2	Canonical sum-of-products form	Simplified function
0	0	0	0	0	0	0
1	0	0	0	1	x_1x_2	x_1x_2
1	0	0	1	0	x_1x_2'	x_1x_2'
1	0	1	0	0	$x_1'x_2$	$x_1'x_2$
1	1	0	0	0	$x_1'x_2'$	$x_1'x_2'$
2	0	0	1	1	$x_1x_2' + x_1x_2$	x_1
2	1	1	0	0	$x_1'x_2' + x_1'x_2$	x_1'
2	0	1	0	1	$x_1'x_2 + x_1x_2$	x_2
2	1	0	1	0	$x_1'x_2' + x_1x_2'$	x_2'
2	1	0	0	1	$x_1'x_2' + x_1x_2$	$x_1x_2 + x_1'x_2'$
2	0	1	1	0	$x_1'x_2 + x_1x_2'$	$x_1x_2' + x_1'x_2$
3	0	1	1	1	$x_1'x_2 + x_1x_2' + x_1x_2$	$x_1 + x_2$
3	1	0	1	1	$x_1'x_2' + x_1x_2' + x_1x_2$	$x_1 + x_2'$
3	1	1	0	1	$x_1'x_2' + x_1'x_2 + x_1x_2$	$x_1' + x_2$
3	1	1	1	0	$x_1'x_2' + x_1'x_2 + x_1x_2'$	$x_1' + x_2'$
4	1	1	1	1	$x_1'x_2' + x_1'x_2 + x_1x_2' + x_1x_2$	1

For n greater than 2, such tables increase rapidly in size, with 256 ($= 2^{2^3}$) rows for three variables and 65,536 ($= 2^{2^4}$) rows for four variables.

Table 2.1.1 indicates that there are 16 possible switching functions of two variables in canonical sum-of-products form. The first column shows the number of terms in each form. In the second column, under each of the product terms, a 0 or 1 is entered. If a 0 is entered, it means that the term at the head of that column is not included in the canonical form. If a 1 is entered in a column, it means that the term at the head of that column is included in the canonical form. In each row, the combined 0's and 1's are treated as a binary number, and the rows are listed in the ascending order of the decimal equivalents of the binary numbers. In the third column, 16 Boolean functions in canonical sum-of-products form are listed. Using the axioms of switching algebra, all of them may be simplified to their simplest equivalent forms which are shown in the last column of Table 2.1.1. These 16 functions are, of course, distinct and form a lattice as shown in Fig. 2.1.1.

Since the value of $f(e_1, e_2, \ldots, e_n)$ and $f'(e_1, e_2, \ldots, e_n)$ of a switching function can *only* be either 0 or 1, Eqs. (1.4.13) and (1.4.14) can be expressed as:

THEOREM 2.1.5

(a) *The canonical sum-of-products form of a switching function:*

$$f(x_1, \ldots, x_n) = \sum_{\substack{\text{all the combinations} \\ \text{of the values of } x_1, \ldots, x_n, \\ \text{for which } f(e_1, \ldots, e_n) = 1}} x_1^{e_1} x_2^{e_2} \ldots x_n^{e_n} \tag{2.1.1}$$

(b) *The canonical product-of-sums form of a switching function:*

$$f(x_1, \ldots, x_n) = \prod_{\substack{\text{all the combinations} \\ \text{of the values of } x_1, \ldots, x_n, \\ \text{for which } f(e_1, \ldots, e_n) = 0}} (x_1^{e_1'} + x_2^{e_2'} + \ldots + x_n^{e_n'}) \tag{2.1.2}$$

where $x_i^0 = x_i'$ and $x_i^1 = x_i$.

Example 2.1.3

The canonical forms of the switching function

$$f(x_1, x_2, x_3) = x_1' x_2 + x_1 x_2' x_3 + x_2 x_3 \tag{2.1.3}$$

are

$$f(x_1, x_2, x_3) = x_1' x_2 x_3' + x_1' x_2 x_3 + x_1 x_2' x_3 + x_1 x_2 x_3 \tag{2.1.4}$$

and

$$f(x_1, x_2, x_3) = (x_1 + x_2 + x_3)(x_1 + x_2 + x_3')(x_1' + x_2 + x_3')(x_1' + x_2' + x_3) \tag{2.1.5}$$

Each term of the canonical sum-of-products of Eq. (2.1.1) is called a *minterm* and each term of the canonical product-of-sums of Eq. (2.1.2) is called a *maxterm*. It is interesting to note that from the definitions of canonical forms of a switching function [Eqs. (2.1.1) and (2.1.2)], the minterms of the canonical sum-of-products form of a switching function can be directly obtained from the rows of the truth table of the function that are mapped into 1, and the

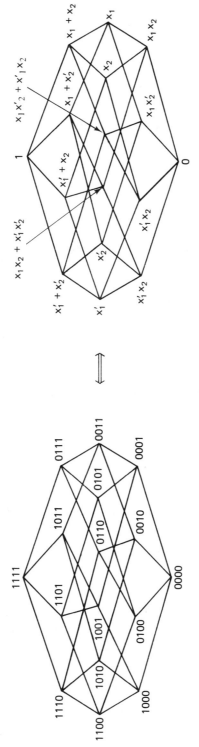

Fig. 2.1.1 Lattice representation of the sixteen simplified functions given in Table 2.1.1.

maxterms of the canonical product-of-sums form of a switching function can be obtained from the rows that are mapped into 0.

Example 2.1.4

Consider the function of Eq. (2.1.3) of Example 2.1.3. The minterms and the maxterms of this function are given in Table 2.1.2. The reason that the row, say 010, corresponds and

TABLE 2.1.2

Row number r	Minterms and maxterms	x_1 x_2 x_3	$f(x_1, x_2, x_3)$ of Eq. (2.1.3)
0	$M_0 = x_1 + x_2 + x_3$	0 0 0	0
1	$M_1 = x_1 + x_2 + x_3'$	0 0 1	0
2	$m_2 = x_1'x_2x_3'$	0 1 0	1
3	$m_3 = x_1'x_2x_3$	0 1 1	1
4	$m_4 = x_1x_2'x_3'$	1 0 0	1
5	$M_5 = x_1' + x_2 + x_3'$	1 0 1	0
6	$M_6 = x_1' + x_2' + x_3$	1 1 0	0
7	$m_7 = x_1x_2x_3$	1 1 1	1

only corresponds to the minterm $x_1'x_2x_3'$ is because $x_1'x_2x_3' = 1$ iff $x_1 = 0$, $x_2 = 1$, and $x_3 = 0$. The reasons the other three rows, 011, 100, and 111, describe the minterms $x_1'x_2x_3$, $x_1x_2'x_3'$, and $x_1x_2x_3$ are similar. Thus, we find that:

RULE 2.1.1

All the minterms of the canonical sum-of-products form of a switching function can be obtained from the rows of the truth table that are mapped into 1. Each 0-value of a variable indicates the complemented form of the variable, and each 1-value of a variable indicates the uncomplemented form of the variable.

Dually, the reason the row 000 corresponds and only corresponds to the maxterm $x_1 + x_2 + x_3$ is because $x_1 + x_2 + x_3 = 0$ iff $x_1 = 0$, $x_2 = 0$, and $x_3 = 0$. The reasons the other three rows, 001, 101, and 110, describe the maxterms $x_1 + x_2 + x_3'$, $x_1' + x_2 + x_3'$, and $x_1' + x_2' + x_3$ are similar. Therefore, we have:

RULE 2.1.2

All the maxterms of the canonical product-of-sums form of a switching function can be obtained from the rows of the truth table that are mapped into 0. Each 0-value of a variable indicates the uncomplemented form of the variable, and each 1-value of a variable indicates the complemented form of the variable.

Because of this one-to-one correspondence between the minterms and maxterms and the rows of the values of the variables of a truth table, we can have a simple representation of minterms and maxterms. It is customary to use m and M to denote minterms and maxterms, respectively. Instead of indicating rows of the truth table by the binary numbers (000, 001,

010, etc.), we use their corresponding decimal numbers (0, 1, 2, etc.). Therefore, every minterm and maxterm can be represented by an m and M letter with a subscript r which is called a *row number*, the decimal-number representation of the binary numbers denoting the row to which it corresponds (see columns 1 and 2 of Table 2.1.2). The above two canonical forms can then be represented by

$$f(x_1, x_2, x_3) = m_2 + m_3 + m_4 + m_7 \tag{2.1.6}$$

and

$$f(x_1, x_2, x_3) = M_0 + M_1 + M_5 + M_6 \tag{2.1.7}$$

which can be further simplified by

$$f(x_1, x_2, x_3) = \sum (2, 3, 4, 7)$$

and

$$f(x_1, x_2, x_3) = \prod (0, 1, 5, 6)$$

respectively. Note that the expression $\sum (2, 3, 4, 7)$ indicates that the rows whose row numbers are not in this summation expression, rows 0, 1, 5, and 6, are mapped into 0, and similarly, the expression $\prod (0, 1, 5, 6)$ indicates that the rows whose row numbers are not in this product expression, rows 2, 3, 4, and 7, are mapped into 1. Also, notice that a row, if it describes a minterm (maxterm), cannot also describe a maxterm (minterm). In other words, the set of rows describing the minterms of completely specified switching functions are mutually exclusive with the set of rows describing the maxterms. These two sets complement each other. Therefore, if the canonical sum-of-products (product-of-sums) form of a switching function is known, the canonical product-of-sums (sum-of-products) form of the function can be obtained from all the rows that are not in the set of rows that correspond to the minterms (maxterms) and application of Rule 2.1.2 (Rule 2.1.1).

Exercise 2.1

1. Find the complement of each of the following functions.
   (a) $f(x_1, x_2, x_3, x_4) = x_1 x_2 [x_3' + x_4'(x_2 x_4 + x_1' x_3')]$
   (b) $f(x_1, x_2, x_3, x_4, x_5, x_6) = x_1 x_3' \{x_2' x_5 [x_4 x_6' + x_1' x_3'(x_2' x_4 + x_5 x_6')]\}$

2. Define the COINCIDENCE operation $\odot$ as $x_1 \odot x_2 = \overline{x_1 \oplus x_2}$. Extend Shannon's theorem to include the operations $\odot$ and $\oplus$, and apply it to the following functions.
   (a) $f(x_1, x_2, x_3) = (x_1 x_2 \oplus x_2' x_3) \odot x_1' x_3'$
   (b) $f(x_1, x_2, x_3, x_4) = [(x_1 + x_2) \oplus (x_3 + x_4)] \odot [(x_1' + x_3') \oplus (x_2' + x_4')]$

3. Expand each of the following functions about its variables.
   (a) $f(x_1, x_2, x_3) = x_1 x_2' + x_2 x_3' + x_3 x_1'$
   (b) $f(x_1, x_2, x_3) = (x_1 + x_2')(x_2 + x_3')(x_3 + x_1')$
   (c) $f(x_1, x_2, x_3) = x_1 x_2 \oplus x_2' x_3' \oplus x_3 x_1$
   (d) $f(x_1, x_2, x_3) = (x_1 + x_2) \odot (x_2' + x_3') \odot (x_3 + x_1)$

4. Find the minterms and maxterms of each of the following functions.
   (a) $f(x_1, x_2, x_3, x_4) = x_1 x_3 + x_2' x_4' + x_1' x_2' x_3' + x_2 x_3 x_4$
   (b) $f(x_1, x_2, x_3, x_4) = (x_1 + x_4)(x_2' + x_3')(x_1 + x_2 + x_3)(x_2' + x_3' + x_4)$
   (c) $f(x_1, x_2, x_3, x_4, x_5) = x_1 x_2' x_3 \oplus x_2 x_4' x_5' \oplus x_3' x_4' x_5'$

2.2   The Generalized Complement and the Rotation Operator

Since every element of a Boolean algebra must have a *unique* complement, a Boolean algebra must have an even number of elements (Theorem 1.3.1). Moreover, since every Boolean algebra is isomorphic to a power set $P(A)$ of a set $A = \{a_1, \ldots, a_n\}$ (Theorem 1.3.4) and there are 2^n elements in $P(A)$ (Theorem 1.1.1), every Boolean algebra has 2^n elements (Corollary 1.3.1). Let B_{2^n} denote the Boolean algebra containing 2^n elements. The B_2, B_4, B_8, and B_{16} are shown in Fig. 2.2.1(a). The elements cover the zero element and are the atoms of the algebra (Definition 1.3.2). For example, I of B_2; a and b of B_4; a, b, and c of B_8; and a, b, c, and d of B_{16} are atoms of their respective algebra. It is seen that B_{2^n} has exactly n atoms (Corollary 1.3.3).

The relation between a 2^n-element Boolean algebra B_{2^n}, and a Cartesian product of n 2-element Boolean algebras, denoted by B_2^n, were given in Corollary 1.3.5, which states: Any Boolean algebra B is isomorphic to a Cartesian product of n B_2, namely, B_2^n for some integer n. For example, $B_2^2 = \{00, 01, 10, 11\}$, $B_2^3 = \{000, 001, 010, 100, 011, 101, 110, 111\}$, and $B_2^4 = \{0000, 0001, 0010, 0100, 1000, 0011, \ldots, 1111\}$ of Fig. 2.2.1(b) are isomorphic to the B_{2^2}, B_{2^3}, and B_{2^4} of Fig. 2.2.1(a), respectively. The isomorphism relationship (denoted by $\sim$) between B_{2^n} and B_2^n may be described by their atoms as follows:

$B_2 \sim B_2^1$	$B_4 \sim B_2^2$	$B_8 \sim B_2^3$	$B_{16} \sim B_2^4$
$I = 1$	$a = 01$	$a = 001$	$a = 0001$
	$b = 10$	$b = 010$	$b = 0010$
		$c = 100$	$c = 0100$
			$d = 1000$

If decimal numbers are used to represent the elements of B_2^n, $n = 1, 2, 3$, and 4, they are shown in Fig. 2.2.1(c).

In ordinary Boolean algebra, one type of complement is defined, namely, the total complement. For example, in B_2^3, the (total) complements of the elements are as follows:

X	$\bar{X}$ (total complement)
000	111
001	110
010	101
100	011
011	100
101	010
110	001
111	000

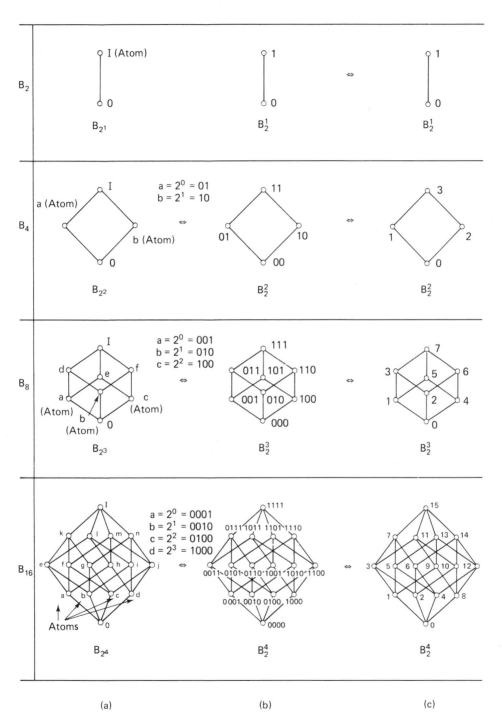

Fig. 2.2.1 B_2, B_4, B_8, and B_{16} Boolean algebras.

The total complement operation $\bar{X}$ is the one that complements every component of X.

In binary-vector switching algebra, the ordinary complement or total complement can be generalized into a more general operation.

DEFINITION 2.2.1

In B_2^n, the *generalized complement* X^P, $P \in B_2^n$, of a variable X is defined to be the element obtained from complementing the components of X according to the value of the corresponding component of P; x_i is complemented or uncomplemented if p_i is 0 or 1, respectively, where x_i and p_i designate the ith component of X and P.

For example, the generalized complements of the elements of B_2^3, for $P = 010$, are as follows:

X	X^P
000	101
001	100
010	111
100	001
011	110
101	000
110	011
111	010

It is important to point out that the generalized complement includes the uncomplement (true form) and the total complement as special cases.

Now we shall introduce another operation, called the *rotation operation*, as follows.

DEFINITION 2.2.2

Let $X = (x_1, x_2, \ldots, x_n)$ be a variable in B_2^n. The *right X* and the *left X rotation* of X are defined as

$$\overset{\rightharpoonup}{X} = (x_n, x_1, x_2, \ldots, x_{n-1})$$

and

$$\overset{\leftharpoonup}{X} = (x_2, x_3, \ldots, x_n, x_1)$$

When the arrow is omitted (i.e., $\tilde{X}$), it means that it can either be the right or the left rotation operation.

DEFINITION 2.2.3

$\overset{(m)}{\tilde{X}} = (((\overset{\frown}{(X)})))$. The operations $\sim$ in this definition are either all right-rotation operations or all left-rotation operations.

Now we are in the position to define a vector switching algebra.

DEFINITION 2.2.4

B_2^n with the three operations $+$, $\cdot$, and $'$ defined in Corollary 1.3.5, plus the generalized complement and the rotation operation defined on it, is called a *vector switching algebra*.

Before studying the properties of vector switching algebra, which will be presented in Section 2.3, we first present some basic properties of the generalized complement and the rotation operation. The following are several basic properties of the generalized complement.

PROPERTY 1

$X^O = \bar{X}$ and $X^I = X$, where $O = (0, \ldots, 0)$ and $I = (1, \ldots, 1)$.

PROPERTY 2

(a) $P^P = I$.
(b) $\bar{P}^P = P^{\bar{P}} = \bar{P}^{\bar{P}} = O$.

PROPERTY 3

$X^P = X \oplus \bar{P}$.

PROPERTY 4

$P_1 = \bar{P}_2$ iff $X^{P_1} = \overline{X^{P_2}}$.

PROPERTY 5

$X^{PP} = (X^P)^P = X^{(PP)} = X$.

PROPERTY 6

$X^{P^{2n+1}} = X^P$ for any positive integer n, where $X^{P^m} \triangleq ((X^{\overbrace{P})\ldots)^{P}}^{m}$.

PROPERTY 7

(a) $P^{P^m} = \begin{cases} I & \text{if } m \text{ is odd} \\ P & \text{if } m \text{ is even.} \end{cases}$

(b) $P^{\bar{P}^m} = \begin{cases} O & \text{if } m \text{ is odd} \\ \bar{P} & \text{if } m \text{ is even.} \end{cases}$

PROPERTY 8

(a) $\overline{(X_1^{P_1} + X_2^{P_2})} = \overline{X_1^{P_1}} \cdot \overline{X_2^{P_2}} = X_1^{\bar{P}_1} \cdot X_2^{\bar{P}_2} = \bar{X}_1^{P_1} \cdot \bar{X}_2^{P_2}$.
(b) $\overline{(X_1^{P_1} \cdot X_2^{P_2})} = \overline{X_1^{P_1}} + \overline{X_2^{P_2}} = X_1^{\bar{P}_1} + X_2^{\bar{P}_2} = \bar{X}_1^{P_1} + \bar{X}_2^{P_2}$.

The proofs for Properties 1–8 are left to the reader as an exercise (problem 1).

PROPERTY 9

(a) $(X_1 + X_2)^P = (\bar{X}_1 \cdot \bar{X}_2)^{\bar{P}}$.
(b) $(X_1 \cdot X_2)^P = (\bar{X}_1 + \bar{X}_2)^{\bar{P}}$.

Proof: First, we prove (a).

$$
\begin{aligned}
(X_1 + X_2)^P &= (X_1 + X_2) \oplus \bar{P} && \text{by Property 3} \\
&= (X_1 + X_2) \oplus P \oplus I && \text{since } \bar{P} = P \oplus I \\
&= (X_1 + X_2) \oplus I \oplus P && \text{since } \oplus \text{ is commutative} \\
&= \overline{(X_1 + X_2)} \oplus P && \text{by Property 3} \\
&= (\bar{X}_1 \cdot \bar{X}_2) \oplus P && \text{by Theorem 1.3.6} \\
&= (\bar{X}_1 \cdot \bar{X}_2)^{\bar{P}} && \text{by Property 3}
\end{aligned}
$$

The proof of (b) is similar and may thus be omitted.   ∎

PROPERTY 10

(a) $(X_1^{P_1} + X_2^{P_2})^{P_3} = (X_1^{\bar{P}_1} \cdot X_2^{\bar{P}_2})^{\bar{P}_3}$.

(b) $(X_1^{P_1} \cdot X_2^{P_2})^{P_3} = (X_1^{\bar{P}_1} + X_2^{\bar{P}_2})^{\bar{P}_3}$.

The proof of Property 10 is similar to that of Property 9. We leave it to the reader as an exercise (problem 2).

The following are basic properties of the rotation operation.

PROPERTY 11

If $X \in B_2^n$, $\overset{(n)}{\widetilde{X}} = X$.

PROPERTY 12

(a) $\widetilde{(X_1 \cdot X_2)} = \widetilde{X}_1 \cdot \widetilde{X}_2$.

(b) $\widetilde{(X_1 + X_2)} = \widetilde{X}_1 + \widetilde{X}_2$.

PROPERTY 13

$\widetilde{X^P} = \tilde{X}^{\hat{P}}$.

The proofs of Properties 11–13 are left to the reader (problem 3).

It is easy to show the following theorem.

THEOREM 2.2.1

In B_2^n,

(a) $(X^P)\overset{}{\widetilde{(X^P)}} \ldots \overset{(n-1)}{\widetilde{(X^P)}} = \begin{cases} I & \text{if } X = P. \\ O & \text{if } X \neq P. \end{cases}$

(b) $X^{\bar{P}} + \overset{}{\widetilde{(X^{\bar{P}})}} + \ldots + \overset{(n-1)}{\widetilde{(X^{\bar{P}})}} = \begin{cases} O & \text{if } X = P. \\ I & \text{if } X \neq P. \end{cases}$

Proof: It is easily seen that if $X = P$, by Property 2(a) the expression above is true. The reason for that, if $X \neq P$, $(X^P)\overset{(n-1)}{\widetilde{(X^P)}} \ldots \overset{}{\widetilde{(X^P)}} = 0$, is that there will be at least one component of X which is 0. This 0 is rotated in X by the $\sim$ operation. The presence of the n terms $(X^P), \ldots, \overset{(n-1)}{\widetilde{(X^P)}}$ ensures that there will be one 0 in every position of the vector X which is in (at least) one of the n terms. This, in turn, ensures that the product of the n terms is 0. Hence, Theorem 2.2.1(a) is proved. Similarly, Theorem 2.2.1(b) can be proved by using Property 2(b). ∎

Exercise 2.2

1. Prove Properties 1–8.

2. Prove Property 10.

3. Prove Properties 11–13.

2.3 Properties of Vector Switching Algebra and Vector Switching Function

First we define the vector switching function.

DEFINITION 2.3.1

Let $X_1, X_2, \ldots, X_n$ be variables whose values lie in a given vector switching algebra. The *vector switching function* is defined by a recursive definition similar to that of the ordinary Boolean function given in Definition 1.4.3 except that all the variables and functions are replaced by vector variables and vector functions, respectively. To be more specific, it is defined as follows:

1. Let A be a constant vector on B_2^n. $F_1(X_1, \ldots, X_m) = A$ and $F_2(X_1, \ldots, X_m) = X_i$ are vector switching functions. The former is called the *constant (vector) function* and the latter is called the *projection (vector) function*.

2. If $F(X_1, \ldots, X_m)$ is a vector switching function, then $F^P(X_1^{P_1}, \ldots, X_m^{P_m})$, for any values of P_i, is a vector switching function.

3. If $F(X_1, \ldots, X_m)$ is a vector switching function, then $F(\overset{n_1}{\widetilde{X_1}}, \ldots, \overset{n_m}{\widetilde{X_m}})$ is a vector switching function.

4. If $F_1(X_1, \ldots, X_m)$ and $F_2(X_1, \ldots, X_m)$ are vector switching functions, then $F_1(X_1, \ldots, X_m) + F_2(X_1, \ldots, X_m)$ and $F_1(X_1, \ldots, X_m) \cdot F_2(X_1, \ldots, X_m)$ are vector switching functions.

5. Any vector switching function that can be constructed by a finite number of applications of the above rules is a vector switching function.

For convenience, from now on we shall use B_2^n to denote a vector switching algebra. Since B_2^n is a generalization of ordinary switching algebra, all the axioms, properties, and theorems of the latter are satisfied by the former. Moreover, all the important theorems in ordinary switching algebra can now be generalized.

THEOREM 2.3.1   Generalized De Morgan's Theorem

(a) $(X_1^{P_1} \cdot X_2^{P_2} \cdot \ldots \cdot X_m^{P_m})^P = (X_1^{\bar{P}_1} + X_2^{\bar{P}_2} + \ldots + X_m^{\bar{P}_m})^{\bar{P}}.$

(b) $(X_1^{P_1} + X_2^{P_2} + \ldots + X_m^{P_m})^P = (X_1^{\bar{P}_1} \cdot X_2^{\bar{P}_2} \cdot \ldots \cdot X_m^{\bar{P}_m})^{\bar{P}}.$

THEOREM 2.3.2 Generalized Shannon's Theorem

$$F^P(X_1^{P_1}, X_2^{P_2}, \ldots, X_m^{P_m}; +, \cdot) = F^{\bar{P}}(X_1^{\bar{P_1}}, X_2^{\bar{P_2}}, \ldots, X_m^{\bar{P_m}}; \cdot, +)$$

The proofs of these theorems are evident from the properties of the generalized complement and the rotation operation described above; they are left to the reader as an exercise (problem 1).

Exercise 2.3

1. Prove Theorems 2.3.1 and 2.3.2.
2. Find the generalized complement of each of the following vector switching functions.
    (a) $F(X_1, X_2, X_3, X_4) = X_1^{P_1} X_2^{P_2}(X_3 + X_4) + X_3^{P_3} X_4^{P_4}(X_1 + X_2)$.
    (b) $F(X_1, X_2, X_3, X_4) = X_1 X_3^{P_1} + X_4(X_2^{P_2} X_3^{P_3} + X_1^{P_4} X_4^{P_5})$.
3. Extend Theorem 2.3.1 to include the **EXCLUSIVE-OR** and **COINCIDENCE** operations.

2.4 Canonical Forms

Consider a general combinational circuit of Fig. 2.4.1, where x_{ij}, $i = 1, \ldots, m$ and $j = 1, \ldots, n$, are binary variables and f_k, $k = 1, \ldots, n$, are binary functions. The n output functions of this circuit are

$$f_1(x_{11}, \ldots, x_{1n}, x_{21}, \ldots, x_{2n}, \ldots, x_{m1}, \ldots, x_{mn})$$

$$\vdots$$

$$f_n(x_{11}, \ldots, x_{1n}, x_{21}, \ldots, x_{2n}, \ldots, x_{m1}, \ldots, x_{mn})$$

In this section, we show that these n functions may be described by a single vector output function of m n-tuple binary-vector variables. Moreover, two canonical forms of it are presented.

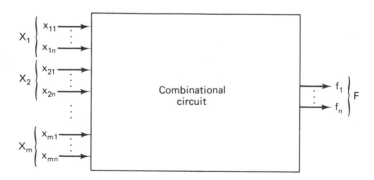

Fig. 2.4.1 General combinational circuit.

THEOREM 2.4.1 Canonical Forms I

Let F be a vector switching function of m binary-vector variables $X_1, X_2, \ldots,$ X_m. Then:

(a) The canonical sum-of-products form of F is

$$F(X_1, \ldots, X_m) = \sum F(P_1, \ldots, P_m)[X_1^{P_1}\overbrace{(X_1^{P_1})\ldots(X_1^{P_1})}^{(n-1)}]\ldots[X_m^{P_m}\overbrace{(X_m^{P_m})\ldots(X_m^{P_m})}^{(n-1)}]$$

(2.4.1)

(b) The canonical product-of-sums form of F is

$$F(X_1, \ldots, X_m) = \prod \{F(P_1, \ldots, P_m) + [X_1^{\bar{P}_1} + \overbrace{(X_1^{\bar{P}_1}) + \ldots + (X_1^{\bar{P}_1})}^{(n-1)}]$$
$$+ \ldots + [X_m^{\bar{P}_m} + \overbrace{(X_m^{\bar{P}_m}) + \ldots + (X_m^{\bar{P}_m})}^{(n-1)}]\}$$

(2.4.2)

where $P_1, \ldots, P_m$ take on values $0, 1, \ldots,$ and $2^n - 1$ in binary form.

Proof: First consider the proof of (a). The function F can be described by a truth table as in Table 2.4.1.

TABLE 2.4.1

X_1	X_2	X_m	F
$x_{11} \ldots x_{1n}$	$x_{21} \ldots x_{2n}$ $\cdots$ $x_{m1} \ldots x_{mn}$		$f_1 \ldots f_n$
$0 \ldots 0$	$0 \ldots 0$	$0 \ldots 0$	$f_{10} \cdots f_{n0}$
$0 \ldots 0$	$0 \ldots 0$	$0 \ldots 1$	$f_{11} \cdots f_{n1}$
.	.	.	.
.	.	.	.
.	.	.	.
$1 \ldots 1$	$1 \ldots 1$	$1 \ldots 1$	$f_{12^{mn}} \ldots f_{n2^{mn}}$

By Theorem 2.2.1(a),

$$F(P_1, \ldots, P_m)[X_1^{P_1}\overbrace{(X_1^{P_1})\ldots(X_1^{P_1})}^{(n-1)}]\ldots[X_m^{P_m}\overbrace{(X_m^{P_m})\ldots X_m^{P_m}}^{(n-1)}]$$
$$= \begin{cases} F(P_1, \ldots, P_m) & \text{if } X_1 = P_1, \ldots, X_m = P_m \\ 0 & \text{otherwise} \end{cases}$$

Since the summation of Eq. (2.4.1) is over all possible values of $P_1, \ldots, P_m$, the equation above ensures that the function value of F of each row of the truth table is described by one and only one term of Eq. (2.4.1). Therefore, Eq. (2.4.1) describes precisely the truth table of F (thus, the function F). Theorem 2.4.1(b) may be proved by using Theorem 2.2.1(b) and a similar argument. ∎

Example 2.4.1

Consider a simple vector switching function described in Table 2.4.2. The canonical

TABLE 2.4.2

X_1	X_2	F
0	0	1
0	1	3
1	2	2
2	3	1
Otherwise		0

X_1	X_2	F
00	00	01
00	01	11
01	10	10
10	11	01
Otherwise		00

(a)   Multivalued logic truth table   (b)   Binary-coded truth table of (a)

sum-of-products form of this function is

$$F(X_1, X_2) = (01)X_1^{00}(\widetilde{X_1^{00}})X_2^{00}(\widetilde{X_2^{00}})$$
$$+ (11)X_1^{00}(\widetilde{X_1^{00}})X_2^{01}(\widetilde{X_2^{01}})$$
$$+ (10)X_1^{01}(\widetilde{X_1^{01}})X_2^{10}(\widetilde{X_2^{10}})$$
$$+ (01)X_1^{10}(\widetilde{X_1^{10}})X_2^{11}(\widetilde{X_2^{11}}) \tag{2.4.3}$$

Now we can generalize Theorem 2.1.3 to vector switching function by applying Theorem 2.4.1.

THEOREM 2.4.2   Generalized Expansion Theorem

(a)

$$F(X_1, X_2, \ldots, X_m) = \sum_{\substack{\text{over every} \\ \text{element } a \in B_2^n}} X_1^a \widetilde{X_1^a} \ldots \overset{(n-1)}{\widetilde{X_1^a}} F(X_1 = a, X_2, \ldots, X_m) \tag{2.4.4a}$$

(b)

$$F(X_1, X_2, \ldots, X_m) = \prod_{\substack{\text{over every} \\ \text{element } a \in B_2^n}} [X_1^a + \widetilde{X_1^a} + \ldots + \overset{(n-1)}{\widetilde{X_1^a}} + F(X_1 = \bar{a}, X_2, \ldots, X_m)]$$

$$\tag{2.4.4b}$$

Example 2.4.2

We can expand vector switching function of Example 2.4.1 about X_1 as:

$$F(X_1, X_2) = X_1^{00}\widetilde{X_1^{00}}F(00, X_2) + X_1^{01}\widetilde{X_1^{01}}F(01, X_2)$$
$$+ X_1^{10}\widetilde{X_1^{10}}F(10, X_2) + X_1^{11}\widetilde{X_1^{11}}F(11, X_2)$$

Besides the canonical forms described in Theorem 2.4.1, we want to show there exist other canonical forms. The one described below is of special interest, because it will be useful for our later discussions.

DEFINITION 2.4.1

Referring to Table 2.4.1, define

$$F_i(X_1, \ldots, X_m) = \sum_{\substack{\text{all the combinations of} \\ \text{the values of } X_1, \ldots, X_m \\ \text{for which } f_{ij}(X_1, \ldots, X_m) = 1; i = 1, 2, \ldots, n \\ \text{for } j = 0, 1, \ldots, (2^{mn} - 1)}} (X_1^{E_{11}} X_1^{\widetilde{E}_{12}} \ldots \overbrace{X_1^{\widetilde{E}_{1,n-1}}}^{(n-1)}) \ldots (X_m^{E_{m1}} X_m^{\widetilde{E}_{m2}} \ldots \overbrace{X_m^{\widetilde{E}_{m,n-1}}}^{(n-1)}) \quad (2.4.5)$$

where E_{ij} take on the values O and I. $E_{ij} = O$ and I for $x_{ij} = 0$ and 1, respectively. Note that

$$X_k^{E_{ij}} = \begin{cases} \bar{X}_k & \text{if } E_{ij} = O \\ X_k & \text{if } E_{ij} = I \end{cases}$$

Example 2.4.3

From Definition 2.4.1, the F_i, $i = 1, 2$, of the vector switching function F of Table 2.4.2(b) are found to be

$$F_1 = \bar{X}_1 \widetilde{\bar{X}}_1 \bar{X}_2 \widetilde{X}_2 + \bar{X}_1 \widetilde{X}_1 X_2 \widetilde{\bar{X}}_2$$
$$F_2 = \bar{X}_1 \widetilde{\bar{X}}_1 \bar{X}_2 \widetilde{\bar{X}}_2 + \bar{X}_1 \widetilde{\bar{X}}_1 \bar{X}_2 \widetilde{X}_2 + X_1 \widetilde{\bar{X}}_1 X_2 \widetilde{X}_2$$

COROLLARY 2.4.1

For any set of X_k's, $k = 1, \ldots, m$, for which $f_{ij}(X_1, X_2, \ldots, X_m) = 1$, we have

$$(X_1^{E_{11}} X_1^{\widetilde{E}_{12}} \ldots \overbrace{X_1^{\widetilde{E}_{1,n}}}^{(n-1)}) \ldots (X_m^{E_{m,1}} X_m^{\widetilde{E}_{m,2}} \ldots \overbrace{X_m^{\widetilde{E}_{m,n}}}^{(n-1)})$$
$$= \begin{cases} (1, 1, \ldots, 1) & \text{if all } X_k\text{'s are either 0 or 1} \\ (0, 0, \ldots, 0) & \text{otherwise} \end{cases}$$

Example 2.4.4

Consider the function of Table 2.4.2. The values of the product terms of F_1 and F_2 found in Example 2.4.3, for which $f_{ij}(X_1, X_2, \ldots, X_m) = 1$ are as follows:

X_1	X_2	f_{ij}	$(X_1^{E_{11}} X_1^{\widetilde{E}_{12}} \ldots \overbrace{X_1^{E_{1,n}}}^{(n-1)}) \ldots (X_m^{E_{m,1}} X_m^{\widetilde{E}_{m,2}} \ldots \overbrace{X_m^{E_{m,n}}}^{(n-1)})$
00	01	$f_{11} = 1$	$\overline{(00)} \widetilde{(00)} \overline{(01)} \widetilde{(01)} = (10)$
01	10	$f_{12} = 1$	$\overline{(01)} \widetilde{(01)} (10) \widetilde{\overline{(10)}} = (10)$
00	00	$f_{20} = 1$	$\overline{(00)} \widetilde{(00)} \overline{(00)} \widetilde{(00)} = (11)$
00	01	$f_{21} = 1$	$\overline{(00)} \widetilde{(00)} \overline{(01)} \widetilde{(01)} = (10)$
10	11	$f_{23} = 1$	$(10) \widetilde{\overline{(10)}} (11) \widetilde{(11)} = (10)$

From the above discussions and Corollary 2.4.1, we have the following corollary.

COROLLARY 2.4.2

If $f_i(X_1, X_2, \ldots, X_m) \neq 0$, then F_i, for $i = 1, 2, \ldots, n$, are either equal to (1, 1, ..., 1) or (1, 0, ..., 0).

DEFINITION 2.4.2

Define δ_i's as the *atoms* of B_2^n, i.e., $\delta_i = (0, \ldots, 0, \underset{i}{1}, 0, \ldots, 0)$.

COROLLARY 2.4.3

$$\delta_i \overset{(i-1)}{\widetilde{F_i}} = \begin{cases} \delta_i & \text{for all } f_{ij} = 1 \\ 0 & \text{for all } f_{ij} = 0 \end{cases}$$

Following this corollary, it is easily seen that we have mapped function f_i of Table 2.4.1 to the ith element of the vector $\delta_i \overset{(i-1)}{\widetilde{F_i}}$.

Following this discussion, we have

THEOREM 2.4.3

Let $F(X_1, \ldots, X_m) = (f_1(X_1, \ldots, X_m), \ldots, f_n(X_1, \ldots, X_m))$ be a vector switching function. Then

$$\delta_i f_i = \delta_i \overset{(i-1)}{\widetilde{F_i}} \qquad i = 1, 2, \ldots, n$$

Following from Theorem 2.4.3, we have

THEOREM 2.4.4 Canonical Form II

Any vector switching function can be expressed as

$$F(X_1, \ldots, X_m) = \begin{cases} \sum\limits_{i=1}^{n} \delta_i f_i = \sum\limits_{i=1}^{n} \delta_i \overset{(i-1)}{\widetilde{F_i}} & (2.4.6) \\ \overset{n}{\underset{i=1}{\boxed{\Sigma}}} \delta_i f_i = \overset{n}{\underset{i=1}{\boxed{\Sigma}}} \delta_i \overset{(i-1)}{\widetilde{F_i}} & (2.4.6a) \end{cases}$$

where $\boxed{\Sigma}$ denotes the modulo-2 sum.

This canonical form is illustrated by the following example.

Example 2.4.5

Consider the same function described in Table 2.4.2. The canonical form of Eq. (2.4.4a) of this function is

$$F(X_1, X_2) = \delta_1(\bar{X}_1\tilde{\bar{X}}_1\bar{X}_2\tilde{X}_2 + \bar{X}_1\tilde{X}_1 X_2\tilde{\bar{X}}_2)^{0}$$

$$+ \delta_2(\bar{X}_1\tilde{\bar{X}}_1\bar{X}_2\tilde{\bar{X}}_2 + \bar{X}_1\tilde{\bar{X}}_1\bar{X}_2\tilde{X}_2 + X_1\tilde{\bar{X}}_1 X_2\tilde{X}_2)^{1} \quad (2.4.7)$$

It is important to point out that the product of every pair of terms of F_i, $i = 1$, $2, \ldots, n$, of Eq. (2.4.6a) is 0. Therefore, Eq. (2.4.7) can be rewritten as

$$F(X_1, X_2) = \delta_1(\bar{X}_1\tilde{\bar{X}}_1\bar{X}_2\tilde{X}_2 \oplus \bar{X}_1\tilde{X}_1 X_2\tilde{\bar{X}}_2)^{0}$$

$$\oplus \delta_2(\bar{X}_1\tilde{\bar{X}}_1\bar{X}_2\tilde{\bar{X}}_2 \oplus \bar{X}_1\tilde{\bar{X}}_1\bar{X}_2\tilde{X}_2 \oplus X_1\tilde{\bar{X}}_1 X_2\tilde{X}_2)^{1} \quad (2.4.7a)$$

In the sequel, it will be shown that the computation of Boolean derivatives of a vector switching function when expressed in the form of Eq. (2.4.7a) becomes very simple.

Applications to Logic Design

Since the advent of integrated-circuit digital components, many basic digital design philosophies have been changed. For example, the switching-function minimization is now no longer important in the design of digital systems, since digital design using ready-made components, SSI, MSI, and LSI, has become today's standard procedure of designing digital systems.

Equations (2.4.1) and (2.4.2) suggest another possible method of designing digital systems using "component modules." The generalized components involved in these equations may be realized by using EXCLUSIVE-OR gates, since

$$X^P = X \oplus \bar{P} = X \oplus P \oplus I \quad \text{(Property 3)}$$

where $X_1 \oplus X_2 = \bar{X}_1 X_2 + X_1 \bar{X}_2$. For example, the first term of Eq. (2.4.3) expressed in terms of EXCLUSIVE-OR operation is

$$(01)(X_1 \oplus \overline{00})\overbrace{(X_1 \oplus \overline{00})}(X_2 \oplus \overline{00})\overbrace{(X_2 \oplus \overline{00})}$$

A general product term of Eq. (2.4.1) is

$$F(P_1 \ldots P_m)[(X_1 \oplus \bar{P}_1) \ldots \overbrace{(X_1 \oplus \bar{P}_1)}^{(n-1)}] \ldots [(X_m \oplus \bar{P}_m) \ldots \overbrace{(X_m \oplus \bar{P}_m)}^{(n-1)}]$$

may be realized by the gate circuit shown in Fig. 2.4.2(a), which may be symbolically represented by the block diagram shown in Fig. 2.4.2(b), in which the symbol $\frown$ denotes that the order of the ordered wires is rotated. The complete realization of

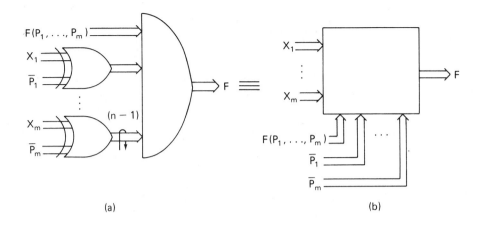

(a)   (b)

Fig. 2.4.2   (a) Gate realization of a products term of Eq. (2.4.1);
(b) its symbolic representation.

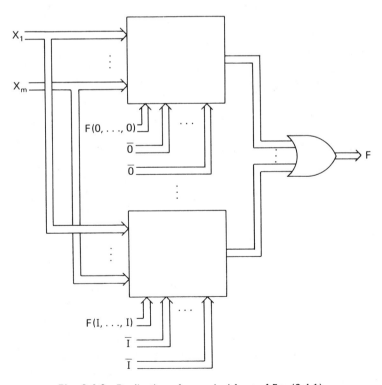

Fig. 2.4.3   Realization of canonical form of Eq. (2.4.1).

Eq. (2.4.1) is presented in the diagram given in Fig. 2.4.3. For example, the realization of Eq. (2.4.3) is shown in Fig. 2.4.4. The realization of the canonical product-of-sums form of Eq. (2.4.2) and the realization of the canonical form of Eq. (2.4.6) can be derived in a similar way. For example, the realization of Eq. (2.4.7) is given in Fig. 2.4.5(a). The function of Eq. (2.4.7) can be realized using programmable ROM arrays, namely, two AND–OR ROM pairs. This is shown in Fig. 2.4.5(b).

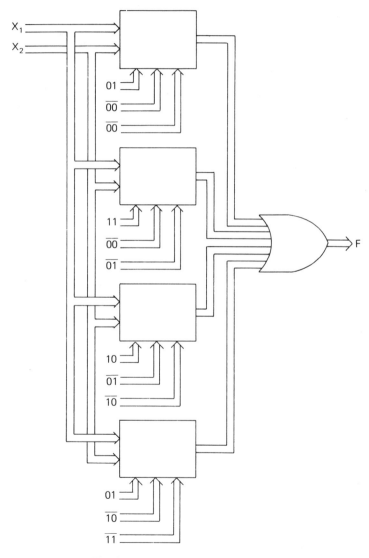

Fig. 2.4.4 Realization of Eq. (2.4.3).

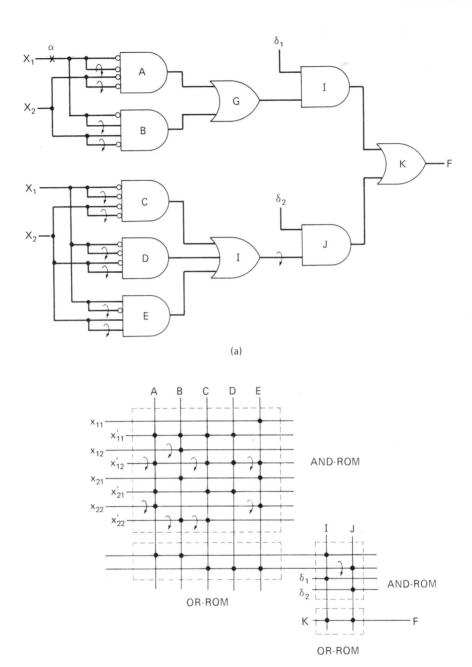

(a)

(b)

Note: The symbol ⌐ in the diagram indicates that the order of the multi-bus wires has been rotated.

Fig. 2.4.5 (a) Gate realization of Eq. (2.4.7); (b) ROM array realization of Eq. (2.4.7).

Exercise 2.4

1. Determine which of the following vector switching functions are in canonical form II in B_2^2.

    (a) $F(X_1, X_2) = \delta_1(X_1 \tilde{X}_1 X_2 \tilde{X}_2 \oplus \bar{X}_1 \tilde{\bar{X}}_1 \bar{X}_2 \tilde{\bar{X}}_2)^{\underline{0}}$

    $\oplus\, \delta_2(X_1 \tilde{X}_1 X_2 \tilde{X}_2 \oplus \bar{X}_1 \tilde{\bar{X}}_1 \bar{X}_2 \tilde{\bar{X}}_2)^{\underline{1}}$

    (b) $F(X_1, X_2) = \delta_1(X_1 \tilde{X}_1 X_2 \tilde{X}_2 \oplus X_1 \tilde{X}_1 X_2 \tilde{X}_2)^{\underline{0}}$

    $\oplus\, \delta_2(\bar{X}_1 \tilde{\bar{X}}_1 \bar{X}_2 \tilde{\bar{X}}_2 \oplus \bar{X}_1 \bar{X}_1 \tilde{\bar{X}}_2 \tilde{\bar{X}}_2)^{\underline{1}}$

2. Find the truth table of the following vector switching function.

$$F(X_1, X_2, X_3) = \delta_1(X_1 \overset{1}{\tilde{\bar{X}}}_1 \overset{2}{\tilde{X}}_1 X_2 \overset{1}{\tilde{\bar{X}}}_2 \overset{2}{\tilde{X}}_2 \bar{X}_3 \overset{1}{\tilde{\bar{X}}}_3 \overset{2}{\tilde{X}}_3)^{\underline{0}}$$

$$\oplus\, \delta_2(X_1 \overset{1}{\tilde{X}}_1 \overset{2}{\tilde{\bar{X}}}_1 X_2 \overset{1}{\tilde{X}}_2 \overset{2}{\tilde{\bar{X}}}_2 X_3 \overset{1}{\tilde{X}}_3 \overset{2}{\tilde{\bar{X}}}_3)^{\underline{1}}$$

$$\oplus\, \delta_3(\bar{X}_1 \overset{1}{\tilde{\bar{X}}}_1 \overset{2}{\tilde{\bar{X}}}_1 \bar{X}_2 \overset{1}{\tilde{\bar{X}}}_2 \overset{2}{\tilde{\bar{X}}}_2 \bar{X}_3 \overset{1}{\tilde{\bar{X}}}_3 \overset{2}{\tilde{\bar{X}}}_3)^{\underline{2}}$$

3. Find the canonical forms I and II of the vector switching function described by the table of Table P2.4.3.

TABLE P2.4.3

X_1	X_2	X_3	F	
			f_1	f_2
00	00	00	0	0
01	01	01	0	1
10	10	10	1	0
11	11	11	1	1
	Otherwise		*	*

*Denotes don't care.

Bibliographical Remarks

This chapter is, in part, based on a paper published by Lee [10]. The material on switching algebra and switching functions can be found in references 1–7. The realization of switching functions using ROM's and PLA's (programmable logic arrays) can be found in references 8–10.

References

1. BOOLE, G., *An Investigation of the Laws of Thought*, Dover, New York, 1854/1940/1958.

2. BERNSTEIN, B. A., "Modular Representations of Finite Algebras," *Proc. 7th Int. Congr.*

Mathematicians, Vol. I, Toronto, Ontario, University of Toronto Press, 1928, pp. 207–216.

3. LOWENSCHUSS, O., "Non-binary Switching Theory," *Proc. IRE Natl. Conv. Rec.*, Vol. 6, No. 4, 1958, pp. 305–317.

4. FLEGG, H. G., *Boolean Algebra and Its Applications*, Wiley, 1964.

5. KORFHAGE, R. R., *Logic and Algorithms*, Wiley, New York, 1965.

6. MacLANE, S., and G. BIRKHOFF, *Algebra*, Macmillan, New York, 1967.

7. WOOD, P. E., *Switching Theory*, McGraw-Hill, New York, 1968.

8. CARR, W. N., and J. P. MIZE, *MOS/LSI Design and Application*, McGraw-Hill, New York, 1972.

9. LUECKE, G., J. P. MIZE, and W. N. CARR, *Semiconductor Memory Design and Application*, McGraw-Hill, New York, 1973.

10. LEE, S. C., "Vector Boolean Algebra and Calculus," *IEEE Trans. Computers*, Vol. C-25, September 1976, pp. 865–874.

3

Boolean Differential Calculus

This chapter introduces the partial derivative, the partial differential, the total differential, and the total variation of a Boolean function. Many properties of these differential and variational operators of Boolean functions are presented. Based on the partial derivative of a Boolean function, the MacLaurin expansion and the Taylor expansion of a Boolean function are derived, which are analogous to those of real functions of real variables.

Two methods for computing Boolean derivatives and differentials of a switching function are presented. One is a graphical method using the Karnaugh map; and the other is a computer algorithm, which is designed for computing derivatives and differentials of functions with a large number of variables on a digital computer.

The following notations are used in this chapter:

Boolean exponentiation:

$$x^{(e)} = \begin{cases} x' & \text{if } e = 0 \\ x & \text{if } e = 1 \end{cases}$$

$$\boldsymbol{x}^{(e)} = x_0^{(e_0)} x_1^{(e_1)} \ldots x_{n-1}^{(e_{n-1})}$$

$$x^e = \begin{cases} 1 & \text{if } e = 0 \\ x & \text{if } e = 1 \end{cases}$$

$$\boldsymbol{x}^e = x_0^{e_0} x_1^{e_1} \ldots x_{n-1}^{e_{n-1}}$$

If D is an operator:

$$D\boldsymbol{x} = Dx_0, Dx_1, \ldots, Dx_{n-1}$$

$$D\boldsymbol{x}^e = Dx_0^{e_0} x_1^{e_1} \ldots x_{n-1}^{e_{n-1}}$$

$$(D\boldsymbol{x})^{(e)} = (Dx_0)^{(e_0)}(Dx_1)^{(e_1)} \ldots (Dx_{n-1})^{(e_{n-1})}$$

$$(D\boldsymbol{x})^e = (Dx_0)^{e_0}(Dx_1)^{e_1} \ldots (Dx_{n-1})^{e_{n-1}}$$

3.1   The Partial Derivative

No one will dispute that the biggest advance in the history of mathematics, which set the foundation for the beginning of the modern mathematics era, was the discovery of calculus by Newton and Leibniz. The basic and most important concept in calculus is that of the derivative of a function,

$$\frac{df(x)}{dx} = \lim_{\Delta x \to 0} \frac{\Delta f(x)}{\Delta x}$$

where $f(x)$ is a real function of a real variable x. In words, it means "the rate of change of $f(x)$ with respect to an infinitesimal change of x." In this section, we shall apply this concept to Boolean functions. More specifically, we shall define the derivative of a Boolean function and investigate its properties.

DEFINITION 3.1.1

Let $f(x)$ be a Boolean function of a Boolean variable x. The *derivative* (or Boolean difference) of f is defined as

$$\frac{df(x)}{dx} = f(x) \oplus f(x') \qquad (3.1.1)$$

In this definition it is assumed that the variation of x is from x to x' and that $dx = \Delta x = x \oplus x' = 1$ and $df(x) = \Delta f(x) = f(x) \oplus f(x')$. Note that in a discrete mathematics such as Boolean algebra, the concept of limit simply does not exist; therefore, dx and Δx must be the same.

Example 3.1.1

Consider the function $f(x) = ax + bx'$ in $B_4 = \{0, a, b, 1\}$, where $b = a'$. The derivative of $f(x)$ with respect to x is

$$\frac{df(x)}{dx} = (ax + bx') \oplus (ax' + bx)$$

$$= (ax + bx')(ax' + bx)' + (ax + bx')'(ax' + bx)$$

$$= 1$$

This result may be verified as follows:

x	$f(x)$	$f(x')$	$f(x) \oplus f(x')$
0	b	a	1
a	1	0	1
b	0	1	1
1	a	b	1

When $df(x)/dx = 1$, it means that $f(x)$ and $f(x')$ are complementary to each other; that is, $f(x') = \overline{f(x)}$.

Definition 3.1.1 may be generalized into the following definition.

DEFINITION 3.1.2

Let $f(x_1, \ldots, x_n)$ be a Boolean function of n variables $x_1, \ldots, x_n$. The *partial derivative* of f with respect to x_i, $1 \leq i \leq n$, is defined as

$$\frac{\partial f}{\partial x_i} = f(x_1, \ldots, x_i, \ldots, x_n) \oplus f(x_1, \ldots, x'_i, \ldots, x_n) \qquad (3.1.2)$$

Example 3.1.2

The partial derivative of the function $f(x_1, x_2, x_3) = x_1 x'_2 + x_1 x_3$ in B_2 with respect to x_1, is

$$\frac{\partial f}{\partial x_1} = (x_1 x'_2 + x_1 x_3) \oplus (x'_1 x'_2 + x'_1 x_3) \qquad (3.1.3)$$

A convenient way to obtain $\partial f/\partial x_i$ in a minimized sum-of-products form is to represent $f(x_1, \ldots, x_i, \ldots, x_n)$ and $f(x_1, \ldots, x'_i, \ldots, x_n)$ by Karnaugh maps and then to find $f(x_1, \ldots, x_i, \ldots, x_n) \oplus f(x_1, \ldots, x'_i, \ldots, x_n)$ graphically. For example, the $\partial f/\partial x_1$ of Eq. (3.1.3) may be simplified as follows:

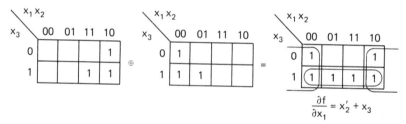

$$\frac{\partial f}{\partial x_1} = x'_2 + x_3$$

THEOREM 3.1.1

The partial derivative of Eq. (3.1.2) is equivalent to the following definition:

$$\frac{\partial f}{\partial x_i} = f(x_1, \ldots, 1, \ldots, x_n) \oplus f(x_1, \ldots, 0, \ldots, x_n) \qquad (3.1.4)$$

Proof: By the expansion theorem [Theorem 2.1.3(a)],

$$f(x_1, \ldots, x_i, \ldots, x_n) = x_i f(x_1, \ldots, 1, \ldots, x_n) + x'_i f(x_1, \ldots, 0, \ldots, x_n)$$
$$= x_i f(x_1, \ldots, 1, \ldots, x_n) \oplus x'_i f(x_1, \ldots, 0, \ldots, x_n)$$

Applying Definition 3.1.2 to this equation, we have

$$= [x_i f(x_1, \ldots, 1, \ldots, x_n) \oplus x'_i f(x_1, \ldots, 0, \ldots, x_n)]$$
$$\oplus [x'_i f(x_1, \ldots, 1, \ldots, x_n) \oplus x_i f(x_1, \ldots, 0, \ldots, x_n)]$$
$$= (x_i \oplus x'_i)(f(x_1, \ldots, 1, \ldots, x_n) \oplus f(x_1, \ldots, 0, \ldots, x_n))$$
$$= f(x_1, \ldots, 1, \ldots, x_n) \oplus f(x_1, \ldots, 0, \ldots, x_n) \quad \blacksquare$$

Example 3.1.3

Find the derivatives of the functions of Examples 3.1.1 and 3.1.2 using Eq. (3.1.4).
(a)   For $f(x) = ax + bx'$ of Example 3.1.1,

$$\frac{df(x)}{dx} = (a \cdot 1 + b \cdot 0) \oplus (a \cdot 0 + b \cdot 1) = 1$$

(b)   For $f(x_1, x_2, x_3) = x_1 x_2' + x_1 x_3$ of Example 3.1.2,

$$\frac{df}{dx} = (1 \cdot x_2' + 1 \cdot x_3) \oplus (0 \cdot x_2' + 0 \cdot x_3) = x_2' + x_3$$

From the examples above, we see that in general the expression of the derivative of a function obtained by using Eq. (3.1.4) is simpler.

Definition 3.1.1 may be further generalized into the following definition.

DEFINITION 3.1.3

The *multiple partial derivative* of a function $f(x_1, \ldots, x_n)$ is defined as

$$\frac{\partial^m f}{\partial x_1 \, \partial x_2 \ldots \partial x_m} = \frac{\partial}{\partial x_1} \left(\frac{\partial}{\partial x_2} \left(\cdots \left(\frac{\partial f}{\partial x_m} \right) \cdots \right) \right) \tag{3.1.5}$$

Example 3.1.4

Evaluate $\partial^2 f / \partial x_1 \, \partial x_2$ of the function of Example 3.1.2.

$$\frac{\partial^2 f}{\partial x_1 \, \partial x_2} = \frac{\partial}{\partial x_1} \left(\frac{\partial f}{\partial x_2} \right)$$

$$= \frac{\partial}{\partial x_1} [(x_1 \cdot 0 + x_1 x_3) \oplus (x_1 \cdot 1 + x_1 x_3)]$$

$$= \frac{\partial}{\partial x_1} (x_1 x_3') = 1 \cdot x_3' \oplus 0 \cdot x_3' = x_3' \tag{3.1.6}$$

The following are some of the basic properties of the derivative of a function. Let $f(x_1, \ldots, x_n)$ and $g(x_1, \ldots, x_n)$ be Boolean functions.

PROPERTY 1

$$\frac{\partial \bar{f}}{\partial x_i} = \frac{\partial f}{\partial x_i}$$

PROPERTY 2

$$\frac{\partial^2 f}{\partial x_i \, \partial x_j} = \frac{\partial^2 f}{\partial x_j \, \partial x_i}$$

PROPERTY 3

$$\frac{\partial^2 f}{\partial x_i \, \partial x_i} = 0$$

PROPERTY 4

$$\frac{\partial (f \cdot g)}{\partial x_i} = f \cdot \frac{\partial g}{\partial x_i} \oplus \frac{\partial f}{\partial x_i} \cdot g \oplus \frac{\partial f}{\partial x_i} \cdot \frac{\partial g}{\partial x_i}$$

PROPERTY 5

$$\frac{\partial (f + g)}{\partial x_i} = \bar{f} \cdot \frac{\partial g}{\partial x_i} \oplus \frac{\partial f}{\partial x_i} \cdot \bar{g} \oplus \frac{\partial f}{\partial x_i} \cdot \frac{\partial g}{\partial x_i}$$

PROPERTY 6

$$\frac{\partial (f \oplus g)}{\partial x_i} = \frac{\partial f}{\partial x_i} \oplus \frac{\partial g}{\partial x_i}$$

The proofs of these properties are left to the reader as an exercise (problem 1).

Example 3.1.5

Consider the function $f(x_1, x_2, x_3) = x_1 x_2' + x_1 x_3$ of Example 3.1.2. Let us illustrate the above six properties by the following examples.

(a) $\dfrac{\partial \bar{f}}{\partial x_1} = \dfrac{\partial f}{\partial x_1}$

$$\frac{\partial \bar{f}}{\partial x_1} = \overline{(1 \cdot x_2' + 1 \cdot x_3)} \oplus \overline{(0 \cdot x_2' + 0 \cdot x_3)} = \overline{(x_2' + x_3)} \oplus 1 = x_2' + x_3$$

which is the same as $\partial f / \partial x_1$ obtained in Example 3.1.2.

(b) $\dfrac{\partial^2 f}{\partial x_1 \, \partial x_2} = \dfrac{\partial^2 f}{\partial x_2 \, \partial x_1}$

$$\frac{\partial^2 f}{\partial x_2 \, \partial x_1} = \frac{\partial}{\partial x_2}(x_2' + x_3) = (0 + x_3) \oplus (1 + x_3) = x_3'$$

which is the same as $\partial^2 f / \partial x_1 \, \partial x_2$ obtained in Eq. (3.1.6).

(c) $\dfrac{\partial^2 f}{\partial x_1 \, \partial x_1} = 0$

$$\frac{\partial^2 f}{\partial x_1 \, \partial x_1} = \frac{\partial}{\partial x_1}(x_2' + x_3) = (x_2' + x_3) \oplus (x_2' + x_3) = 0$$

Let $g(x_1, x_2, x_3) = x_2 x_3'$

(d) $\dfrac{\partial (f \cdot g)}{\partial x_2} = f \cdot \dfrac{\partial g}{\partial x_2} \oplus \dfrac{\partial f}{\partial x_2} \cdot g \oplus \dfrac{\partial f}{\partial x_2} \cdot \dfrac{\partial g}{\partial x_2}$

$$\frac{\partial (f \cdot g)}{\partial x_2} = \frac{\partial}{\partial x_2}(0) = 0$$

$$f \cdot \frac{\partial g}{\partial x_2} \oplus \frac{\partial f}{\partial x_2} \cdot g \oplus \frac{\partial f}{\partial x_2} \cdot \frac{\partial g}{\partial x_2}$$

$$= [(x_1 x_2' + x_1 x_3) \cdot x_3'] \oplus [(x_1 x_3') \cdot x_2 x_3'] \oplus [(x_1 x_3') x_3']$$

$$= x_1 x_2' x_3' \oplus x_1 x_2 x_3' \oplus x_1 x_3'$$

$$= x_1 x_3' \oplus x_1 x_3'$$

$$= 0$$

(e) $\dfrac{\partial(f+g)}{\partial x_2} = \bar{f} \cdot \dfrac{\partial g}{\partial x_2} \oplus \dfrac{\partial f}{\partial x_2} \cdot g \oplus \dfrac{\partial f}{\partial x_2} \cdot \dfrac{\partial g}{\partial x_2}$

$$\begin{aligned}
\frac{\partial(f+g)}{\partial x_2} &= \frac{\partial}{\partial x_2}(x_1 x_2' + x_1 x_3 + x_2 x_3')\\
&= (x_1 \cdot 0 + x_1 x_3 + 1 \cdot x_3') \oplus (x_1 \cdot 1 + x_1 x_3 + 0 \cdot x_3')\\
&= (x_1 + x_3') \oplus x_1\\
&= x_1' x_3'
\end{aligned}$$

$$\begin{aligned}
\bar{f} \cdot \frac{\partial g}{\partial x_2} &\oplus \frac{\partial f}{\partial x_2} \bar{g} \oplus \frac{\partial f}{\partial x_2} \cdot \frac{\partial g}{\partial x_2}\\
&= [\overline{(x_1 x_2' + x_1 x_3)} \cdot x_3'] \oplus [(x_1 x_3') \cdot \overline{x_2 x_3'}] \oplus [(x_1 x_3') x_3']\\
&= (x_1' + x_2) x_3' \oplus x_1 x_2' x_3' \oplus x_1 x_3'
\end{aligned}$$

This expression can be minimized to $x_1' x_3'$, as seen from the following Karnaugh maps.

(f) $\dfrac{\partial(f \oplus g)}{\partial x_2} = \dfrac{\partial f}{\partial x_2} \oplus \dfrac{\partial g}{\partial x_2}$

$$\begin{aligned}
\frac{\partial(f \oplus g)}{\partial x_2} &= \frac{\partial}{\partial x_2}[(x_1 x_2' + x_1 x_3) \oplus x_2 x_3']\\
&= [(x_1 \cdot 0 + x_1 x_3) \oplus 1 \cdot x_3'] \oplus [(x_1 \cdot 1 + x_1 x_3) \oplus 0 \cdot x_3']\\
&= (x_1 + x_3') \oplus x_1\\
&= x_1' x_3'
\end{aligned}$$

$$\begin{aligned}
\frac{\partial f}{\partial x_2} \oplus \frac{\partial g}{\partial x_2} &= x_1 x_3' \oplus x_3'\\
&= x_1' x_3'
\end{aligned}$$

Exercise 3.1

1. Prove Properties 1–6 of the partial derivative.

2. The partial p- and q-derivatives of a function $f(x_0, \ldots, x_{n-1})$ are defined as follows:

$$\frac{qf}{qx_i} = f(x_0, \ldots, x_i, \ldots, x_{n-1}) + f(x_0, \ldots, x_i', \ldots, x_{n-1})$$

$$\frac{pf}{px_i} = f(x_0, \ldots, x_i, \ldots, x_{n-1}) \cdot f(x_0, \ldots, x_i', \ldots, x_{n-1})$$

Let $x = x_0, x_1, \ldots, x_{n-1}$. If x_0, x_1 is a partition of x, the multiple partial q- and p-derivatives of the function f with respect to the m variables in x_0 will be denoted by qf/qx_0 and pf/px_0 and are defined as follows:

$$\frac{qf}{qx_0} = \frac{q}{qx_0}\left(\frac{q}{qx_1}\left(\cdots \frac{qf}{qx_{m-1}}\right)\cdots\right)$$

$$\frac{pf}{px_0} = \frac{p}{px_0}\left(\frac{p}{px_1}\left(\cdots \frac{pf}{px_{m-1}}\right)\cdots\right)$$

Prove that:

(a) $\dfrac{qf}{qx_0} = \sum\limits_{e_0} f(e_0, x_1)$

(b) $\dfrac{pf}{px_0} = \prod\limits_{e_0} f(e_0, x_1)$

where $0 \le e_0 \le 2^m - 1$

3. (a) Show that the function pf/px_0 is the sum of the prime implicants of f independent of x_0.
    (b) Verify part (a) by an example. Assume that $f(x_0, x_1) = x_1 x_2' + x_1 x_3$, where $x_0 = (x_1, x_2)$ and $x_1 = x_3$.

4. (a) The sum terms of a product-of-sums of a Boolean function f are called the *prime implicates* of f. Show that the function qf/qx_0 is the product of the prime implicates of f independent of x_0.
    (b) Verify part (a) by an example.

5. A Boolean function $f(x_1, \ldots, x_n)$ is said to be *independent* of a variable x_i iff $f(x_1, \ldots, x_n)$ is logically invariant under complementation of x_i, that is, iff $f(x_1, \ldots, x_i, \ldots, x_n) = f(x_1, \ldots, x_i', \ldots, x_n)$. Show that a necessary and sufficient condition that a function $f(x_1, \ldots, x_n)$ be independent of a variable x_i is that $\partial f/\partial x_i = 0$.

3.2 Series Expansions of a Boolean Function

Having defined the Boolean partial derivative similar to ordinary partial derivative, it is natural to ask if it is possible to expand a Boolean function in terms of its partial derivatives. The answer is yes. The following is the MacLaurin expansion of a Boolean function.

THEOREM 3.2.1

Any Boolean function $f(x_1, x_2)$ may be expanded as

$$f(x_1, x_2) = \left.\frac{\partial f}{\partial x_1^0}\right|_{x_1=0} x_1^0 \oplus \left.\frac{\partial f}{\partial x_1^1}\right|_{x_1=0} x_1^1 \oplus \cdots \oplus \left.\frac{\partial f}{\partial x_1^{2^m-1}}\right|_{x_1=0} x_1^{2^m-1} \tag{3.2.1}$$

$$= \sum\limits_e \left(\frac{\partial f}{\partial x_1^e}\right)_{x_1=0} x_1^e, \qquad 0 \le e \le 2^m - 1$$

where $\partial f/\partial x_1^0 \triangleq f$ and $2^m - 1 \triangleq \underbrace{(1, 1, \cdots, 1)}_{m}$.

Example 3.2.1

Given $f(x_1, x_2) = x_1 x_2' + x_1 x_3$, where $x_1 = (x_1, x_2)$ and $x_2 = x_3$, the above theorem states that f may be expressed as

$$f(x_1, x_2) = \frac{\partial f}{\partial x_1^0}\bigg|_{x_1=0} x_1^0 \oplus \frac{\partial f}{\partial x_1^1}\bigg|_{x_1=0} x_1^1 \oplus \frac{\partial f}{\partial x_1^2}\bigg|_{x_1=0} x_1^2 \oplus \frac{\partial f}{\partial x_1^3}\bigg|_{x_1=0} x_1^3$$

$$= f\bigg|_{x_1=00} x_1^{00} \oplus \frac{\partial f}{\partial x_1^{01}}\bigg|_{x_1=00} x_1^{01} \oplus \frac{\partial f}{\partial x_1^{10}}\bigg|_{x_1=00} x_1^{10} \oplus \frac{\partial f}{\partial x_1^{11}}\bigg|_{x_1=00} x_1^{11}$$

where

$$f\bigg|_{x_1=00} = 0$$

$$\frac{\partial f}{\partial x_1^{01}}\bigg|_{x_1=00} = \frac{\partial^2 f}{\partial x_1^0 \partial x_2^1}\bigg|_{x_1=00} = \frac{\partial f}{\partial x_2}\bigg|_{x_1=00} = x_1 x_3'\bigg|_{x_1=0, x_2=0} = 0$$

$$\frac{\partial f}{\partial x_1^{10}}\bigg|_{x_1=00} = \frac{\partial^2 f}{\partial x_1^1 \partial x_2^0}\bigg|_{x_1=00} = \frac{\partial f}{\partial x_1}\bigg|_{x_1=00} = x_2' + x_3\bigg|_{x_1=0, x_2=0} = 1$$

$$\frac{\partial f}{\partial x_1^{11}}\bigg|_{x_1=00} = \frac{\partial^2 f}{\partial x_1^1 \partial x_2^1}\bigg|_{x_1=00} = \frac{\partial}{\partial x_1}\left(\frac{\partial f}{\partial x_2}\right)\bigg|_{x_1=00} = \frac{\partial}{\partial x_1}(x_1 x_3')\bigg|_{x_1=00}$$

$$= x_3'\bigg|_{x_1=0, x_2=0} = x_3'$$

The MacLaurin expansion of the function $f(x_1, x_2) = x_1 x_2' + x_1 x_3$ therefore is

$$f(x_1, x_2) = 0 \cdot x_1^0 x_2^0 \oplus 0 \cdot x_1^0 x_2^1 \oplus 1 \cdot x_1^1 x_2^0 \oplus x_3' \cdot x_1^1 x_2^1 \qquad (3.2.2)$$

where $x_1^0 \triangleq 1$, $x_2^0 \triangleq 1$, $x_1^1 \triangleq x_1$, and $x_2^1 \triangleq x_2$. It may be easily verified that this is equal to the original function.

Proof of Theorem 3.2.1: We shall prove this theorem by mathematical induction. Let m be the number of x_i's in x_1.

(1) First, we shall prove the validity of Eq. (3.2.1) for $m = 1$. From the expansion theorem (Theorem 2.1.3), we have

$$f(x_1, x_2) = (x_1, x_2) = x_1 \cdot f(x_1, x_2)|_{x_1=1} \oplus x_1' \cdot f(x_1, x_2)|_{x_1=0} \qquad (3.2.3)$$

Since x_1 has only one variable in it, we can apply the definition of partial derivative given in Theorem 3.1.1 to this function and obtain

$$\frac{\partial f}{\partial x_1} = f(x_1, x_2)\bigg|_{x_1=0} \oplus f(x_1, x_2)\bigg|_{x_1=1} \qquad (3.2.4)$$

Addition of the term $f(x_1, x_2)_{x_1=0}$ to both sides of Eq. (3.2.4) yields

$$f(x_1, x_2)\bigg|_{x_1=0} \oplus \frac{\partial f}{\partial x_1} = f(x_1, x_2)\bigg|_{x_1=0} \oplus f(x_1, x_2)\bigg|_{x_1=0} \oplus f(x, x_2)\bigg|_{x_1=1} \qquad (3.2.5)$$

By using the properties of the EXCLUSIVE-OR operation: $a \oplus 0 = a$ and $a \oplus a = 0$, Eq. (3.2.5) can be rewritten as

$$f(x_1, x_2)\Big|_{x_1=0} \oplus \frac{\partial f}{\partial x_1} = f(x_1, x_2)\Big|_{x_1=1} \tag{3.2.6}$$

Substitution of Eq. (3.2.6) into Eq. (3.2.3) gives

$$f(x_1, x_2) = x_1 \left(f(x_1, x_2)\Big|_{x_1=0} \oplus \frac{\partial f}{\partial x_1} \right) \oplus x_1' \cdot f(x_1, x_2)\Big|_{x_1=0}$$

$$= [f(x_1, x_2)|_{x_1=0} \cdot (x_1 \oplus x_1')] \oplus x_1 \frac{\partial f}{\partial x_1} \quad \text{(since } x_1 \oplus x_1' = 1)$$

$$= f(x_1, x_2)|_{x_1=0} \oplus x_1 \frac{\partial f}{\partial x_1}$$

$$= \frac{\partial f}{\partial x_1^0}\Big|_{x_1=0} x_1^0 \oplus \frac{\partial f}{\partial x_1^1}\Big|_{x_1=0} x_1^1 \tag{3.2.7}$$

In this expression, the reason that $\partial f/\partial x_1 = \partial f/\partial x_1^1|_{x_1=0}$ is because, for $m=1$, $\partial f/\partial x_1$ is independent of x_1.

(2) Assuming that Eq. (3.2.1) holds for $m=M$, we want to prove that the theorem holds for $m = M+1$. Without loss of generality, we let the function be $f(z, x_1, x_2)$, where x_1 has M variables and z is a single variable. Expanding the function $f(z, x_1, x_2)$ with respect to x_1 and the single variable z, we have

$$f(z, x_1, x_2) = \sum_e \left(\frac{\partial f}{\partial x_1^e} \right)_{x_1=0} x_1^e, \quad 0 \le e \le 2^M - 1 \tag{3.2.8}$$

and

$$f(z, x_1, x_2) = \frac{\partial f}{\partial z^0}\Big|_{z=0} z^0 \oplus \frac{\partial f}{\partial z^1}\Big|_{z=0} z^1 \tag{3.2.9}$$

respectively. The validity of Eq. (3.2.8) is based on our assumption, and that of Eq. (3.2.9) was proved in (1). Substituting Eq. (3.2.8) into Eq. (3.2.9),

$$f(x_1, z, x_2) = \frac{\partial}{\partial z^0} \left[\left(\sum_e \frac{\partial f}{\partial x_1^e} \right)_{x_1=0} x_1^e \right]\Big|_{z=0} z^0 \oplus \frac{\partial}{\partial z^1} \left[\left(\sum_e \frac{\partial f}{\partial x_1^e} \right)_{x_1=0} x_1^e \right]\Big|_{z=0} z^1$$

$$0 \le e \le 2^M - 1 \tag{3.2.10}$$

Let $x_1^* = zx_1$. The first term on the right side of Eq. (3.2.10) can be written as

$$\frac{\partial}{\partial z^0} \left[\left(\sum_e \frac{\partial f}{\partial x_1^e} \right)_{x_1=0} x_1^e \right]\Big|_{z=0} z^0 = \left(\sum_e \frac{\partial f}{\partial x_1^e} \right)_{x_1=0} x_1^e \Big|_{z=0} \cdot 1 \quad \left[\text{since } \frac{\partial g(z)}{\partial z^0} = g(z) \right]$$

$$= \sum_e \left(\frac{\partial f}{\partial x_1^e} \right)_{x_1^*=0} x_1^e, \quad 0 \le e \le 2^M - 1 \tag{3.2.11}$$

$$= \sum_e \left(\frac{\partial f}{\partial x_1^{*e}} \right)_{x_1^*=0} x_1^{*e}, \quad 0 \le e \le 2^M - 1$$

Note that these are the first 2^M terms of the expansion of $f(x_1^*, x_2)$ with respect to x_1^*, namely, for $x_1^* = 0x_1$ (since $z = 0$).

Examining the second term on the right-hand side of Eq. (3.2.10), we find that

$$\frac{\partial}{\partial z^1}\left[\left(\sum_e \frac{\partial f}{\partial x_1^e}\right)_{x_1=0} x_1^e\right]\Bigg|_{z=0} z^1 = \sum_e \left(\frac{\partial f}{\partial x_1^{*e}}\right)_{x_1=0} x_1^{*e}, \quad 2^M \leq e \leq 2^{M+1}-1 \qquad (3.2.12)$$

Combining Eqs. (3.2.11) and (3.2.12), we have

$$f(x_1^*, x_2) = \sum_e \left(\frac{\partial f}{\partial x_1^{*e}}\right)_{x_1^*=0} x_1^{*e}, \quad 0 \leq e \leq 2^{M+1}-1 \quad \blacksquare \qquad (3.2.13)$$

As an extension of the MacLaurin expansion, we present the Taylor expansion of a Boolean function as follows.

THEOREM 3.2.2

Any Boolean function $f(x_1, x_2)$ may be expanded as

$$f(x_1, x_2) = \frac{\partial f}{\partial x_1^0}\bigg|_{x_1=h_1} (x_1 \oplus h_1)^0 \oplus \frac{\partial f}{\partial x_1^1}\bigg|_{x_1=h_1} (x_1 \oplus h_1)^1 \oplus \cdots \oplus \frac{\partial f}{\partial x_1^{2^m-1}}\bigg|_{x_1=h_1}$$

$$(x_1 \oplus h_1)^{2^m-1}$$

$$= \sum_e \left(\frac{\partial f}{\partial x_1^e}\right)\bigg|_{x_1=h_1} (x_1 \oplus h_1)^e, \quad 0 \leq e \leq 2^m-1 \qquad (3.2.14)$$

Example 3.2.2

Expand the function $f(x_1, x_2) = x_1 x_2' + x_1 x_3$, where $x_1 = (x_1, x_2)$ and $x_2 = x_3$ in a Taylor series expansion of Eq. (3.2.14). Since, in this example, $m = 2$,

$$f(x_1, x_2) = \frac{\partial f}{\partial x_1^0}\bigg|_{x_1=h_1} (x_1 \oplus h_1)^0 \oplus \frac{\partial f}{\partial x_1^1}\bigg|_{x_1=h_1} (x_1 \oplus h_1)^1 \oplus \frac{\partial f}{\partial x_1^2}\bigg|_{x_1=h_1} (x_1 \oplus h_1)^2$$

$$\oplus \frac{\partial f}{\partial x_1^3}\bigg|_{x_1=h_1} (x_1 \oplus h_1)^3, \text{ where } h_1 = (h_1, h_2)$$

then

$$\frac{\partial f}{\partial x_1^0}\bigg|_{x_1=h_1} = f\bigg|_{\substack{x_1=h_1 \\ x_2=h_2}} = h_1 h_2' + h_1 x_3$$

$$\frac{\partial f}{\partial x_1^1}\bigg|_{x_1=h_1} = \frac{\partial f}{\partial x_2}\bigg|_{\substack{x_1=h_1 \\ x_2=h_2}} = x_1 x_3'\bigg|_{\substack{x_1=h_1 \\ x_2=h_2}} = h_1 x_3'$$

$$\frac{\partial f}{\partial x_1^2}\bigg|_{x_1=h_1} = \frac{\partial f}{\partial x_1}\bigg|_{\substack{x_1=h_1 \\ x_2=h_2}} = (x_2' + x_3)\bigg|_{\substack{x_1=h_1 \\ x_2=h_2}} = h_2' + x_3$$

$$\frac{\partial f}{\partial x_1^3}\bigg|_{x_1=h_1} = \frac{\partial^2 f}{\partial x_1 \partial x_2}\bigg|_{\substack{x_1=h_1 \\ x_2=h_2}} = x_3'$$

The Taylor expansion of the function is

$$
\begin{aligned}
f(x_1, x_2) = &[(h_1 h_2' + h_1 x_3)(x_1 \oplus h_1)^0 (x_2 \oplus h_2)^0] \\
&\oplus [h_1 x_3'(x_1 \oplus h_1)^0 (x_2 \oplus h_2)^1] \\
&\oplus [(h_2' + x_3)(x_1 \oplus h_1)^1 (x_2 \oplus h_2)^0] \qquad (3.2.15)\\
&\oplus [x_3'(x_1 \oplus h_1)^1 (x_2 \oplus h_2)^1] \\
= &f_{00} \oplus f_{01} \oplus f_{10} \oplus f_{11}
\end{aligned}
$$

To verify that this expansion is indeed equal to the given function, we first simplify the four terms $f_{00}, f_{01}, f_{10},$ and f_{11} as follows:

$$
\begin{aligned}
f_{00} &= h_1 h_2' + h_1 x_3 \\
f_{01} &= h_1 x_3'(x_2 \oplus h_2) = h_1 h_2' x_2 x_3' + h_1 h_2 x_2' x_3' \\
f_{10} &= (h_2' + x_3)(x_1 \oplus h_1) = h_1' h_2' x_1 + h_1' x_1 x_3 + h_1 h_2' x_1' + h_1 x_1' x_3 \\
f_{11} &= x_3'(x_1 \oplus h_1)(x_2 \oplus h_2) = h_1' h_2' x_1 x_2 x_3' + h_1 h_2' x_1' x_2 x_3' + h_1' h_2 x_1 x_2' x_3' + h_1 h_2 x_1' x_2' x_3'
\end{aligned}
$$

The Karnaugh maps of these four functions and the EXCLUSIVE-OR of them are shown on the following page.

LEMMA 3.2.1

$$
\frac{\partial f(x_1 \oplus h_1)}{\partial x_1}\bigg|_{x_1=0} = \frac{\partial f(x_1)}{\partial x_1}\bigg|_{x_1=h_1} \qquad (3.2.16)
$$

Proof: From the definition of the partial derivative, we have

$$
\begin{aligned}
\frac{\partial f(x_1 \oplus h_1)}{\partial x_1}\bigg|_{x_1=0} &= (f(x_1 \oplus h_1) \oplus f(x_1' \oplus h_1))|_{x_1=0} \\
&= f(h_1) \oplus f(h_1') \\
&= (f(x_1) \oplus f(x_1'))|_{x_1=h_1} \\
&= \frac{\partial f(x_1)}{\partial x_1}\bigg|_{x_1=h_1} \quad \blacksquare
\end{aligned}
$$

LEMMA 3.2.2

Let $x_1 = x_1 x_2 \ldots \ldots x_m$. Then

$$
\frac{\partial f(x_1 \oplus h_1, x_2)}{\partial x_1}\bigg|_{x_1=0} = \frac{\partial f(x_1, x_2)}{\partial x_1}\bigg|_{x_1=h_1} \qquad (3.2.17)
$$

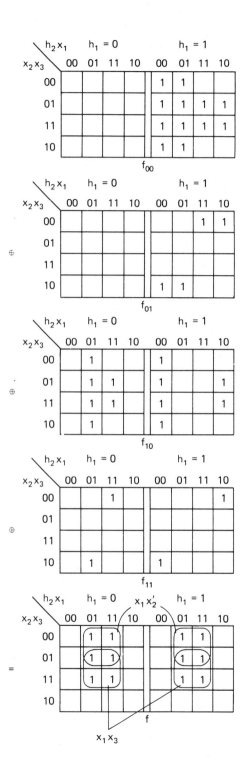

Proof: By Definition 3.1.3,

$$\left.\frac{\partial f(x_1 \oplus h_1, x_2)}{\partial x_1}\right|_{x_1=0} = \frac{\partial}{\partial x_m}\frac{\partial}{\partial x_{m-1}}\cdots\frac{\partial}{\partial x_2}\frac{\partial}{\partial x_1}$$

$$\left.(f(x_1 \oplus h_1, x_2 \oplus h_2, \ldots, x_m \oplus h_m, x_2))\right|_{x_1=0,\ldots,x_m=0}$$

$$= \frac{\partial}{\partial x_m}\frac{\partial}{\partial x_{m-1}}\cdots\frac{\partial}{\partial x_2}$$

$$\left.\left(\frac{\partial}{\partial x_1}(f(x_1, x_2 \oplus h_2, \ldots, x_m \oplus h_m, x_2))\right)\right|_{x_1=h_1, x_2=0,\ldots,x_m=0}$$

$$= \frac{\partial}{\partial x_m}\left(\frac{\partial}{\partial x_{m-1}}\cdots\left(\frac{\partial}{\partial x_2}\frac{(f(x_1, x_2, \ldots, x_m, x_2))}{x_1}\right)\cdots\right)\Bigg|_{x_1=h_1,\ldots,x_m=h_m}$$

$$= \left.\frac{\partial f(x_1, x_2)}{\partial x_1}\right|_{x_1=h_1} \quad\blacksquare \tag{3.2.18}$$

Following Lemma 3.2.2, we have

LEMMA 3.2.3

$$\left.\frac{\partial f(x_1 \oplus h_1, x_2)}{\partial x_1^e}\right|_{x_1=0} = \left.\frac{\partial f(x_1, x_2)}{\partial x_1^e}\right|_{x_1=h_1}, \quad 0 \le e \le 2^m - 1$$

Proof of Theorem 3.2.2: The MacLaurin series expansion of $f(x_1 \oplus h_1, x_2)$ is

$$f(x_1 \oplus h_1, x_2) = \sum_e \left(\frac{\partial f(x_1 \oplus h_1, x_2)}{\partial x_1^e}\right)_{x_1=0} x_1^e, \quad 0 \le e \le 2^m - 1$$

By Lemma 3.2.3, we have

$$f(x_1 \oplus h_1, x_2) = \sum_e \left(\frac{\partial f(x_1, x_2)}{\partial x_1^e}\right)_{x_1=h_1} x_1^e, \quad 0 \le e \le 2^m - 1 \tag{3.2.19}$$

Now if we replace all the x_1 in Eq. (3.2.19) by $x_1 \oplus h_1$, since $x_1 \oplus h_1 \oplus h_1 = x_1$ and $\partial f(x_1, x_2)/\partial x_1^e|_{x_1=h_1}$ is independent of x_1, the equation above becomes Eq. (3.2.14). Hence, the theorem is proved. $\blacksquare$

Some other expansions arise from searching to know if a Boolean function is completely determined by means of the set of all its simple derivatives. An answer to this question is given by means of the following theorem.

THEOREM 3.2.3

A Boolean function $f(x)$, $x = (x_0, x_1, \ldots, x_{n-1})$, can be expanded in the form

$$f(x) = f(h) \oplus \sum_{i=0}^{n-1} x_i^{(h_i')}\left(\frac{\partial f}{\partial x_i}\right)_{x_j=h_j}, \, j > i \tag{3.2.20}$$

where $h = (h_0, h_1, \ldots, h_{n-1})$.

Before proving this theorem, we present the following lemma.

LEMMA 3.2.4

Another type of MacLaurin series expansion of a function $f(x_1, x_2)$ with respect to a single variable x_1 is

$$f(x_1, x_2) = f(x_1, x_2)|_{x_1=1} \oplus x_1' \frac{\partial f}{\partial x_1}$$

$$= \frac{\partial f}{\partial x_1^0}\bigg|_{x_1=1} (x_1')^0 \oplus \frac{\partial f}{\partial x_1^1}\bigg|_{x_1=1} (x_1')^1 \tag{3.2.21}$$

Proof: From Eq. (3.2.4),

$$f(x_1, x_2)|_{x_1=0} = f(x_1, x_2)|_{x_1=1} \oplus \frac{\partial f}{\partial x_1}$$

Substituting this equation into Eq. (3.2.3), we have

$$f(x_1, x_2) = x_1 \cdot f(x_1, x_2)|_{x_1=1} \oplus x_1' \cdot \left[f(x_1, x_2)|_{x_1=1} \oplus \frac{\partial f}{\partial x_1} \right]$$

$$\doteq [f(x_1, x_2)|_{x_1=1} \cdot (x_1 \oplus x_1')] \oplus x_1' \frac{\partial f}{\partial x_1}$$

$$= f(x_1, x_2)|_{x_1=1} \oplus x_1' \frac{\partial f}{\partial x_1}$$

This completes the proof. ∎

Proof of Theorem 3.2.3: Let $f(x)$ be a Boolean function of $x = (x_0, x_1, \ldots, x_{n-1})$ variables. Expanding the function with respect to x_{n-1} using Eqs. (3.2.7) and (3.2.21), we have

$$f(x) = f(x_0, x_1, \ldots, x_{n-2}, 0) \oplus x_{n-1} \frac{\partial f}{\partial x_{n-1}} \tag{3.2.22}$$

and

$$f(x) = f(x_0, x_1, \ldots, x_{n-2}, 1) \oplus x_{n-1}' \frac{\partial f}{\partial x_{n-1}} \tag{3.2.23}$$

respectively. Combining these two equations, we have

$$f(x) = f(x_0, x_1, \ldots, x_{n-2}, h_{n-1}) \oplus x_{n-1}^{(h'_{n-1})} \frac{\partial f}{\partial x_{n-1}} \tag{3.2.24}$$

where h_{n-1} is a Boolean constant.

Repeating the process above with respect to x_{n-2}, x_{n-3}, etc., we obtain

$$f(x_0, x_1, \ldots, x_{n-3}, x_{n-2}, h_{n-1}) = f(x_0, x_1, \ldots, h_{n-2}, h_{n-1}) \oplus x_{n-2}^{(h'_{n-2})} \frac{\partial f}{\partial x_{n-2}}$$

$$\vdots \tag{3.2.25}$$

$$f(x_0, h_1, \ldots, h_{n-1}) = f(h_0, h_1, \ldots, h_{n-2}, h_{n-1}) \oplus x_0^{(h'_0)} \frac{\partial f}{\partial x_0}$$

Next, define $h = (h_0, h_1, \ldots, h_{n-1})$. By substituting the equations above into Eq. (3.2.24), we obtain

$$f(x) = f(h) \oplus \sum_{i=0}^{n-1} x_i^{(h'_i)} \left(\frac{\partial f}{\partial x_i} \right)_{x_j = h_j}, \qquad j > i$$

The theorem is thus proved. ∎

Example 3.2.3

The expansion of the function $f(x) = x_1 x_2' + x_1 x_3$ using Eq. (3.2.20) is

$$f(x) = (h_1 h_2' + h_1 h_3) \oplus x_3^{(h'_3)}(x_1 x_2) \oplus x_2^{(h'_2)}(x_1 h_3') \oplus x_1^{(h'_1)}(h_2' + h_3) \qquad (3.2.26)$$

For example, for $h_1 = h_2 = 1$ and $h_3 = 0$,

$$f(x) = (1 \cdot 0 + 1 \cdot 0) \oplus x_3^{(1)}(x_1 x_2) \oplus x_2^{(0)}(x_1 \cdot 1) \oplus x_1^{(0)}(0 + 0)$$
$$= x_1 x_2 x_3 \oplus x_1 x_2' = x_1 x_2' + x_1 x_3$$

The complete verification of Eq. (3.2.26) involves the substitutions of all eight possible values of $h = (h_1, h_2, h_3)$, which is left to the reader as an exercise (problem 4).

THEOREM 3.2.4

Let h be an increment of x. Then

$$f(x) \oplus f(x \oplus h) = \sum_e \left(\frac{\partial f}{\partial x^e} \right) h^e, \qquad 0 \leq e \leq 2^m - 1 \qquad (3.2.27)$$

Proof: From Theorem 3.2.2, we have

$$f(x) = \sum_e \left(\frac{\partial f}{\partial x^e} \right)_{x_1 = h_1} (x_1 \oplus h_1)^e, \qquad 0 \leq e \leq 2^m - 1$$

Substituting x by $x \oplus h$ in the equation above,

$$f(x \oplus h) = \sum_e \left(\frac{\partial f(x \oplus h)}{\partial (x \oplus h^e)} \right)_{x_1 \oplus h_1} x_1^e$$
$$= \sum_e \left(\frac{\partial f(x)}{\partial x} \right)_{x = h} x_1^e, \qquad 0 \leq e \leq 2^m - 1$$

By Lemma 3.2.1, we have

$$f(x \oplus h) = \sum_e \frac{\partial f(h)}{\partial h^e} x^e, \qquad 0 \leq e \leq 2^m - 1$$
$$= f(h) \oplus \sum_e \frac{\partial f(x)}{\partial x^e}\bigg|_{x = h} x^e, \qquad 0 \leq e \leq 2^m - 1$$

Adding the term $f(h)$ to both sides of the equation and interchanging the roles of x and h we obtain

$$f(x) \oplus f(x \oplus h) = \sum_e \frac{\partial f(h)}{\partial h^e}\bigg|_{h = x} h^e$$
$$= \sum_e \frac{\partial f(x)}{\partial x} h^e, \qquad 0 \leq e \leq 2^m - 1$$

This completes the proof of this theorem. ∎

Example 3.2.4

Suppose that $f(x) = x_1 x_2'$.

$$\frac{\partial f}{\partial x^0} = f(x) = x_1 x_2'$$

$$\frac{\partial f}{\partial x^1} = \frac{\partial f}{\partial x_2} = x_1$$

$$\frac{\partial f}{\partial x^2} = \frac{\partial f}{\partial x_1} = x_2'$$

$$\frac{\partial f}{\partial x^3} = \frac{\partial^2 f}{\partial x_1 \, \partial x_2} = 1$$

Substituting these equations into the right-hand side of Eq. (3.2.27), we have

$$\underset{e}{\Sigma} \left(\frac{\partial f}{\partial x^e} \right) h^e = h_2 x_1 \oplus h_1 x_2' \oplus h_1 h_2 \qquad (3.2.28)$$

The left-hand side of Eq. (3.2.27) is

$$f(x) \oplus f(x \oplus h) = x_1 x_2' \oplus (x_1 \oplus h_1)(x_2 \oplus h_2)'$$
$$= x_1 x_2' \oplus \{h_1' h_2 x_1 x_2 + k_1' h_2' x_1 x_2' + h_1 h_2 x_1' x_2 + h_1 h_2' x_1' x_2'\} \qquad (3.2.29)$$

Equations (3.2.28) and (3.2.29) can be computed using Karnaugh maps as follows:

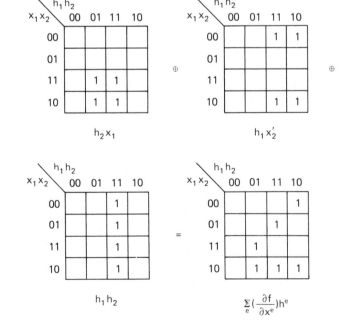

and

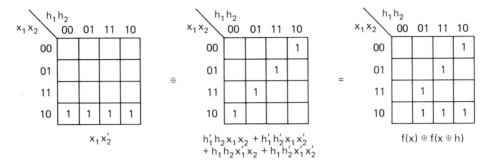

It is seen that Theorem 3.2.4 holds.

Exercise 3.2

1. Expand the function $f(x_1, x_2) = x_1'x_3'x_4' + x_2x_4' + x_3'x_4$ in the MacLaurin series by letting
    (a) $x_1 = x_1$ and $x_2 = (x_2, x_3, x_4)$
    (b) $x_1 = (x_1, x_2)$ and $x_2 = (x_3, x_4)$
    (c) $x_1 = (x_1, x_2, x_3)$ and $x_2 = x_4$
    Be sure to verify your results.

2. Repeat problem 1 for Taylor series expansion.

3. Repeat problem 1 for the series expansion of Eq. (3.2.20).

4. Complete the verification of Eq. (3.2.26) of Example 3.2.3.

5. (a) Find $f(x) \oplus f(x \oplus h)$ for each of the following functions:
        (1) $f(x) = x_1 + x_2'$
        (2) $f(x) = x_1x_2' + x_1x_3$
    (b) Verify your results obtained in part (a) using Theorem 3.2.4.

3.3 The Total Differential and the Total Variation

In this section, two important concepts of the differential and variation are introduced. First, we study the differential.

DEFINITION 3.3.1

The *differential* dx_i of the variable x_i is the increment of x_i.

Let us recall that the increment in a two-valued Boolean algebra is obtained by performing the EXCLUSIVE-OR operation; the differential of a Boolean variable is thus itself a Boolean variable.

DEFINITION 3.3.2

The *partial differential* $d_{x_i} f$ of the function $f(x)$ with respect to a variable x_i is the increment of f due to the increment of x_i. Mathematically, it may be expressed as

$$d_{x_i} f = f(x_1, \ldots, x_{i-1}, x_i, x_{i+1}, \ldots, x_n) \oplus f(x_1, \ldots, x_{i-1}, x_i \oplus d_{x_i}, x_{i+1}, \ldots, x_n)$$

$$(3.3.1)$$

THEOREM 3.3.1

$$d_{x_i} f = \frac{\partial f}{\partial x_i} \, dx_i$$

The proof of this theorem is left to the reader as an exercise (problem 1).

Example 3.3.1

Consider the function $f(x) = x_1 x_2' + x_1 x_3$. The partial differential $d_{x_1} f$ obtained by using Definition 3.3.2 is

$$d_{x_1} f = f(x_1, x_2, x_3) \oplus f(x_1 \oplus dx_1, x_2, x_3)$$
$$= (x_1 x_2' + x_1 x_3) \oplus [(x_1 \oplus dx_1) x_2' + (x_1 \oplus dx_1) x_3]$$
$$= (x_1 x_2' + x_1 x_3) \oplus [x_1 (dx_1)' x_2' + x_1' (dx_1) x_2' + x_1 (dx_1)' x_3 + x_1' (dx_1) x_3] \qquad (3.3.2)$$

Since dx_1 can be 0 or 1, it may be considered as a variable; the above expression can be simplified using four-variable Karnaugh maps as follows:

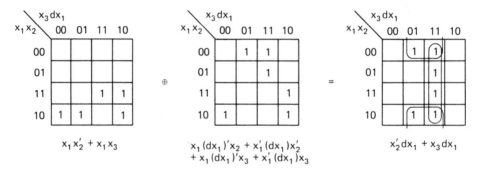

$$x_1 x_2' + x_1 x_3 \qquad\qquad \begin{array}{l} x_1 (dx_1)' x_2 + x_1' (dx_1) x_2' \\ + x_1 (dx_1)' x_3 + x_1' (dx_1) x_3 \end{array} \qquad\qquad x_2' dx_1 + x_3 dx_1$$

Thus, we find that

$$d_{x_1} f = x_2' \, dx_1 + x_3 \, dx_1$$
$$= (x_2' + x_3) \, dx_1$$

From the result of Example 3.1.2, we see that

$$d_{x_1} f = \frac{\partial f}{\partial x_1} \, dx_1$$

The partial differential is thus a parametric form of the partial derivative, the parameter being the differential dx_i, and consequently enjoys similar properties.

DEFINITION 3.3.3

The *multiple partial differential* $d_{x_1} f$ of f with respect to the m variables in x_1 is the function

$$d_{x_1}f = d_{x_0}(d_{x_1}(\ldots d_{x_{m-1}}f)\ldots)$$

$$= \frac{\partial f}{\partial x_1} \prod dx_i, \qquad x_i \in x_1 = x_0, x_1, \ldots, x_{m-1}$$

(3.3.3)

Example 3.3.2

Let $x_1 = (x_1, x_2)$. By Theorem 3.3.1 and Definition 3.3.3, the $d_{x_1}f$ of $f(x) = x_1 x_2' + x_1 x_3$ is found to be

$$d_{x_1}f = d_{x_1}(d_{x_2}f)$$

$$= d_{x_1}\left(\frac{\partial f}{\partial x_2} dx_2\right)$$

$$= d_{x_1}(x_1 x_3' \, dx_2)$$

$$= \frac{\partial}{\partial x_1}(x_1 x_3' \, dx_2) \, dx_1$$

$$= x_3' \, dx_1 \, dx_2$$

DEFINITION 3.3.4

The *total differential df* of a function $f(x)$ is the increment of this function due to the increment dx of x. Mathematically, it may be expressed as

$$df = f(x) + f(x + dx)$$

(3.3.4)

THEOREM 3.3.2

$$df = \sum_e \frac{\partial f}{\partial x^e}(dx)^e, \qquad 0 < e \le 2^n - 1$$

(3.3.5)

$$= \sum_e d_{x^e}f, \qquad 0 < e \le 2^n - 1$$

(3.3.6)

The proof of this theorem follows directly from Eqs. (3.2.27), (3.3.3), and (3.3.4).

The total differential operator has the following basic properties. Let a be a constant.

PROPERTY 1

$$da = 0$$

PROPERTY 2

$$d(af) = a\,df$$

PROPERTY 3

$$d(\bar{f}) = df$$

PROPERTY 4

$$d(df) = 0$$

PROPERTY 5

$$d(fg) = f\,dg \oplus g\,df \oplus df\,dg$$

PROPERTY 6

$$d(f + g) = f'\,dg \oplus g'df \oplus df\,dg$$

PROPERTY 7

$$d(f \oplus g) = df \oplus dg$$

The proofs of these properties are left to the reader as an exercise (problem 2).

DEFINITION 3.3.5

The *sensitivity function* Sf/Sx_1 of a function $\mathbf{x}(\mathbf{x}_1, \mathbf{x}_2)$ with respect to the m variables in $\mathbf{x}_1'$ is defined as

$$\frac{Sf}{Sx_1} = f(\mathbf{x}_1, \mathbf{x}_2) \oplus f(\mathbf{x}_1', \mathbf{x}_2) \tag{3.3.7}$$

where $f(\mathbf{x}_1')$ is the value of f for $\mathbf{x}_1$ substituted by $\mathbf{x}_1'$, the vector of complemented literals.

It is easy to show that the sensitivity operator S/Sx_1 also satisfies Properties 1–7 of the total differential operator (problem 3). Furthermore, it can be easily shown that the total differential operator and the sensitivity operator are related by the following relation:

$$\frac{Sf}{Sx_1} = (df)_{dx_1=1, dx_2=0} \tag{3.3.8}$$

The differential operator of functions of functions is defined as follows:

DEFINITION 3.3.6

Let $f[\mathbf{x}, \mathbf{y}(\mathbf{x})]$ be a function where $\mathbf{x}$ and $\mathbf{y}$ are n- and m-dimensional vectors, respectively. The *total derivative* of f with respect to a variable x_i is denoted by df/dx_i and is defined as follows:

$$\frac{df}{dx_i} = f[x_i, \mathbf{y}(x_i)] \oplus f[x_i', \mathbf{y}(x_i')] \tag{3.3.9}$$

DEFINITION 3.3.7

Let $\mathbf{x}_1, \mathbf{x}_2$ be a partition of $\mathbf{x}$. The *multiple total derivative* of the function f with respect to the m variables in $\mathbf{x}_1$, denoted by $df/d\mathbf{x}_1$, is defined as follows:

$$\frac{df}{d\mathbf{x}_1} = \frac{d}{dx_0}\left(\frac{d}{dx_1}\left(\cdots \frac{df}{dx_{m-1}}\right)\cdots\right), \qquad \mathbf{x}_1 = (x_0, x_1, \ldots, x_{m-1}) \tag{3.3.10}$$

DEFINITION 3.3.8

The *total sensitivity function* Sf/Sx_1 of the function f with respect to the m variables in x_1 is the following function:

$$\frac{Sf}{Sx_1} = f[x_1, x_2, y(x_1, x_2)] \oplus f[x_1', x_2, y(x_1', x_2)] \qquad (3.3.11)$$

Let $\phi(x)$ be a function such that

$$f[x, y(x)] = \phi(x)$$

holds for any value of the vector x. Then

$$\frac{df}{dx_1} = \frac{\partial\phi}{\partial x_1}$$

$$\frac{Sf}{Sx_1} = \frac{S\phi}{Sx_1} \qquad (3.3.12)$$

$$df = d\phi \qquad (3.3.13)$$

Now we shall study various types of variable and functional variations and the relations among them.

DEFINITION 3.3.9

The *variation* ∂x_i of an independent variable is the increment of x_i: namely, $\partial x_i = dx_i$.

DEFINITION 3.3.10

The *total variation* δf of a function $f(x)$ is the maximal increment of this function due to the total variation δx of x.

The following theorem relates the variation δx and the total variation δf.

THEOREM 3.3.3

Let $f(x)$ be a function of independent vector variable x. Then

$$\delta f = \sum_e \left(\frac{\partial f}{\partial x^e}\right)(\delta x)^e, \qquad 0 < e \leq 2^n - 1 \qquad (3.3.14)$$

Proof: From Definitions 3.3.1 and 3.3.2, we have

$$\delta f = \sum_{e_1} [f(x) \oplus f(x \oplus e_1 \,\delta x)], \qquad 0 < e_1 \leq 2^n - 1$$

$$= \sum_{e_1} \sum_e \left(\frac{\partial f}{\partial x^e}(e_1 \,\delta x)^e\right), \qquad 0 < e \leq 2^n - 1$$

$$= \sum_e \left(\frac{\partial f}{\partial x^e}\right)(\delta x)^e \quad \blacksquare$$

DEFINITION 3.3.11

The Δ-*operator* $\Delta f/\Delta x_1$ of the function $f(x_1, x_2)$ with respect to the m variables in x_1 is defined as

$$\frac{\Delta f}{\Delta x_1} = \sum_e \frac{\partial f}{\partial x_1^e}, \qquad 0 < e \le 2^m - 1 \tag{3.3.15}$$

Following Theorem 3.3.3 and Definition 3.3.11, we immediately have

COROLLARY 3.3.1

$$\frac{\Delta f}{\Delta x_1} = (\delta f)_{\delta x_1 = 1, \delta x_2 = 0} \tag{3.3.16}$$

Proof:

$$(\delta f)_{\delta x_1 = 1, \delta x_2 = 0} = \sum_e \frac{\partial f}{\partial x_1^e}, \qquad 0 < e \le 2^m - 1$$

$$= \frac{\Delta f}{\Delta x_1} \quad \blacksquare$$

Two important theorems about the Δ-operator are given below.

THEOREM 3.3.4

$\Delta f(x_1, x_2)/\Delta x_1$ is independent of x_1.

Proof: Let $x_1 = (x_1, x_2, \ldots, x_m)$. We must show that

$$\frac{\partial[\Delta f(x_1, x_2)/\Delta x_1]}{\partial x_i} = 0, \qquad \text{for } i = 1, 2, \ldots, m.$$

We shall prove this by mathematical induction on m. If $m = 1$,

$$\frac{\Delta f(x_1, x_2)}{\Delta x_1} = \frac{\partial f}{\partial x_1}$$

which is independent of x_1, since $\partial^2 f/\delta x_1 \, \partial x_1 = 0$.
    Assume that

$$\frac{\partial[\Delta f(x_1, x_2)/\Delta x_1]}{\partial x_i} = 0$$

for $i = 1, 2, \ldots, M$. Consider x_{M+1} and denote it by y. Then

$$\frac{\Delta f((x_1 y), x_2)}{\Delta(x_1 y)} = \frac{\Delta f((x_1 y), x_2)}{\Delta x_1} + \frac{\Delta[\partial f((x_1 y), x_2)/\partial y]}{\Delta x_1} + \frac{\partial f((x_i y), x_2)}{\partial y}$$

By hypothesis, the partial derivatives of the first two terms with respect to any x_i are 0. Hence, by Property 5 of the partial derivative presented in Section 3.1,

$$\frac{\partial[\Delta f((x_1 y), x_2)/\Delta(x_1 y)]}{\partial x_i} = \left(\frac{\Delta f}{\Delta x_1} + \frac{\Delta[\partial f/\partial y]}{\Delta x_1} \right)' \cdot \frac{\partial^2 f}{\partial x_i \, \partial y}$$

for any i. But the first factor is a product of terms among which is $(\partial^2 f/\partial x_i \, \partial y)'$. Thus,

$$\frac{\partial[\Delta f/\Delta(x_1 y)]}{\partial x_i} = 0$$

for all i. This completes the proof. ∎

THEOREM 3.3.5

$\Delta f(x)/\Delta x$ is 0 or 1 and is 0 iff f is itself 0 or 1.

Proof: The fact that $\Delta f(x)/\Delta x$ must be 0 or 1 follows from Theorem 3.3.4. If f itself is 0 or 1, all its differences will be 0, and hence $\Delta f(x)/\Delta x$ will be 0. If f is not 0 or 1, it must be a dependent function of at least one basic variable and its partial derivative with respect to that variable will not be 0. Thus, $\Delta f(x)/\Delta x$ will not be 0, and hence it must be 1. ∎

The total variation operator δ has the following basic properties. Let a be a constant.

PROPERTY 1

$$\delta a = 0$$

PROPERTY 2

$$\delta(af) = a\,\delta f$$

PROPERTY 3

$$\delta(fg) = f\,\delta g + g\,\delta f + \delta f\,\delta g$$

PROPERTY 4

$$\delta(f + g) = f'\,\delta g + g'\,\delta f + \delta f\,\delta g$$

PROPERTY 5

$$\delta(f \oplus g) = \delta f + \delta g$$

The proofs of these properties are left to the reader as an exercise (problem 7). In addition, if $f = f[x, y(x)]$, then

$$\delta f = \sum_e \frac{\partial f}{\partial(xy)^e}[\delta(x, y)]^e, \qquad 0 < e \le 2^{m+n} - 1 \qquad (3.3.17)$$

where x and y have m and n variables, respectively. The proof of this theorem is evident from Definition 3.3.10.

THEOREM 3.3.6

For any functions $\phi(x)$ and $f[x, y(x)]$ that are such that $\phi(x) = f[x, y(x)]$,

$$d\phi = df \le \delta\phi \le \delta f$$

The proof of this theorem is also left to the reader as an exercise (problem 8).
The following are basic properties of the total Δ-operator. Let $f = f(x_1, x_2)$ and a be a constant.

PROPERTY 1

$$\frac{\Delta a}{\Delta x_1} = 0$$

PROPERTY 2

$$\frac{\Delta af}{\Delta x_1} = a\frac{\Delta f}{\Delta x_1}$$

PROPERTY 3

$$\frac{\Delta f'}{\Delta x_1} = \frac{\Delta f}{\Delta x_1}$$

PROPERTY 4

$$\frac{\Delta}{\Delta x_1}\frac{\Delta f}{\Delta x_1} = 0$$

PROPERTY 5

$$\frac{\Delta}{\Delta x_1}(fg) = f\frac{\Delta g}{\Delta x_1} + g\frac{\Delta f}{\Delta x_1} + \frac{\Delta f}{\Delta x_1}\frac{\Delta g}{\Delta x_1}$$

PROPERTY 6

$$\frac{\Delta}{\Delta x_1}(f + g) = f'\cdot\delta g + g'\cdot\delta f + \frac{\Delta f}{\Delta x_1}\frac{\Delta g}{\Delta x_1}$$

PROPERTY 7

$$\frac{\Delta}{\Delta x_1}(f \oplus g) = \frac{\Delta f}{\Delta x_1} + \frac{\Delta g}{\Delta x_1}$$

In addition, the total variation of a function of independent variables may also be expressed as a function of all the Δ-operators as

$$\delta f = \sum_e \left(\frac{\Delta f}{\Delta x^e}\right)(\delta x)^{(e)}, \qquad 0 < e \leq 2^n - 1 \qquad\qquad (3.3.18)$$

Exercise 3.3

1. Prove Theorem 3.3.1.

2. Prove Properties 1–7 of the total differential operator.

3. Prove that the sensitivity S/Sx_1 also satisfies Properties 1–7 of the total differential operator.

4. Show that

$$\frac{df}{dx_i} = \frac{\partial f}{\partial x_i} \oplus \sum_e \left(\frac{\partial f}{\partial y^e} \oplus \frac{\partial f}{\partial(x_i y^e)}\right)\left(\frac{\partial y}{\partial x_i}\right)^e, \qquad 0 < e \leq 2^m - 1$$

where

$$\frac{\partial y}{\partial x_t} = \left(\frac{\partial y_0}{\partial x_t}, \frac{\partial y_1}{\partial x_t}, \ldots, \frac{\partial y_{m-1}}{\partial x_t} \right)$$

5. The total q-differential qf (p-differential pf) of a function $f(x)$ is the maximum (minimum) of f in the subcube generated by the variation dx of x. Let a be a constant. Prove the following basic properties of the q- and p-differentials.

(1) $qa = a$
(2) $q(af) = aqf$
(3) $qf' = (pf)'$
(4) $q(qf) = qf$
(5) $q(f + g) = qf + qg$
(6) $p(qf) = qf$

(1') $pa = a$
(2') $p(af) = apf$
(3') $pf' = (qf)'$
(4') $p(pf) = pf$
(5') $p(fg) = pfpg$
(6') $q(pf) = pf$

6. (a) Show that

$$qf = \sum_e f(x \oplus e \, dx), \qquad 0 \leq e \leq 2^n - 1$$

$$= \sum_e \frac{qf}{qx^e} (dx)^e$$

(b) Show that

$$pf = \prod_e f(x \oplus e \, dx), \qquad 0 \leq e \leq 2^n - 1$$

$$= \prod_e \left(\frac{pf}{px^e} + [1 \oplus (dx)^e] \right)$$

7. Prove the five basic properties of the total variation described in this section.

8. Prove Theorem 3.3.6.

3.4 Graphical Method and Computer Algorithm for Computing Boolean Derivatives and Differentials

In the foregoing sections, we have discussed Boolean derivatives and differentials of switching functions. Now, we would like to present a convenient graphical method for computing them using the Karnaugh map. For a given function $f(x)$, the partial derivative $\partial f / \partial x_i$ of f with respect to a variable $x_i \in x$ can be conveniently found by the graphical method described below.

Step 1 Display the given function on a Karnaugh map.

Step 2 Locate the x_i axis (or axes) and rotate the map with respect to the x_i axis. Every 1 in the map (a minterm of the function) will either generate a new 1 (a new minterm) or eliminate a 1 in its mirror-image cell about the x_i axis. A 1 is entered to the cell if no 1 has occupied it; and a 1 is deleted from the cell if it has already a 1. For differentiating the 1's generated by different x_i's, we shall use a bold number "*i*" to denote the 1's that are generated by the rotation of the map with respect to the x_i axis.

Step 3   Find the minimal expression of $\partial f/\partial x_i$ from the remaining 1's and the newly created i's (which are considered to be 1's) of the map.

This method is best illustrated by examples.

Example 3.4.1

Consider the function $f(x_1, x_2, x_3) = x_1 x_2' + x_1 x_3$. The $\partial f/\partial x_i$, $i = 1, 2, 3$, can be found by the graphical method shown in Fig. 3.4.1(a). By extending this method, we can obtain

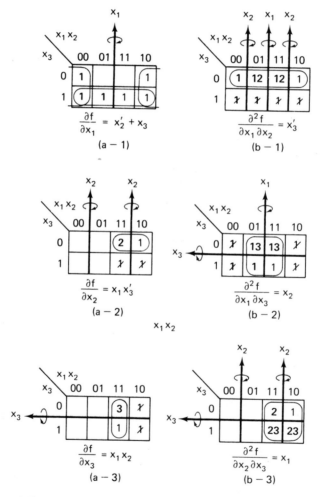

Note: 1. The minterms indicated by bold number i (ij) are the ones generated by rotating the map with respect to the x_i-axis (first with respect to the x_i-axis, then the x_j-axis).
2. A slash / drawn on a minterm indicator means the deletion of that minterm.

Fig. 3.4.1   Example 3.4.1.

$\partial^2 f/\partial x_i \partial x_j, 1 \leq i \leq j \leq 3$, which are depicted in Fig. 3.4.1(b). Note that $\partial^3 f/\partial x_1 \partial x_2 \partial x_3 = 1$.

After all the partial derivatives of the function are found, from Theorems 3.3.2 and 3.3.3, we obtain the total differential and the total variation of the function,

$$df = (x_2' + x_3)\, dx_1 \oplus x_1 x_3'\, dx_2 \oplus x_1 x_2\, dx_3 \oplus x_3'\, dx_1\, dx_2 \oplus x_2\, dx_1\, dx_3 \oplus x_1\, dx_2\, dx_3$$
$$\oplus\, dx_1\, dx_2\, dx_3$$

and

$$\delta f = (x_2' + x_3)\, \delta x_1 + x_1 x_3'\, \delta x_2 + x_1 x_2\, \delta x_3 + x_3'\, \delta x_1\, \delta x_2 + x_2\, \delta x_1\, \delta x_3 + x_1\, \delta x_2\, \delta x_3$$
$$+\, \delta x_1\, \delta x_2\, \delta x_3$$

Example 3.4.2

Consider the function

$$f(x_1, \ldots, x_5) = \sum (4, 5, 6, 7, 8, 10, 12, 15, 20, 21, 24, 27, 29, 30)$$

Find $\partial f/\partial x_1$ using the graphical method.

With reference to Fig. 3.4.2, $\partial f/\partial x_1$ is found to be

$$\frac{\partial f}{\partial x_1} = x_2 x_3 + x_2 x_4 + x_3 x_4$$

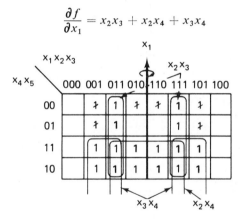

Fig. 3.4.2 Example 3.4.2.

The graphical method becomes more and more laborious and thus less and less attractive as the number of variables gets large. In the following, a computer algorithm for computing $\partial f/\partial x_i$ and multiple partial derivatives is presented which is essentially the same as the graphical method described above, but it requires no maps and is designed to be implemented on a digital computer. The algorithm is described and illustrated by an example.

Example 3.4.3

For convenience, let us consider the function with six variables,

$$f(x_1, \ldots, x_6) = \sum (0, 2, 5, 6, 7, 13, 21, 23, 28, 30, 33, 35, 36, 47, 49, 51, 60, 62)$$

TABLE 3.4.1 Example 3.4.3

(a) Minterms of the function

$x_1x_2x_3x_4x_5x_6$	$x_1x_2x_3x_4x_5x_6$
0 0 0 0 0 0	1 0 0 0 0 0
0 0 0 0 1 0	1 0 0 0 1 0
0 0 0 1 0 1	1 0 0 1 0 1
0 0 0 1 1 0	1 0 0 1 1 0
0 0 0 1 1 1	1 1 0 0 0 1
0 0 1 1 0 1	1 1 0 0 1 1
~~0 1 1 1 0 0~~	~~1 1 1 1 0 0~~
~~0 1 1 1 1 0~~	~~1 1 1 1 1 0~~
0 1 0 1 0 1	
0 1 0 1 1 1	

(b) Minterms created by changing x_1 of the minterms of (a) to x_1' for computing $\partial f/\partial x_1$

$x_1x_2x_3x_4x_5x_6$	$x_1x_2x_3x_4x_5x_6$
0 0 0 0 0 0	1 0 0 0 0 0
0 0 0 0 1 1	1 0 0 0 1 0
0 0 0 1 0 0	1 0 0 1 0 1
0 0 1 1 1 0	1 0 0 1 1 0
0 1 0 0 0 1	1 0 0 1 1 1
0 1 0 0 1 1	1 0 1 1 0 1
~~0 1 1 1 0 0~~	~~1 1 1 1 0 0~~
~~0 1 1 1 1 0~~	~~1 1 1 1 1 0~~
	1 1 0 1 0 1
	1 1 0 1 1 1

(c) Minterms created by changing x_2 of the minterms of (a) and (b) to x_2' for computing $\partial^2 f/\partial x_2 \partial x_1$.

$x_1x_2x_3x_4x_5x_6$	$x_1x_2x_3x_4x_5x_6$	$x_1x_2x_3x_4x_5x_6$	$x_1x_2x_3x_4x_5x_6$
0 1 0 0 0 0	1 1 0 0 0 1	1 1 0 0 0 0	0 1 0 0 0 1
0 1 0 0 1 0	1 1 0 0 1 1	1 1 0 0 1 0	0 1 0 0 1 1
0 1 0 1 0 1	1 1 0 1 0 0	1 1 0 1 0 1	0 1 0 1 0 0
0 1 0 1 1 0	1 1 1 1 1 1	1 1 0 1 1 0	0 1 1 1 1 1
0 1 0 1 1 1	1 0 0 0 0 1	1 1 0 1 1 1	0 0 0 0 0 1
0 1 1 1 0 1	1 0 0 0 1 1	1 1 1 1 0 1	0 0 0 0 1 1
0 0 0 1 0 1		1 0 0 1 0 1	
0 0 0 1 1 1		1 0 0 1 1 1	

The $\partial f/\partial x_1$ can be computed as follows:

Step 1   Find the binary representations of the minterms. They are shown in Table 3.4.1(a).

Step 2   Create new minterms by changing x_1 of the minterms to x_1' and delete those (the original and the minterm generated by it) which appear twice. They are shown in Table 3.4.1(b). Note that the minterms 011100, 011110, 111100, and 111110 appear twice and thus are deleted from both Tables 3.4.1(a) and (b).

Step 3   Find the minimal expression of $\partial f/\partial x_1$ using the Quine–McCluskey method. The minimized $\partial f/\partial x_1$ is found to be

$$\frac{\partial f}{\partial x_1} = x_2'x_3' + x_2'x_4x_6 + x_2x_3'x_6$$

Now suppose that we want to find $\partial^2 f/\partial x_2\,\partial x_1$. The new minterms, created by changing x_2 of the minterms of Tables 3.4.1(a) and (b) to x_2', are shown in Table 3.4.3(c). There are 16 minterms (indicated by frame) appearing twice in Table 3.4.1. After having them deleted from the table, we obtain the minimal form of $\partial^2 f/\partial x_2\,\partial x_1$:

$$\frac{\partial^2 f}{\partial x_2\,\partial x_1} = x_3'x_6' + x_3x_4x_6$$

The corresponding graphical representation of the algorithm is depicted in Fig. 3.4.3.

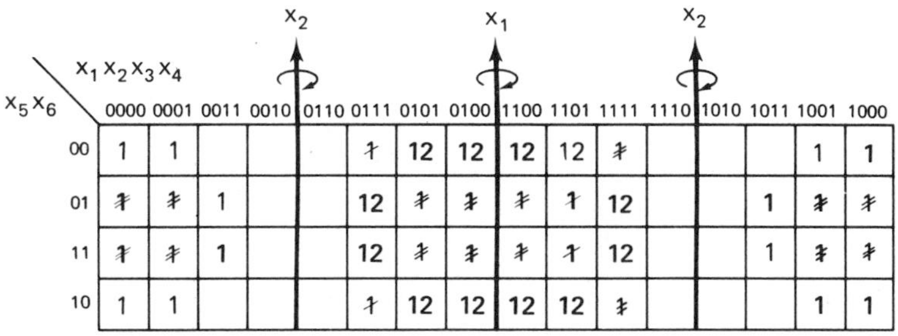

x_5x_6 \ $x_1x_2x_3x_4$	0000	0001	0011	0010	0110	0111	0101	0100	1100	1101	1111	1110	1010	1011	1001	1000
00	1	1			✝	12	12	12	12	✝					1	1
01	✝	✝	1		12	✝	✝	✝	✝	12			1	✝	✝	
11	✝	✝	1		12	✝	✝	✝	✝	12			1	✝	✝	
10	1	1			✝	12	12	12	12	✝					1	1

Note: 1 and ✝   The 1's created and eliminated, respectively, by changing x_1 of the minterms of the function to x_1'.

12 and   The 1's created and eliminated, respectively, by changing x_2 of the
(✱ and ✝)   minterms of the function plus the minterms created by changing x_1 of the original minterms to x_1', to x_2'.

Fig. 3.4.3   Example 3.4.3.

Exercise 3.4

1. Compute $\partial f/\partial x_i$ and all the multiple partial derivatives of each of the following functions using the graphical method.
    (a) $f(x_1, x_2, x_3) = x_1'x_2' + x_2x_3 + x_1x_3$
    (b) $f(x_1, x_2, x_3, x_4) = x_1x_3 + x_2'x_4 + x_1'x_3'x_4 + x_2x_3'$

(c) $f(x_1, \ldots, x_5) = x_1'x_4 + x_2'x_3 + x_3'x_5 + x_2x_3'x_4'$

(d) $f(x_1, \ldots, x_6) = x_1x_4'x_6 + x_3'x_4x_5' + x_1x_3'x_5'x_6$

2. Repeat problem 1 using the computer algorithm.

Bibliographical Remarks

The concept of partial derivative of a Boolean function was first introduced by Akers [1] and the concept of sensitivity function was also introduced by Akers [1] and later developed by others [2, 3]. Akers [1] also introduced the Δ-operator, which is a measure of the invariance of a Boolean function on a subcube. The total differential and the total variation were defined by Thayse [4]. Much advanced research on Boolean differential calculus and its applications has been published by Thayse, Davio, and Deschamps [5–20]. All these references are recommended.

References

1. AKERS, S. B., JR., "On a Theory of Boolean Functions," *J. Soc. Indust. Appl. Math.* Vol. 7, No. 4, December 1959.

2. SELLERS, F. F., M. Y. HSIAO, and L. W. BEARNSON, "Analyzing Errors with the Boolean Difference," *IEEE Trans. Computers*, Vol. C-17, July 1968, pp. 676–683.

3. SELLERS, F. F., M. Y. HSIAO, and L. W. BEARNSON, *Error Detecting Logic for Digital Computers*, McGraw-Hill, New York, 1968, pp. 17–37.

4. THAYSE, A., "Boolean Differential Calculus," *Philips Res. Rept.*, Vol. 26, 1971, pp. 229–246.

5. THAYSE, A., "A Variational Diagnosis Method for Stuck-Faults in Combinational Networks," *Philips Res. Rept.*, Vol. 27, 1972, pp. 82–98.

6. THAYSE, A., "Testing of Asynchronous Sequential Switching Circuits," *Philips Res. Rept.*, Vol. 27, 1972, pp. 99–106.

7. THAYSE, A., "A Fast Algorithm for the Proper Decomposition of Boolean Functions," *Philips Res. Rept.*, Vol. 27, 1972, pp. 140–150.

8. THAYSE, A., "Multiple-Fault Detection in Large Logic Networks," *Philips Res. Rept.*, Vol. 27, 1972, pp. 582–602.

9. THAYSE, A., "Disjunctive and Conjunctive Operators for Boolean Functions," *Philips Res. Rept.*, Vol. 28, 1973, pp. 1–16.

10. THAYSE, A., "On Some Iterative Properties of Boolean Functions," *Philips Res. Rept.*, Vol. 28, 1973, pp. 107–119.

11. DAVIO, M., "Taylor Expansions of Symmetric Boolean Functions," *Philips Res. Rept.*, Vol. 28, 1973, pp. 466–474.

12. THAYSE, A., and M. DAVIO, "Boolean Differential Calculus and Its Application to Switching Theory," *IEEE Trans. Computers*, Vol. C-22, 1973, pp. 409–420.

13. DESCHAMPS, J. P., and A. THAYSE, "Applications of Discrete Functions, Part I. Transient Analysis of Combinational Networks," *Philips Res. Rept.*, Vol. 28, 1973, pp. 497–529.

14. THAYSE, A., "Applications of Discrete Functions, Part II. Transient Analysis of Asynchronous Switching Networks," *Philips Res. Rept.*, Vol. 29, 1974, pp. 155–192.

15. THAYSE, A., "Applications of Discrete Functions, Part III. The Use of Functions of Functions in Switching Circuits," *Philips Res. Rept.*, Vol. 29, 1974, pp. 429–452.

16. DESCHAMPS, J. P., and A. THAYSE, "On a Theory of Discrete Functions, Part I. The Lattice Structure of Discrete Functions," *Philips Res. Rept.*, Vol. 28, 1973, pp. 397–423.

17. THAYSE, A., and J. P. DESCHAMPS, "On a Theory of Discrete Functions, Part II. The Ring and Field Structures of Discrete Functions," *Philips Res. Rept.*, Vol. 28, 1973, pp. 424–465.

18. DESCHAMPS, J. P., "On a Theory of Discrete Functions, Part III. Decomposition of Discrete Functions," *Philips Res. Rept.*, Vol. 29, 1974, pp. 193–213.

19. THAYSE, A., "On a Theory of Discrete Functions, Part IV. Discrete Functions of Functions," *Philips Res. Rept.*, Vol. 29, 1974, pp. 305–329.

20. THAYSE, A., "Differential Calculus for Functions from $(GF(p))^n$ into $GF(p)$," *Philips Res. Rept.*, Vol. 29, 1974, pp. 560–586.

4

Special Switching Functions

In Section 2.1, we studied general properties of switching functions. In this chapter, we study four special classes of switching functions: (1) monotonic functions, (2) threshold functions, (3) symmetric functions, and (4) functionally complete functions. Many properties of these functions are discussed. Algorithms for determining them are presented. The relationships among them are explored.

4.1 Monotonic Functions

We begin this study with monotonic functions. First we review the ordering relation among the elements of the Cartesian product of two-element Boolean algebras B_2, namely, $B_2^n = \underbrace{B_2 \times B_2 \times \ldots \times B_2}_{n}$. Let x and y be two elements of B_2^n.

$$x = (a_1, a_2, \ldots, a_n)$$
$$y = (b_1, b_2, \ldots, b_n)$$

where $a_i, b_i \in B_2$. We define

$$x \leq y \qquad \text{iff } a_i \leq b_i \tag{4.1.1}$$

Otherwise, they are unrelated.

Example 4.1.1

Let $x = (1, 0, 0, 0)$, $y = (1, 0, 1, 0)$, and $z = (0, 1, 1, 1)$. Then

$$x \leq y, \qquad x \nleq z, \qquad y \nleq z$$

This may be seen from B_2^4 in Fig. 1.3.2(b).

DEFINITION 4.1.1

Let $f(x_1, x_2, \ldots, x_n)$ be a switching function in canonical sum-of-products form, and let $x = (a_1, a_2, \ldots, a_n)$ and $y = (b_1, b_2, \ldots, b_n)$ be the binary representations of two terms of f. We define that the term x is less than or equal to y by Eq. (4.1.1).

DEFINITION 4.1.2

Let x and y be any two terms of the canonical sum-of-products form of a switching function $f(x_1, x_2, \ldots, x_n)$. f is said to be *monotonically increasing* iff $x \leq y$ implies $f(x) \leq f(y)$ and is said to be *monotonically decreasing* iff $x \leq y$ implies that $f(x) \geq f(y)$.

From this definition, we immediately have:

THEOREM 4.1.1

Let $f(x_1, x_2, \ldots, x_n)$ be a switching function of n variables. f is monotonically increasing (decreasing) iff the complement of f is monotonically decreasing (increasing).

Example 4.1.2

Referring to the 16 simplified functions of two variables in Fig. 2.1.1, close observation reveals that the six functions on the far left are monotonically increasing, and the six functions on the far right are monotonically decreasing (see Fig. 4.1.1). They are verified in Table 4.1.1.

Observe that all the six monotonically increasing functions of two variables are in the form of a sum-of-products of the x_i, and all six monotonically decreasing functions of two variables are in the form of the sum-of-products of x_i'. As a matter of fact, this observation is true in general. Before proving this, we need the following definition.

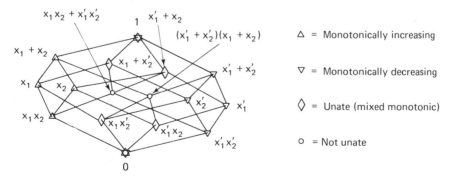

Fig. 4.1.1 Classification of the 16 simplified functions.

**TABLE 4.1.1 Mapping of the Monotonically
Increasing and Decreasing Functions
of Two Variables**

x_1	x_2	Six monotonically increasing functions					
		0	$x_1 x_2$	x_1	x_2	$x_1 + x_2$	1
0	0	0	0	0	0	0	1
0	1	0	0	0	1	1	1
1	0	0	0	1	0	1	1
1	1	0	1	1	1	1	1
Complements of monotonically increasing functions		$0'$	$x_1' + x_2'$	x_1'	x_2'	$x_1' + x_2'$	$1'$
			Six monotonically decreasing functions				

DEFINITION 4.1.3

Let $f(x_1, x_2, \ldots, x_n)$ be a monotonically increasing function. Let $x = (x_1, x_2, \ldots, x_n)$ be a generic element in the domain of f. Define the set $U = \{x \,|\, f(x) = 1\}$. An element $x^i \in U$ is called a *minimal element* in U if there does not exist an element x^j in U such that $x^j < x^i$.

Example 4.1.3

Consider the monotonically increasing function with three variables whose mapping is given in Table 4.1.2. Since the domain of this function forms an eight-element lattice, it is convenient to indicate the mapping directly on this lattice underneath the 3-tuples. The set U of this function is

$$U = \{(0, 1, 0), (0, 0, 1), (1, 1, 0), (0, 1, 1), (1, 0, 1), (1, 1, 1)\}$$

**TABLE 4.1.2 Mapping of a Monotonically
Increasing Function**

x_1	x_2	x_3	$f(x_1, x_2, x_3)$
0	0	0	0
0	0	1	1
0	1	0	1
0	1	1	1
1	0	0	0
1	0	1	1
1	1	0	1
1	1	1	1

The two minimal elements of U are $(0, 1, 0)$ and $(0, 0, 1)$,since there is no element in U which is less than either of them. Written in terms of the variables, the set $U = \{x_1'x_2x_3', x_1'x_2'x_3, x_1x_2x_3', x_1'x_2x_3, x_1x_2'x_3, x_1x_2x_3\}$, and the two minimal elements are $x_1'x_2x_3'$ and $x_1'x_2'x_3$.

THEOREM 4.1.2

Let x^i be a minimal element of the set U of a monotonically increasing switching function $f(x_1, x_2, \ldots, x_n)$. Define U_i to be the set

$$U_i = \{x \mid x \geq x^i \quad \text{and} \quad f(x) = 1\}$$

Then U_i forms a sublattice of the 2^n-element lattice.

Proof: In view of the fact that f is a monotonically increasing function, and by the way U_i is defined, it is clear that U is partially ordered. Now let x^j and x^k be two elements in U_i. $f(x^j) = f(x^k) = 1$. The sum and product of x^j and x^k are greater than or equal to x^i. Thus, they are all in U_i.   ∎

Example 4.1.4

The two U_i's of Example 4.1.3, corresponding to the minimum elements $x_1'x_2x_3'$ and $x_1'x_2'x_3$, are

$$U_1 = \{x_1'x_2x_3', x_1x_2x_3', x_1'x_2x_3, x_1x_2x_3\}$$
$$U_2 = \{x_1'x_2'x_3, x_1x_2'x_3, x_1'x_2x_3, x_1x_2x_3\}$$

The two sublattices are shown in Fig. 4.1.2.

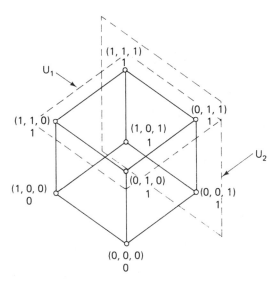

Fig. 4.1.2   Display of the mapping of the switching function of Table 4.1.2.

Now we prove the following important theorem.

THEOREM 4.1.3

A function $f(x_1, x_2, \ldots, x_n)$ is monotonically increasing (decreasing) iff it is expressible as a sum-of-products of the $x_i(x_i')$.

Proof: The proof of the necessity follows from the fact that x_i is monotonically increasing, the product of monotonically increasing functions $x_i x_j \ldots x_n$ is a monotonically increasing function, and the sum of monotonically increasing functions is a monotonically increasing function.

Now we want to prove the sufficiency. Before proving it in general, let us first examine an example. Consider the monotonically increasing function of Example 4.1.3. The canonical form of this function is

$$f(x_1, x_2, x_3) = x_1 x_2 x_3 + x_1 x_2 x_3' + x_1 x_2' x_3 + x_1' x_2 x_3 + x_1' x_2 x_3' + x_1' x_2' x_3 \qquad (4.1.2)$$

The two minimal elements found in Example 4.1.3 were $x_1' x_2 x_3'$ and $x_1' x_2' x_3$. Two important properties of U_i's were found in Example 4.1.4:

(1) $U_1 \cup U_2 = U$.

(2)   The sum of the elements of U_i is a sum-of-products of the x_i; in other words, no variable appears complemented in this sum.

The sum of the elements of U_1   *The sum of the elements of U_2*

$$\left. \begin{array}{l} x_1' x_2 x_3' \\ x_1 x_2 x_3' \end{array} \right\} x_2 x_3' \\ \left. \begin{array}{l} x_1' x_2 x_3 \\ x_1 x_2 x_3 \end{array} \right\} x_2 x_3 \right\} x_2 \qquad \left. \begin{array}{l} x_1' x_2' x_3 \\ x_1 x_2' x_3 \end{array} \right\} x_2' x_3 \\ \left. \begin{array}{l} x_1' x_2 x_3 \\ x_1 x_2 x_3 \end{array} \right\} x_2 x_3 \right\} x_3$$

Therefore, the function is $f(x_1, x_2, x_3) = x_2 + x_3$, in which no variables appear complemented.

Now extend the example above to the general case. Suppose $f(x_1, x_2, \ldots, x_n)$ is a switching function of n variables $x_1, x_2, \ldots, x_n$. The domain of f forms a 2^n-element lattice, say, there are m ($< n$) minimal elements in the set U of f. Now we want to show that

(1) $\displaystyle \bigcup_{i=1}^{m} U_i = U$

(2)   The sum of all the elements of U_i is a sum-of-products of the x_i.

Proof of (1):   Since U is defined as $U = \{x \mid f(x) = 1\}$, every element in $\displaystyle \bigcup_{i=1}^{m} U_i$ is an element in U; thus, $\displaystyle \bigcup_{i=1}^{m} U_i \subseteq U$. Conversely, let x be an element of U. Then either x is a minimal element in U, or there exists an element $y \in U$, $y < x$, because U is a partially ordered set. If it is of the former case, then x is in one of the U_i's; if it is of the latter case, then either y is a minimal element in U, or there exists an element $z \in U$ such that $z < y$, and so on. Suppose after repeating this process l times, we have w, of which there does not exist an element in U that is less than w. Then by definition of U_i, w is a minimal element of U and belongs to one of the U_i's, say U_k. Since $f(x) = 1$ and $x > w$, x is in U_k. Of course, x may be in more than one U_i; thus, $\displaystyle \bigcup_{i=1}^{m} U_i \supseteq U$. Hence, $\displaystyle \bigcup_{i=1}^{m} U_i = U$.

Proof of (2):   Suppose that x^i is the minimal element of U_i. If we write x^i in such a way that the variables in x^i without complementation appear before all of those with complementation, x^i may be expressed as

$$x^i = x_{j_1}x_{j_2} \ldots x_{j_p} \; x'_{k_1}x'_{k_2} \ldots x'_{k_q}, \qquad p + q = n \qquad (4.1.3)$$

$$\underbrace{\phantom{x_{j_1}x_{j_2} \ldots x_{j_p}}}_{\substack{\text{variables} \\ \text{without} \\ \text{complementation}}} \underbrace{\phantom{x'_{k_1}x'_{k_2} \ldots x'_{k_q}}}_{\substack{\text{variables} \\ \text{with} \\ \text{complementation}}}$$

We can show that all the elements

$$x^1 = x_{j_1}x_{j_2} \cdots x_{j_p}x_{k_1}x'_{k_2}x'_{k_3} \cdots x'_{k_q}$$
$$x^2 = x_{j_1}x_{j_2} \cdots x_{j_p}x'_{k_1}x_{k_2}x'_{k_3} \cdots x'_{k_q}$$
$$\cdot$$
$$\cdot$$
$$\cdot$$
$$x^{2^q-1} = x_{j_1}x_{j_2} \cdots x_{j_p}x_{k_1}x_{k_2}x_{k_3} \cdots x_{k_q}$$

are in U_i. Observe that if

$$x^i = x_{j_1}x_{j_2} \ldots x_{j_p}x'_{k_1}x'_{k_2} \ldots x'_{k_q}$$

is in U_i, then the element

$$x^1 = x_{j_1}x_{j_2} \ldots x_{j_p}x_{k_1}x'_{k_2} \ldots x'_{k_q}$$

is mapped to 1, since $x^i < x^1$ and f is monotonically increasing. By induction, $x^2, \ldots,$ x^{2^q-1} are all in U_i. In fact, $U_i = \{x^i, x^1, x^2, \ldots, x^{2^q-1}\}$. By repeatedly applying the property that the sum of two terms which are different in value of only one variable results in the elimination of one variable, it is easy to show that the sum s_i of the elements $x^i, x^1, \ldots, x^{2^q-1}$ is $s_i = x_{j_1}x_{j_2} \ldots x_{j_p}$, in which no variable appears complemented. Since f is equal to the sum of s_i's, f is expressible as a sum of products of the x_i. Hence, the theorem is proved. ∎

Following from this theorem, we immediately have

THEOREM 4.1.4

Let $f(x_1, x_2, \ldots, x_n)$ be a monotonically increasing function. Then in the expression

$$f(x_1, x_2, \ldots, x_n) = s_1 + s_2 + \ldots + s_m \qquad (4.1.4)$$

no product s_i includes another product s_j, for all i and j.

Proof: We shall prove it by contradiction. Suppose that there is a s_i which includes s_j. Suppose that

$$s_i = x_{j_1}x_{j_2} \ldots x_{j_p}$$
$$s_j = x_{j_1}x_{j_2} \ldots x_{j_r}, \qquad j_r < j_p$$

From the definitions of s_i and s_j,

$$f(s_i) = 1$$
$$f(s_j) = 1$$

By Theorem 4.1.3, s_i is a minimal element in U_i. But by hypothesis, $s_j < s_i$ and $f(s_j) = 1$, which implies that s_i is not a minimal element. This is a contradiction. Hence, in the expression of Eq. (4.1.4), no s_i includes s_j for all i, j.   ∎

THEOREM 4.1.5

The expression of Eq. (4.1.4) is unique.

Proof: Let us suppose that there exist two such expressions for f,

$$f(x_1, x_2, \ldots, x_n) = s_1 + s_2 + \ldots + s_m = r_1 + r_2 + \ldots + r_l$$

Now we want to show that there exists a one-to-one correspondence between the two expressions. Take an element from s_i. Suppose that

$$s_i = x_{j_1} x_{j_2} \ldots x_{j_p}$$

Define

$$x^i = x_{j_1} x_{j_2} \ldots x_{j_p} x'_{k_1} x'_{k_2} \ldots x'_{k_q}$$

If we assign 1 to all x_j's and 0 to all x_k's, we find that with this assignment $\underbrace{(1, 1, \ldots, 1,}_{p}$

$\underbrace{0, 0, \ldots, 0)}_{q}$, all the terms of the first expression are 0 except the term s_i. Suppose that this is not true; that is, there is a term s_j which is not equal to 0, namely, 1. This cannot be the case, since, otherwise, either s_i would include s_j or vice versa, which contradicts Theorem 4.1.4.

Let us consider under this assignment what happens in the second expression. Since

$$f(\underbrace{1, 1, \ldots, 1,}_{p} \quad \underbrace{0, 0, \ldots, 0}_{q}) = 1$$

there must exist one term in the second expression which is 1 under this assignment. As a matter of fact, for the same reason just stated, there is *only* one term, say r_j, which is 1; the rest are all 0.

We want to show that $s_i = r_j$. Since s_i is a minimal element of f, by Theorem 4.1.3, s_i cannot include r_j, so it must be equal to r_j or be included in r_j. Similarly, by the same arguments, r_j must be equal to s_i or be included in s_i. Therefore, $s_i = r_j$. The theorem is thus proved.   ∎

We now introduce a class of functions having a property that is a mixture of monotonically increasing and decreasing.

DEFINITION 4.1.4

A function $f(x_1, x_2, \ldots, x_n)$ is said to be *mixed monotonic* or *unate* if there exists a subset $S = \{x_{i_1}, x_{i_2}, \ldots, x_{i_m}\}$ of $X = \{x_1, x_2, \ldots, x_n\}$ of variables such that by replacing all x_{i_j} by x'_{i_j} or all x'_{i_j} by x_{i_j} in the expression of f, f becomes either monotonically increasing or monotonically decreasing.

Example 4.1.5

Consider a simple example

$$f(x_1, x_2, x_3) = x'_1 + x'_1 x_2 + x'_1 x_3$$

If we replace x'_1 by x_1, f becomes monotonically increasing; or if we replace x_2 by x'_2 and x_3 by x'_3, f becomes monotonically decreasing. Thus, f is unate.

For any function with three variables or less, the determination of whether the function is monotonically increasing or decreasing, or unate, may be found either by inspection or by mapping the function on a lattice as we did in Example 4.1.3. As the number of elements increases, the problem becomes complicated. Therefore, we need to find some general procedure by which we can always determine whether a given function is monotonically increasing, monotonically decreasing, or unate.

The procedure given below is a direct application of the Karnaugh map and the Quine–McCluskey methods. Recall that the latter contains two parts. The first part is finding all the prime implicants of the function, and the second is a tabular procedure to minimize redundant prime implicants, that is, to find a minimal set of prime implicants which cover the function. Before presenting the procedure, we first need to prove the following theorem.

THEOREM 4.1.6

Let $f(x_1, x_2, \ldots, x_n)$ be a monotonically increasing, decreasing, or unate function. It is necessary that the minimal coverage of f obtained from either the Karnaugh map or the Quine–McCluskey method be a sum of essential prime implicants.

Proof: To prove this theorem, we need to prove the following:
(1)   Every minimal element x^i must be covered by *only* one prime implicant, A_i.
(2)   The union of A_i covers all the canonical product terms of f; that is, $f = s_1 + s_2 + \ldots + s_m$, where s_i is the sum of all elements of A_i.
Since the proofs for the cases where f are monotonically increasing, monotonically decreasing, and unate functions are similar, the proof for the case where f is monotonically increasing should suffice. Let x^i be a minimal element. Suppose that it is covered by two prime implicants A_i and A_j. Since A_i and A_j are prime implicants and cover x^i, they should contain all the elements which are greater than x^i. Moreover, since f is monotonically increasing, every element of A_i should also be in A_j and vice versa. Hence, $A_i = A_j$.

Now we prove that the union of A_i covers all the canonical product terms of f. Choose a term x^j in the canonical sum-of-products form. Since f is a monotonically increasing function, and $f(x^j) = 1$, x^j must be greater than some minimal element, say, x^i, that is, $x^j \geq x^i$; and therefore it is in A_i. Hence, the union of A_i covers all the canonical product terms of f. ∎

Based on this theorem, a procedure for determining the monotonic characteristics of a switching function may be derived which is described in Fig. 4.1.3.

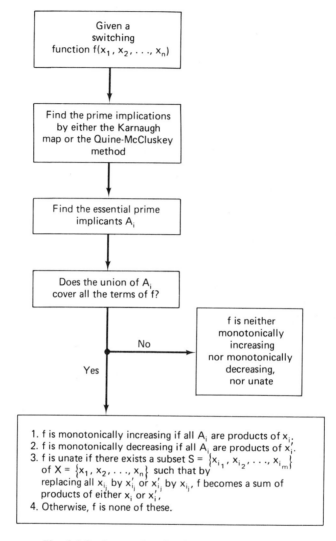

Fig. 4.1.3 A procedure for determining the monotonic characteristics of a function.

Example 4.1.6

Consider the function

$$f(x_1, x_2, x_3, x_4) = \sum (0, 5, 7, 8, 9, 10, 11, 14, 15)$$

The four essential prime implicants of this function are $x_1 x_2'$, $x_1 x_3$, $x_1' x_2 x_4$, and $x_2' x_3' x_4'$

which are shown in the following Karnaugh map:

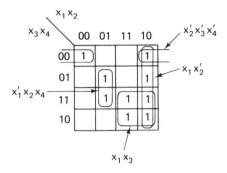

The sum of them does cover f. However, it is neither monotonically increasing, nor mono-tonically decreasing, nor unate, because it is not a sum-of-products of the x_i or x_i' or subject to $x_i \longleftrightarrow x_i'$ modifications to obtain either of the two properties.

TABLE 4.1.3 **Finding the Prime Implicants and Essential Prime Implicants by the Quine-McCluskey Method**

	Decimal number	Binary representation of each term	Decimal numbers	Binary representation of each term	Decimal numbers	Binary representation of each term
Index 1	{ 4	000100 ✓	4, 20	0–0100 ✓	4, 20, 36, 52	––0100 A
Index 2	⎰20	010100 ✓	4, 36	–00100 ✓	4, 36, 20, 52	––0100
	⎱36	100100 ✓	20, 21	01010– ✓	20, 21, 52, 53	–1010– B
Index 3	⎰21	010101 ✓	20, 52	–10100 ✓	20, 52, 21, 53	–1010–
	⎱52	110100 ✓	36, 52	1–0100 ✓	52, 53, 54, 55	1101–– C
Index 4	⎰53	110101 ✓	21, 53	–10101 ✓	52 54, 53, 55	1101––
	⎱54	110110 ✓	52, 53	11010– ✓		
Index 5	55	110111 ✓	52, 54	1010–0 ✓		
			53, 55	1101–1 ✓		
			54, 55	11011– ✓		

(a) Table for obtaining the prime implicants

	4	20	21	36	52	53	54	55
*A	⊗	×		⊗	×			
*B		×	⊗		×	×		
*C					×	×	×	⊗

(b) Table for reducing redundant prime implicants

Example 4.1.7

Consider the function

$$f(x_1, x_2, x_3, x_4, x_5, x_6) = \sum (4, 20, 21, 36, 52, 53, 54, 55)$$

Using the Quine–McCluskey method, the three prime implicants $A = x_3'x_4x_5'x_6'$, $B = x_2x_3x_4x_5$, and $C = x_1x_2x_3x_4$ are obtained which are shown in Table 4.1.3(a). In Table 4.1.3(b), it is seen that all three of these prime implicants are essential; hence, it satisfies the necessary condition for a switching function being monotone. Close examination reveals that if we make the following changes:

$$x_3' \longrightarrow x_3$$
$$x_5' \longrightarrow x_5$$
$$x_6' \longrightarrow x_6$$

the function becomes

$$f_1(x_1, x_2, x_3, x_4, x_5, x_6) = x_1x_2x_3x_4 + x_2x_3x_4x_5 + x_3x_4x_5x_6$$

which is clearly a monotonically increasing function. Hence, the given function is a unate function.

Exercise 4.1

1. A switching function $f(x_1, x_2, \ldots, x_n)$ is called *positive* in x_1 if f can be expressed as

$$f(x_1, x_2, \ldots, x_n) = x_1g_1(x_2, \ldots, x_n) + g_2(x_2, \ldots, x_n)$$

without x_1' appearing, and is called *negative* in x_1 if f can be expressed as

$$f(x_1, x_2, \ldots, x_n) = g_1(x_2, \ldots, x_n) + x_1'g_2(x_2, \ldots, x_n)$$

without x_1 appearing.
Show that the following definitions are equivalent to the definitions of monotonically increasing, monotonically decreasing, and unate given in this section.
(a) A function $f(x_1, \ldots, x_n)$ is monotonically increasing if it is positive in each x_i.
(b) A function $f(x_1, \ldots, x_n)$ is monotonically decreasing if it is negative in each x_i.
(c) A function $f(x_1, \ldots, x_n)$ is unate if, given any x_i, $f(x_1, \ldots, x_n)$ is either positive in x_i or negative in x_i.

2. An *n-ordering* of the vertices of an *n*-cube is a partial ordering for which
(1) There is a least vertex $a = (a_1, \ldots, a_n)$ such that if $b = (b_1, \ldots, b_n)$ is any other vertex $a < b$.
(2) If b and c are any pair of vertices $b \leq c$ iff either $b_i = a_i$ or $c_i = a_i'$.
Let the points of the 3-cube be given an *n*-ordering in which $(0, 1, 1)$ is the least vertex. Draw a lattice diagram for this *n*-ordering.

3. Prove that f is unate iff it is monotonically increasing for some *n*-ordering.

4. Let $f(x_1, x_2, \ldots, x_n)$ be a switching function of n variables. Let $y_1, y_2, \ldots, y_n$ be a set of variables, where y_i is equal to x_i or equal to its complement. Show that if $f(x_1, x_2, \ldots, x_n)$ is unate, then $f(y_1, y_2, \ldots, y_n)$ is unate.

5. A switching function $f(x_1, x_2, \ldots, x_n)$ is said to be *vacuous* in an independent variable x_i, if for all possible values of $x_1, x_2, \ldots, x_{i-1}, x_{i+1}, \ldots, x_n$,

$$f(x_1, x_2, \ldots, x_{i-1}, 0, x_{i+1}, \ldots, x_n) = f(x_1, x_2, \ldots, x_{i-1}, 1, x_{i+1}, \ldots, x_n)$$

How many switching functions of n variables are not vacuous?

6. A switching function $f(x_1, x_2, \ldots, x_n)$ is *nondegenerate* iff it is not vacuous in all its n variables. Prove that every nondegenerate monotonically increasing function of n variables defines 2^n unate functions.

7. Show that the number of nondegenerate unate functions of n variables is equal to 2^n times the number of nondegenerate monotonically increasing functions.

8. A function $f(x_1, \ldots, x_n)$ is monotonically increasing in an n-ordering if given that $c_i \le b_i$, for $i = 1, 2, \ldots, n$, in the ordering, then

$$f(b_1, \ldots, b_n) \ge f(c_1, \ldots, c_n)$$

Show that the set of monotonically increasing functions form a lattice with 0 and 1. Is it distributive? Complemented?

9. Let f have the following Karnaugh map. Is f monotonic? Is f unate?

	$x_1 x_2$			
	00	01	11	10
$x_3 x_4$ 00			1	
01	1	1	1	1
11	1	1		1
10		1		1

10. Let $f(x_1, x_2, x_3, x_4) = x_1' x_2' + x_1' x_2 x_3' + x_1' x_3 x_4' + x_1 x_2' x_3' + x_1 x_2' x_3 x_4'$. Is f monotonic? Is f unate?

11. Consider the following functions:
    (a) $f(x_1, x_2, x_3, x_4, x_5) = \sum (1, 3, 4, 5, 7, 8, 11, 15, 19, 23)$
    (b) $f(x_1, x_2, x_3, x_4, x_5, x_6) = \sum (8, 9, 10, 14, 26, 45, 68, 80, 84, 96)$
    Find a minimal sum of f. Is f monotonic?

12. Show that the set of monotonically increasing functions forms a lattice with 0 and 1. Is it distributive? Complemented?

4.2 Threshold Functions

The second class of special functions we would like to investigate is the threshold function.

DEFINITION 4.2.1

A switching function $f(x_1, x_2, \ldots, x_n)$ is a *threshold function* iff there exists a set of real numbers $w_1, w_2, \ldots, w_n$ and T such that

$$\sum_{i=1}^{n} w_i x_i \geq T \qquad \text{if } f(x_1, x_2, \ldots, x_n) = 1$$

$$\sum_{i=1}^{n} w_i x_i < T \qquad \text{if } f(x_1, x_2, \ldots, x_n) = 0$$

$(4.2.1)$

where $w_1, w_2, \ldots, w_n$ are called the *weights* of the variables $x_1, x_2, \ldots, x_n$, and T is called the *threshold* of the function.

Example 4.2.1

Consider the function

$$f(x_1, x_2, x_3) = x_1 + x_2 + x_3' \qquad (4.2.2)$$

whose mapping may be compressed to four lines as shown in Table 4.2.1, where the dash "–" means that the variable at the top of that column may either be 1 or 0. With reference to

TABLE 4.2.1 Compressed Mapping of
$f(x_1, x_2, x_3) = x_1 + x_2 + x_3'$

x_1	x_2	x_3	$f(x_1, x_2, x_3) = x_1 + x_2 + x_3'$
1	–	–	1
–	1	–	1
–	–	0	1
0	0	1	0

Table 4.2.1, if $f(x_1, x_2, x_3)$ is a threshold function, there must exist a set of values of the weights $w_1, w_2, \ldots, w_n$ which satisfy the following inequalities:

$$\left. \begin{array}{c} w_1 \\ w_1 \overset{\wedge}{+} w_2 \\ w_1 \overset{\wedge}{+} w_3 \\ w_2 \\ w_2 \overset{\wedge}{+} w_3 \\ w_1 \overset{\wedge}{+} w_2 \overset{\wedge}{+} w_3 \\ 0 \end{array} \right\} \geq T > w_3 \qquad (4.2.3)$$

The sign $\overset{\wedge}{+}$ is used for ordinary arithmetic addition to distinguish it from the Boolean addition. We find that when $w_1 = w_2 = -w_3 = 1$ and $T = 0$, the inequalities above are all satisfied, which implies that the conditions of Eq. (4.2.1) are satisfied. By Definition 4.2.1, f is a threshold function.

Example 4.2.2

As another simple example of a threshold function, consider

$$f(x_1, x_2, x_3) = x_1 x_2 + x_3$$

From the truth table of this function, the following inequalities may be obtained:

$$\left.\begin{array}{l} w_1 \stackrel{\wedge}{+} w_2 \\ w_1 \stackrel{\wedge}{+} w_2 \stackrel{\wedge}{+} w_3 \\ w_3 \\ w_1 \stackrel{\wedge}{+} w_3 \\ w_2 \stackrel{\wedge}{+} w_3 \end{array}\right\} \geq T > \left\{\begin{array}{l} w_1 \\ w_2 \\ 0 \end{array}\right.$$

It is easy to see that when $w_1 = w_2 = 1$, $w_3 = 2$, and $T = 2$, all the above inequalities are satisfied. Hence, the function is a threshold function.

Now we may ask: Are all the switching functions of two-valued variables threshold functions? The answer is no. The following is an example of a nonthreshold function.

Example 4.2.3

Consider

$$f(x_1, x_2) = x_1 x_2' + x_1' x_2$$

By inspection, the truth inputs of this function are 10 and 01, and the false inputs are 11 and 00. Therefore, we obtain the following inequalities:

$$\left.\begin{array}{l} w_1 \\ w_2 \end{array}\right\} \geq T > \left\{\begin{array}{l} 0 \\ w_1 \stackrel{\wedge}{+} w_2 \end{array}\right.$$

which implies that

$$\left.\begin{array}{l} w_1 > 0 \\ w_1 > w_1 \stackrel{\wedge}{+} w_2 \end{array}\right\} \Longrightarrow w_2 < 0$$

and

$$\left.\begin{array}{l} w_2 > 0 \\ w_2 > w_1 \stackrel{\wedge}{+} w_2 \end{array}\right\} \Longrightarrow w_1 < 0$$

Thus, there does not exist a set of weights which will satisfy the conditions of Eq. (4.2.1). Hence, the function is *not* a threshold function.

DEFINITION 4.2.2

Let

$$F = \sum_{i=1}^{n} w_i x_i$$

where w_i, $i = 1, \ldots, n$ is a set of weights which is valid for the threshold function

$f(x_1, x_2, \ldots, x_n)$. We define the *gap* $G = \bar{F} - \underline{F}$, where $\bar{F}$ is the least value of F over all truth inputs of f and $\underline{F}$ is the greatest value of F over all false inputs of f.

THEOREM 4.2.1

A function is a threshold function if there exists a set of weights $w_1, w_2, \ldots, w_n$ such that $G > 0$.

Proof: The proof of this theorem is quite simple. From the previous examples, we saw that for any given function, we could derive a set of inequalities from its truth table as follows:

$$\left. \begin{array}{c} F_i(w_1, \ldots, w_n) \\ \text{over all} \\ \text{truth inputs} \end{array} \right\} \geq T > \left\{ \begin{array}{c} F_j(w_1, \ldots, w_n) \\ \text{over all} \\ \text{false inputs} \end{array} \right.$$

If there exists a set of weights $w_1, w_2, \ldots, w_n$ such that

$$G = \min_i F_i(w_1, \ldots, w_n) - \max_j F_j(w_1, \ldots, w_n) > 0$$

and if T has any value in the interval

$$[\min_i F_i(w_1, \ldots, w_n), \max_j F_j(w_1, \ldots, w_n)]$$

then the inequalities of Eq. (4.2.1) are satisfied by this set of weights and T. By Definition 4.2.1, the function is a threshold function. ∎

For example, in Example 4.2.1, for $w_1 = w_2 = -w_3 = 1$,

$$G = \min [w_1, w_1 \overset{\frown}{+} w_2, w_1 \overset{\frown}{+} w_3, w_2, w_2 \overset{\frown}{+} w_3, w_1 \overset{\frown}{+} w_2 \overset{\frown}{+} w_3, 0] - \max [w_3]$$
$$= 0 - (-1) = 1 > 0$$

and in Example 4.2.2, for $w_1 = w_2 = 1$, and $w_3 = 2$,

$$G = \min [w_1 \overset{\frown}{+} w_2, w_1 \overset{\frown}{+} w_2 \overset{\frown}{+} w_3, w_3, w_1 \overset{\frown}{+} w_3, w_2 \overset{\frown}{+} w_3] - \max [w_1, w_2, 0]$$
$$= 2 - 1 = 1 > 0$$

Hence, they are threshold functions. On the other hand, in Example 4.2.3, we cannot find a set of weights w_1 and w_2 such that

$$\min [w_1, w_2] - \max [0, w_1 \overset{\frown}{+} w_2] > 0$$

Thus, it is *not* a threshold function.

THEOREM 4.2.2

Any threshold function is unate.

Proof: Let $f(x_1, x_2, \ldots, x_n)$ be a threshold function; that is, there exists a set of real numbers $w_1, w_2, \ldots, w_n, T$ such that

$$F = \sum_{i=1}^{n} w_i x_i \geq T \quad \text{if } f(x_1, \ldots, x_n) = 1 \tag{4.2.4}$$

and

$$F = \sum_{i=1}^{n} w_i x_i < T \quad \text{if } (x_1, \ldots, x_n) = 0 \tag{4.2.5}$$

Now consider the following three cases:

CASE 1 $w_i = 0$

When $w_i = 0$, then F is independent of x_i. Equations (4.2.4) and (4.2.5) imply that f is also independent of x_i. We can consider this case as f is vacuous† in x_i.

CASE 2 $w_i > 0$

In this case, consider the two subcases.
(2-1) If $f(x_1, \ldots, x_{i-1}, 0, x_{i+1}, \ldots, x_n) = 0$, f is unate.
(2-2) If $f(x_1, \ldots, x_{i-1}, 0, x_{i+1}, \ldots, x_n) = 1$, then

$$\sum_{\substack{k=1 \\ k \neq i}}^{n} w_k x_k \geq T \Longrightarrow w_i \overset{\frown}{+} \sum_{\substack{k=1 \\ k \neq i}}^{n} w_k x_k \geq T \Longrightarrow f(x_1, \ldots, x_{i-1}, 1, x_{i+1}, \ldots, x_n) = 1$$

Thus, f is positive in x_1 (see problem 1 of Exercise 4.1).

CASE 3 $w_i < 0$

By an argument similar to the proof of case 2, it can be shown that f is negative for $w_i < 0$. This completes the proof. ∎

THEOREM 4.2.3

There exist unate functions which are not threshold.

Proof: We shall prove this theorem by presenting an example. Consider the function

$$f(x_1, x_2, x_3, x_4) = x_1 x_2 + x_3 x_4$$

which is unate (monotonically increasing). However, it is not threshold. The proof is as follows: For f to be a threshold function, there must exist a set of real numbers $w_1, w_2, w_3, w_4,$ T such that the conditions of Eqs. (4.2.4) and (4.2.5) are satisfied. Now consider that when $x_1 = x_2 = 1$ and $x_3 = x_4 = 0$, $f(1, 1, 0, 0) = 1$, which requires that

$$w_1 \overset{\frown}{+} w_2 \geq T \tag{4.2.6}$$

When $x_1 = x_2 = 0$ and $x_3 = x_4 = 1$, $f(0, 0, 1, 1) = 1$, which requires that

$$w_3 \overset{\frown}{+} w_4 \geq T \tag{4.2.7}$$

†The definition of vacuous function was given in problem 5 of Exercise 4.1.

The sum of Eqs. (4.2.6) and (4.2.7) gives

$$w_1 \hat{+} w_2 \hat{+} w_3 \hat{+} w_4 \geq 2T \qquad (4.2.8)$$

On the other hand, when $x_1 = x_3 = 1$ and $x_2 = x_4 = 0$, and when $x_1 = x_3 = 0$ and $x_2 = x_4 = 1$, $f((1, 0, 1, 0)) = f((0, 1, 0, 1)) = 0$, which requires that

$$w_1 \hat{+} w_3 < T \qquad (4.2.9)$$
$$w_2 \hat{+} w_4 < T \qquad (4.2.10)$$

The sum of the two equations is

$$w_1 \hat{+} w_2 \hat{+} w_3 \hat{+} w_4 < 2T \qquad (4.2.11)$$

Comparing the two inequalities of Eqs. (4.2.8) and (4.2.11), it is seen that there does *not* exist w_1, w_2, w_3, w_4, T for which Eqs. (4.2.4) and (4.2.5) are satisfied; hence, f is not threshold. ∎

THEOREM 4.2.4

If the set of weights $w_1, w_2, \ldots, w_n$ and T is a threshold realization for a non-degenerate† monotonically increasing (decreasing) threshold function $f(x_1, x_2, \ldots, x_n)$, then $w_i > 0$ ($w_i < 0$) for all i.

Proof: Since f is monotonically increasing, f is positive in every x_i, and f may be expressed as

$$f(x_1, \ldots, x_n) = x_i g_1(x_1, \ldots, x_{i-1}, x_{i+1}, \ldots, x_n) + g_2(x_1, \ldots, x_{i-1}, x_{i+1}, \ldots, x_n)$$

which implies that if

$$f(x_1, \ldots, x_{i-1}, 0, x_{i+1}, \ldots, x_n) = 1 \qquad (4.2.12)$$

then

$$f(x_1, \ldots, x_{i-1}, 1, x_{i+1}, \ldots, x_n) = 1 \qquad (4.2.13)$$

This condition, in turn, implies that

$$\sum_{\substack{k=1 \\ k \neq i}}^{n} w_k x_k \geq T \Longrightarrow w_i \hat{+} \sum_{\substack{k=1 \\ k \neq i}}^{n} w_k x_k \geq T \qquad (4.2.14)$$

Thus, w_i must be positive. Hence, the theorem is proved. ∎

THEOREM 4.2.5

If a set of weights and a value T, $\{w_1, \ldots, w_n; T\}$, is a threshold realization for the function $f(x_1, \ldots, x_i, \ldots, x_n)$, then $\{w_1, \ldots, -w_i, \ldots, w_n; T - w_i\}$ is a realization for the function $f(x_1, \ldots, x_i', \ldots, x_n)$.

†The definition of nondegenerate function was given in problem 6 of Exercise 4.1.

Proof: f is a threshold function realized by w_i and T, by definition,

$$f(x_1, \ldots, x_i, \ldots, x_n) = 1 \quad \text{iff } w_1 x_1 \mathbin{\hat{+}} \ldots \mathbin{\hat{+}} w_i x_i \mathbin{\hat{+}} \ldots \mathbin{\hat{+}} w_n x_n \geq T$$

Since $x_i' = 1 - x_i$,

$$f(x_1, \ldots, x_i', \ldots, x_n) = 1 \quad \text{iff } w_1 x_1 \mathbin{\hat{+}} \ldots \mathbin{\hat{+}} w_i(1 - x_i) \mathbin{\hat{+}} \ldots \mathbin{\hat{+}} w_n x_n \geq T$$

or

$$f(x_1, \ldots, x_i', \ldots, x_n) = 1 \quad \text{iff } w_1 x_1 \mathbin{\hat{+}} \ldots \mathbin{\hat{+}} (-w_i)x_i \mathbin{\hat{+}} \ldots \mathbin{\hat{+}} w_n x_n > T - w_i$$

Thus, the theorem is proved. ∎

DEFINITION 4.2.3

Let $f(x_1, \ldots, x_n)$ be a switching function of n variables and $\alpha = (\alpha_{i_1}, \ldots, \alpha_{i_p})$ and $\beta = (\beta_{i_1}, \ldots, \beta_{i_p})$ be binary p-tuples, where $i_1, \ldots, i_p$ are any p of the integers $1, 2, \ldots, n$. Let

$$f(\alpha) = f(x_1, \ldots, x_n)|_{x_{i_j} = \alpha_{i_j}}$$
$$f(\beta) = f(x_1, \ldots, x_n)|_{x_{i_j} = \beta_{i_j}}$$

for $1 \leq j \leq p \leq n$. Then the function f is *k-monotonic* iff $f(\alpha) \geq f(\beta)$ or $f(\beta) \geq f(\alpha)$ for every pair of values α and β on every common set of variables $x_{i_1}, \ldots, x_{i_p}$ for which $p \leq k$. If it is n-monotonic, then f is said to be *completely monotonic*.

THEOREM 4.2.6

A threshold function is completely monotonic.

Proof: Let the set $\{w_1, \ldots, w_n; T\}$ be a realization for the threshold function f. Then

$$F = \sum_{j=1}^{n} w_j x_j \geq T \quad \text{if } f(x_1, \ldots, x_n) = 1 \tag{4.2.15}$$

$$F = \sum_{j=1}^{n} w_j x_j < T \quad \text{if } f(x_1, \ldots, x_n) = 0 \tag{4.2.16}$$

Define

$$F_\alpha = \sum_{j=1}^{p} \alpha_{i_j} w_{i_j} \quad \text{and} \quad F_\beta = \sum_{j=1}^{p} \beta_{i_j} w_{i_j}$$

Then

$$\sum_{\substack{\text{over all} \\ j \neq i_1, \ldots, i_p}} w_j x_j \geq T - F_\alpha \quad \text{iff } f(\alpha) = 1$$

$$\sum_{\substack{\text{over all} \\ j \neq i_1, \ldots, i_p}} w_j x_j \geq T - F_\beta \quad \text{iff } f(\beta) = 1$$

There are three cases:

(1) $T - F_\alpha > T - F_\beta$; then $f(\alpha) > f(\beta)$.
(2) $T - F_\alpha < T - F_\beta$; then $f(\beta) > f(\alpha)$.
(3) $T - F_\alpha = T - F_\beta$; then $f(\alpha) = f(\beta)$.

In view of the fact that the variables $x_{i_1}, \ldots, x_{i_p}$, where $1 \le p \le n$, are completely arbitrary, the function f is completely monotonic. ∎

Furthermore, it has been shown [13] that

THEOREM 4.2.7

A function $f(x_1, \ldots, x_n)$, $n \le 8$, is a threshold function iff it is completely monotonic.

We omit the proof here, which may be found in Paull and McCluskey [13].

Exercise 4.2

1. Determine whether or not the following functions are threshold functions.
   (a) $f_1(x_1, x_2) = \Sigma\ (0, 1, 2)$
   (b) $f_2(x_1, x_2) = \Sigma\ (0, 3)$
   (c) $f_3(x_1, x_2, x_3) = \Sigma\ (1, 2, 3, 4, 5, 6)$
   (d) $f_4(x_1, x_2, x_3) = \Sigma\ (0, 1, 2, 5, 6, 7)$
   (e) $f_5(x_1, x_2, x_3, x_4) = \Sigma\ (2, 3, 4, 5, 6, 10, 12, 13, 14)$
   (f) $f_6(x_1, x_2, x_3, x_4) = \Sigma\ (0, 2, 5, 8, 10, 12, 13)$

2. Let x be a two-valued variable. Prove that if $f(x_1, \ldots, x_n)$ is a threshold function, then
   (a) $x + f$
   (b) xf
   (c) $x' + f$
   (d) $x'f$
   are also threshold functions.

3. (a) Show that $f(x_1, x_2, x_3, x_4) = x_1 + x_2 + x_3 x_4$ is a threshold function.
   (b) Verify problem 2 by this example.
   (c) Verify Theorem 4.2.4 by this example.
   (d) Verify Theorem 4.2.5 by this example.

4. Show that a hyperplane defined by the linear equation

$$x_1 w_1 \stackrel{\wedge}{+} x_2 w_2 \stackrel{\wedge}{+} \ldots \stackrel{\wedge}{+} x_n w_n = T$$

separates the vertices of the n-cube associated with $f^{-1}(1)$ from those associated with $f^{-1}(0)$ iff $f(x_1, x_2, \ldots, x_n)$ is a threshold function.

5. List all two-variable switching functions which are not threshold functions. Check your result with the answers you found in problem 1(a) and (b).

6. List all three-variable switching functions which are not threshold functions. Check your result with the answers you found in problem 1(c) and (d).

7. Prove that if a function is unate, it is 1-monotonic.

8. Show that

(a) 1-monotonicity is sufficient for two-variable and three-variable functions to be threshold functions.

(b) 2-monotonicity is sufficient for four-variable and five-variable functions to be threshold functions.

(c) 3-monotonicity is sufficient for six-variable and seven-variable functions to be threshold functions.

(d) 4-monotonicity is sufficient for eight-variable functions to be threshold functions.

9. Prove, in general, that $n/2$-monotonicity, for even n, and $(n - 1)/2$-monotonicity, for odd n, is sufficient for an n-variable function to be a threshold function.

4.3 Symmetric Functions

The third important special type of switching functions we introduce is the symmetric function. First, let us introduce the partially symmetric function.

DEFINITION 4.3.1

A switching function $f(x_1, \ldots, x_n)$ is *partially symmetric* in x_i and x_j, $i \neq j$, if

$$f(x_1, \ldots, x_i, \ldots, x_j, \ldots, x_n) = f(x_1, \ldots, x_j, \ldots, x_i, \ldots, x_n)$$

Example 4.3.1

The following functions are partially symmetric.

(a) $f(x_1, x_2, x_3) = x_1 x_3 + x_2 x_3$ is partially symmetric in x_1 and x_2.

(b) $f(x_1, x_2, x_3, x_4) = (x_1 + x_3)x_2 x_4$ is partially symmetric in x_1 and x_3, and also x_2 and x_4.

DEFINITION 4.3.2

A function is *totally symmetric* or simply *symmetric* if it is partially symmetric in x_i and x_j, for all i, j.

Example 4.3.2

The following functions are symmetric functions.

(a) $f_1(x_1, x_2, x_3) = (x_1 + x_2)(x_2 + x_3)(x_3 + x_1)$

(b) $f_2(x_1, x_2, x_3) = x_1' x_2' + x_2' x_3' + x_3' x_1'$

(c) $f_3(x_1, \ldots, x_n) = 0$

(d) $f_4(x_1, \ldots, x_n) = 1$

A basic property of symmetric functions is stated as follows.

THEOREM 4.3.1

If $f(x_1, \ldots, x_n)$ is a symmetric function and for $x_1 = x_2 = \ldots = x_k = 1$ and $x_{k+1} = x_{k+2} = \ldots = x_n = 0$,

$$f(\underbrace{1, 1, \ldots, 1}_{k}, \underbrace{0, 0, \ldots, 0}_{n-k}) = 1$$

then f is 1 for any k of the x_i's equal to 1 and the remaining $n - k$ x_i's equal to 0.

Proof: Choose any k distinct $x_{i_1}, x_{i_2}, \ldots, x_{i_k}$ from the set $X = \{x_1, x_2, \ldots, x_n\}$. Since f is symmetric, the function must remain the same under the permutation $x_{i_1} \longrightarrow x_1$, $x_{i_2} \longrightarrow x_2, \ldots, x_{i_k} \longrightarrow x_k$. ∎

Example 4.3.3

Consider the symmetric function $f(x_1, x_2, x_3) = x_1x_2 + x_2x_3 + x_3x_1$. If $x_1 = x_2 = 1$ and $x_3 = 0, f = 1$; and if $x_1 = x_2 = x_3 = 1, f = 1$. Thus, if any *two* of the three variables or all *three* variables are 1, the function is 1.

Define the *characteristic* set A of a symmetric function $f(x_1, \ldots, x_n)$ as $A = \{a \mid a$ is the number of 1's assignments of x_i's for which f is 1$\}$, and define $S_A(x_1, \ldots, x_n)$ for the symmetric function of n variables whose a numbers are elements of A of this function.

COROLLARY 4.3.1

Any symmetric function $f(x_1, \ldots, x_n)$ is completely characterized by $S_A(x_1, \ldots, x_n)$.

Example 4.3.4

The symmetric function of Example 4.3.3 may be represented by $S_{\{2,3\}}(x_1, x_2, x_3)$, and the four functions of Example 4.3.2 may be represented by

(a) $S_{\{2,3\}}(x_1, x_2, x_3)$
(b) $S_{\{0,1\}}(x_1, x_2, x_3)$
(c) $S_{\{\varnothing\}}(x_1, \ldots, x_n)$
(d) $S_{\{0,1,2,\ldots,n\}}(x_1, \ldots, x_n)$

COROLLARY 4.3.2

There are 2^{n+1} symmetric functions of n variables.

Proof: For an n-variable function, the allowable a numbers are $0, 1, \ldots, n$. Therefore, the number of functions is

$$\sum_{k=0}^{n+1} C_k^n = (1+1)^{n+1} = 2^{n+1} \quad \blacksquare$$

COROLLARY 4.3.3

There are $2^{3 \times 2^{n-2}}$ partially symmetric functions in x_i and x_j of n variables.

Proof: Let us partition the set of n variables into four equivalent classes with respect to x_i and x_j in the following way. Those which contain x_ix_j, x_ix_j', $x_i'x_j$, and $x_i'x_j'$, respectively, are grouped into one group, since these four groups have the same number of elements and the total number of minterms of an n-variable function is 2^n. Thus, each group contains $2^{(n-2)}$ elements, as shown in Fig. 4.3.1. Every element in the upper left corner and lower right corner groups is a partially symmetric function in x_i and x_j. The elements in the groups of $x_i'x_j$ and x_ix_j' are not partially symmetric. However, each term in the group of $x_i'x_j$ (for instance, a minterm $x_1, \ldots, x_{i-1}, x_i', x_{i+1}, \ldots, x_{j-1}, x_j, x_{j+1}, \ldots, x_n$) combining with the one in the

	x_j	x_j'
x_i	2^{n-2} Elements	2^{n-2} Elements
x_i'	2^{n-2} Elements	2^{n-2} Elements

Fig. 4.3.1

group of $x_i x_j'$ where all the variables of the minterm are the same except x_i and x_j (e.g., the one in the group of $x_i x_j'$ corresponding to the minterm $x_1, \ldots, x_{i-1}, x_i', x_{i+1}, \ldots, x_{j-1}, x_j,$ $x_{j+1}, \ldots, x_n$ is $x_1, \ldots, x_{i-1}, x_i, x_{i+1}, \ldots, x_{j-1}, x_j', x_{j+1}, \ldots, x_n)$ is partially symmetric in x_i and x_j. Obviously, there are $2^{(n-2)}$ such combinations between the groups $x_i' x_j$ and $x_i x_j'$. Therefore, there are $3 \times 2^{(n-2)}$ number of terms which are partially symmetric in x_i and x_j. Each minterm can be assigned 1 (which means that the minterm is included in the canonical sum-of-products form of the function) or 0 (which means that the minterm is not included in the canonical sum-of-products form). Hence, there are $2^{3 \times 2^{(n-2)}}$ partially symmetric functions in x_i and x_j of n variables. ∎

COROLLARY 4.3.4

There are $n + 2$ symmetric functions of n variables which are monotonically increasing and $n + 2$ symmetric functions of n variables which are monotonically decreasing.

Proof: By Theorem 4.1.3, if $f(x_1, \ldots, x_n)$ is a monotonically increasing (decreasing) function, it is expressible as a sum of products of $x_i(x_i')$. Hence, functions which satisfy both this condition and the condition for symmetry are as follows:

Symmetric functions which are also monotonically increasing	*Symmetric functions which are also monotonically decreasing*
$S_{\{\varnothing\}}(x_1, \ldots, x_n) = 0$ $S_{\{0,1,\ldots,n\}}(x_1, \ldots, x_n) = 1$ $S_{\{1,\ldots,n\}}(x_1, \ldots, x_n) = \sum_{i=1}^{n} x_i$ $S_{\{2,\ldots,n\}}(x_1, \ldots, x_n) = \sum_{1 \leq i < j \leq n} x_i x_j$ $\cdots\cdots\cdots\cdots\cdots\cdots$ $S_{\{n\}}(x_1, \ldots, x_n) = x_1 x_2 \ldots x_n$ $\Big\} \, n+2$	$S_{\{\varnothing\}}(x_1, \ldots, x_n) = 0$ $S_{\{0,1,\ldots,n\}}(x_1, \ldots, x_n) = 1$ $S_{\{0,1,\ldots,n-1\}}(x_1, \ldots, x_n) = \sum_{i=1}^{n} x_i'$ $S_{\{0,1,\ldots,n-2\}}(x_1, \ldots, x_n) = \sum_{1 \leq i < j \leq n} x_i' x_j'$ $\cdots\cdots\cdots\cdots\cdots\cdots$ $S_{\{0\}}(x_1, \ldots, x_n) = x_1' x_2' \ldots x_n'$ $\Big\} \, n+2$

Hence, the corollary is proved.

It is clear that

COROLLARY 4.3.5

There are $2n + 2$ symmetric functions of n variables which are monotonically increasing and decreasing functions.

Moreover, we can prove the following:

THEOREM 4.3.2

Let $f(x_1, \ldots, x_n)$ be a symmetric function. f is monotonic iff it is threshold.

Proof: When f is a constant function (i.e., 0 or 1), the theorem is obvious. When it is a nonconstant function, the proof is as follows.

(1) *Proof of the Necessity:* Suppose that f is symmetric and monotonic. Consider

$$S_{\{k, \ldots, n\}}(x_1, \ldots, x_n) = \sum_{1 \leq i_1 \leq i_2 \leq \ldots < i_k \leq n} x_{i_1}, x_{i_2}, \ldots, x_{i_k}$$

In accordance with Corollary 4.3.4, it is monotonically increasing.
Let $w_i = 1/k$ for $i = 1, 2, \ldots, n$, and $T = 1$; then

$$\sum_{i=1}^{n} w_i x_i = \frac{1}{k} \sum_{i=1}^{n} x_i \geq T = 1$$

iff among the assignments of x_i there are at least k 1's. Consider the symmetric and monotonically decreasing function

$$S_{\{0, 1, \ldots, n-k\}}(x_1, \ldots, x_n) = \sum_{i \leq i_1 < i_2 < \ldots < i_k \leq n} x'_{i_1} x'_{i_2} \ldots x'_{i_k}$$

Letting $w_i = -1/(n-k)$ for $i = 1, 2, \ldots, n$, and $T = -1$,

$$\sum_{i=1}^{n} w_i x_i = \frac{-1}{n-k} \sum_{i=1}^{n} x_i \geq T = -1$$

iff among the assignments of x_i there are at least k 1's. Hence, the necessity is proved.

(2) *Proof of the Sufficiency:* Now, if f is threshold from Theorem 4.2.7, it is unate. Suppose that f is monotonically increasing in $x_{i_1} x_{i_2} \ldots x_{i_k}$ and monotonically decreasing in the rest $x_{i_{k+1}} x_{i_{k+2}} \ldots x_{i_n}$ variables. Suppose that we interchange the variables x_{i_1} and $x_{i_{k+1}}$. By hypothesis, f is symmetric which implies that f should be monotonically increasing in $x_{i_{k+1}}$ and monotonically decreasing in x_{i_1}. This is a contradiction. Hence, the theorem is proved. ∎

LEMMA 4.3.1

Let A_1 and A_2 be the characteristic sets of the symmetric functions $f_1(x_1, \ldots, x_n)$ and $f_2(x_1, \ldots, x_n)$. Then

(a) $S_{A_1}(x_1, \ldots, x_n) + S_{A_2}(x_1, \ldots, x_n) = S_{A_1 \cup A_2}(x_1, \ldots, x_n)$.
(b) $S_{A_1}(x_1, \ldots, x_n) \cdot S_{A_2}(x_1, \ldots, x_n) = S_{A_1 \cap A_2}(x_1, \ldots, x_n)$.
(c) The complement of $S_{A_1}(x_1, \ldots, x_n)$ is equal to $S_{A_1'}(x_1, \ldots, x_n)$, where $A_1 = Z_{n+1} - A_1'$ and $Z_{n+1} = \{0, 1, \ldots, n\}$.

Proof: (a) Since

$$S_A(x_1, \ldots, x_n) = S_{A_1}(x_1, \ldots, x_n) + S_{A_2}(x_1, \ldots, x_n) = 1$$

iff

$$S_{A_1}(x_1, \ldots, x_n) = 1 \quad \text{or} \quad S_{A_2}(x_1, \ldots, x_n) = 1$$

then an element of A is either in set A_1, in set A_2, or in both. By definition of union, $A = A_1 \cup A_2$.

(b)   Dually, consider that

$$S_A(x_1, \ldots, x_n) = S_{A_1}(x_1, \ldots, x_n) \cdot S_{A_2}(x_1, \ldots, x_n) = 1$$

iff both

$$S_{A_1}(x_1, \ldots, x_n) = 1 \quad \text{and} \quad S_{A_2}(x_1, \ldots, x_n) = 1$$

Therefore, an element of A must be in both sets A_1 and A_2; that is, $A = A_1 \cap A_2$.

(c)   Let $S_X(x_1, \ldots, x_n)$ be the complement of $S_{A_1}(x_1, \ldots, x_n)$. From the definition of complement,

$$S_{A_1}(x_1, \ldots, x_n) + S_X(x_1, \ldots, x_n) = S_{Z_{n+1}}(x_1, \ldots, x_n)$$
$$S_{A_1}(x_1, \ldots, x_n) \cdot S_X(x_1, \ldots, x_n) = S_{\varnothing}(x_1, \ldots, x_n)$$

From (a) and (b), it follows that

$$A_1 \cup X = Z_{n+1}$$

and

$$A_1 \cap X = \varnothing$$

which imply that $X = Z_{n+1} - A_1 = A_1'$. ∎

Following from the lemma, we have

THEOREM 4.3.3

The set of symmetric switching functions of n variables forms a sub-Boolean algebra of the Boolean algebra of all the switching functions of n variables.

Now we introduce the expansion theorem for symmetric functions.

THEOREM 4.3.4

Let $f(x_1, \ldots, x_n)$ be a symmetric function of n variables described by $S_{\{a\}}(x_1, \ldots, x_n)$. Then

$$S_{\{a\}}(x_1, \ldots, x_n) = x_i \cdot S_{\{a-1\}}(x_1, \ldots, x_{i-1}, x_{i+1}, \ldots, x_n)$$
$$+ x_i' \cdot S_{\{a\}}(x_1, \ldots, x_{i-1}, x_{i+1}, \ldots, x_n), \qquad a > 0 \quad (4.3.1)$$

More generally, if a symmetric function $f(x_1, \ldots, x_n)$ is described by $S_{\{a_1, \ldots, a_m\}}(x_1, \ldots, x_n)$, then

$$S_{\{a_1, \ldots, a_m\}}(x_1, \ldots, x_n) = x_i \cdot S_{\{a_1-1, \ldots, a_m-1\}}(x_1, \ldots, x_{i-1}, x_{i+1}, \ldots, x_n)$$
$$+ x_i' \cdot S_{\{a_1, \ldots, a_m\}}(x_1, \ldots, x_{i-1}, x_{i+1}, \ldots, x_n)$$
$$a_i > 0, \quad i = 1, 2, \ldots, m \quad (4.3.2)$$

Proof: The proof of this theorem follows directly from the general expansion theorem,

$$f(x_1, \ldots, x_n) = x_i \cdot f(x_1, \ldots, x_{i-1}, 1, x_{i+1}, \ldots, x_n) + x_i' \cdot f(x_1, \ldots, x_{i-1}, 0, x_{i+1}, \ldots, x_n)$$

$$(4.3.3)$$

Since $f(x_1, \ldots, x_n)$ is described by $S_{\{a\}}(x_1, \ldots, x_n)$, $f(x_1, \ldots, x_{i-1}, 1, x_{i+1}, \ldots, x_n)$ must be described by $S_{\{a-1\}}(x_1, \ldots, x_{i-1}, x_{i+1}, \ldots, x_n)$. This is because x_i has been made equal to 1. On the other hand, $f(x_1, \ldots, x_{i-1}, 0, x_{i+1}, \ldots, x_n)$ is $S_{\{a\}}(x_1, \ldots, x_{i-1}, x_{i+1}, \ldots, x_n)$; it has the same a-numbers, since no variables are made equal to 1. Thus, Eq. (4.3.3) may be rewritten as

$$S_{\{a\}}(x_1, \ldots, x_n) = x_i \cdot S_{\{a-1\}}(x_1, \ldots, x_{i-1}, x_{i+1}, \ldots, x_n)$$
$$+ x_i' \cdot S_{\{a\}}(x_1, \ldots, x_{i-1}, x_{i+1}, \ldots, x_n)$$

Now suppose that f is described by more than one a-number; that is, $S_{\{a_1, \ldots, a_m\}}(x_1, \ldots, x_n)$. By Lemma 4.3.1(a),

$$S_{\{a_1, \ldots, a_m\}}(x_1, \ldots, x_n) = S_{\{a_1\}}(x_1, \ldots, x_n) + S_{\{a_2\}}(x_1, \ldots, x_n) + \ldots + S_{\{a_m\}}(x_1, \ldots, x_n)$$
$$= x_i \cdot S_{\{a_1-1\}}(x_1, \ldots, x_{i-1}, x_{i+1}, \ldots, x_n)$$
$$+ x_i' \cdot S_{\{a_1\}}(x_1, \ldots, x_{i-1}, x_{i+1}, \ldots, x_n)$$
$$+ x_i \cdot S_{\{a_2-1\}}(x_1, \ldots, x_{i-1}, x_{i+1}, \ldots, x_n)$$
$$+ x_i' \cdot S_{\{a_2\}}(x_1, \ldots, x_{i-1}, x_{i+1}, \ldots, x_n)$$
$$+ \ldots\ldots\ldots\ldots\ldots\ldots\ldots\ldots\ldots\ldots\ldots\ldots$$
$$+ x_i \cdot S_{\{a_m-1\}}(x_1, \ldots, x_{i-1}, x_{i+1}, \ldots, x_n)$$
$$+ x_i' \cdot S_{\{a_m\}}(x_1, \ldots, x_{i-1}, x_{i+1}, \ldots, x_n)$$
$$= x_i \cdot \{S_{\{a_1-1\}}(x_1, \ldots, x_{i-1}, x_{i+1}, \ldots, x_n)$$
$$+ \ldots + S_{\{a_m-1\}}(x_1, \ldots, x_{i-1}, x_{i+1}, \ldots, x_n)\}$$
$$+ x_i' \cdot \{S_{\{a_1\}}(x_1, \ldots, x_{i-1}, x_{i+1}, \ldots, x_n)$$
$$+ \ldots + S_{\{a_m\}}(x_1, \ldots, x_{i-1}, x_{i+1}, \ldots, x_n)\}$$

Using Lemma 4.3.1(a) again,

$$S_{\{a_1, \ldots, a_m\}} = x_i \cdot S_{\{a_1-1, \ldots, a_m-1\}}(x_1, \ldots, x_{i-1}, x_{i+1}, \ldots, x_n)$$
$$+ x_i' \cdot S_{\{a_1, \ldots, a_m\}}(x_1, \ldots, x_{i-1}, x_{i+1}, \ldots, x_n)$$

Hence, the theorem is proved. ∎

Example 4.3.5

Consider the first symmetric function of Example 4.3.2:

$$f(x_1, x_2, x_3) = (x_1 + x_2)(x_2 + x_3)(x_3 + x_1) \qquad (4.3.4)$$

which may be described as

$$f(x_1, x_2, x_3) = S_{\{2, 3\}}(x_1, x_2, x_3) \qquad (4.3.5)$$

Applying the expansion theorem above to symmetric functions, say, $x_i = x_1$,

$$S_{\{2,3\}}(x_1, x_2, x_3) = x_1 \cdot S_{\{1,2\}}(x_2, x_3) + x_1' \cdot S_{\{2,3\}}(x_2, x_3) \tag{4.3.6}$$

Let us verify Eq. (4.3.6) by substituting

$$S_{\{1,2\}}(x_2, x_3) = S_{\{2,3\}}(1, x_2, x_3) = (x_2 + x_3)$$
$$S_{\{2,3\}}(x_2, x_3) = S_{\{2,3\}}(0, x_2, x_3) = x_2 x_3 (x_2 + x_3) = x_2 x_3$$

into the right-hand side of Eq. (4.3.6), we obtain

$$x_1 \cdot S_{\{1,2\}}(x_2, x_3) + x_1' \cdot S_{\{2,3\}}(x_2, x_3)$$
$$= x_1 \cdot (x_2 + x_3) + x_1' x_2 x_3 (x_2 + x_3)$$
$$= x_1 x_2 x_3 + x_1 x_2 x_3' + x_1 x_2' x_3 + x_1' x_2 x_3$$

which is symmetric and, in fact, is $S_{\{2,3\}}(x_1, x_2, x_3)$.

Now we come to the problem of identification of symmetric functions. In practice, one of the advantages of knowing that a given function is symmetric is the potential economy offered by the realization of such a function using smaller number of components. There are three methods for determining whether a given function is symmetric: (1) the simplified permutation method, (2) the graphical pattern-recognition method, and (3) the McCluskey method. The three methods are described below.

1. Simplified Permutation Method

This method is based on the following theorem.

THEOREM 4.3.5

A function $f(x_1, \ldots, x_n)$ is *symmetric* iff f satisfies the following two properties: (1) f is partially symmetric in x_1 and x_2; that is,

$$f(x_1, x_2, \ldots, x_n) = f(x_2, x_1, \ldots, x_n) \tag{4.3.7}$$

(2) $\qquad\qquad f(x_1, x_2, x_3, \ldots, x_n) = f(x_1, x_3, \ldots, x_n, x_2) \tag{4.3.8}$

Proof: The proof of the necessity follows from the definition of symmetric function. We need only prove the sufficiency.

First, we want to show that for any $x_i, x_j, f(x_1, \ldots, x_n)$ is partially symmetric in x_1 and x_i and x_2 and x_j. The proof that f is partially symmetric in x_1 and x_i is as follows. After applying Eq. (4.3.8) to f for $(i - 2)$ times, we have

$$f(x_1, x_2, x_3, \ldots, x_n) = f(x_1, x_i, \ldots, x_{i-2}, x_{i-1})$$

By Eq. (4.3.7),

$$f(x_1, x_2, x_3, \ldots, x_n) = f(x_i, x_1, \ldots, x_{i-2}, x_{i-1})$$

By repeatedly applying Eq. (4.3.8) to the above equation until x_2 is at the position next to x_i, we obtain

$$f(x_1, x_2, x_3, \ldots, x_n) = f(x_i, x_2, \ldots, x_1, \ldots, x_n)$$

Thus, f is partially symmetric in x_1 and x_i. That f is partially symmetric in x_2 and x_j may be proved in a similar way. By hypothesis, f is partially symmetric in x_1 and x_2; and now we have shown that f is partially symmetric in x_1 and x_i and in x_2 and x_j. Therefore, f is partially symmetric in x_i and x_j, where x_i and x_j are any two variables of the n-variables; hence, f is a symmetric function. ∎

Example 4.3.6

Consider the switching function

$$f(x_1, x_2, x_3, x_4) = x_1 x_2 x_3' x_4' + x_1 x_2' x_3 x_4' + x_1 x_2' x_3' x_4 + x_1' x_2 x_3 x_4'$$
$$+ x_1' x_2 x_3' x_4 + x_1' x_2' x_3 x_4 + x_1 x_2 x_3 x_4' + x_1 x_2 x_3' x_4$$
$$+ x_1 x_2' x_3 x_4 + x_1' x_2 x_3 x_4 \tag{4.3.9}$$

First, we test whether f is partially symmetric in x_1 and x_2. After interchanging x_1 and x_2, we have

$$f_{x_1 \leftrightarrow x_2}(x_1, x_2, x_3, x_4) = x_2 x_1 x_3' x_4' + x_2 x_1' x_3 x_4' + x_2 x_1' x_3' x_4$$
$$+ x_2' x_1 x_3 x_4' + x_2' x_1 x_3' x_4 + x_2' x_1' x_3 x_4 + x_2 x_1 x_3 x_4'$$
$$+ x_2 x_1 x_3' x_4 + x_2 x_1' x_3 x_4 + x_2' x_1 x_3 x_4 \tag{4.3.10}$$

A close examination of Eqs. (4.3.9) and (4.3.10) confirms that they are the same; thus, f satisfies the first condition of Theorem 4.3.5.

Next, we perform the permutation: $x_2 \longrightarrow x_3$, $x_3 \longrightarrow x_4$, and $x_4 \longrightarrow x_2$ on the variables of the function of Eq. (4.3.9),

$$f(x_1, x_2, x_3, x_4) = x_1 x_3 x_4' x_2' + x_1 x_3' x_4 x_2' + x_1 x_3' x_4' x_2 + x_1' x_3 x_4 x_2'$$
$$+ x_1' x_3 x_4' x_2 + x_1' x_3' x_4 x_2 + x_1 x_3 x_4 x_2' + x_1 x_3 x_4' x_2$$
$$+ x_1 x_3' x_4 x_2 + x_1' x_3 x_4 x_2 \tag{4.3.11}$$

Again, we find that this equation is the same as Eq. (4.3.9). Hence, f is symmetric. ∎

2. Graphical Pattern-Recognition Method

The recognition of symmetric functions of four variables or less is easily accomplished by observing their pattern when they are entered in a Karnaugh map. Since any symmetric function with three variables or less may be recognized by inspection, only functions with four variables will be discussed. An extended Karnaugh map of four variables is shown in Fig. 4.3.2. The number indicated in each cell of the map is the a-number of that term. For instance, in the cells corresponding to $x_1'x_2x_3'x_4$ are 2's and in the cells corresponding to $x_1'x_2'x_3'x_4'$ are 0's. The patterns of symmetric functions, $S_i(x_1, x_2, x_3, x_4)$, $i = 0, 1, 2, 3, 4$, are shown in Fig. 4.3.2. Therefore, a symmetric function should contain terms which form these patterns, no more and no less.

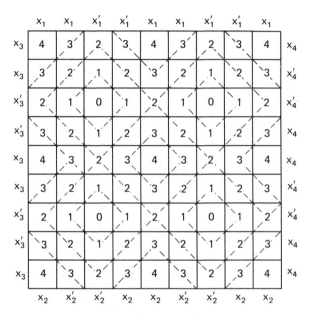

Fig. 4.3.2 An extended Karnaugh map indicating the patterns of symmetric functions.

Example 4.3.7

Consider the function

$$f(x_1, x_2, x_3, x_4) = x_1x_2'x_4 + x_1x_3'x_4 + x_1'x_2x_3 + x_1'x_3x_4 + x_1'x_2x_4 + x_2x_3x_4'$$

which we cannot tell by inspection whether it is symmetric. However, if we write down the canonical sum form of this function,

$$f = x_1x_2x_3'x_4' + x_1x_2'x_3x_4' + x_1x_2'x_3'x_4 + x_1'x_2x_3x_4' + x_1'x_2x_3'x_4 + x_1'x_2'x_3x_4$$
$$+ x_1x_2x_3x_4' + x_1x_2x_3'x_4 + x_1x_2'x_3x_4 + x_1'x_2x_3x_4$$

we can, with the aid of the Karnaugh map shown in Fig. 4.3.3, easily determine that the function is symmetric.

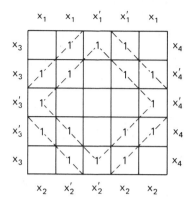

Fig. 4.3.3 The extended Karnaugh map of the function of Example 4.3.6.

3. McCluskey Method

When the number of variables is more than four, it is frequently necessary to use the McCluskey method to determine whether a function is symmetric. The method for identifying symmetric functions using the McCluskey tabular procedure is best illustrated by the use of examples.

Example 4.3.8

Consider

$$f(x_1, x_2, x_3, x_4, x_5) = \Sigma(3, 5, 6, 7, 9, 10, 11, 12, 13, 14, 15, 17, 18, 19, 20, 21, 22,$$
$$23, 24, 25, 26, 27, 28, 29, 30)$$

First, construct the table as shown in Fig. 4.3.4. The first and second columns are the decimal- and binary-number representation of each term of the given function. The a-number for each term is shown in the third column. The number of terms of each a-number is given in the fourth column.

From the definition of a-number and Theorem 4.3.1, a function is symmetric iff it satisfies the condition that the number of terms m (indicated in the fourth column) must be equal to $C_{a_i}^n$. In this example, $a_1 = 2$, $a_2 = 3$, and $a_3 = 4$. We find that

$$C_2^5 = \frac{5!}{2!(5-2)!} = 10, \qquad C_3^5 = \frac{5!}{3!(5-2)!} = 10, \qquad C_4^5 = \frac{5!}{4!(5-4)!} = 5$$

which are, respectively, the same as the numbers indicated in the fourth column. Hence, the function is symmetric. It may be represented by $S_{\{2,3,4\}}(x_1, x_2, x_3, x_4, x_5)$.

Example 4.3.9

Consider the function

$$f(x_1, x_2, x_3, x_4, x_5, x_6) = \Sigma(4, 11, 16, 21, 22, 28, 35, 41, 42, 43, 47, 52, 59)$$

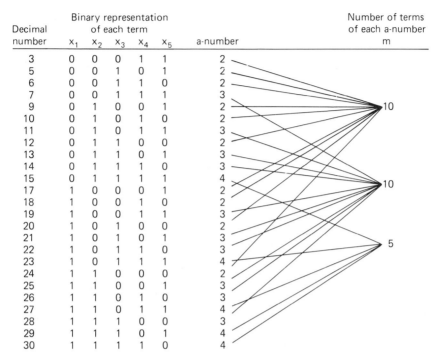

Decimal number	x_1	x_2	x_3	x_4	x_5	a-number	Number of terms of each a-number m
3	0	0	0	1	1	2	
5	0	0	1	0	1	2	
6	0	0	1	1	0	2	
7	0	0	1	1	1	3	
9	0	1	0	0	1	2	10
10	0	1	0	1	0	2	
11	0	1	0	1	1	3	
12	0	1	1	0	0	2	
13	0	1	1	0	1	3	
14	0	1	1	1	0	3	
15	0	1	1	1	1	4	
17	1	0	0	0	1	2	10
18	1	0	0	1	0	2	
19	1	0	0	1	1	3	
20	1	0	1	0	0	2	
21	1	0	1	0	1	3	
22	1	0	1	1	0	3	
23	1	0	1	1	1	4	5
24	1	1	0	0	0	2	
25	1	1	0	0	1	3	
26	1	1	0	1	0	3	
27	1	1	0	1	1	4	
28	1	1	1	0	0	3	
29	1	1	1	0	1	4	
30	1	1	1	1	0	4	

Fig. 4.3.4 The table for identifying symmetric functions using the McCluskey method.

Referring to Fig. 4.3.5, the a-numbers are 1, 3, 4, and 5. Thus, we check the number of terms of each a-number with the formula $C_{a_i}^n$, namely,

$$C_1^6 = 6, \quad C_3^6 = \frac{6!}{3!\,3!} = 20, \quad C_4^6 = \frac{6!}{4!\,2!} = 15, \quad C_5^6 = \frac{6!}{5!\,1!} = 6$$

It is seen that none of the numbers given in the fourth column agrees with the number obtained by the formula $C_{a_i}^n$. Hence, the function is not a symmetric function in x_1, x_2, x_3, x_4, x_5, and x_6.

However, this does not rule out the possibility that the function is symmetric in the complements of some of the six variables. Define the r-number of a variable x as the ratio of 1's to the number of terms:

$$r\text{-number} = \frac{\text{number of 1's in } x \text{ column}}{\text{number of terms}}$$

Then, a necessary condition for this to happen is that the r-numbers of each pair of variables should be either equal to or complement each other. The r-numbers of the six variables are shown at the bottom of each column in Fig. 4.3.5. The r-numbers of x_2 and x_4 are complements of the others. Replacing x_2 and x_4 by their complements, the table is shown in Fig. 4.3.6.

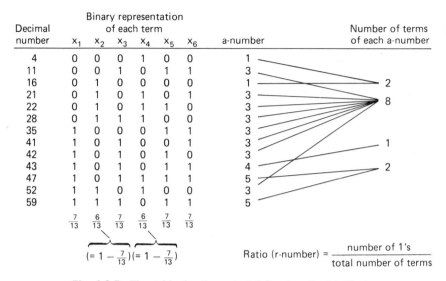

Decimal number	Binary representation of each term						a-number	Number of terms of each a-number
	x_1	x_2	x_3	x_4	x_5	x_6		
4	0	0	0	1	0	0	1	
11	0	0	1	0	1	1	3	
16	0	1	0	0	0	0	1	2
21	0	1	0	1	0	1	3	
22	0	1	0	1	1	0	3	8
28	0	1	1	1	0	0	3	
35	1	0	0	0	1	1	3	
41	1	0	1	0	0	1	3	1
42	1	0	1	0	1	0	3	
43	1	0	1	0	1	1	4	2
47	1	0	1	1	1	1	5	
52	1	1	0	1	0	0	3	
59	1	1	1	0	1	1	5	
	$\frac{7}{13}$	$\frac{6}{13}$	$\frac{7}{13}$	$\frac{6}{13}$	$\frac{7}{13}$	$\frac{7}{13}$		

$$(= 1 - \tfrac{7}{13})(= 1 - \tfrac{7}{13})$$

$$\text{Ratio (r-number)} = \frac{\text{number of 1's}}{\text{total number of terms}}$$

Fig. 4.3.5 The table for Example 4.3.8 using the McCluskey method.

Decimal number	Binary representation of each term						a-number	Number of terms of each a-number
	x_1	x_2'	x_3	x_4'	x_5	x_6		
16	0	1	0	0	0	0	1	
31	0	1	1	1	1	1	5	
4	0	0	0	1	0	0	1	
1	0	0	0	0	0	1	1	6
2	0	0	0	0	1	0	1	
8	0	0	1	0	0	0	1	
55	1	1	0	1	1	1	5	
61	1	1	1	1	0	1	5	6
62	1	1	1	1	1	0	5	
63	1	1	1	1	1	1	6	1
59	1	1	1	0	1	1	5	
32	1	0	0	0	0	0	1	
47	1	0	1	1	1	1	5	
r-number	$\frac{7}{13}$	$\frac{7}{13}$	$\frac{7}{13}$	$\frac{7}{13}$	$\frac{7}{13}$	$\frac{7}{13}$		

Fig. 4.3.6 The table of Fig. 4.3.5 with x_2 and x_4 replaced by x_2' and x_4'.

In this table it is seen that the a-numbers are 1, 5, and 6. The numbers of terms of the three a-numbers are 6, 6, and 1, respectively, which are exactly equal to

$$C_1^6 = 6, \quad C_5^6 = 6, \quad \text{and} \quad C_6^6 = 1$$

Therefore, this function is symmetric in $x_1, x_2', x_3, x_4', x_5$, and x_6, or we may write it as $S_{\{1,5,6\}}(x_1, x_2', x_3, x_4', x_5, x_6)$.

Exercise 4.3

1. Let $f(x_1, x_2, x_3, x_4) = x_3x_4' + x_1x_3' + x_1x_2' + x_3'x_4 + x_2x_3'x_4'$.
   (a) Construct a Karnaugh map for f.
   (b) Is f a symmetric function? If so, what are its a-numbers?

2. Let $f(x_1, x_2, x_3, x_4) = x_2'x_4' + x_2x_4 + x_1x_3 + x_1'x_3'$.
   (a) Give a Karnaugh map for f.
   (b) Name explicitly the permutations leaving f fixed.

3. Let $f(x_1, x_2, x_3, x_4) = S_{\{2,3\}}(x_1, x_2, x_3, x_4)$.
   (a) Draw a Karnaugh map for f.
   (b) Find all the prime implicants of f.
   (c) Is f a threshold function? Prove your answer.
   (d) How many symmetric functions of four variables are threshold?

4. Let $f(x, y, z, w) = x'y' + x'yz + x'zw + xz'w' + xy'z' + xy'zw'$.
   (a) Draw a Karnaugh map for f.
   (b) Is f a symmetric function? Prove your answer.
   (c) Give a minimum product-of-sums expression for f.
   (d) How many symmetric functions of four variables are there? How many of them are monotone-increasing? Enumerate them.

5. Devise a product-of-sums Karnaugh map method to tell by inspection if a function of four variables is symmetric.

6. How many functions of three variables are partially symmetric in x_1 and x_2? How many of these functions are not partially symmetric in x_2 and x_3?

7. Let $f(x_1, x_2, x_3, x_4)$ have the following Karnaugh map:

	00	01	11	10
00			1	
01		1	1	1
11	1	1		1
10		1	1	1

Is f symmetric? Is f monotonic? Is f unate? Give brief reasons for your answers.

8. Give a switching function for each of the following cases:
   (a) It is monotonically increasing, but not threshold and symmetric.
   (b) It is threshold, but not monotonic and symmetric.
   (c) It is symmetric, but not monotonic and threshold.
   (d) It is monotonic and threshold, but not symmetric.

9. Show that any symmetric function of n variables can be written as

$$S_A(x_1, \ldots, x_n) = x_1 x_2 S_{A_1}(x_3, \ldots, x_n) + x_1 x_2' S_{A_2}(x_3, \ldots, x_n)$$
$$+ x_1' x_2 S_{A_3}(x_3, \ldots, x_n) + x_1' x_2' S_{A_4}(x_3, \ldots, x_n)$$

and determine A_i, $i = 1, 2, 3, 4$, in terms of A, where $A, A_1, \ldots, A_4$ are the sets of a-numbers.

10. A function $f(x_1, \ldots, x_n)$ is said to be *antisymmetric* in x_1 and x_2 if

$$f(x_1, x_2, x_3, \ldots, x_n) = (f(x_2, x_1, x_3, \ldots, x_n))'$$

How many functions of n variable are antisymmetric in x_1 and x_2?

11. Derive a formula similar to Lemma 4.3.1 for $S_{A_1}(x_1, \ldots, x_n) \oplus S_{A_2}(x_1, \ldots, x_n)$.

12. Let S_4 be the group of 1-1 transformations on the set $\{1, 2, 3, 4\}$.
(a) How many elements does S_4 contain?
(b) Give a noncommutative subgroup of S_4 or show that none exists.

13. Give a characterization of symmetric monotonically increasing switching functions in terms of their a-numbers. Let $\mathcal{F}_n$ be the space of binary switching functions from $\{0, 1\}^n \longrightarrow \{0, 1\}$. How many symmetric functions in $\mathcal{F}_n$ are monotonically increasing? How many are unate?

14. Let $f_1(x_1, \ldots, x_n)$ and $f_2(x_1, \ldots, x_n)$ be totally symmetric functions and let $\phi(x_1, x_2)$ be totally symmetric. Then prove that $\phi(f_1(x_1, \ldots, x_n) \cdot f_2(x_1, \ldots, x_n))$ is totally symmetric in $x_1, \ldots, x_n$.

15. Let ϕ_1 be the function $\phi_1(x_1, x_2) = x_1 \cdot x_2$. Find the a-numbers of $\phi_1(h_1(x_1, \ldots, x_n), h_2(x_1, \ldots, x_n))$ in terms of the a-numbers of h_1 and h_2, where h_1 and h_2 are totally symmetric. Same for $\phi_2(x_1, x_2) = x_1 + x_2$.

16. Determine which of the following functions are symmetric using the simplified permutation method.
(a) $f(x_1, x_2, x_3, x_4) = x_1' x_2' + x_1' x_2 x_3' + x_1' x_3 x_4' + x_1 x_3' x_4' + x_1 x_2' x_3' + x_1 x_2' x_3 x_4'$
(b) $f(x_1, x_2, x_3, x_4) = x_1 x_3 + x_2 x_3' + x_1 x_2' x_4' + x_1' x_3' x_4 + x_2' x_3' x_4'$
(c) $f(x_1, x_2, x_3, x_4) = x_3 x_4' + x_1' x_3 + x_1 x_2' + x_3' x_4 + x_2 x_3' x_4'$
(d) $f(x_1, x_2, x_3, x_4) = x_1 x_2 x_3 x_4 + x_1 x_2' x_3' x_4' + x_1' x_2 x_3' x_4' + x_1' x_2' x_3 x_4' + x_1' x_2' x_3' x_4$

17. Repeat problem 16 by using the graphical pattern-recognition method.

18. Using the McCluskey method, determine which of the following are symmetric or mixed-symmetric functions.
(a) $f_1(x_1, x_2, x_3, x_4, x_5) = \Sigma(1, 2, 3, 4, 5, 6, 7, 8, 9, 10, 11, 12, 13, 14, 16, 17, 18, 19, 20,$ $21, 22, 24, 25, 26, 28)$
(b) $f_2(x_1, x_2, x_3, x_4, x_5) = \Sigma(0, 1, 3, 4, 5, 6, 8, 9, 10, 12, 14, 16, 17, 20, 23, 27)$
(c) $f_3(x_1, x_2, x_3, x_4, x_5) = \Sigma(1, 3, 8, 10, 13, 15, 16, 18, 21, 23, 28, 30)$
(d) $f_4(x_1, x_2, x_3, x_4, x_5, x_6) = \Sigma(0, 1, 2, 4, 8, 16, 31, 32, 47, 55, 59, 61, 62, 63)$
(e) $f_5(x_1, x_2, x_3, x_4, x_5, x_6) = \Sigma(2, 13, 21, 25, 28, 31, 32, 34, 35, 38, 42, 50, 61)$
(f) $f_6(x_1, x_2, x_3, x_4, x_5, x_6, x_7) = \Sigma(0, 63, 95, 111, 119, 123, 125, 126, 127)$
(g) $f_7(x_1, x_2, x_3, x_4, x_5, x_6, x_7) = \Sigma(8, 32, 40, 41, 42, 44, 56, 87, 104)$

19. How many mixed-symmetric functions of n variables are there?

20. Let $f(y_1, y_2)$ be a function of two variables y_1 and y_2, and let $S_{A_1}(x_1, \ldots, x_n)$ and $S_{A_2}(x_1, \ldots, x_n)$ be two symmetric functions of n variables. Show that

$$f(S_{A_1}(x_1, \ldots, x_n), \quad S_{A_2}(x_1, \ldots, x_n))$$

is a symmetric function.

21. Let $p(n)$ be the number of n variable switching functions which are partially symmetric in at least one pair of variables. Show that

$$\lim_{n \to \infty} \left[\frac{p(n)}{2^{2^n}} \right] = 0$$

4.4 Functionally Complete Functions

An important concept in switching theory is that of functional completeness.

DEFINITION 4.4.1

Let $\psi = \{\phi_1(x_1, \ldots, x_n), \ldots, \phi_m(x_1, \ldots, x_n)\}$ be a set of operations of n variables on B_2. The set ψ is said to be *functionally complete* if any switching function on B_2 is expressible as a *finite* composition of the ϕ_i's.

It has been shown that every switching function can be constructed from the variables $x_1, x_2, \ldots, x_n$ and the operations $+$, $\cdot$, and $'$. Moreover, by De Morgan's laws, it has been shown that only the complement operation plus either the OR or the AND operation are actually needed. We define the sets of operations

$$\psi_1 = \{+, '\} \quad \text{and} \quad \psi_2 = \{\cdot, '\}$$

as the *fundamental sets* of operations, which are, of course, functionally complete.

In proving that a given set of operations is functionally complete, all we need to prove is that they can realize either ψ_1 or ψ_2. This is illustrated by the following example.

Example 4.4.1

Show that the set of operations $\psi = \{\supset, \not\subset\}$ is functionally complete.

From Theorem 1.4.2, $x_1 \supset x_2 = x_1' + x_2$ and $x_1 \not\subset x_2 = x_1' \cdot x_2$. It is obvious that the fundamental set ψ_1 can be realized by these two operations:

$$x_1 \supset (x_1 \not\subset x_1) = x_1'$$
$$x_1' \subset x_2 = x_1 + x_2$$

and also, they can realize ψ_2:

$$x_1 \supset (x_1 \not\subset x_1) = x_1'$$
$$x_1' \not\subset x_2 = x_1 \cdot x_2$$

It should be clear that if a set of operations realizes ψ_1, it must also be able to realize ψ_2. Hence, $\psi = \{\supset, \nsubseteq\}$ is functionally complete.

Moreover, we can show that there are more pairs of operations of (a)–(h) of Theorem 1.4.2 which form functionally complete sets.

THEOREM 4.4.1

The following pairs of operations constitute functionally complete sets.

$$\psi_a = \{\supset, \nsubseteq\} \qquad \psi_e = \{\supset, \oplus\}$$
$$\psi_b = \{\subset, \nsupseteq\} \qquad \psi_f = \{\subset, \oplus\}$$
$$\psi_c = \{\supset, \nsupseteq\} \qquad \psi_g = \{\equiv, \nsubseteq\}$$
$$\psi_d = \{\subset, \nsubseteq\} \qquad \psi_h = \{\equiv, \nsupseteq\}$$

Since the proofs for ψ_b–ψ_h are similar to the proof of ψ_a which was given above, they are left to the reader as an exercise.

Before introducing the next theorem, we need the following definition.

DEFINITION 4.4.2

The *weight* of a switching function f is defined to be the number of elements of $\{0 \text{ (false)}, 1 \text{ (true)}\}^n$ which are mapped into 1 (true). The weight of f is denoted by $w(f)$.

Now we have

THEOREM 4.4.2

Given a switching function $f(x_1, x_2)$ of two variables and the negation operation, f is functionally complete iff $w(f)$ is odd.

Proof: (1) *The proof of Sufficiency:* Suppose that $w(f)$ is odd. With reference to Table 4.1.4, the functions with odd weights are

$$\vee, \downarrow, \wedge, |, \supset, \nsupseteq, \subset, \nsubseteq$$

It is clear that $\vee$ and $\wedge$, together with the negation operation, form ψ_1 and ψ_2. Hence, they are functionally complete, since, as easily seen from equations (a), (b), (c), (d), (g), and (h) of Theorem 1.4.2, that we can obtain either the operation $+$ or the operation $\cdot$. The $+$ operation may be obtained by $\supset$, $\subset$, and | together with the negation operation

$$(x_1' \supset x_2) = (x_1 + x_2)$$
$$(x_1 \subset x_2') = (x_1 + x_2)$$
$$(x_1' \,|\, x_2') = (x_1 + x_2)$$

and the $\cdot$ operation may be obtained by $\nsupseteq$, $\nsubseteq$, and $\downarrow$ together with the negation operation

$$(x_1 \nsupseteq x_2') = x_1 \cdot x_2$$

$$(x_1' \not\subset x_2) = x_1 \cdot x_2$$
$$(x_1' \downarrow x_2') = x_1 \cdot x_2$$

(2) *Proof of the Necessity:* Now suppose that f is functionally complete. By examining all the functions (operations) in Fig. 1.4.2(a) the functions with even weights can neither form the $+$ operation nor the $\cdot$ operation. This is obvious for the functions x_1, x_2, x_1', x_2', true, and false. But it is less obvious for the operations $\equiv$ and $\not\equiv$. The proofs for the latter are given as follows. First, let us prove that the $\equiv$ and the negation operation can never realize either the operation $+$ or the operation $\cdot$. It can easily be shown that there are *only eight distinct* functions that may be constructed by the operations $\equiv$ and the negation. They are†: $x_1 \equiv x_2, x_1 \oplus x_2, 1, 0, x_1, x_1', x_2$, and x_2'. It is seen that none of them has formed the $+$ operation. (None of them is odd.) Hence, the theorem is proved. ∎

COROLLARY 4.4.1

Given a function $f(x_1, x_2)$ of two variables with even weights and the negation operation, we can, at most, construct eight distinct functions.

Proof: From Fig. 1.4.2(a), there are eight functions of odd weights and eight functions of even weights. Now suppose that we can construct more than eight distinct functions. Then among them there must be at least one function which is of odd weight. This contradicts Theorem 4.4.2. Hence, the corollory is proved. ∎

COROLLARY 4.4.2

The functions obtained from even weights and the negation operation are always even.

Proof: By Theorem 4.4.2, the set of a function of even weight and the negation operation (which is even) are not functionally complete. Therefore, if a function of odd weight is obtained from operations of even weights, these operations, with the negation, will form a functionally complete set which contradicts Theorem 4.4.2. Hence, the corollary is proved. ∎

Furthermore, Theorem 4.4.2 can be extended to the case where n is greater than 2. First, we want to prove the following lemma.

LEMMA 4.4.1

For any function of n variables of odd weight, there exists at least one way of reducing it into a function of two variables of odd weight.

Proof: To illustrate the proof, let us first consider a function of four variables of odd weight which is mapped on a Karnaugh map as shown in Fig. 4.4.1(a), where the number of 1's is odd. Now if we *partition* the weights of this function into two disjoint submaps by the

†The reader who is interested in this result may verify it by constructing C_2^8 functions.

equivalence relation defined as

$$x_1 \, R \, x_2 \qquad \text{iff } x_1 = x_2 \qquad\qquad (4.4.1)$$

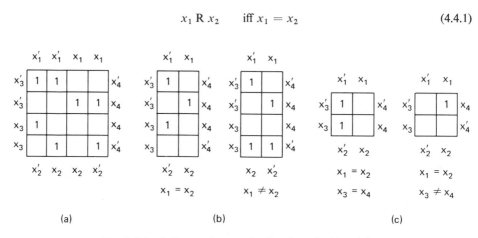

(a)   (b)   (c)

Fig. 4.4.1   A Karnaugh map of a function of odd weight.

then one of the two functions represented by these two submaps must be of odd weight, which
is shown in Fig. 4.4.1(b). This is because the weight of the original function is odd, and odd
can only be the sum of an odd integer and an even integer. We shall disregard the subfunction
of even weight and repartition the weight of the subfunction of odd weight by the relation
defined in Eq. (4.4.1) with respect to another pair of variables, say,

$$x_3 \, R \, x_4 \qquad \text{iff } x_3 = x_4$$

which is shown in Fig. 4.4.1(c). The subfunction corresponding to the rightmost submap is
a function of odd weight. By Theorem 4.4.2, the original function is functionally complete.

Now consider the general case where the function has n variables. Repeating this process
$(n-2)$ times, by the argument stated above, we can obtain a subfunction of two variables of
odd weight. Hence, the lemma is proved. ∎

As a direct consequence of the lemma, we have

THEOREM 4.4.3

(a)   Given a switching function $f(x_1, \ldots, x_n)$, with $n > 2$ and the negation
operation, f is functionally complete if $w(f)$ is odd.

(b)   There are functions of n variables, where $n > 2$, of even weights, which
together with the negation operation form functionally complete sets.

Proof: The proof of (a) follows immediately from the lemma.

In view of the fact that an even positive integer may be expressed as a sum of two odd
positive integers (see Fig. 4.4.2), then by Theorem 4.4.2, functions of even weights, together
with the negation operation, may form functionally complete sets. ∎

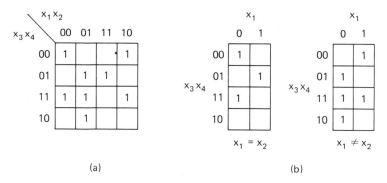

(a)   (b)

Fig. 4.4.2 An example showing a function with even weight can be partitioned into two subfunctions.

In the above we have shown that some pairs of functions form functionally complete sets. We may ask: Is it necessary to require two functions to form a functionally complete set? The answer is no. This is shown in the following.

LEMMA 4.4.2

The Pierce arrow (dagger function)

$$x_1 \downarrow x_2 = x_1' \cdot x_2'$$

is functionally complete, and so is the Sheffer stroke,

$$x_1 \mid x_2 = x_1' + x_2'$$

Proof: First, we prove that the Pierce arrow is functionally complete. If $x_1 = x_2$,

$$x_1 \downarrow x_1 = x_1' \cdot x_1' = x_1'$$

we obtain the complement operation. From

$$(x_1 \downarrow x_1) \downarrow (x_2 \downarrow x_2) = x_1' \downarrow x_2' = (x_1')' \cdot (x_2')' = x_1 \cdot x_2$$

we obtain the AND operation. Thus, it is functionally complete.

Similarly, for the Sheffer stroke,

$$x_1 \mid x_1' = x_1' + x_1' = x_1'$$

and

$$(x_1 \mid x_1) \mid (x_2 \mid x_2) = x_1' \mid x_2' = (x_1')' + (x_2')' = x_1 + x_2$$

Hence, it is functionally complete. ∎

Now we want to show an important theorem concerning the sufficient conditions for a single function forming a functionally complete set. First, we need the following definitions.

DEFINITION 4.4.3

The *dual* of a function $f(x_1, \ldots, x_n)$, denoted by $f_d(x_1, \ldots, x_n)$, is defined by

$$f_d(x_1, \ldots, x_n) = f(x_1', \ldots, x_n')$$

and the *complementary dual* of $f(x_1, \ldots, x_n)$, denoted by $h(x_1, \ldots, x_n)$, is defined by

$$h(x_1, \ldots, x_n) = (f_d(x_1, \ldots, x_n))' = (f(x_1', \ldots, x_n'))'$$

THEOREM 4.4.4

If $f(x_1, \ldots, x_n)$ is a function of n variables and
(1) $f(0, \ldots, 0) = 1$,
(2) $f(1, \ldots, 1) = 0$,
(3) f is not its own complementary dual,
then f is functionally complete.

Proof: Since f satisfies conditions (1) and (2), it is its own dual; hence, it realizes the complement operation.

Condition (3) implies that there exists at least one set of values of $(x_1, \ldots, x_n)$, say, $(\alpha_1, \ldots, \alpha_n)$, such that

$$(f_d(\alpha_1, \ldots, \alpha_n))' \neq (f(\alpha_1', \ldots, \alpha_n'))'$$

With loss of generality, let

$$(f_d(\alpha_1, \ldots, \alpha_n))' = 1 \Longrightarrow f_d(\alpha_1, \ldots, \alpha_n) = 0$$
$$(f(\alpha_1', \ldots, \alpha_n'))' = 0 \Longrightarrow f(\alpha_1', \ldots, \alpha_n') = 1$$

From conditions (1) and (2), we know that there is at least one $\alpha_i = 0$ and one $\alpha_j = 1$. Let $\alpha_{i_1}, \ldots, \alpha_{i_s}$ be the α's which are 0, and $\alpha_{j_1}, \ldots, \alpha_{j_t}$ be the α's which are 1. For instance,

$$f_d(1, 1, \ldots, 0, 0, \ldots, 1, 1, 1) = 0$$
$$f(0, 0, \ldots, 1, 1, \ldots, 0, 0, 0) = 1$$

Since the assignments of these n-variables are restricted to this form and f_d is the dual of f, f may be reduced to a function of two variables

$$\phi(x, y) = f(x, x, \ldots, y, y, \ldots, x, x, x)$$

The truth table of this function is shown in Fig. 4.4.3(a). Conditions (1) and (2) imply that $\phi(1, 1) = 0$ and $\phi(0, 0) = 1$. There are two possible ways to assign the β_1 and β_2 in the truth table of Fig. 4.4.3(a): $\beta_1 = \beta_2 = 1$ and $\beta_1 = \beta_2 = 0$, which are shown in Figs. 4.4.3(b) and (c). It should be noted that we cannot assign either $\beta_1 = 0$ and $\beta_2 = 1$ or $\beta_1 = 1$ and $\beta_2 = 0$, because either of these assignments would contradict condition (3).

The two truth tables of Fig. 4.4.3(b) and (c) indicate that one is the Sheffer stroke $(\,|\,)$ and the other is the Pierce arrow $(\downarrow)$. By Lemma 4.4.2, f is functionally complete. ∎

x	y	$\phi(x, y)$	x	y	$\phi(x, y)$	x	y	$\phi(x, y)$
1	1	0	1	1	0	1	1	0
1	0	β_1	1	0	1	1	0	0
0	1	β_2	0	1	1	0	1	0
0	0	1	0	0	1	0	0	1
		(a)			(b)			(c)

Fig. 4.4.3

Lastly, let us examine the EXCLUSIVE-OR operation, $\oplus$ which has the following properties:

(1) $x \oplus 0 = x$
(2) $x \oplus 1 = x'$
(3) $x \oplus x = 0$
(4) Commutative: $x_1 \oplus x_2 = x_2 \oplus x_1$
(5) Associative: $x_1 \oplus (x_2 \oplus x_3) = (x_1 \oplus x_2) \oplus x_3$

Property (2) indicates that the EXCLUSIVE-OR operation can perform the negation operation if a signal representing a constant 1 is available. Under the condition that a signal representing a constant 1 is available, the sets $\{\oplus, +\}$ and $\{\oplus, \cdot\}$ are complete.

Exercise 4.4

1. Prove or disprove the functional completeness of the following sets:
   (a) $\{x_1 \oplus x_2, x'\}$
   (b) $\{x_1 \oplus x_2, 1\}$
   (c) The set of all symmetric functions of three variables.
   (d) The set of all unate functions of two variables.

2. Which of the following statements are true? Which are false? Give your reasons.
   (a) The complement of a functionally complete function is always functionally complete.
   (b) The complement of a unate function is always unate.
   (c) The complement of a threshold function is always a threshold function.
   (d) All monotone-decreasing functions except 0 and 1 are functionally complete.
   (e) The dual of a functionally complete function is always functionally complete.

3. (a) How many functionally complete functions of n variables are there for $n \geq 2$?
   (b) How many n-variable functions are both monotone-increasing and monotone-decreasing?
   (c) How many functions of n variables are both functionally complete and monotonically increasing?
   (d) How many symmetric functions of n-variables are functionally complete?
   (e) How many n-variable symmetric functions are unate?
   (f) How many k-dimensional subcubes can be formed from the vertices of an n-cube?
   (g) What is the maximum number of prime implicants for an n-variable function?
   (h) How many nondegenerate functions of n variables are there?
   (i) How many nondegenerate symmetric functions of n variables are there?

4. Let $M(n)$ be the fraction of functions of n variables which are functionally complete. Evaluate $M(2)$, $M(3)$, $M(4)$, and $M(\infty)$.

5. Show that the following sets of functions are not functionally complete.
    (a) {true, false}
    (b) $\{\vee, \wedge\}$
    (c) $\{\equiv, \not\equiv\}$

6. Except for the three sets of functions of problem 5, examine carefully the 16 functions of Fig. 1.4.2(a) and list as many pairs of functions that are not functionally complete as you can.

7. Supply the proof for Theorem 4.4.1.

8. (a) A switching function is called *self-dual* if

    $$f(x_1, \ldots, x_n) = (f(x'_1, \ldots, x'_n))'$$

    List all the *self-dual* functions of two variables.
    (b) How many self-dual functions of n variables are there when n is odd?
    (c) How many self-dual functions of n variables are there when n is even?

9. Let $f(x_1, x_2) = x'_1 + x_2$. Show that f is functionally complete when the constants 0 and 1 are available as auxiliary inputs.

10. Is the function $f = x \longrightarrow y = x' + y$ complete; that is, can all functions be obtained using only f? Why?

11. Let n binary variables $x_1, x_2, \ldots, x_n$ be represented by binary signals appearing on n wires. Assume that two additional wires provide the constant signals 0 and 1. The only logical elements available are EXCLUSIVE-OR gates with two inputs.
    (a) How many different Boolean functions can be constructed? (The number of EXCLUSIVE-OR gates is not limited.)
    (b) Show all the functions that can be constructed in part (a) for the case $n = 2$.
    (c) Do the functions in part (b) form a subalgebra of the Boolean algebra of all functions of two variables? Prove your answer.

12. Let F_n be the set of all Boolean functions of n variables. Let E be a subset of F_n and let $\{E, \oplus\}$ be the set of all functions obtainable as the EXCLUSIVE-OR of a finite number of functions in E.
    (a) Show that $f \equiv 0$ (f is identically 0) is in $\{E, \oplus\}$ iff E is nonempty.
    (b) Give a set E having the minimum number of members such that $\{E, \oplus\}$ is the set of symmetric functions of n variables or prove that no such E exists.
    (c) Give a set E having the minimum number of members such that $\{E, \oplus\}$ is the set of monotonically increasing functions of n variables or prove that no such E exists.
    (d) Prove that the number of functions in $\{E, \oplus\}$ is a power of 2 provided that E is nonempty.

Bibliographical Remarks

Monotonic function is discussed in references 3 (Chapter 6) and 8. Threshold gates and properties and realizations of threshold functions are well documented in several books [2, 4–6]. Thus, only relevant and recent publications on the subject are included in these references. An extensive discussion on symmetric functions is contained in reference 1 (Chapter 7). References 20 and 21 are the original papers on the identification of symmetric functions.

References

Books

1. CALDWELL, S. H., *Switching Circuits and Logical Design*, McGraw-Hill, New York, 1959.

2. DERTOUZOS, M. L., *Threshold Logic*, MIT Press, Cambridge, Mass., 1965.

3. HARRISON, M. A., *Introduction to Switching and Automata Theory*, McGraw-Hill, New York, 1965.

4. HU, S. T., *Threshold Logic*, University of California Press, Berkeley, Calif., 1965.

5. LEWIS, P. M., II and C. L. COATES, *Threshold Logic*, Wiley, New York, 1967.

6. SHENG, C. L., *Threshold Logic*, Academic Press, New York, 1969.

7. WOOD, P. E., JR., *Switching Theory*, McGraw-Hill, New York, 1968.

Papers

Monotonic Functions

8. NCNAUGHTON, R., "Unate Truth Function," *IRE Trans. Electronic Computers*, Vol. EC-10, March 1961, pp. 1–6.

Threshold Functions

9. CHOW, C. K., "On the Characterization of Threshold Functions," *Switching Circuit Theory and Logical Design*, AIEE Spec. Publ. S134, 1961, pp. 34–38.

10. COATES, C. L., and P. M. LEWIS, "Linearly Separable Switching Functions," *J. Franklin Inst.*, Vol. 272, November 1961, pp. 360–410.

11. MUROGA, S., I. TODA, and S. TAKASU, "Theory of Majority Decision Elements," *J. Franklin Inst.*, Vol. 271, May 1961, pp. 376–418.

12. MUROGA, S., T. TSUBOI, and C. R. BAUGH, "Enumeration of Threshold Functions of Eight Variables," *IEEE Trans. Computers*, Vol. C-19, September 1970, pp. 818–825.

13. PAULL, M. C., and E. J. McCLUSKEY, "Boolean Functions Realizable with Single Threshold Devices," *Proc. IRE*, Vol. 48, No. 7, 1960, pp. 1335–1337.

14. WINDER, R. O., "Threshold Functions Through $n = 7$," *Air Force Cambridge Res. Lab. Rept.*, 1964, pp. 64–925.

Symmetric Functions

15. ARNOLD, R. F., and M. A. HARRISON, "Algebraic Properties of Symmetric and Partially Symmetric Boolean Functions," *IEEE Trans. Electronic Computers*, Vol. EC-12, June 1963, pp. 244–251.

16. BORN, R. C., and A. K. SCIDMORE, "Transformation of Switching Functions to Completely Symmetric Switching Functions," *IEEE Trans. Computers*, Vol. C-17, June 1968, pp. 596–599.

17. DAHLBERY, B., "On Symmetric Functions with Redundant Variables—Weighted Functions," *IEEE Trans. Computers*, Vol. C-22, May 1973, pp. 450–458.

18. DAS, S. R., and C. L. SHENG, "On Detecting Total or Partial Symmetry of Switching Functions," IEEE Trans. Computers, Vol. C-20, March 1971, pp. 352–355.

19. DIETMEYER, D. L., and P. R. SCHNEIDER, "Identification of Symmetric, Redundancy and Equivalence of Boolean Functions," *IEEE Trans. Electronic Computers*, Vol. EC-16, December 1967, pp. 804–817.

20. MARCUS, M. P., "The Detection and Identification of Symmetric Switching Functions with the Use of Tables of Combinations," *IRE Trans. Electronic Computers*, Vol. EC-5, December 1956, pp. 237–239.

21. McCLUSKEY, E. J., JR., "Detection of Group Invariance or Total Symmetry of a Boolean Function," *Bell System Tech. J.*, Vol. 35, No. 6, November 1956, pp. 1445–1453.

22. MUKHOPADHYAY, A., "Detection of Total or Partial Symmetry of a Switching Function with the Use of Decomposition Charts," *IEEE Trans. Electronic Computers*, Vol. EC-12, October 1963, pp. 553–557.

23. SHENG, C. L., "Detection of Totally Symmetric Boolean Functions," *IEEE Trans. Electronic Computers*, Vol. EC-14, December 1965, pp. 924–926.

24. YAU, S. S., and Y. S. TANG, "On Identification of Redundancy and Symmetry of Switching Functions," *IEEE Trans. Computers*, Vol. C-20, December 1971, pp. 1609–1613.

25. YAU, S. S., and Y. S. TANG, "Transformation of an Arbitrary Switching Function to a Totally Symmetric Function," *IEEE Trans. Computers*, Vol. C-20, December 1971, pp. 1606–1609.

5

Multivalued Logic

Vector switching algebra, discussed in Chapter 2, is, in a sense, a type of multi-valued logic system, since every element of it is an n-dimensional binary vector which corresponds to a (logic) value between 0 and $2^n - 1$. Hence, it is a 2^n-valued logic system.

In this chapter, two additional types of multivalued logic system are introduced. The first type is the general *m-valued logic* (m need not be equal to 2^n). Most definitions and theorems about two-valued special switching functions discussed in the previous chapter may be extended to the m-valued case. However, even for special functions, the multivalued functions in general are far more difficult to analyze than the two-valued functions, with the exception of symmetric functions. Because of their symmetry property, the representation and realization of multivalued symmetric functions are simple; and therefore it is convenient to use it whenever it is applicable.

The second type of multivalued logic system is the *m-class logic*. This logic is derived from the fuzzy logic, which is an infinitely many valued logic. In this logic the AND and OR operations are defined by the minimum and maximum operations, respectively. The analysis and realization of m-class logic functions are discussed in detail.

5.1 *m*-Valued Logic Functions

Just as a two-valued switching function, a multivalued logic function can be described by either a truth table or by a function that is a union of sets of functions. This is best described by use of an example. Consider a simple example of a three-variable three-valued logic function described by the truth table of Table 5.1.1. There are 27 rows in this table, each of which is assigned a truth value 0, 1, or 2. We can therefore divide them into three groups and denote them by 0-group, 1-group, and 2-group as shown in the table. This table may be represented by an expression using the decimal representations of the rows and the union operation:

$$f(x_1, x_2, x_3) = \Sigma_1(1, 4, 5, 6, 11, 12, 13, 14, 19, 24) \cup \Sigma_2(2, 3, 7, 17, 18, 21, 25, 26)$$

TABLE 5.1.1   A Typical Multivalued Logic Function Described by a Truth Table

Decimal representation	Ternary representation x_1 x_2 x_3	$f(x_1, x_2, x_3)$
0	0 0 0	0
1	0 0 1	1
2	0 0 2	2
3	0 1 0	2
4	0 1 1	1
5	0 1 2	1
6	0 2 0	1
7	0 2 1	2
8	0 2 2	0
9	1 0 0	0
10	1 0 1	0
11	1 0 2	1
12	1 1 0	1
13	1 1 1	1
14	1 1 2	1
15	1 2 0	0
16	1 2 1	0
17	1 2 2	2
18	2 0 0	2
19	2 0 1	1
20	2 0 2	0
21	2 1 0	2
22	2 1 1	0
23	2 1 2	0
24	2 2 0	1
25	2 2 1	2
26	2 2 2	2

(0-group, 1-group, 2-group)

with the understanding that the remaining rows 0, 8, 9, 10, 15, 16, 20, 22, and 23 are mapped to the truth value 0.

It should be clear that the truth table of an n-variable, m-valued function contains m^n rows and that there are m^{m^n} different functions. This number increases rapidly with the increase of m. For instance, for two variables (i.e., $n = 2$) there are only $2^{2^2} = 16$ two-valued functions, but there are $3^{3^2} = 19{,}683$ three-valued functions. For m-valued ($m > 2$) functions, they are not only large in number, but also of widely different algebraic structure. This is the reason why the multivalued ($m > 2$) functions are in general much more difficult to analyze than the two-valued functions, with the exception of the class of multivalued symmetric functions.

DEFINITION 5.1.1

Let $f(x_1, \ldots, x_n)$ be an n-variable m-valued logic function. f is *symmetric* if

$$f(x_1, \ldots, x_i, \ldots, x_j, \ldots, x_n) = f(x_1, \ldots, x_j, \ldots, x_i, \ldots, x_n) \quad (5.1.1)$$

for all x_i and x_j.

From the definition above, for a given function, we need to check $C_2^n = n(n-1)/2$ functions. However, only n tests,

$$f(x_1, x_2, \ldots, x_n) = f(x_2, x_1, \ldots, x_{n-1}, x_n)$$
$$f(x_1, x_2, x_3, \ldots, x_n) = f(x_3, x_2, x_1, \ldots, x_{n-1}, x_n)$$
$$\cdot$$
$$\cdot \qquad\qquad\qquad\qquad\qquad\qquad (5.1.2)$$
$$\cdot$$
$$f(x_1, x_2, \ldots, x_n) = f(x_n, x_2, \ldots, x_{n-1}, x_1)$$

suffice. Moreover, we can prove that

THEOREM 5.1.1

A function $f(x_1, \ldots, x_n)$ is *symmetric* iff it satisfies the following conditions:
(a) $f(x_1, x_2, x_3, \ldots, x_n) = f(x_2, x_1, x_3, \ldots, x_n)$
(b) $f(x_1, x_2, \ldots, x_n) = f(x_1, x_3, x_4, \ldots, x_n, x_2)$ $\qquad$ (5.1.3)

The proof is similar to that of the corresponding theorem of two-valued symmetric switching function (Theorem 4.3.5).

DEFINITION 5.1.2

For each assignment, we define an **a**-*vector* $\mathbf{a} = (a_0, a_1, \ldots, a_{m-1})$ associated with this assignment, where a_i is the number of i's in the assignment.

THEOREM 5.1.2

A function is *symmetric* iff the values of the function for all the assignments having the same **a**-vector are the same.

Proof: Suppose that $f(x_1, x_2, \ldots, x_n)$ is an m-valued symmetric function. Consider any assignment of the function

$$f(\underbrace{0, 0, \ldots, 0}_{a_0}, \underbrace{1, 1, \ldots, 1}_{a_1}, \ldots, \underbrace{(m-1), (m-1), \ldots, (m-1)}_{a_{m-1}}) = t_i$$

where $\sum_{i=0}^{m-1} a_i = n$. By definition of symmetry, the value of the function $f(x_1, x_2, \ldots, x_n)$ remains the same for any interchange of variables x_i and x_j. This implies that the values of the function for all the assignments having the same **a**-vector are the same. Conversely, suppose that the values of $f(x_1, \ldots, x_i, \ldots, x_j, \ldots, x_n)$ for all the assignments having the same **a**-vector are the same. Suppose that we now interchange two variables, x_i and x_j, namely, $f(x_1, \ldots, x_j, \ldots, x_i, \ldots, x_n)$. Since the **a**-vectors of all the assignments of $f(x_1, \ldots, x_i,$ $\ldots, x_j, \ldots, x_n)$ are the same as those of $f(x_1, \ldots, x_j, \ldots, x_i, \ldots, x_n)$ and since the values of $f(x_1, \ldots, x_i, \ldots, x_j, \ldots, x_n)$ for all the assignments having the same **a**-vector are the same, $f(x_1, \ldots, x_i, \ldots, x_j, \ldots, x_n) = f(x_1, \ldots, x_j, \ldots, x_i, \ldots, x_n)$ for all x_i and x_j. By Definition 5.1.1, f is symmetric. ∎

Example 5.1.1

Now let us consider a simple example of two-variable three-valued symmetric function described in Table 5.1.2. The function is symmetric because $f(x_1, x_2) = f(x_2, x_1)$. Observe

TABLE 5.1.2 A 2-Variable 3-Valued Symmetric Function

Decimal representation	x_1	x_2	a-vector	$f(x_1, x_2)$
0	0	0	(2, 0, 0)	0
1	0	1	(1, 1, 0) ⎤ 1	1
2	0	2	(1, 0, 1) ⎤	2
3	1	0	(1, 1, 0) ⎦	1
4	1	1	(0, 2, 0) ⊢ 2	0
5	1	2	(0, 1, 1) ⎤	2
6	2	0	(1, 0, 1) ⎦ 2	2
7	2	1	(0, 1, 1) ⎦	2
8	2	2	(0, 0, 2)	0

0-group, 1-group, 2-group

that the values of this function for all the assignments having the same a-vector are the same. This function may be represented by

$$f(x_1, x_2) = \Sigma_1(1, 3) \cup \Sigma_2(2, 5, 6, 7)$$

The following theorem concerns the number of symmetric functions of n variables.

THEOREM 5.1.3

There are $m^{C_{m-1}^{n+m-1}}$ n-variable m-valued symmetric functions.

Proof: The number of distinct symmetric functions is equal to the number of different a-vectors which is calculated by C_{m-1}^{n+m-1}. The proof is given as follows: An a-vector has m components

$$(a_0, a_1, \ldots, a_{m-1})$$

and each component a_i denotes the number of variables assigned with truth value i and thus can be $0, 1, \ldots,$ or n. The sum of all a_i's must be equal to n. First, let us consider the simple case $n = 1$. For m-valued logic, it is clear that the different a-vectors are

$$(1, 0, \ldots, 0)$$
$$(0, 1, \ldots, 0)$$
$$\cdots\cdots\cdots$$
$$(0, 0, \ldots, 1)$$

For the case $n = 2$, the different a-vectors are

$$(2, 0, \ldots, 0)$$
$$(0, 2, \ldots, 0)$$
$$\cdots\cdots\cdots$$
$$(0, 0, \ldots, 2)$$

the a-vectors corresponding to the rows of the truth table with two variables having the same truth value

$$
\left.
\begin{array}{l}
(1, 1, 0, \ldots, 0) \\
(1, 0, 1, \ldots, 0) \\
\cdots\cdots\cdots \\
(1, 0, 0, \ldots, 1) \\
(0, 1, 1, \ldots, 0) \\
\cdots\cdots\cdots \\
(0, 0, \ldots, 1, 1)
\end{array}
\right\}
$$
the **a**-vectors corresponding to the rows of the truth table with two variables having different truth values

If we represent a^i of each **a**-vector using *unary* representation, then the first group of **a**-vectors becomes

$$
\overbrace{n\,1\text{'s}}\;\overbrace{(m-1)\,0\text{'s}}
$$
$$
(11, 0, \ldots, 0)
$$
$$
(0, 11, \ldots, 0)
$$
$$
\cdots\cdots\cdots
$$
$$
(0, 0, \ldots, 11)
$$

The two examples above illustrate that the number of different **a**-vectors for n variables and m-valued logic may be seen as the number of different $(m-1)$-tuples whose components are unary that can be formed from n 1's and $(m-1)$ 0's, which is readily seen to be C_{m-1}^{n+m-1}. There are m ways to assign values to each **a**-vector; hence, there are $m^{C_{m-1}^{n+m-1}}$ different functions. ∎

For example, there are $3^{C_2^4} = 729$ two-variable three-valued symmetric functions.

Because of their symmetry property, the representation, realization, and computational complexity in time of symmetric functions are simple compared to nonsymmetric functions, and because of these advantages of symmetric functions, it is often desirable to know whether a given function is symmetric or not. An algorithm is presented in this section for identifying symmetric and mixed symmetric functions. First we define

DEFINITION 5.1.3

Define the operation $\sim x$ as follows:

x	$\sim x$
0	1
1	2
2	3
.	.
.	.
.	.
$m-2$	$m-1$
$m-1$	0

and define $\sim^{(k)} x = \underbrace{\sim(\sim(\ldots(\sim x)\ldots))}_{k \text{ times}}$.

DEFINITION 5.1.4

A function $f(x_1, \ldots, x_n)$ is *mixed-symmetric* if there exist integers $k_1, k_2, \ldots,$ k_n, $0 \leq k_i \leq m - 1$, where not all k_i are equal to 0, such that

$$f(\sim^{(k_1)} x_1, \ldots, \sim^{(k_n)} x_n)$$

is symmetric.

DEFINITION 5.1.5

Define the *j-group* as the set of all the rows of the truth table which are mapped to the truth value j. For symmetric functions, this set is equivalent to the set of rows with the same **a**-vector, which are mapped to the truth value j.

DEFINITION 5.1.6

For each *j*-group we define a *ratio vector* or simply **r**-*vector* for each column i, $\mathbf{r}_i^{(j)} = (r_{i0}^{(j)}, r_{i1}^{(j)}, \ldots, r_{im-1}^{(j)})$, where $r_{il}^{(j)}$ is the number of l's in the ith column in the jth group.

DEFINITION 5.1.7

Define $M_{\mathbf{a}_i}^{(j)}$ to be the number of rows of a truth table having the same $\mathbf{a}_i^{(j)}$-vector and in the same *j*-group.

The algorithm is as follows:

1. Determine *simultaneously* the n **r**-vectors for the n columns for each group.
2. For each *j*-group ($j = 0, 1, \ldots, m - 1$), arbitrarily choose an $\mathbf{r}^{(j)}$-vector, say $\mathbf{r}_1^{(j)}$. If $\mathbf{r}_1^{(j)} = \mathbf{r}_i^{(j)}$ for $i = 2, 3, \ldots, n$ and $j = 0, 1, \ldots, m - 1$, then the given function is a candidate for a symmetric function. (Otherwise, it is not a symmetric function.)
3. Next, examine the equation

$$M_{\mathbf{a}_i^{(j)}} = C_{a_0^{(j)}}^n C_{a_1^{(j)}}^{n - a_0^{(j)}} \ldots C_{a_{m-1}^{(j)}}^{n - \sum\limits_{k=0}^{m-2} a_k^{(j)}}$$

(which gives the number of the same $\mathbf{a}_i^{(j)}$ as it should be if the given function is symmetric) and see whether it holds for all $\mathbf{a}_i^{(j)}$ and $j = 0, 1, \ldots, m - 1$.

4. If in 3, the answer is yes, then it is a symmetric function.
5. If in 3, the answer is no, then fix the first column, replace the second column of x_2 by that of $\sim x_2$ and repeat 1–3.
6. Repeat 5 for all possible combinations of $\sim^{(k_2)} x_2, \sim^{(k_3)} x_3, \ldots, \sim^{(k_n)} x_n$, $0 \leq k_i \leq m - 1$ and see whether there is a set of $x_1, \sim^{(k_2)} x_2, \sim^{(k_3)} x_3, \ldots, \sim^{(k_n)} x_n$ for which 2 and 3 are satisfied. If so, it is a mixed-symmetric function; otherwise, it is not a mixed-symmetric function.

Example 5.1.2

Identify whether the two-variable three-valued function

$$f(x_1, x_2) = \Sigma_1(1, 3) \cup \Sigma_2(2, 5, 6, 7)$$

is a symmetric function.

The algorithm for identifying a symmetric function is as follows: ($n = 2, m = 3$)

1. Find the **r**-vectors. They are shown in Table 5.1.3.

TABLE 5.1.3 An Example of Symmetric Function

Decimal representation	Ternary representation x_1 x_2	$a_i^{(j)} = (a_1^{(i)}, a_2^{(i)}, a_3^{(i)})$	$M_{a_i}(j)$ (No. of the same $a_i^{(i)}$ in j-group)	$M_{a_i}(j) = C^n_{a_0^{(i)}} C^{n-a_0^{(i)}}_{a_1^{(i)}} C^{n-a_0^{(i)}-a_1^{(i)}}_{a_2^{(i)}}$
0-group ⎰ 0	0 0	$a_1^{(0)} = (2, 0, 0)$	1	$C^2_2 \, C^{2-2}_0 \, C^{2-2}_0 = 1$
0-group ⎱ 4	1 1	$a_2^{(0)} = (0, 2, 0)$	1	$C^2_0 \, C^{2-0}_2 \, C^{2-2}_0 = 1$
8	2 2	$a_3^{(0)} = (0, 0, 2)$	1	$C^2_0 \, C^2_0 \, C^2_2 = 1$
r-vector	(1, 1, 1) (1, 1, 1)			
1-group ⎰ 1	0 1	$a_1^{(1)} = (1, 1, 0)$	\} 2	$C^2_1 \, C^{2-1}_1 \, C^{2-1-1}_0 = 2$
1-group ⎱ 3	1 0	$a_2^{(1)} = (1, 1, 0)$		
r-vector	(1, 1, 0) (1, 1, 0)			
2	0 2	$a_1^{(2)} = (1, 0, 1)$	\} 2	$C^2_1 \, C^{2-1}_0 \, C^{2-1}_1 = 2$
2-group ⎰ 5	1 2	$a_2^{(2)} = (0, 1, 1)$		
2-group ⎱ 6	2 0	$a_3^{(2)} = (1, 0, 1)$	\} 2	$C^2_0 \, C^{2-0}_1 \, C^{2-1}_1 = 2$
7	2 1	$a_4^{(2)} = (0, 1, 1)$		
r-vector	(1, 1, 2) (1, 1, 2)			

2. See whether all the **r**-vectors of each *j*-group are the same. If so, it is a candidate for a symmetric function.

3. Calculate $M_{a_i}(j)$. This is shown in the fifth column of the table.

4. Compare $M_{a_i}(j)$ obtained from the function (in the fourth column) with $M_{a_i}(j)$ calculated (in the fifth column) and see whether they agree. For this example, we see that it is so; hence, the function is symmetric.

Example 5.1.3

Consider a two-variable three-valued function

$$f(x_1, x_2) = \Sigma_1(2, 4) \cup \Sigma_2(0, 3, 7, 8)$$

The **r**-vectors are found and shown in the first column of the Table 5.1.4. It is seen that the **r**-vectors of the 1-group and 2-group are not the same. So it is not a symmetric function.

TABLE 5.1.4 Example of Mixed-Symmetric Functions

	x_1	x_2	a	x_1	$\sim x_2$	a
0-group	0	1	(1, 1, 0)	0	2	(1, 0, 1)
	1	2	(0, 1, 1)	1	0	(1, 1, 0)
	2	0	(1, 0, 1)	2	1	(0, 1, 1)
r-vector	(1, 1, 1)	(1, 1, 1)		(1, 1, 1)	(1, 1, 1)	
1-group	0	2	(1, 0, 1)	0	0	(2, 0, 0)
	1	1	(0, 2, 0)	1	2	(0, 1, 1)
r-vector	(1, 1, 0)	(0, 1, 1)		(1, 1, 0)	(1, 0, 2)	
2-group	0	0	(2, 0, 0)	0	1	(1, 1, 0)
	1	0	(1, 1, 0)	1	1	(0, 2, 0)
	2	1	(0, 1, 1)	2	2	(0, 0, 2)
	2	2	(0, 0, 2)	2	0	(1, 0, 1)
r-vector	(1, 1, 2)	(2, 1, 1)		(1, 1, 2)	(1, 2, 1)	

	x_1	$\sim(\sim x_2)$	a
	0	0	(2 0, 0)
	1	1	(0, 2, 0)
	2	2	(0, 0, 2)
r-vector	(1, 1, 1)	(1, 1, 1)	
	0	1	(1, 1, 0)
	1	0	(1, 1, 0)
r-vector	(1, 1, 0)	(1, 1, 0)	
	0	2	(1, 0, 1)
	1	2	(0, 1, 1)
	2	0	(1, 0, 1)
	2	1	(0, 1, 1)
r-vector	(1, 1, 2)	(1, 1, 2)	

Now we want to determine whether it is a mixed-symmetric function. By replacing x_2 by $\sim x_2$, which is shown in the second column, we obtain a new set of r-vectors. We find that the condition in the second step of the algorithm is still not met; we then try $\sim(\sim x_2)$, which is shown in the third column. This time we find that the condition in the second step of the algorithm is satisfied. This shows that $f(x_1, \sim(\sim x_2))$ is a candidate of a symmetric function. A close observation indicates that this function is the same as that in Example 5.1.2, which has been shown to be symmetric. Therefore, $f(x_1, x_2)$ is a mixed-symmetric function.

It was pointed out that the values of the function of two assignments with the same a-vector are the same. This property of symmetric functions permits us to greatly reduce the size of the truth table (m^n rows) of the function to an equivalent truth table whose rows are the C_{m-1}^{n+m-1} a-vectors (see the proof of Theorem 5.1.3), as shown in Table 5.1.5.

TABLE 5.1.5 Condensed Truth Table Using a-Vectors

	a-*number*	$f(x_1, \ldots, x_n)$
0-group	$\mathbf{a}_1^{(0)}$ . . . $\mathbf{a}_{l_0}^{(0)}$	0 . . . 0
1-group	$\mathbf{a}_1^{(1)}$ . . . $\mathbf{a}_{l_1}^{(1)}$	1 . . . 1
	. . .	. . .
(*m* − 1)-group	$\mathbf{a}_1^{(m-1)}$ . . . $\mathbf{a}_{l_{m-1}}^{(m-1)}$	*m* − 1 . . . *m* − 1

For example, the function of Example 5.1.1 may be represented by using **a**-vectors as follows:

a-vector	$f(x_1, x_2)$
(2, 0, 0)	0
(0, 2, 0)	0
(0, 0, 2)	0
(1, 1, 0)	1
(1, 0, 1)	2
(0, 1, 1)	2

The *reduction factor*, r.f., of the size of the truth table by such a representation is

$$\text{r.f.} = \frac{C_{m-1}^{n+m-1}}{m^n} = \frac{(n+m-1)!}{(m-1)!\,n!\,m^n}$$

For any fixed finite *N*,

$$\lim_{m \to \infty} \text{r.f.} = 0$$

and likewise,

$$\lim_{n \to \infty} \text{r.f.} = 0$$

for any fixed finite *M*. Furthermore, r.f. is monotonically decreasing as *n* and *m* increase.

This may be shown by examining the ratios

$$\frac{\text{r.f.}\,(m = M, n = N + 1)}{\text{r.f.}\,(m = M, n = N)} = \frac{N + M}{(N + 1)\cdot M} \le 1$$

and

$$\frac{\text{r.f.}\,(m = M + 1, n = N)}{\text{r.f.}\,(m = M, n = N)} = \frac{N + M}{M\cdot(1 + 1/M)^N} \le 1$$

for any positive integers M and N. The values of r.f. for some values of n and m are tabulated in Table 5.1.6. From Table 5.1.6, it is seen that the size of the truth table is

TABLE 5.1.6   Values of r.f. for Some Values of m and n

m	n	$r.f.\,(m, n)$
3	2	0.67
3	3	0.37
4	3	0.31
3	4	0.19
4	4	0.14
4	5	0.055
5	5	0.044

reduced greatly as m and n increase.

Exercise 5.1

1. Give a (nontrivial) example of multivalued symmetric function with
    (a) $n = 2$ and $m = 3$ (d) $n = 4$ and $m = 3$
    (b) $n = 3$ and $m = 3$ (e) $n = 4$ and $m = 4$
    (c) $n = 3$ and $m = 4$ (f) $n = 5$ and $m = 5$

2. Prove the following three-variable three-valued function:

$$f(x_1, x_2, x_3) = \Sigma_1(1, 3, 5, 7, 9, 11, 13, 14, 15, 16, 19, 21, 22)$$
$$\cup\ \Sigma_2(0, 2, 4, 6, 10, 12, 18, 26)$$

    is a symmetric function using
    (a) Eq. (5.1.1)
    (b) Eq. (5.1.2)
    (c) Eq. (5.1.3)

3. How many n-variable m-valued mixed-symmetric functions are there?

4. Which of the following two-variable three-valued functions are symmetric? Which are mixed-symmetric?

(a) $\Sigma_1(1, 3) \cup \Sigma_2(2, 5, 6, 7)$

(b) $\Sigma_1(0, 2, 6) \cup \Sigma_2(1, 3)$

(c) $\Sigma_1(2, 4, 6) \cup \Sigma_2(5, 7, 8)$

(d) $\Sigma_1(0, 2, 4, 5, 7) \cup \Sigma_2(3, 6, 8)$

(e) $\Sigma_1(0, 1, 5, 8) \cup \Sigma_2(4, 6)$

(f) $\Sigma_1(0, 2, 5) \cup \Sigma_2(6, 8)$

5. Which of the following three-variable three-valued functions are symmetric? Which are mixed-symmetric?

(a) $\Sigma_1(0, 4, 10, 12, 13, 14, 16, 22, 26) \cup \Sigma_2(1, 3, 5, 7, 9, 11, 15, 17, 19, 21, 23, 25)$

(b) $\Sigma_1(0, 2, 6, 11, 15, 17) \cup \Sigma_2(4, 8, 9, 13, 19, 21, 22, 23, 25)$

(c) $\Sigma_1(11, 15, 17, 18, 20, 24, 26) \cup \Sigma_2(1, 3, 5, 7, 10, 12, 14, 16)$

(d) $\Sigma_1(1, 3, 4, 5, 7, 13, 17, 18, 22) \cup \Sigma_2(0, 1, 6, 8, 10, 12, 14, 16, 19, 21, 23, 25)$

6. Any *n*-variable *m*-valued symmetric function can be represented by the union of the functions

$$S^{(0)}_{\{a_1^{(0)}, a_2^{(0)}, \ldots, a_{t_0}^{(0)}\}}(x_1, \ldots, x_n)$$

$$S^{(1)}_{\{a_1^{(1)}, a_2^{(1)}, \ldots, a_{t_1}^{(1)}\}}(x_1, \ldots, x_n)$$

$$\ldots \ldots \ldots \ldots \ldots \ldots \ldots \ldots \ldots \ldots \ldots$$

$$S^{(m-1)}_{\{a_1^{(m-1)}, a_2^{(m-1)}, \ldots, a_{t_{m-1}}^{(m-1)}\}}(x_1, \ldots, x_n)$$

It is clear that only $m - 1$ among them are linearly independent. Represent the symmetric and mixed-symmetric functions of problems 4 and 5 by the **a**-vector representation shown above.

7. By properly assigning values to the "don't cares," show that the following two-variable three-valued functions can be made symmetrical.

(a) $\Sigma_1(0, 8) \cup \Sigma_2(2, 6) \cup \Sigma_d(4, 5, 7)$

(b) $\Sigma_1(1, 3, 5, 6) \cup \Sigma_2(4, 8) \cup \Sigma_d(2, 7)$

(c) $\Sigma_1(0, 1, 2) \cup \Sigma_2(4, 8) \cup \Sigma_d(3, 5, 6, 7)$

(d) $\Sigma_1(0, 1, 3, 4) \cup \Sigma_2(8) \cup \Sigma_d(2, 5, 6, 7)$

8. Assign the values of "don't cares" in each of the following three-variable three-valued functions so that it becomes symmetric.

(a) $\Sigma_1(1, 3, 9, 11, 15) \cup \Sigma_2(2, 6, 17) \cup \Sigma_d(5, 7, 18, 19, 21, 23, 25)$

(b) $\Sigma_1(0, 1, 2, 3) \cup \Sigma_2(4, 5, 6) \cup \Sigma_d(6, 9, 10, 12, 18, 19, 23, 25, 8, 20, 24)$

(c) $\Sigma_1(1, 3, 9, 16, 22) \cup \Sigma_2(5, 7, 11, 15, 17, 23) \cup \Sigma_d(0, 13, 14, 19, 21, 25, 26)$

(d) $\Sigma_1(0, 7, 13, 25) \cup \Sigma_2(1, 8, 14, 23) \cup \Sigma_d(3, 5, 9, 11, 14, 15, 16, 17, 19, 20, 21, 22, 24, 25)$

5.2 Fuzzy Logic and *m*-Class Logic

In ordinary set theory, the sets considered are abstract sets which are defined as collections of objects having some very general property P; nothing special is assumed or considered about the nature of the individual objects. For example, we define a set A as the set of cars. Symbolically, $A = \{x \,|\, x \text{ is a car}\}$.

Now what about the "class of *new* cars"? First, is it a set in the ordinary sense?

Before we answer, we may first ask: How "new" is a new car? Is a one-year-old car a new car? If so, then, is there any difference between a half-year-old new car and one-year-old new car? And so on. Frankly, we do not know how to adequately answer these questions from the information "*new* cars." Because the class of "new cars" does not possess a *sharply defined* boundary, it does not constitute a set in the usual sense. Sets of this nature very often involve some not sharply defined adjectives, verbs, adverbs, and some combination thereof in their descriptions. For example:

  1. The class of *short* men. (Am I a member of this set?)
  2. The class of *high* buildings. (Is the United Nations Building a member of this set?)
  3. The class of all real numbers which are *much* greater than 1. (Is 25 a member of this set?)

Numerous other examples may be found in almost every branch of science and engineering, as well as in writings and daily conversations. In fact, most of the classes of objects encountered in the real world are of this *fuzzy, not sharply defined* type. They do not have precisely defined criteria of membership. In such classes, an object need not necessarily either belong to or not belong to a class; there may be intermediate grade of membership. This is the concept of a fuzzy set, which is a "class" with a *continuum* of grades of membership.

DEFINITION 5.2.1

A *fuzzy* set A in the object space X is characterized by a membership (characteristic) *function* with respect to certain properties of x of interest, $p_1, p_2, \ldots, p_n$, denoted by $f_A(x = (p_1, p_2, \ldots, p_n))$, which is a *functional* mapping from the property space defined by the object space X into the interval [0, 1]. The value of $f_A(x = (p_1, p_2, \ldots, p_n))$ at x represents the *grade* of *membership* of x in A which ranges from 0 to 1, where 0 and 1 indicate nonmembership and full membership, respectively.

A fuzzy set is *empty* iff its membership function is identically zero on X. A fuzzy set is *universal* iff its membership function is identically unity on X. Two fuzzy sets A and B are *equal* iff $f_A(x) = f_B(x)$ for all x in X.

DEFINITION 5.2.2

The *union* of two fuzzy sets A and B with respective membership functions $f_A(x)$ and $f_B(x)$ is a fuzzy set C, written as $C = A \cup B$, whose membership function is related to those of A and B by

$$f_C(x) = \max [f_A(x), f_B(x)], \qquad x \in X$$

The *intersection* of A and B, written as $D = A \cap B$, is a fuzzy set D whose membership function is

$$f_D(x) = \min [f_A(x), f_B(x)], \qquad x \in X$$

The *complement* of A, written A', is a fuzzy set whose membership function is

$$f_{A'}(x) = 1 - f_A(x), \qquad x \in X$$

Based on the ordinary (or abstract) set theory, the two-valued logic is derived in which every variable can take on the value of 1 or 0 and every proposition that can be described by it must be of yes-or-no type or all-or-none type. For the same reasons cited above, in the physical world this assumption is not always adequate, for the attributes of the system variable are often ambiguously and subjectively defined. Such problems arise in the fields of pattern classification, information processing, control, system identification, artificial intelligence, and, more generally, decision processes involving incomplete or ill-defined data.

In contrast to ordinary logic (two-valued or multivalued logic) in which a given proposition is ascribed *objectively* using either *deterministic* or *probabilistic* approaches, fuzzy logic deals with propositions which may *subjectively* be ascribed values between falsehood and truth (zero and one) either in a *continuous* or a *discrete* fashion. A nonfuzzy statement is as follows:

If John passes both examination *A* and examination *B*, then the company he works with will pay his tuition iff the company has money available and his boss approves it.

An appropriate logical analysis of this statement might begin with the symbolism

$$a \wedge b \supset (p \equiv m \wedge f) \tag{5.2.1}$$

where the letters denote the following statements:

a: John passes examination *A*.   *m*: The company has money available.
b: John passes examination *B*.   *f*: His boss approves it.
p: The company he works with will pay his tuition.

and the logic connectives $\wedge$, $\supset$, and $\equiv$ denote the conjunction, conditional, and equivalence, respectively.

The proposition in Eq. (5.2.1) can be synthesized by an ordinary logic circuit. However, if the statement above is changed to

If John passes both examination *A* and examination *B satisfactorily*, the company he works with will pay a *large* portion of his tuition iff the company has a *large* amount of uncommitted money and his boss approves it without *too much* reservation.

it becomes a fuzzy statement. Even though the logical expression of this statement is identical to that of the previous statement, namely Eq. (5.2.1), the basic building blocks, as well as the method of synthesizing this expression, are completely different.

In contrast to ordinary Boolean algebra, we define a fuzzy Boolean algebra as follows:

DEFINITION 5.2.3

A *fuzzy algebra* is an algebraic system $(B, +, \cdot, ')$, which consists of (1) a fuzzy set B (the grade of membership of every member of B is between 0 and 1); (2) two binary operations $+$ and $\cdot$, which are defined as: If x and y are in B, $x + y = \max (x, y)$ and $x \cdot y = \min (x, y)$; and (3) a unary operation $'$ defined as: If x is in B, $x' = 1 - x$.

Let B be a fuzzy algebra and x, x_1, x_2, and x_3 be elements in B. From the above definition, we immediately have the following properties about the three basic fuzzy operations:

$$x + x = x \qquad\qquad (x_1 + x_2)' = x_1' x_2' \text{ (De Morgan's law)}$$
$$x \cdot x = x \qquad\qquad (x_1 x_2)' = x_1' + x_2' \text{ (De Morgan's law)}$$
$$(x_1 x_2) x_3 = x_1 (x_2 x_3) \qquad\qquad x + x' = \max (x, x') \qquad\qquad (5.2.2)$$
$$(x_1 + x_2) + x_3 = x_1 + (x_2 + x_3) \qquad\qquad x \cdot x' = \min (x, x')$$
$$x_1 (x_2 + x_3) = x_1 x_2 + x_1 x_3 \qquad x_1 (x_1 + x_2) = x_1$$

We define

DEFINITION 5.2.4

Let B be a fuzzy algebra. *A fuzzy logic function*, or simply *fuzzy function*, defined on B is a mapping from B to B using exclusively the three operations $+$, $\cdot$, and $'$.

For example,

$$f(x_1, x_2, x_3) = x_1 x_2' x_3 + x_1 x_2' x_3 x_3' + x_1' x_2 x_2'$$

is a fuzzy function which may be simplified as

$$f(x_1, x_2, x_3) = x_1 x_2' x_3 (1 + x_3') + x_1' x_2 x_2'$$
$$= x_1 x_2' x_3 + x_1' x_2 x_2'$$
$$= x_2' (x_1 x_3 + x_1' x_2)$$

It should be noted that terms such as $x_1 x_2' x_3 x_3'$ and $x_1' x_2 x_2'$ which reduce to zero in ordinary two-valued logic are not necessarily zero in fuzzy logic. Similarly, factors such as $x_1 (x_2 + x_2')$ cannot, in general, be reduced to x_1. However, the following inequalities hold for any fuzzy variable x,

$$x + x' \geq 0.5$$
$$x \cdot x' \leq 0.5 \qquad\qquad (5.2.3)$$

When these inequalities are applied to the function

$$f(x_1, x_2) = x_1 x_1' x_2 + x_1 x_1' x_2'$$

we obtain the following simple function,

$$f(x_1, x_2) = x_1 x_1' (x_2 + x_2') = x_1 x_1'$$

In summary, by utilizing the above equalities and inequalities, fuzzy functions can generally be simplified in a similar manner as ordinary switching functions. For example,

$$
\begin{aligned}
f(x_1, x_2, x_3) &= x_1 x_2' x_3 + x_1 x_1' x_3' + x_1 x_2' + x_1 x_2 + x_1 x_1' x_2 \\
&\quad + x_1 x_2 x_3 + x_1 x_1' x_3 \\
&= x_1 x_1' (x_3' + x_3) + x_1 x_2' (x_3 + 1) + x_1 x_2 (1 + x_1' + x_3) \\
&= x_1 x_1' + x_1 x_2' + x_1 x_2
\end{aligned}
$$

If for some reason $x_2 \geq x_1$, then the above function can be further simplified as

$$
\begin{aligned}
f(x_1, x_2, x_3) &= x_1 x_1' + x_1 x_2' + x_1 \\
&= x_1 x_1' + x_1 (x_2' + 1) \\
&= x_1 x_1' + x_1 \\
&= x_1 (x_1' + 1) \\
&= x_1
\end{aligned}
$$

It is seen from the above example that certain conditions on the fuzzy variables may be used very advantageously in the simplification of fuzzy functions. Such restrictions on the variables may be introduced intentionally in some cases without impairing the practical usefulness of the function.

The fuzzy function defined above does not provide us with a "decision mechanism" in the sense that a two-valued (or *m*-valued) logic function does with its "one–zero," "yes–no," or "true–false" states. To know, for instance, that a fuzzy function, $f(x_1, x_2)$, has the value $f(x_1, x_2) = 0.8$ does not have much significance unless it is related to some outcome which is dependent upon the fact that $f(x_1, x_2) = 0.8$. A possible way for introducing this quality into a fuzzy system is by reintroducing the abandoned concept of "belonging to a set" but in a rather qualified sense. That is, we agree to have the continuous range of membership grades [0, 1] subdivided into a finite number of classes in the following manner:

$$
\begin{array}{ll}
\text{class } 1: & \alpha_1 < x \leq 1 \\
\text{class } 2: & \alpha_2 < x \leq \alpha_1 \\
\quad \cdot & \quad \cdot \\
\quad \cdot & \quad \cdot \\
\quad \cdot & \quad \cdot \\
\text{class } m: & 0 \leq x \leq \alpha_{m-1}
\end{array}
\tag{5.2.4}
$$

where $1 > \alpha_1 > \alpha_2 > \ldots > \alpha_{m-1} > 0$. A fuzzy variable or a fuzzy function can now be identified with one of these *m* classes according to the value they assume in the closed region [0, 1]. This subdivision of the region [0, 1] into a finite number of classes enables one to utilize the properties of *m*-valued logic in the treatment of fuzzy logic systems. For example, for $m = 3$,

$$\text{class 1:}\quad \alpha_1 < x \leq 1$$
$$\text{class 2:}\quad \alpha_2 < x \leq \alpha_1 \qquad\qquad (5.2.5)$$
$$\text{class 3:}\quad 0 \leq x \leq \alpha_2$$

We may now ascribe certain meanings to the different classes. For instance, we may assume that an object x, (a) belongs to a set if it is in "class 1," (b) does not belong to a set if it is in "class 3," and (c) its status remains undecided if it is in "class 2."

DEFINITION 5.2.5

An *m-class algebra* is a fuzzy algebra (Definition 5.2.3), except the variable values are specified by "classes" rather than by specific values.

DEFINITION 5.2.6

An *m-class logic function* is a fuzzy logic function (Definition 5.2.4), except whose variable values and function value are specified by "classes" rather than by specific values.

The next two sections are devoted to the analysis and realization of *m*-class logic functions.

5.3 Analysis of *m*-Class Logic Functions

In the analysis of *m*-class logic functions, we consider two functional representations: the "sum-of-products" and "product-of-sums" forms. Any other functional form would be a combination of these two forms.

A. Sum-of-Products Form

The analysis of *m*-class logic functions of sum-of-products form is best illustrated by use of examples.

Example 5.3.1

Consider a simple function

$$f(x_1, x_2, x_3) = x_1 + x_2 + x_3' \qquad\qquad (5.3.1)$$

In analyzing the function, one is interested in determining the conditions that must be satisfied by fuzzy variables x_1, x_2, and x_3 in order for the function to belong to a certain class M as defined in Eq. (5.2.4).

Let $f(x_1, x_2, x_3)$ belong to class M or, equivalently, let

$$a_M \leq f(x_1, x_2, x_3) < a_{M-1} \qquad\qquad (5.3.2)$$

which is

$$x_1 + x_2 + x_3' \geq a_M \tag{5.3.3a}$$

and

$$x_1 + x_2 + x_3' < a_{M-1} \tag{5.3.3b}$$

From Eq. (5.3.3a), which is

$$\max(x_1, x_2, x_3') \geq a_M$$

we have

$$x_1 \geq a_M$$

or

$$x_2 \geq a_M$$

or

$$x_3' \geq a_M, \qquad x_3 \leq 1 - a_M$$

From Eq. (5.3.3b), which is

$$\max(x_1, x_2, x_3') < a_{M-1}$$

we have

$$x_1 < a_{M-1}$$

and

$$x_2 < a_{M-1}$$

and

$$x_3' < a_{M-1}, \qquad x_3 > 1 - a_{M-1}$$

In summary, in order for the function of Eq. (5.3.1) to belong to the class $a_M \leq f(x_1, x_2, x_3)$ $< a_{M-1}$, x_1, x_2, and x_3 must satisfy the following conditions:

$$\text{group 1} = [x_1 \geq a_M \text{ or } x_2 \geq a_M \text{ or } x_3 \leq 1 - a_M] \tag{5.3.4a}$$

and

$$\text{group 2} = [x_1 < a_{M-1} \text{ and } x_2 < a_{M-1} \text{ and } x_3 > 1 - a_{M-1}] \tag{5.3.4b}$$

where group 1 is the set of conditions on x_1, x_2, and x_3 so that $f(x_1, x_2, x_3) \geq a_M$, and group 2 is the set of conditions on x_1, x_2, and x_3 so that $f(x_1, x_2, x_3) < a_{M-1}$. Note that group 1 and group 2 are dual in nature.

Example 5.3.2

Suppose that the function is

$$f(x_1, x_2, x_3) = x_1' x_2 x_3' + x_1 x_2' x_3 \tag{5.3.5}$$

We want to find the conditions that must be satisfied by the variables x_1, x_2, and x_3 so that

$a_M \leq f(x_1, x_2, x_3) < a_{M-1}$. From the condition $a_M \leq f(x_1, x_2, x_3)$, we immediately have

$$\text{group 1} = \left[\left\{ \begin{matrix} x_1 \leq 1 - a_M \\ \text{and } x_2 \geq a_M \\ \text{and } x_3 \leq 1 - a_M \end{matrix} \right\} \quad \text{or} \quad \left\{ \begin{matrix} x_1 \geq a_M \\ \text{and } x_2 \leq 1 - a_M \\ \text{and } x_3 \geq a_M \end{matrix} \right\} \right] \qquad (5.3.6a)$$

and from the condition

$$x_1' x_2 x_3' + x_1 x_2' x_3 < a_{M-1}$$

by taking the complement of both sides of this inequality and applying De Morgan's law, we get

$$(x_1 + x_2' + x_3) \cdot (x_1' + x_2 + x_3') > 1 - a_{M-1}$$

which means that

$$\min \left[(x_1 + x_2' + x_3), \quad (x_1' + x_2 + x_3') \right] > 1 - a_{M-1}$$

That is,

$$x_1 + x_2' + x_3 > 1 - a_{M-1}$$

and

$$x_1' + x_2 + x_3' > 1 - a_{M-1}$$

From these two inequalities, we obtain

$$\text{group 2} = \left[\left\{ \begin{matrix} x_1 > 1 - a_{M-1} \\ \text{or } x_2 < a_{M-1} \\ \text{or } x_3 > 1 - a_{M-1} \end{matrix} \right\} \quad \text{and} \quad \left\{ \begin{matrix} x_1 < a_{M-1} \\ \text{or } x_2 > 1 - a_{M-1} \\ \text{or } x_3 < a_{M-1} \end{matrix} \right\} \right] \qquad (5.3.6b)$$

B. Product-of-Sums Form

Next, consider the product-of-sums form.

Example 5.3.3

Consider a simple function of product-of-sums form,

$$f(x_1, x_2, x_3) = (x_1 + x_2' + x_3)(x_1' + x_2 + x_3') \qquad (5.3.7)$$

Following the same procedure presented in A, we have that in order for $f(x_1, x_2, x_3)$ to belong to class M or, equivalently,

$$a_M \leq f(x_1, x_2, x_3) < a_{M-1}$$

the variables x_1, x_2, and x_3 must satisfy the conditions given as

$$\text{group 1} = \left[\left\{\begin{array}{l} x_1 \geq a_M \\ \text{or } x_2 \leq 1 - a_M \\ \text{or } x_3 \geq a_M \end{array}\right\} \text{ and } \left\{\begin{array}{l} x_1 \leq 1 - a_M \\ \text{or } x_2 \geq a_M \\ \text{or } x_3 \leq 1 - a_M \end{array}\right\}\right] \qquad (5.3.8a)$$

and

$$\text{group 2} = \left[\left\{\begin{array}{l} x_1 < a_{M-1} \\ \text{and } x_2 > 1 - a_{M-1} \\ \text{and } x_3 < a_{M-1} \end{array}\right\} \text{ or } \left\{\begin{array}{l} x_1 > 1 - a_{M-1} \\ \text{and } x_2 < a_{M-1} \\ \text{and } x_3 > 1 - a_{M-1} \end{array}\right\}\right] \qquad (5.3.8b)$$

where groups 1 and 2 are defined as before.

From the above examples, we find that not only are groups 1 and 2 dual in nature, but the conditions on the variables for the two forms are also dual in nature, as seen from Eqs. (5.3.7) and (5.3.8). More specifically, the rules in constructing sets of conditions groups 1 and 2 *directly* from the function, which is either of the sum-of-products form or the product-of-sums form, may be stated as follows. For both forms:

1. The variables of group 1 associated with inequalities directed to the left appear in their complemented form in the term they enter, while those associated with inequalities directed to the right appear in their uncomplemented form.

2. The variables of group 2 associated with inequalities directed to the left appear in their uncomplemented form in the term they enter, while those associated with inequalities directed to the right appear in their complemented form.

3. The operations "and" and "or" represent the logic operations of intersection and union, respectively, when associated with group 1. The same operations are logically reversed if associated with group 2. Thus, in the latter case, "or" and "and" mean intersection and union, respectively.

Example 5.3.4

We want to find the conditions that must be satisfied by the variables x_1, x_2, x_3, and x_4 of the function

$$f(x_1, x_2, x_3, x_4) = x_1' x_2 (x_3' + x_4) + (x_1' + x_2) \cdot (x_3 + x_4') \qquad (5.3.9)$$

so that $a_M \leq f(x_1, x_2, x_3, x_4) < a_{M-1}$. This function obviously does not belong to either one of the two forms examined previously. But the sets of conditions, groups 1 and 2 can still be obtained by directly applying the rules stated above. The result is

$$\text{group 1} = \left[\left\{\begin{array}{l} x_1 \leq 1 - \alpha_M \\ \text{and } x_2 \geq \alpha_M \\ \text{and } (x_3 \leq 1 - \alpha_M \text{ or } x_4 \leq 1 - \alpha_M) \end{array}\right\}\right.$$
$$\left. \text{or } \left\{\begin{array}{l} (x_1 \leq 1 - \alpha_M \text{ or } x_2 \geq \alpha_M) \\ \text{and } (x_3 \geq 1 - \alpha_M \text{ or } x_4 \leq 1 - \alpha_M) \end{array}\right\}\right] \qquad (5.3.10a)$$

and

$$\text{group 2} = \left[\left[\begin{cases} x_1 > 1 - \alpha_{M-1} \\ \text{or } x_2 < \alpha_{M-1} \\ \text{or } (x_3 > 1 - \alpha_{M-1} \text{ and } x_4 > 1 - \alpha_{M-1}) \end{cases}\right.\right.$$

$$\text{and } \begin{cases} (x_1 > 1 - \alpha_{M-1} \text{ and } x_2 < \alpha_{M-1} \\ \text{or } (x_3 < 1 - \alpha_{M-1} \text{ and } x_4 > 1 - \alpha_{M-1}) \end{cases}\right]\right] \tag{5.3.10b}$$

Exercise 5.3

1. Minimize the following m-class logic functions:
   (a) $f(x_1, x_2, x_3) = x_1 x_2 x_3 + x_1 x_2 + x_1 x_3' + x_2$
   (b) $f(x_1, x_2, x_3) = x_1 x_2 x_2' x_3 + x_1 x_2 + x_1 x_2' x_3 + x_1 x_2 x_2'$
   (c) $f(x_1, x_2, x_3, x_4) = x_1 x_2' x_3 x_4' + x_2 x_3 x_3' x_4 + x_3 x_4 x_4' + x_2 x_3 + x_1 x_2' + x_3 x_4' + x_4 x_4'$
   (d) $f(x_1, x_2, x_3, x_4) = x_2 x_3 x_3' x_4 x_4' + x_1' x_3 x_4' + x_1 x_2' x_4 + x_2 x_3 x_3' x_4 + x_1 x_4 + x_1' x_4'$
2. Simplify the functions above under the condition $x_1 \geq x_2 \geq x_3 \geq x_4$.
3. Find the sets of conditions, groups 1 and 2, for each of the following m-class logic functions to belong to the class $a_M \leq f < a_{M-1}$.
   (a) $f(x_1, x_2, x_3) = (x_1' + x_2 + x_3')(x_1 + x_3)$
   (b) $f(x_1, x_2, x_3) = x_1 x_2' + x_2 x_3' + x_3 x_1'$
   (c) $f(x_1, x_2, x_3, x_4) = x_1 x_2 x_3 (x_1' + x_2' + x_3') + x_1 x_3 + x_2' x_4'$
   (d) $f(x_1, x_2, x_3, x_4) = x_2 x_4' (x_1 + x_2' x_3 + x_4 x_4')(x_1' + x_2' + x_3 x_4)$

5.4   Realization of m-Class Logic Functions

Before discussing realization of m-class logic functions, first let us examine the electronic implementation of the three m-class logic operations. The implementation of these operations can be easily carried out by utilizing, for instance, diode transistor logic. The operations $x_1 \cdot x_2$ and $x_1 + x_2$ have exactly the same electronic implementation as the AND and OR operations of two-valued logic. The logic operation complementation is easily implemented by making use of a transistor inverter as shown in Fig. 5.4.1. In designing the inverter, care must be exercised to ensure that a set of input–output characteristics can be approximated fairly closely by biasing the transistor properly. It is now apparent that when $V_x = 0$, which corresponds to a membership grade of $g = 0$, the output voltage $V_0 = V_{cc} - V_x$, which is equivalent to $x' = 1 - x$.

These three operations can also be implemented digitally using the circuitry shown in Fig. 5.4.2(b). This circuitry has been designed for realizing fuzzy AND-, OR-, and NOT-gates which will perform the max-, min-, and not-operations on two digitized n-bit (for simplicity, we use 4 bits) binary variables. This device performs fuzzy gating operations, hence the name "fuzzy gate" or "maxmin gate." It is designed to determine

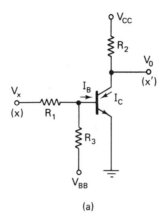

(a)

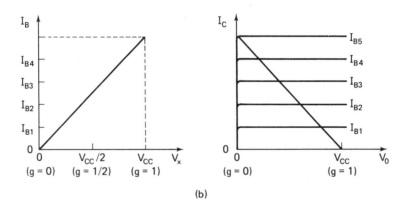

(b)

Fig. 5.4.1   Electronic implementation of the fuzzy not-operation.

which of the two binary inputs is larger and to gate out the maximum or minimum of the two or the inverse of one, depending on the control setting [Fig. 5.4.2(a)]. Table 5.4.1 shows which input will be gated given the control inputs. This is accomplished

TABLE 5.4.1

G_0	G_1	Output (C)
0	1	max (A, B)
1	1	min (A, B)
$\times$	0	$1 - A$

Note: $\times$ denotes don't care.

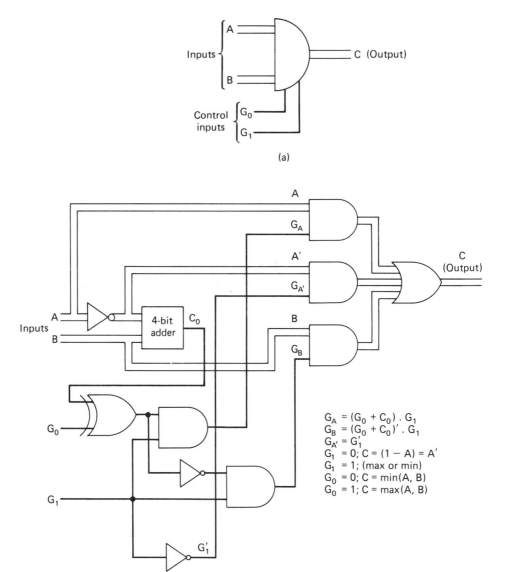

$$G_A = (G_0 + C_0) \cdot G_1$$
$$G_B = (G_0 + C_0)' \cdot G_1$$
$$G_{A'} = G_1'$$
$$G_1 = 0; C = (1 - A) = A'$$
$$G_1 = 1; \text{(max or min)}$$
$$G_0 = 0; C = \min(A, B)$$
$$G_0 = 1; C = \max(A, B)$$

Fig. 5.4.2 A digital implementation of maxmin gate.

by the use of a 4-bit binary adder, one of whose inputs is directly added to the complement of the other. The carry from this adder is then used with the control inputs to determine which input should be output, and gating is appropriately carried out.

The realization problem involves determination of fuzzy function describing a process from a set of specifications such as those given in Eq. (5.3.4), (5.3.6), (5.3.8), or (5.3.10).

Example 5.4.1

A certain process with attributes of interest $x_1, x_2,$ and x_3 is recognizable when its describing function $f(x_1, x_2, x_3) \geq \alpha_1$. Obtain the describing function, $f(x_1, x_2, x_3)$, if $f(x_1, x_2, x_3) \geq \alpha_1$, when

$$\left\{ \begin{array}{l} x_1 \geq \alpha_1 \\ \text{and } x_2 \leq 1 - \alpha_1 \\ \text{and } x_3 \geq \alpha_1 \end{array} \right\} \quad \text{or} \quad \left\{ \begin{array}{l} x_1 \geq \alpha_1 \\ \text{and } x_2 \geq \alpha_1 \\ \text{and } x_3 \leq 1 - \alpha_1 \end{array} \right\} \quad \text{or} \quad \left\{ \begin{array}{l} x_1 \leq 1 - \alpha_1 \\ \text{and } x_2 \leq 1 - \alpha_1 \\ \text{and } x_3 \leq 1 - \alpha_1 \end{array} \right\} \quad (5.4.1)$$

The required function is easily obtained by utilizing the procedure described above in reverse. The function is

$$f(x_1, x_2, x_3) = x_1 x_2' x_3 + x_1 x_2 x_3' + x_1' x_2' x_3' \qquad (5.4.2)$$

The realization of this function is shown in Fig. 5.4.3, in which the symbol $x \rightarrow \textcircled{\alpha_1} \rightarrow z$ represents a threshold gate whose operation is $z = +1$ and -1 if $x \geq \alpha_1$ and $x < \alpha_1$, respectively. (The general threshold gate is described in Section 4.2.) If we let the two classes be denoted by

$$\text{class } 1 = \{(x_1, x_2, x_3) | f(x_1, x_2, x_3) \geq \alpha_1\}$$

and

$$\text{class } 2 = \{(x_1, x_2, x_3) | f(x_1, x_2, x_3) < \alpha_1\}$$

then (x_1, x_2, x_3) belongs to class 1 if $z = +1$; otherwise, it belongs to class 2.

Example 5.4.2

If $f(x_1, x_2, x_3) \geq \alpha_1$ when

$$\left\{ \begin{array}{l} x_1 \geq \alpha_{11} \\ \text{and } x_2 \leq \alpha_{12} \end{array} \right\} \quad \text{or} \quad \left\{ \begin{array}{l} x_1 \leq \alpha_{12} \\ \text{and } x_2 \geq \alpha_{13} \end{array} \right\} \quad \text{or} \quad \left\{ \begin{array}{l} x_1 \geq \alpha_{11} \\ \text{and } x_2 \geq \alpha_{13} \\ \text{and } x_3 \leq \alpha_{14} \end{array} \right\} \quad (5.4.3)$$

find the expression $f(x_1, x_2, x_3)$.

The essential difference between the above example and this one lies in the fact that the ranges of interest for $x_1, x_2,$ and x_3 are not necessarily related to the critical value α_1. Instead, arbitrary values such as $\alpha_{11}, \alpha_{12}, \alpha_{13},$ etc., were selected. This, however, does not pose any problems if we keep in mind that an inequality of the type $x_1 \geq \alpha_{1N}$ for its implementation

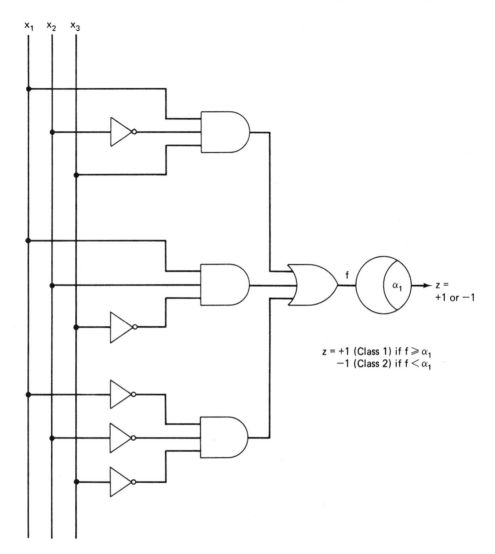

Fig. 5.4.3 The realization of fuzzy function of Eq. (5.4.2).

will depend on α_1, while one of the form $x_1 \leq \alpha_{1M}$ will be dependent on $(1 - \alpha_1)$. This, then, implies that α_{1N} and α_{1M} must be properly multiplied (amplified) to attain the levels α_1 and $(1 - \alpha_1)$, respectively, prior to using them in an actual implementation. The required function is

$$f(x_1, x_2, x_3) = (k_{11}x_1)\cdot(k_{12}x_2') + (k_{12}x_1')\cdot(k_{13}x_2) + (k_{11}x_1)\cdot(k_{13}x_2)\cdot(k_{14}x_3') \qquad (5.4.4)$$

where k_{ij} represents the multiplication factor to be associated with the various variables. More specifically,

k_{11} in $(k_{11}x_1)$ is equal to $\dfrac{\alpha_1}{\alpha_{11}}$

k_{12} in $(k_{12}x_2')$ is equal to $\dfrac{1 - \alpha_1}{\alpha_{12}}$

k_{12} in $(k_{12}x_1')$ is equal to $\dfrac{1 - \alpha_1}{\alpha_{12}}$ $\qquad$ (5.4.5)

k_{13} in $(k_{13}x_2)$ is equal to $\dfrac{\alpha_1}{\alpha_{13}}$

k_{14} in $(k_{14}x_3')$ is equal to $\dfrac{1 - \alpha_1}{\alpha_{14}}$

Figure 5.4.4 shows the electronic implementation of the function of Eq. (5.4.4).

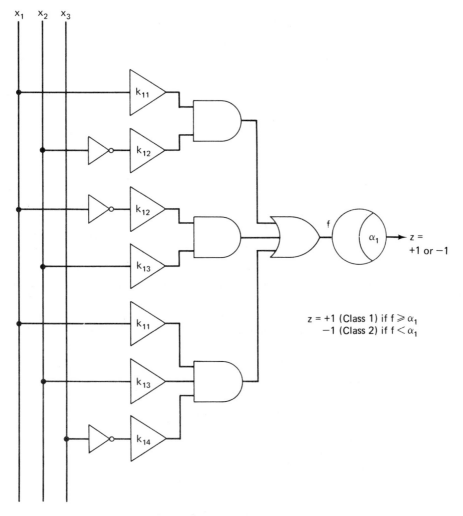

Fig. 5.4.4   The realization of the function of Eq. (5.4.4).

Example 5.4.3

Synthesize and implement electronically the function $f(x_1, x_2, x_3)$ so that $a_M \leq f(x_1, x_2, x_3) < a_{M-1}$ when the variables x_1, x_2, and x_3 satisfy the conditions:

$$\text{group 1} = \left[\left\{ \begin{array}{l} x_1 \leq \alpha_1 \\ \text{and } x_3 \geq \alpha_2 \end{array} \right\} \quad \text{or} \quad \left\{ \begin{array}{l} x_1 \geq \alpha_3 \\ \text{and } x_2 \geq \alpha_4 \\ \text{and } x_3 \leq \alpha_5 \end{array} \right\} \right] \tag{5.4.6a}$$

and

$$\text{group 2} = \left[\left\{ \begin{array}{l} x_1 > \alpha_6 \\ \text{or } x_3 < \alpha_7 \end{array} \right\} \quad \text{and} \quad \left\{ \begin{array}{l} x_1 < \alpha_8 \\ \text{or } x_2 < \alpha_9 \\ \text{or } x_3 > \alpha_{10} \end{array} \right\} \right] \tag{5.4.6b}$$

It should be noted that groups 1 and 2 are dual in nature as defined earlier. Making use of either group 1 or group 2, along with the analytical rules established above, we have that

$$f(x_1, x_2, x_3) = x_1' x_3 + x_1 x_2 x_3' \tag{5.4.7}$$

Since the choice of the constants α_k was arbitrary, the electronic implementation of the function of Eq. (5.4.7) cannot be carried out in the usual conventional manner. One must first properly scale the various fuzzy variables and then develop the terms $x_1' x_3$ and $x_1 x_2 x_3'$. Proper scaling is necessary in order for the fuzzy system to offer a response, that is, $a_M \leq f(x_1, x_2, x_3) < a_{M-1}$, whenever the conditions of Eqs. (5.4.6a) and (5.4.6b) are satisfied. The scaling factors sought must be such that Eqs. (5.4.6a) and (5.4.6b) become

$$\text{group 1} = \left[\left\{ \begin{array}{l} x_1 \leq 1 - a_M \\ \text{and } x_3 \geq a_M \end{array} \right\} \quad \text{or} \quad \left\{ \begin{array}{l} x_1 \geq a_M \\ \text{and } x_2 \geq a_M \\ \text{and } x_3 \leq 1 - a_M \end{array} \right\} \right] \tag{5.4.8a}$$

and

$$\text{group 2} = \left[\left\{ \begin{array}{l} x_1 > 1 - a_{M-1} \\ \text{or } x_3 < a_{M-1} \end{array} \right\} \quad \text{and} \quad \left\{ \begin{array}{l} x_1 < a_{M-1} \\ \text{or } x_2 < a_{M-1} \\ \text{or } x_3 > 1 - a_{M-1} \end{array} \right\} \right] \tag{5.4.8b}$$

Thus, the scaling factors for the variables in groups 1 and 2 are as follows:

Group 1:

$$\alpha_1 k_{11} = 1 - a_M \quad \text{or} \quad k_{11} = \frac{1 - a_M}{\alpha_1}$$

$$\alpha_2 k_{12} = a_M \quad \text{or} \quad k_{12} = \frac{a_M}{\alpha_2}$$

$$\alpha_3 k_{13} = a_M \quad \text{or} \quad k_{13} = \frac{a_M}{\alpha_3} \tag{5.4.9a}$$

$$\alpha_4 k_{14} = a_M \quad \text{or} \quad k_{14} = \frac{a_M}{\alpha_4}$$

$$\alpha_5 k_{15} = 1 - a_M \quad \text{or} \quad k_{15} = \frac{1 - a_M}{\alpha_5}$$

Group 2:

$$\alpha_6 k_{21} = 1 - a_{M-1} \quad \text{or} \quad k_{21} = \frac{1 - a_{M-1}}{\alpha_6}$$

$$\alpha_7 k_{22} = a_{M-1} \quad \text{or} \quad k_{22} = \frac{a_{M-1}}{\alpha_7}$$

$$\alpha_8 k_{23} = a_{M-1} \quad \text{or} \quad k_{23} = \frac{a_{M-1}}{\alpha_8} \qquad (5.4.9b)$$

$$\alpha_9 k_{24} = a_{M-1} \quad \text{or} \quad k_{24} = \frac{a_{M-1}}{\alpha_9}$$

$$\alpha_{10} k_{25} = 1 - a_{M-1} \quad \text{or} \quad k_{25} = \frac{1 - a_{M-1}}{\alpha_{10}}$$

In electronic parlance, the scale factors k_{ij} are known as analog multipliers. Each variable must be scaled by its respective scale factor prior to being operated upon logically. Thus, in forming the product $x_1' x_3$ as required by group 1 of Eq. (5.4.6a), one must first multiply x_1 and x_3 by k_{11} and k_{12}, respectively, subsequently complement x_1, and then form the product $x_1' x_3$; similarly, for the term $x_1 x_2 x_3'$. A more detailed description of the electronic implementation of this system is given in Fig. 5.4.5. It is clear from Fig. 5.4.5 that

$$z_1 = \begin{cases} +1 & \text{if } f \geq a_M \\ -1 & \text{if } f < a_M \end{cases}$$

$$z_2 = \begin{cases} +1 & \text{if } f \geq a_{M-1} \\ -1 & \text{if } f < a_{M-1} \end{cases}$$

$$z = \begin{cases} +1 & \text{if both } z_1 = +1 \text{ and } z_2' = +1 \text{ (i.e., } a_M \leq f < a_{M-1}) \\ -1 & \text{otherwise} \end{cases}$$

That is, an output $+1$ is generated iff $a_M \leq f(x_1, x_2, x_3) < a_{M-1}$. Note that the two threshold elements T_1 and T_2 are adjusted so that they respond when $f(x_1, x_2, x_3) \geq a_M$ and $f(x_1, x_2, x_3) < a_{M-1}$, respectively, except in the case in which $a_{M-1} = 1$. In that case, threshold element T_2 is not needed, and thus may be removed.

Class selection may be varied by varying the boundaries a_M and a_{M-1} or, equivalently, the settings on the threshold elements T_1 and T_2. These changes may be carried out either manually or automatically through program control. It is also conceivable that one might outline a set of n specifications such as those given in Eqs. (5.4.6a) and (5.4.6b), in which case the previous synthesis procedure must be repeated n times.

In the above example, groups 1 and 2 are dual in nature. Duality in this example simply implies that groups 1 and 2 are from the same logical function. However, it is not necessary that this be so. The conditions on the fuzzy variables expressed by group 1 may be satisfied by $f_1 \geq a_M$ and those of group 2 by $f_2 \leq a_{M-1}$, where f_1 and f_2 are different functions. The functions f_1 and f_2 in such a case would be the system describing functions, and their electronic implementation is analogous to that given in Fig. 5.4.6.

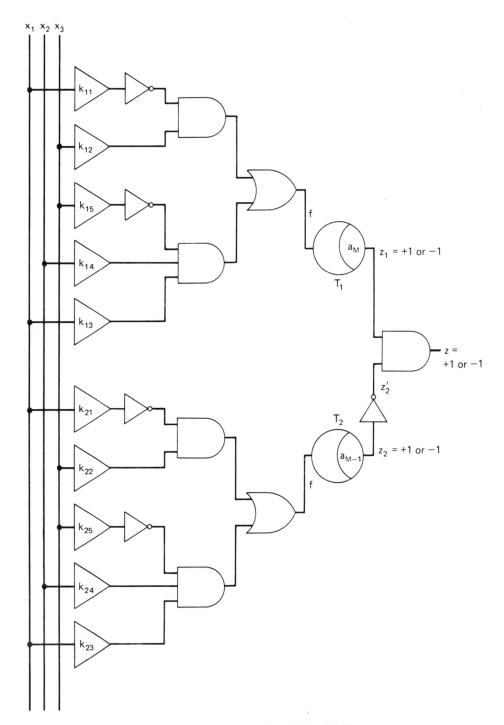

Fig. 5.4.5 The realization of Eq. (5.4.6).

174

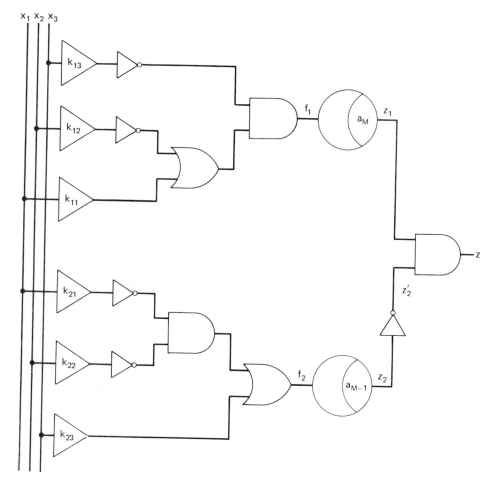

Fig. 5.4.6 The realization of Eqs. (5.4.11) and (5.4.12).

Example 5.4.4

Design a system that generates an output $a_M \leq z < a_{M-1}$ when

$$\text{group } 1 = \left[\left\{ \begin{array}{c} x_1 \geq b_1 \\ \text{or } x_2 \leq b_2 \end{array} \right\} \quad \text{and} \quad \left\{ x_3 \leq b_3 \right\} \right] \qquad (5.4.10a)$$

and

$$\text{group } 2 = \left[\left\{ \begin{array}{c} x_1 \geq b_4 \\ \text{or } x_2 \geq b_5 \end{array} \right\} \quad \text{and} \quad \left\{ x_3 \leq b_6 \right\} \right] \qquad (5.4.10b)$$

Groups 1 and 2 of Eqs. (5.4.10a) and (5.4.10b) are not dual in nature, and therefore each group must be associated with a different function. Let f_1 and f_2 be the two functions synthesized from groups 1 and 2, respectively. These functions are such that $f_1 \geq a_M$ and $f_2 \leq a_{M-1}$ when the conditions of Eqs. (5.4.10a) and (5.4.10b) are satisfied.

Using the analytical rules stated previously, we can easily verify that

$$f_1(x_1, x_2, x_3) = (x_1 + x_2') \cdot x_3' \qquad (5.4.11)$$

and

$$f_2(x_1, x_2, x_3) = x_1' x_2' + x_3 \qquad (5.4.12)$$

The scaling factors involved in this case are as follows.

Group 1	*Group 2*	
$k_{11} = \dfrac{a_M}{b_1}$	$k_{21} = \dfrac{1 - a_{M-1}}{b_4}$	
$k_{12} = \dfrac{1 - a_M}{b_2}$	$k_{22} = \dfrac{1 - a_{M-1}}{b_5}$	(5.4.13)
$k_{13} = \dfrac{1 - a_M}{b_3}$	$k_{23} = \dfrac{a_{M-1}}{b_6}$	

The electronic implementation of this system is given in Fig. 5.4.6.

In the above, when the ranges of interest for the variables were not related to the critical values a_M and a_{M-1}, multiplicative scaling factors were used to scale the variables to appropriate levels prior to using them in an actual implementation. In the following, it will be seen that the scaling of variables may also be attained by using additive scaling factors. This is illustrated by the following example.

Example 5.4.5

Repeat the design problem of Example 5.4.4 using additive scaling factors. From Eqs. (5.4.11) and (5.4.12), using additive scaling factors K_{ij}, we obtain

$$f_1(x_1, x_2, x_3) = ((x_1 \mathbin{\widehat{+}} K_{11}) + (x_2' \mathbin{\widehat{+}} K_{12})) \cdot (x_3' \mathbin{\widehat{+}} K_{13}) \qquad (5.4.14)$$
$$f_2(x_1, x_2, x_3) = ((x_1' \mathbin{\widehat{+}} K_{21})(x_2' \mathbin{\widehat{+}} K_{22}) + (x_3 \mathbin{\widehat{+}} K_{23}) \qquad (5.4.15)$$

The additive scaling factors for variables in groups 1 and 2 are found to be

$$
\begin{aligned}
(x_1 \mathbin{\widehat{+}} K_{11}) &\geq a_M \quad \text{for } x_1 \geq b_1; \quad \text{or } K_{11} = a_M - b_1 \\
(x_2' \mathbin{\widehat{+}} K_{12}) &\geq a_M \quad \text{for } x_2' \geq b_2'; \quad \text{or } K_{12} = a_M - b_2' \\
(x_3' \mathbin{\widehat{+}} K_{13}) &\geq a_M \quad \text{for } x_3' \geq b_3'; \quad \text{or } K_{13} = a_M - b_3' \\
(x_1' \mathbin{\widehat{+}} K_{21}) &\geq a_{M-1} \quad \text{for } x_1' \geq b_4'; \quad \text{or } K_{21} = a_{M-1} - b_4' \\
(x_2' \mathbin{\widehat{+}} K_{22}) &\geq a_{M-1} \quad \text{for } x_2' \geq b_5'; \quad \text{or } K_{22} = a_{M-1} - b_5' \\
(x_3 \mathbin{\widehat{+}} K_{23}) &\geq a_{M-1} \quad \text{for } x_3 \geq b_6; \quad \text{or } K_{23} = a_{M-1} - b_6
\end{aligned}
\qquad (5.4.16)
$$

where the symbol $\widehat{+}$ represents the arithmetic addition. The "head" $\wedge$ is used to distinguish it from the fuzzy OR operation $+$. A realization of Eqs. (5.4.10a) and (5.4.10b) using additive scaling factors is shown in Fig. 5.4.7.

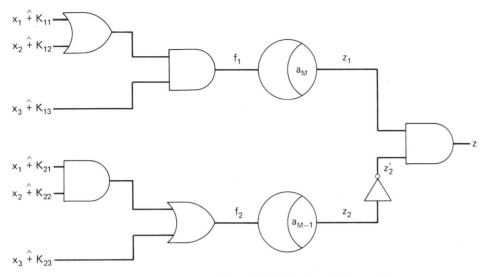

Fig. 5.4.7 A realization of Eqs. (5.4.10) using additive scaling factors.

From an implementation point of view, sometimes it is desirable to make all the K_{ij} be positive whenever possible so that no subtraction operations will be required in the implementation. This is illustrated in the following example.

Example 5.4.6

Consider the same problem as that in Example 5.4.2 but using additive scaling factors. Define $f^*(x_1, x_2, x_3) = f(x_1, x_2, x_3) + K_0$, where $K_0 \geq 0$. Then the problem is equivalent to the following problem: If $f^*(x_1, x_2, x_3) \geq \alpha_1^* = \alpha_1 + K_0$,

$$\left\{ \begin{array}{l} x_1 \geq \alpha_{11} \\ \text{and } x_2' \geq \alpha_{12}' \end{array} \right\} \quad \text{or} \quad \left\{ \begin{array}{l} x_1' \geq \alpha_{21}' \\ \text{and } x_2 \geq \alpha_{22} \end{array} \right\} \quad \text{or} \quad \left\{ \begin{array}{l} x_1 \geq \alpha_{31} \\ \text{and } x_2 \geq \alpha_{32} \\ \text{and } x_3' \geq \alpha_{33}' \end{array} \right\} \quad (5.4.17)$$

From these conditions, it is readily seen that

$$\begin{aligned} f^*(x_1, x_2, x_3) &= [(x_1 + K_1) \cdot (x_2' + K_2)] + [(x_1' + K_3) \cdot (x_2 + K_4)] \\ &\quad + [(x_1 + K_5) \cdot (x_2 + K_6) \cdot (x_3' + K_7)] \end{aligned} \quad (5.4.18)$$

Solve for the K_i's such that

$$\begin{array}{ll} (x_1 + K_1) \cdot (x_2' + K_2) \geq \alpha_1 + K_0 & \text{for } (x_1 \geq \alpha_{11}) \text{ and } (x_2' \geq \alpha_{12}') \\ (x_1' + K_3) \cdot (x_2 + K_4) \geq \alpha_1 + K_0 & \text{for } (x_1' \geq \alpha_{21}') \text{ and } (x_2 \geq \alpha_{22}) \\ (x_1 + K_5) \cdot (x_2 + K_6) \cdot (x_3' + K_7) \geq \alpha_1 + K_0 & \text{for } (x_1 \geq \alpha_{31}) \text{ and } (x_2 \geq \alpha_{32}) \\ & \quad \text{and } (x_3' \geq \alpha_{33}') \end{array} \quad (5.4.19)$$

First letting $K_0 = 0$, we find that

(a) $(x_1 + K_1) \geq \alpha_1$ for $x_1 \geq \alpha_{11}$; or $K_1 = \alpha_1 - \alpha_{11}$

(b) $(x_2' + K_2) \geq \alpha_1$ for $x_2' \geq \alpha_{12}'$; or $K_2 = \alpha_1 - \alpha_{12}'$

(c) $(x_1' + K_3) \geq \alpha_1$ for $x_1' \geq \alpha_{21}'$; or $K_3 = \alpha_1 - \alpha_{21}'$

(d) $(x_2 + K_4) \geq \alpha_1$ for $x_2 \geq \alpha_{22}$; or $K_4 = \alpha_1 - \alpha_{22}$ (5.4.20)

(e) $(x_1 + K_5) \geq \alpha_1$ for $x_1 \geq \alpha_{31}$; or $K_5 = \alpha_1 - \alpha_{31}$

(f) $(x_2 + K_6) \geq \alpha_1$ for $x_2 \geq \alpha_{32}$; or $K_6 = \alpha_1 - \alpha_{32}$

(g) $(x_3' + K_7) \geq \alpha_1$ for $x_3' \geq \alpha_{33}'$; or $K_7 = \alpha_1 - \alpha_{33}'$

Should any K_i be negative, let K_0 be any number equal to or greater than the absolute value of the smallest negative K_i and recalculate K_i's from Eq. (5.4.20). Otherwise, $K_0 = 0$. For example, let

$$\alpha_{11} = 4, \quad \alpha_{21} = 7, \quad \alpha_{31} = 4, \quad \alpha_{33} = 6$$
$$\alpha_{12} = 7, \quad \alpha_{22} = 9, \quad \alpha_{32} = 9, \quad \alpha_1 = 5$$
                                                                                    (5.4.21)

from which the K_i's are found to be

$$K_1 = \alpha_1 - \alpha_{11} = 5 - 4 = 1$$
$$K_2 = \alpha_1 - \alpha_{12}' = 5 - 8 = -3$$
$$K_3 = \alpha_1 - \alpha_{21}' = 5 - 8 = -3$$
$$K_4 = \alpha_1 - \alpha_{22} = 5 - 9 = -4 \qquad (5.4.22)$$
$$K_5 = \alpha_1 - \alpha_{31} = 5 - 4 = 1$$
$$K_6 = \alpha_1 - \alpha_{32} = 5 - 9 = -4$$
$$K_7 = \alpha_1 - \alpha_{33}' = 5 - 9 = -4$$

In order to make all K_i nonnegative, we find that $K_0 \geq |-4|$. Let $K_0 = 4$; the corresponding recalculated K_i are

$$K_1 = 5, \quad K_2 = 1, \quad K_3 = 1, \quad K_4 = 0, \quad K_5 = 5, \quad K_6 = 0, \quad K_7 = 0 \qquad (5.4.23)$$

and $\alpha_1^* = 9$. Thus,

$$f^*(x_1, x_2, x_3) = [(x_1 \mathbin{\widehat{+}} 5) \cdot (x_2' \mathbin{\widehat{+}} 1)] + [(x_1' \mathbin{\widehat{+}} 1) \cdot (x_2)] + [(x_1 \mathbin{\widehat{+}} 5) \cdot x_2 \cdot x_3'] \qquad (5.4.24)$$

The realization of this function is shown in Fig. 5.4.8.

The preceding examples are offered as illustrations of analyzing and realizing m-class logic functions. As to the application of fuzzy logic, pattern recognition and classification, prediction making, decision making, and formal languages are potential users of m-class logic. Patterns, for instance, may be classified by identifying certain pattern attributes, and then proceeding to classify these patterns according to well-

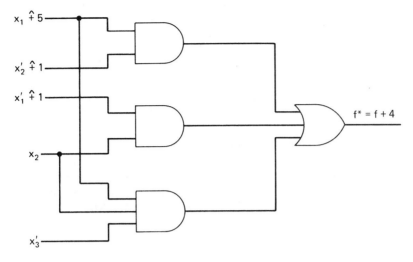

Fig. 5.4.8 Realization of Example 5.4.5.

defined conditions that are satisfied by their respective attributes (pattern-describing variables).

Exercise 5.4

1. Realize the following three-variable fuzzy functions with the conditions described below.

(a)
$$\left\{\begin{matrix} x_1 \le 1 - \beta \\ \text{and } x_2 \ge \beta \\ \text{and } x_3 \le 1 - \beta \end{matrix}\right\} \quad \text{or} \quad \left\{\begin{matrix} x_1 \le 1 - \beta \\ \text{and } x_2 \le 1 - \beta \\ \text{and } x_3 \ge \beta \end{matrix}\right\} \quad \text{or} \quad \left\{\begin{matrix} x_1 \ge \beta \\ \text{and } x_3 \le 1 - \beta \end{matrix}\right\}$$

(b)
$$\left\{\begin{matrix} x_1 \ge 0.25 \\ \text{or } x_2 \le 0.75 \\ \text{or } x_3 \ge 0.25 \end{matrix}\right\} \quad \text{and} \quad \left\{\begin{matrix} x_1 \le 0.75 \\ \text{or } x_2 \ge 0.25 \end{matrix}\right\} \quad \text{and} \quad \left\{\begin{matrix} x_2 \ge 0.25 \\ \text{or } x_3 \le 0.75 \end{matrix}\right\}$$

(c)
$$\left\{\begin{matrix} x_1 \ge 0.3 \\ \text{or } x_2 \le 0.8 \end{matrix}\right\} \quad \text{and} \quad \left\{\begin{matrix} x_2 \ge 0.5 \\ \text{or } x_3 \le 0.4 \end{matrix}\right\} \quad \text{and} \quad \left\{\begin{matrix} x_3 \ge 0.2 \\ \text{or } x_1 \le 0.9 \end{matrix}\right\}$$

(d)
$$\text{group 1} = \left[\left\{\begin{matrix} x_1 \le 0.7 \\ \text{or } x_2 \ge 0.2 \\ \text{or } x_3 \ge 0.3 \end{matrix}\right\} \quad \text{and} \quad \left\{\begin{matrix} x_1 \le 0.9 \\ \text{or } x_2 \le 0.4 \\ \text{or } x_3 \ge 0.2 \end{matrix}\right\} \right]$$

and

$$\text{group 2} = \left[\left\{\begin{matrix} x_1 > 0.2 \\ \text{and } x_2 < 0.6 \\ \text{and } x_3 < 0.8 \end{matrix}\right\} \quad \text{or} \quad \left\{\begin{matrix} x_1 < 0.9 \\ \text{and } x_2 > 0.5 \\ \text{and } x_3 < 0.7 \end{matrix}\right\} \right]$$

(e)

$$\text{group } 1 = \left[\left\{ \begin{array}{l} x_1 \leq 0.2 \\ \text{or } x_2 \geq 0.6 \\ \text{or } x_3 \leq 0.7 \end{array} \right\} \quad \text{and} \quad \left\{ \begin{array}{l} x_1 \geq 0.3 \\ \text{or } x_3 \leq 0.6 \end{array} \right\} \right]$$

and

$$\text{group } 2 = \left[\left\{ \begin{array}{l} x_1 \geq 0.3 \\ \text{and } x_2 \leq 0.4 \end{array} \right\} \quad \text{or} \quad \left\{ \begin{array}{l} x_2 \leq 0.8 \\ \text{and } x_3 \geq 0.6 \end{array} \right\} \quad \text{or} \quad \left\{ \begin{array}{l} x_3 \geq 0.3 \\ \text{and } x_1 \leq 0.7 \end{array} \right\} \right]$$

Bibliographical Remarks

This chapter is, in part, based on references 1 and 2. The design of nonbinary electronic switching circuits can be found in references 3–10. References 11–13 are relevant original papers on fuzzy sets and fuzzy systems. References 14–21 are excellent references on multivalued logic.

References

1. LEE, S. C., and E. T. LEE, "On Multivalued Symmetric Functions," *IEEE Trans. Computers*, Vol. C-21, March 1972, pp. 312–317.

2. MARINOS, P. N., "Fuzzy Logic and Its Application to Switching Systems," *IEEE Trans. Computers*, Vol. C-18, April 1969.

3. POST, E. L., "Introduction to a General Theory of Elementary Propositions," *J. Amer. Math.*, Vol. 43, 1921, pp. 163–185.

4. LEE, C. Y., and W. H. CHEN, "Several-Valued Combinational Switching Circuits," *AIEE Trans. (Commun. Electron.)*, Vol. 75, July 1956, pp. 278–283.

5. BERLIN, R. D., "Synthesis of *N*-Valued Switching Circuits," *IRE Trans. Electron. Comput.*, Vol. EC-7, March 1958, pp. 52–56.

6. MUEHLDORF, E., "Multivalued Switching Algebras and Application to Digital Systems," *Proc. NEC*, Vol. 15, October 1959.

7. SNARE, R. C., "Ternary Memory Store, Data Register, and Arithmetic Unit," Ohio State University, Columbus, Ohio, Res. Rept.

8. WOLF, J. K., and W. R. RICHARD, "Binary to Ternary Conversion by Linear Filtering," Griffiss Air Force Base, Rome, N.Y., Rept. RADC-TRD-62-230, May 1962.

9. KEIR, Y. A., "Algebraic Properties of Three-Valued Compositions," *IEEE Trans. Electron. Comput.* (Short Notes), Vol. EC-13, October 1964, pp. 635–639.

10. YOELI, M., and G. ROSENFELD, "Logical Design of Ternary Switching Circuits," *IEEE Trans. Electron. Comput.*, Vol. EC-14, February 1965, pp. 19–29.

11. ZADEH, L. A., "Fuzzy Sets," *Information and Control*, Vol. 8, 1965, pp. 338–353.

12. ZADEH, L. A., "Shadows of Fuzzy Sets," *Problems in Transmission and Information*, Vol. 2, 1966, pp. 37–44.

13. ZADEH, L. A., "Fuzzy Sets and Systems," *Proc. Symp. Systems Theory*, Polytechnic Press of Polytechnic Institute of Brooklyn, N.Y., 1965, pp. 29–37.

14. POST, E. L., "Introduction to a General Theory of Elementary Propositions," *Amer. J. Math.*, Vol. 43, 1921, pp. 163–185.

15. EPSTEIN, G., "The Lattice Theory of Post Algebra," *Trans. Amer. Math Soc.*, Vol. 95, No. 2, May 1960, pp. 300–317.

16. TACZYK, T., "Axioms and Some Properties of Post Algebras," *Colloquium Mathematicum*, Vol. 10, 1963, pp. 193–209.

17. BRADDOCK, R. C., G. EPSTEIN, and H. YAMANAKS, "Multiple-Valued Logic Design and Applications," 1971 Symposium on Multiple-Valued Logic Design, Buffalo, N.Y., pp. 13–35.

18. SU, S. Y. H., and A. A. SARRIS, "The Relationship Between Multi-valued Switching Algebra and Boolean Algebra Under Different Definitions of Complement," *IEEE Trans. Computers*, Vol. C-21, No. 5, May 1972.

19. Special issue of *Computer* on Multiple-Valued Logic, September 1974.

20. *Proceedings of the Sixth Internal Symposium on Multiple-Valued Logic,* 1976.

21. RINE, D. C., (ed.), *Computer Science and Multiple-Valued Logic: Theory and Applications,* North-Holland, Amsterdam, 1977.

6

Fault Detection in Combinational Circuits

In this chapter, various methods for deriving fault-detection tests for single and multiple permanent logical faults (stuck-at-zero and stuck-at-one faults) in combinational circuits are presented. The need for conceptually simple and straightforward ways of deriving tests is the impetus behind the Boolean derivative (Boolean difference) method. The derivation of a complete set of tests (the set of all possible tests) for detecting any single logical fault in any combinational circuit is presented. To show how easily a minimal complete test set (experiment) could be obtained, the derivation of a minimal complete experiment for the class of monotonic two-level circuits is included. It is followed by an extensive discussion of multiple fault detection, including methods for deriving tests for detecting multiple faults in various types of combinational circuits.

For fault detection in large combinational circuits two computer-oriented algorithms, D-ALGorithm version II (DALG-II) and TEST-DETECT, are presented. The former computes a test to detect a failure and the latter ascertains all failures detected by a given test set. Both are based upon the utilization of a "calculus of D-cubes" that provides the means for effectively performing the necessary computations for large combinational circuits. The two algorithms have been programmed in APL (A Programming Language) and run on the APL interpreter at the IBM Watson Research Center since 1966.

6.1 Test Generation by Boolean Derivative (Difference)

One of the applications of Boolean differential calculus discussed in Chapter 3 to switching circuit theory is the use of Boolean derivatives (differences) to the derivation of fault-detection tests for combinational circuits. Recall that Boolean derivatives were defined as being the EXCLUSIVE-OR operation between two Boolean functions, one representing the normal circuit and the other representing the faulty circuit. Thus, if the Boolean derivative is a 1, a fault is indicated.

Consider a line j of a combinational circuit C. To derive a test of a fault on line

j, line j is "cut" and is considered an applied "pseudo-input" x_j. The primary output z is then expressed in terms of the primary input variables $x_1, \ldots, x_n$ and this pseudo-input x_j; that is, $z_{x_j} = F(x_1, \ldots, x_n, x_j)$. It is noted that the actual value on line j depends on the values $x_1, x_2, \ldots, x_n$, and we let this dependence relation be denoted by the logic function $X_j(x_1, \ldots, x_n)$. To find a test of the stuck-at-zero (s-a-0) or stuck-at-one (s-a-1) fault on line j, we must specify an input pattern applied to the primary input terminals such that the value of f depends on the value of x_j and that $X_j = 1$ (or 0) under the fault-free condition. Let a_i be the binary value of the primary input variable x_i. Then, it is obvious that $(a_1, \ldots, a_n)$ is a test of the s-a-0 fault on line j iff

$$X_j(x_1, \ldots, x_n) \frac{dF(x_1, \ldots, x_n, x_j)}{dx_j}\bigg|_{a_1, \ldots, a_n} = 1 \qquad (6.1.1)$$

Similarly, $(a_1, \ldots, a_n)$ is a test of the s-a-1 fault on line j iff

$$\overline{X_j(x_1, \ldots, x_n)} \frac{dF(x_1, \ldots, x_n, x_j)}{dx_j}\bigg|_{a_1, \ldots, a_n} = 1 \qquad (6.1.2)$$

Thus, we have the following theorem:

THEOREM 6.1.1

Let $x_i^0 = x_i'$ and $x_i^1 = x_i$. Let a_i be the binary value of input variable x_i. Then $(a_1, \ldots, a_n)$ is a test of the s-a-0 (or s-a-1) fault on line j iff $x_1^{a_1} \ldots x_n^{a_n}$ is a minterm in the canonical sum-of-products form of

$$X_j(x_1, \ldots, x_n) \frac{dF(x_1, \ldots, x_n, x_j)}{dx_j} \qquad \left[\text{or } \overline{X_j(x_1, \ldots, x_n)} \frac{dF(x_1, \ldots, x_n, x_j)}{dx_j} \right]$$

Following the theorem above, we have

COROLLARY 6.1.1

The s-a-0 (or s-a-1) fault on line j is undetectable iff

$$X_j(x_1, \ldots, x_n) \frac{dF(x_1, \ldots, x_n, x_j)}{dx_j} = 0 \qquad \left[\text{or } \overline{X_j(x_1, \ldots, x_n)} \frac{dF(x_1, \ldots, x_n, x_j)}{dx_j} = 0 \right]$$

Having presented the theorem and the corollary, we would now like to apply them to some of the circuits considered in previous sections by other methods, so that we can compare them.

First, we find that it is convenient to determine whether a fault of a circuit is detectable or not by using Corollary 6.1.1.

Example 6.1.1

Consider the circuit of Fig. 6.1.1. The output functions expressed in terms of primary input variables and pseudo-inputs e and h are

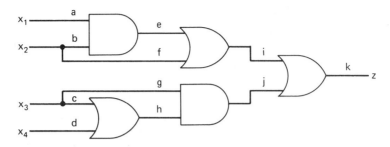

Fig. 6.1.1 Circuit of Example 6.1.1.

$$F(x_3, x_4, e) = (e + x_2) + x_3(x_3 + x_4)$$

and

$$F(x_1, x_2, h) = (x_1 x_2 + x_2) + h x_3$$

respectively, and the Boolean differences with respect to e and h are

$$\frac{dF(x_3, x_4, e)}{de} = [(1 + x_2) + x_3] \oplus [(0 + x_2) + x_3] = x_2' x_3'$$

and

$$\frac{dF(x_1, x_2, h)}{dh} = [x_2 + 1 \cdot x_3] \oplus [x_2 + 0 \cdot x_3] = x_2' x_3$$

It is easy to see that $X_e = x_1 x_2$ and $\bar{X}_h = x_3' x_4'$. Thus,

$$X_e \frac{dF(x_3, x_4, e)}{de} = 0 \quad \text{and} \quad \bar{X}_h \frac{dF(x_1, x_2, h)}{dh} = 0$$

According to Corollary 6.1.1, e_0 and h_1 are undetectable faults.

$$\bar{X}_e \frac{dF(x_3, x_4, e)}{de} = (x_1 x_2)' x_2' x_3' = x_2' x_3'$$

and

$$X_h \frac{dF(x_1, x_2, h)}{dh} = (x_3 + x_4) x_2' x_3 = x_2' x_3$$

Thus, the complete test sets for detecting e_1 and h_0 are {0000, 0001, 1000, 1001} and {0010, 0011, 1010, 1011}, respectively.

A detailed treatment of this subject can be found in Section 7.6 of reference 4.

6.2 Derivation of Minimal Fault-Detection Experiments for Monotonic Two-Level Circuits

Recall that there are three types of monotonic functions: monotonically increasing, monotonically decreasing, and unate. Accordingly, we have three types of monotonic two-level circuits: monotonically increasing two-level circuits, monotonically decreasing two-level circuits, and unate two-level circuits. For example, the circuits of Figs. 6.2.1(a), (b), and (c) realize, respectively, the functions:

$$\text{(a) } z = f(x_1, x_2, x_3, x_4) = x_1 x_3 + x_2 x_4 + x_3 x_4$$
$$= (x_1 + x_4)(x_2 + x_3)(x_3 + x_4)$$
$$\text{(b) } z = f(x_1, x_2, x_3, x_4) = x_1' x_3' + x_1' x_4' + x_2' x_3' + x_3' x_4'$$
$$= (x_1' + x_3')(x_3' + x_4')(x_1' + x_2' + x_4')$$
$$\text{(c) } z = f(x_1, x_2, x_3, x_4) = x_1' x_4 + x_1' x_3 + x_2' x_3$$
$$= (x_1' + x_2')(x_3 + x_4)(x_1' + x_3)$$

which are monotonically increasing, monotonically decreasing, and unate functions. The methods for determining minimal fault-detection experiments for monotonically increasing and monotonically decreasing circuits will be presented first. The determination of minimal fault-detection experiments for unate circuits follows directly from those for monotonically increasing and monotonically decreasing circuits.

DEFINITION 6.2.1

A *zero-bias function* is the function realized by a two-level AND–OR or OR–AND circuit when a line is s-a-0, and a *one-bias function* is the function realized by a two-level AND–OR or OR–AND circuit when a line is s-a-1.

It was shown in Theorem 4.1.3 that a monotonically increasing function has only uncomplemented variables in its minimal sum-of-products (or minimal product-of-sums) form. Then it is easy to prove the following theorems.

THEOREM 6.2.1

In an irredundant monotonically increasing two-level AND–OR circuit that realizes $f = f(x_1, x_2, \dots, x_n)$, the zero-bias functions are monotonically increasing and are strictly included in f.

Proof: Assume that line j is s-a-0. Remove the part of the circuit that feeds the stuck line. Call g the function realized by the pruned-out part. Expressing f in normal form as a sum of products and factoring g, we get $f = g F_1 + F_2$, where g, F_1, and F_2 are obviously monotonically increasing functions (only AND and OR gates are present). Then

$$f|_{\text{line } j \text{ s-a-0}} = g|_{=0} \cdot F_1 + F_2 = F_2 \subseteq f$$

Since the circuit is nonredundant, the fault is detectable and $f \neq F_2$; then $F_2 \subset f$. ∎

THEOREM 6.2.2

In an irredundant monotonically increasing circuit that realizes $f = f(x_1, x_2, \dots, x_n)$, the one-bias functions are monotonically increasing and strictly include f.

Proof: This is similar to Theorem 6.2.1, except that f is expressed in canonical form as a product of sums.

THEOREM 6.2.3

Let $f = f(x_1, x_2, \ldots, x_n)$ be a monotonically increasing function, p^i a vertex of the n cube, and

$$T_0 = \{p^i \mid f(p^i) = 1 \quad \text{and} \quad p^i > p^j \Longrightarrow f(p^j) = 0\} \qquad (6.2.1a)$$

Then T_0 is a sufficient set of input combinations to detect all single s-a-0 faults in any irredundant monotonically increasing circuit that realizes f.

Proof: Suppose that $p^j \notin T_0$ detects a s-a-0 fault and g_0 is the zero-bias function corresponding to that fault. By Theorem 6.2.1, $g_0 \subset f$ and consequently $f(p^j) = 1$ and $g_0(p^j) = 0$. Since $f(p^j) = 1$ and $p^j \notin T_0$, there exists a $p^k \in T_0$, $p^k < p^j$ such that $f(p^k) = 1$. Now g_0 is monotonically increasing; thus, $g_0(p^j) = 0$ implies that $g_0(p^k) = 0$. Hence, $p^k \in T_0$ is also a test for that failure. ∎

THEOREM 6.2.4

Let $f = f(x_1, x_2, \ldots, x_n)$ be a monotonically increasing function, p^i a vertex of the n cube, and

$$T_1 = \{p^i \mid f(p^i) = 0 \quad \text{and} \quad p^j > p^i \Longrightarrow f(p^j) = 1\} \qquad (6.2.1b)$$

Then T_1 is a sufficient set of input combinations to detect all single s-a-1 faults in any irredundant monotonically increasing circuit that realizes f.

Proof: Suppose that $p^j \notin T_1$ detects a s-a-1 fault and g_1 is the one-bias function corresponding to that fault. By Theorem 6.2.2, $g \supset f$ and consequently $f(p^j) = 0$ and $g_1(p^j) = 1$. Since $f(p^j) = 0$ and $p^j \notin T_1$, there exists a $p^k \in S_1$, $p^k > p^j$, such that $f(p^k) = 0$. Now, g_1 is monotonically increasing; thus, $g_1(p^j) = 1$ implies that $g_1(p^k) = 1$, and hence $p^k \in T_1$ is also a test for that failure. ∎

In a monotonically increasing function all the prime implicants are essential, and to uniquely determine a monotonically increasing function, it is only necessary to know the smallest vertex (closest to 000 . . . 00) in each prime implicant. This follows from the fact that if $p^j > p^k$ and $f(p^k) = 1$, then $f(p^j) = 1$. These smallest vertices correspond to the elements of T_0, and therefore we have that to each prime implicant there corresponds one and only one element of T_0.

Similarly, in every prime implicate† the largest vertex (the farthest from 000 . . . 0) corresponds to an element of T_1.

To obtain T_0 by this method, substitute any present variable in each prime implicant by a 1 and substitute a 0 for every variable not present. To obtain T_1, in each prime implicate substitute any present variable by a 0 and any variable not present by a 1.

---

†Each term (sum of literals) of the canonical product-of-sums form of a switching function is called an *implicate* of the function.

Example 6.2.1

Consider the monotonically increasing function

$$z = x_1 x_3 + x_2 x_4 + x_3 x_4$$
$$= (x_1 + x_4)(x_2 + x_3)(x_3 + x_4) \qquad (6.2.2)$$

Then,

$$
\begin{array}{c}
x_1\ \ x_3\ \ +\ \ x_2\ \ x_4\ \ +\ \ \ x_3\ x_4 \\
T_0 = \{1\ 0\ 1\ 0\ ,\ 0\ 1\ 0\ 1\ ,\ 0\ 0\ 1\ 1\ \} \\
= \{10, 5, 3\}
\end{array}
\qquad (6.2.3a)
$$

and

$$
\begin{array}{c}
(x_1\ +\ x_4)\cdot\ \ (x_2 + x_3)\ \cdot (\ \ \ x_3 + x_4) \\
T_1 = \{0\ 1\ 1\ 0\ ,\ 1\ 0\ \ \ 0\ 1\ ,\ 1\ 1\ 0\ \ \ 0\} \\
= \{6, 9, 12\}
\end{array}
\qquad (6.2.3b)
$$

It has been shown above that T_0 and T_1 are sufficient to detect all single faults, but there are circuits that can be tested completely using subsets of $T_0 \cup T_1$, while other realizations need all the tests given by T_0 and T_1. In particular, we have the following theorem.

THEOREM 6.2.5

All the elements of $T_0(T_1)$ are necessary to detect all s-a-0 (s-a-1) faults in a two-level irredundant AND–OR (OR–AND) realization of a function f. So the sets T_0 and T_1 are minimal for the class of monotonically increasing realizations of f.

Proof: It has been shown above that every prime implicant is determined uniquely by an element of T_0; that is, each element of T_0 is covered by one and only one prime implicant. An input combination, say, t_0, cannot be a test for single s-a-0 faults in lines feeding different gates A_i and A_j in the first level of an AND–OR circuit: the outputs of A_i and A_j for such t_0 would be 1 if the gates were working properly, and, if any of the input lines to A_i are s-a-0, the output of A_j, and consequently the output of the entire circuit, is still 1, and t_0 cannot be a test for A_i.

Similar reasoning applies for a two-level OR–AND circuit. A two-level AND–OR (OR–AND) circuit may not need all the T_1 (T_0) tests to detect all s-a-1 (s-a-0) faults, as shown in Fig. 6.2.1(a); but $T_0 \cup T_1$ is minimum for the class of realizations of f in the sense that there is no set Q with fewer elements than $T_0 \cup T_1$ that detects all single stuck faults in all irredundant monotonically increasing realizations of a function f. ∎

Thus, the minimal complete fault-detection test set T for the circuits in Figs. 6.2.1 (a-1) and (a-2) is

$$T = T_0 \cup T_1 = \{3, 5, 6, 9, 10, 12\} \qquad (6.2.4)$$

It has also been shown in Theorem 4.1.3 that a monotonically decreasing function has only complemented variables in its minimal sum-of-products (or minimal product-

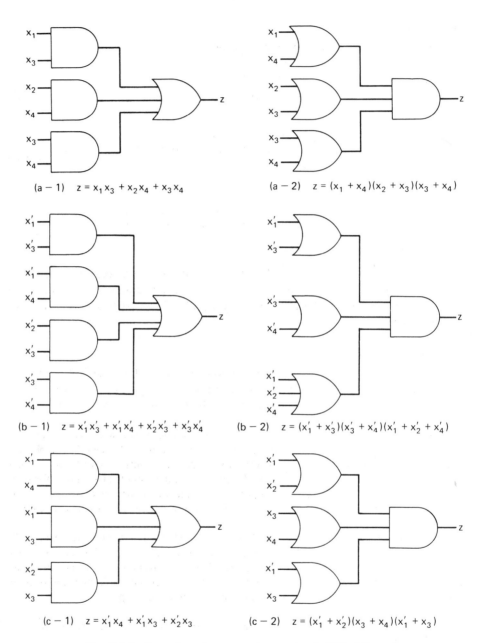

Fig. 6.2.1 Examples of monotonic two-level AND-OR and OR-AND circuits (a) monotonic increasing circuit, (b) monotonic decreasing circuit, and (c) unate circuit.

of-sums) form. Thus, the theory developed so far for monotonically increasing functions applies also to monotonically decreasing functions, with some modifications.

THEOREM 6.2.6

Let $f = f(x_1, x_2, \ldots, x_n)$ be a monotonically decreasing function and let

$$T_0 = \{p^i \,|\, f(p^i) = 1 \quad \text{and} \quad p^j > p^i \Longrightarrow f(p^j) = 0\} \tag{6.2.5a}$$

and

$$T_1 = \{p^i \,|\, f(p^i) = 0 \quad \text{and} \quad p^j < p^i \Longrightarrow f(p^j) = 1\} \tag{6.2.5b}$$

Then T_0 and T_1 are sufficient sets of input combinations to detect all single s-a-0 and s-a-1 faults in any irredundant monotonically decreasing circuit that realizes f.

The proof of this theorem is similar to those of Theorems 6.2.3 and 6.2.4 and can thus be omitted.

Following an argument similar to that given above, T_0 for a monotonically decreasing function is obtained by substituting any present variable in each prime implicant by a 0 and substituting a 1 for every missing variable; and T_1 for a monotonically decreasing function is obtained by substituting any present variable in each prime implicate by a 1 and substituting a 0 for every missing variable.

Example 6.2.2

Consider the monotonically decreasing function

$$
\begin{aligned}
z &= x_1' x_3' + x_1' x_4' + x_2' x_3' + x_3' x_4' \\
&= (x_1' + x_3')(x_3' + x_4')(x_1' + x_2' + x_4')
\end{aligned}
\tag{6.2.6}
$$

Then,

$$
\begin{aligned}
& \qquad x_1'\,x_3' + x_1' \qquad x_4' + \quad x_2'x_3' + \qquad x_3'x_4' \\
T_0 &= \{0\ 1\ 0\ 1\ ,\ 0\ 1\ 1\ 0\ ,\ 1\ 0\ 0\ 1\ ,\ 1\ 1\ 0\ 0\} \\
&= \{5, 6, 9, 12\}
\end{aligned}
\tag{6.2.7a}
$$

and

$$
\begin{aligned}
& \qquad (x_1'\ +\ x_3')\ \cdot(\quad x_3' + x_4')\cdot(x_1' + x_2'\ + x_4') \\
T_1 &= \{1\ 0 \qquad 1\ 0, 0\ 0\ 1 \qquad 1\ ,\ 1 \qquad 1\ 0 \qquad 1\} \\
&= \{10, 3, 13\}
\end{aligned}
\tag{6.2.7b}
$$

THEOREM 6.2.7

All the elements of T_0 (T_1) are necessary to detect all s-a-0 (s-a-1) faults in a two-level irredundant AND–OR (OR–AND) realization of a function f. So the sets T_0 and T_1 are minimal for the class of monotonically decreasing realizations of f.

The proof is similar to that of Theorem 6.2.5. Hence, the minimal complete fault-detection experiment for the circuits of Fig. 6.2.1(b-1) and (b-2) is

$$T = T_0 \cup T_1 = \{3, 5, 6, 9, 10, 12, 13\} \tag{6.2.8}$$

After having the derivations of fault-detection experiments for monotonically increasing and decreasing functions, let us now consider fault detection for unate functions. Every variable in the canonical form of a unate function appears either complemented or uncomplemented, but not both.

Let $f = f(x_1, x_2, \ldots, x_s, y_1, y_2, \ldots, y_{n-s})$ be a unate function where f is monotonically increasing in the set $X = \{x_1, x_2, \ldots, x_s\}$ and monotonically decreasing in the set $Y = \{y_1, \ldots, y_{n-s}\}$ ($x_1, \ldots, x_s$ appear only uncomplemented and $y_1, \ldots, y_{n-s}$ appear only complemented in a minimal sum-of-products form); then Theorems 6.2.1–6.2.5 apply to the set X and Theorems 6.2.6 and 6.2.7 apply to the set Y. In particular, to find the sets T_0 and T_1 for a unate function, the following procedure is used. In each prime implicant of the minimal sum, every x_i present is substituted by 1 and the missing x_i by 0, and every y_i' present by 0 and the missing y_i' by 1. The binary vectors so obtained are the elements of T_0. Similarly, in each prime implicate of the minimal product, every x_i present is substituted by a 0 and the missing x_i by 1, and every y_i' present is substituted by a 1 and the missing y_i' by 0, giving the elements of T_1.

Example 6.2.3

$$z = x_1'x_4 + x_1'x_3 + x_2'x_3 = (x_1' + x_2')(x_3 + x_4)(x_1' + x_3) \qquad (6.2.9)$$

$$
\begin{array}{cccccc}
 & x_1' & x_4 + x_1' & x_3 + & x_2' & x_3 \\
T_0 = \{ 0 & 1 & 0\ 1\ , & 0\ 1\ 1\ 0\ , & 1\ 0 & 1\ 0 \}
\end{array}
$$
$$= \{5, 6, 10\} \qquad (6.2.10a)$$

$$
\begin{array}{cccc}
 & (x_1' + x_2')\cdot(& x_3 + x_4)\cdot(x_1' + x_3 &) \\
T_1 = \{ 1\ 1\ 1\ 1\ , & 0\ 0\ 0 & 0\ ,\ 1\ 0 & 0\ 1 \}
\end{array}
$$
$$= \{0, 9, 15\} \qquad (6.2.10b)$$

THEOREM 6.2.8

The sets T_0 and T_1 defined above are sufficient sets of tests to detect any s-a-0 or s-a-1 fault in any AND–OR (OR–AND) realization of the unate function f.

Proof: If f is monotonically increasing, this theorem is equivalent to Theorems 6.2.3 and 6.2.4, and similar arguments hold if f is monotonically decreasing. For a general unate function, we introduce the relationship $<:$. We say that $s_1 <: s_2$ if s_1 has a 0 in every position corresponding to a variable of X where s_2 has a 0, and s_1 has a 1 in every position corresponding to a variable of Y where s_2 has a 1. For example, suppose that f is monotonically increasing in $X = \{u_1, u_2\}$ and monotonically decreasing in $Y = \{u_3, u_4\}$; then $0010 <: 1010$ and $1001 <: 1100$. Then, $f(s_1) < f(s_2) \longrightarrow s_1 <: s_2$.

Let g_0 be the function realized by a circuit when a line is s-a-0 and $s \notin T_0$ an input combination that detects such a fault; then, clearly, $f(s) = 1$ and $g_0(s) = 0$. But in the prime implicant that includes s, there is another point $t \in T_0$, $t <: s$, such that $f(t) = 1$. Now, since the circuit consists only of AND and OR gates, $g_0(t) = 0$ and $t \in T_0$ is also a test for that fault. Similarly, let g_1 be the function realized by a circuit when a line is s-a-1 and $s \notin T_1$ an input combination that detects such a fault; then $f(s) = 0$ and $g_1(s) = 1$. But in the prime implicate that includes s, there is also another point, $t \in T_1, t :> s$ such that $f(t) = 0$. Now, since the circuit consists only of AND and OR gates, $g_1(t) = 1$ and $t \in T_1$ is also a test for the fault. ∎

Similar reasoning applies for two-level OR–AND circuits.

THEOREM 6.2.9

Let m_0 and m_1 denote the number of prime implicants and the number of prime implicates of a unate function f, respectively. The minimum number of s-a-0 (s-a-1) tests and the minimum number of s-a-1 (s-a-0) tests for a two-level AND–OR (OR–AND) realization of f are m_0 and m_1, respectively.

Proof: Since each prime implicant is realized by an AND gate in the two-level AND–OR realization and each prime implicate is realized by an OR gate in the two-level OR–AND realization and since each of the AND gates in the AND–OR realization has to be tested at least once for s-a-0 faults, and each of the OR gates in the OR–AND realization has to be tested at least once for s-a-1 faults, the minimum number of s-a-0 tests for the two-level AND–OR realization is m_0 and the minimum number of s-a-1 tests for two-level OR–AND realization is m_1. A s-a-0 test for the AND–OR realization is also a s-a-0 test for the OR–AND realization, and a s-a-1 test for the AND–OR realization is also a s-a-1 test for the OR–AND realization. Hence, the minimum numbers of tests for both two-level AND–OR and OR–AND realizations of f to detect any s-a-0 and s-a-1 faults are m_0 and m_1, respectively. ∎

Following from Theorem 6.2.9, we have

COROLLARY 6.2.1

The test sets T_0 and T_1 obtained above for a unate function are minimal for detecting any s-a-0 and s-a-1 faults in the two-level realizations. The union of T_0 and T_1 constitutes a minimal complete test set for the realizations.

Hence, T_0 and T_1 of Eq. (6.2.10) are minimal s-a-0 and s-a-1 test sets for the circuits of Figs. 6.2.1(c-1) and (c-2), and the union of them is a minimal complete fault-detection test set for the circuits.

Exercise 6.2

1. Derive a complete test set for each of the minimal two-level AND–OR realizations of the following functions.
   (a) Monotonically increasing function:

   $$f(x_1\, x_2, x_3, x_4) = x_1 x_2 x_3 + x_2 x_3 x_4 + x_1 x_3 x_4 + x_1 x_2 x_4$$

   (b) Monotonically decreasing function:

   $$f(x_1, x_2, x_3, x_4) = x_1' x_2' + x_3' x_4' + x_1' x_3' + x_2' x_4' + x_1' x_4' + x_2' x_3'$$

   (c) Unate function:

   $$f(x_1, x_2, x_3, x_4, x_5, x_6) = x_1 x_2 x_3' x_4 + x_2 x_3' x_4 x_5' + x_3' x_4 x_5' x_6'$$

6.3 Analysis of Multiple Faults
in Combinational Circuits

The fault-detection methods introduced in the previous sections were designed mainly for the detection of single faults. The assumption that only single faults can occur is reasonably valid for circuits that have been running correctly and for which the most probable equipment failures produce single faults. However, this assumption is invalid for circuits which are undergoing initial testing and circuit technologies, such as large-scale integration (LSI), where a logical failure may cause massive rather than single faults.

In this section, we shall study multiple faults in redundant and irredundant circuits. In redundant circuits, some undesirable features about fault masking among undetectable faults on redundant connections and normally detectable faults (without the presence of undetectable faults) are discussed. It is also shown by example that undetectable faults may make two normally distinguishable faults become indistinguishable. A method for controlling the effect of redundancy by adding test inputs is discussed. Many interesting results on multiple fault detectability in irredundant or redundancy-controlled circuits are presented. Based on these results, it is found that a complete test set for testing any single fault in an irredundant circuit (obtained by any one of the methods described in the previous sections) will detect any multiple fault if it is a two-level circuit, and will detect any fault set of three or fewer elements if it is a treelike circuit. Conditions for two faults to be mutually masking in an irredundant circuit containing internal fan-outs are presented at the end of this section.

A. Multiple Faults in Redundant Circuits

Redundancy is undesirable in combinational circuits from the point of view of testing. There are three main reasons for this.

1. Some faults on the redundant connections in a circuit are undetectable. The faults e s-a-0 and h s-a-1 in the circuit of Fig. 6.1.1 and the faults b s-a-1, c s-a-1, and d s-a-1 in the circuit of Fig. 6.3.1(a) are examples of undetectable faults on the redundant connections. The former group had been explained in Section 6.1. The reason for the latter may be seen from the following output function of the circuit.

$$z = (x_1 x_2 + x_2')'(x_1 x_3 + x_3')'(x_1 x_4 + x_4')'$$

which remains unchanged when the variables x_2, x_3, and x_4 in the product terms $x_1 x_2$, $x_1 x_3$, and $x_1 x_4$ are replaced by a value 1.

2. The presence of undetectable faults on redundant connections may cause other detectable faults to become undetectable. For example, it can easily be shown that in the circuit of Fig. 6.3.1(a), the multiple fault f s-a-0, h s-a-0, and j s-a-0 can be detected by $x_1 = 1$, $x_2 = 0$, $x_3 = 0$, and $x_4 = 0$, but becomes undetectable under the presence of the undetectable faults b s-a-1, c s-a-1, and d s-a-1.

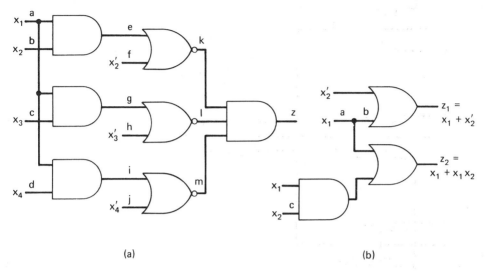

(a)   (b)

Fig. 6.3.1 An example of showing two distinguishable faults may be indistinguishable under the presence of an undetectable fault.

3.   The presence of undetectable faults may make some distinguishable faults become indistinguishable. For example, in the circuit of Fig. 6.3.1(b), the fault a s-a-0 can be detected on output z_1 by the test $x_1 = 1$ and $x_2 = 1$ and on output z_2 by the test $x_1 = 1$ and $x_2 = 0$, and the fault b s-a-0 can be detected only on z_1 by the test $x_1 = 1$ and $x_2 = 1$. Thus, faults a s-a-0 and b s-a-0 are distinguishable. But under the presence of the undetectable fault c s-a-1, the fault a s-a-0 is no longer detectable on output z_2. Consequently, faults a s-a-0 and b s-a-0 become indistinguishable.

Even though redundancy is undesirable from the point of view of fault detection, in some cases the presence of redundancy is intentional and desirable for other reasons, for example, hazard-free realizations (see Section 5.7 of reference 4). If a complete testing capability is desired and the redundancy cannot be removed, additional methods are required, such as inserting test points within the circuit.

It is also possible to control the effect of redundancy by adding test inputs. Consider a node N which has a redundant connection [e.g., Fig. 6.3.2(a)] and think of the function described at the output z of N as described by

$$z = f_1 \odot f_2 \qquad (6.3.1)$$

where the operation $\odot$ is the operation of N, f_1 represents the function comprised of the irredundant inputs, and f_2 represents the redundant connections; f_2 may be considered to be irredundant in its own right. If it is not, multiple redundant connections exist and the one nearest the input must be treated first. Now create a new function with additional test input variables as in Eq. (6.3.2):

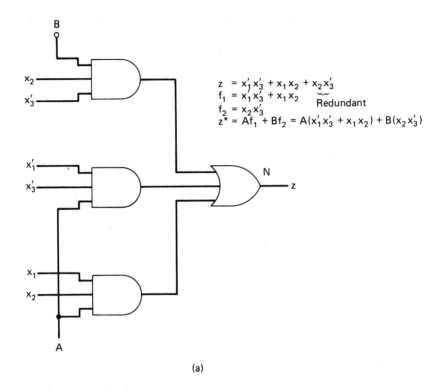

$$z = x_1'x_3' + x_1x_2 + x_2x_3'$$
$$f_1 = x_1'x_3' + x_1x_2 \quad \underbrace{}$$
$$f_2 = x_2x_3' \qquad \text{Redundant}$$
$$z^* = Af_1 + Bf_2 = A(x_1'x_3' + x_1x_2) + B(x_2x_3')$$

(a)

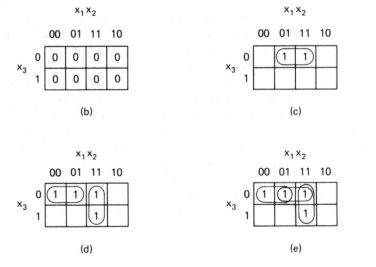

(b)

(c)

(d)

(e)

Fig. 6.3.2 A controlled rendundant function.
     (a) Redundant subcircuit.
     (b) $A = 0, B = 0.$
     (c) $A = 0, B = 1.$
     (d) $A = 1, B = 0.$
     (e) $A = 1, B = 1.$

$$z^* = Af_1 \odot Bf_2 \qquad\qquad (6.3.2)$$

For the simple case of an alternating AND–OR circuit only two test variables will be required. One input is fanned out to all gates which are predecessors of the irredundant connections to N and the other to the redundant predecessor gates. The use of these control inputs is shown in Fig. 6.3.2 considering the case where N is an OR gate and each predecessor is an AND gate.

Under the control input $A = 0$ and $B = 1$ the redundant portion of the circuit is passed as an irredundant function, and faults in f_2 may be sensitized to an observable output [Fig. 6.3.2(c)]. Similarly, $A = 1$ and $B = 0$ passes the irredundant function, which may then be tested in a normal fashion [Fig. 6.3.2(d)]. When no test is being performed, both control inputs must remain at 1 and the original function is passed by node N [Fig. 6.3.2(e)]. This scheme guarantees that the redundancy does not lose its effectiveness and that it does not mask a normally detectable fault and does not make normally distinguishable faults become indistinguishable.

B. Multiple Faults in General Irredundant Circuits

In the previous sections, efforts have been directed to the search for methods for deriving a complete test set for testing any single fault in a combinational circuit. It would be of great interest to know if the same test set can be used to detect certain (if not all) types of multiple faults. Moreover, we want to investigate classes of circuits (if they exist) whose complete test set for testing single faults will also detect multiple faults in the circuit.

The following analysis assumes that the circuit will be irredundant or will contain controlled redundancy. It is further assumed that a set of tests that will detect all single faults is given for the circuit. Such a test set is known to exist if the circuit is irredundant (problem 2).

DEFINITION 6.3.1

A *complete test set T*, or simply a *T set*, is a complete test set that will detect any single fault in an irredundant circuit.

DEFINITION 6.3.2

The *sensitizing line value* for a gate is that value which is applied to all lines when the inputs are all simultaneously under test. For the AND and NAND gates the sensitizing line values are a logic 1. For the OR and NOR gates the sensitizing line values are a logic 0.

DEFINITION 6.3.3

A fault is a *spanning fault* with respect to a gate if it is a member of the fault class which is tested when all inputs to a gate are the sensitizing line value. Notice that there is only one such class for any gate.

DEFINITION 6.3.4

A fault is a *sensitizing fault* with respect to a gate if it is not a member of the spanning fault class for the gate. Notice that there are K such classes for every gate, where K is the number of input connections.

THEOREM 6.3.1

Any set of multiple faults occurring in a single gate will be detected by a T set.

Proof: The only way that a sensitizing fault in a gate may be blocked is by a spanning fault, which will destroy the sensitized configuration. However, any number of spanning faults will be detected by the spanning test, since they are in the same indistinguishability class and cannot be blocked by a sensitizing fault. ∎

DEFINITION 6.3.5

A *convergence point* is a node at which two or more faults interact. These faults may be faults in the convergence point or have propagated from some other node to the convergence point.

DEFINITION 6.3.6

When multiple faults meet at a convergence point, the fault or faults that can propagate through the convergence point for some test in T will be denoted as *dominant faults* for that node. It is clear that whenever there is a spanning fault at a convergence point, it will be dominant.

THEOREM 6.3.2

Any set of multiple faults occurring on a simple path will be detected by a T set.

Proof: As the faults nearest the input are tested they are sensitized so that they propagate to a convergence point with another fault. This condition will look exactly as though there are multiple faults in a single gate. From the results of Theorem 6.3.1, it is clear that some dominant fault is always propagated past this convergence point. A similar argument will apply at each convergence point down the path. The dominant fault nearest the output will always be detected by the T set. ∎

THEOREM 6.3.3

Any set of faults occurring in a two-level irredundant circuit will be detected by a T set.

The proof of this theorem is quite complicated and lengthy and hence is not included. It may be found in references 12 and 13.

DEFINITION 6.3.7

A *closed fault set* is a set of faults such that each element in the set is masked by another element in the set when it is under test.

DEFINITION 6.3.8

A *fault set graph* is a graph with a node for each element in a set of multiple faults and a directed edge between nodes i and j if fault i masks fault j.

It is clear that a closed fault set and a set graph are functions of the T set selected.

THEOREM 6.3.4

A fault set graph for a closed fault set contains at least one directed cycle. The proof of the theorem is obvious from Definitions 6.3.7 and 6.3.8 and the fact that the fault set graph is finite.

DEFINITION 6.3.9

A set of faults is an *undetected fault set* (UFS) if the faults are mutually masking in such a way that the T set will pass the circuit.

THEOREM 6.3.5

A set of faults will be an undetectable fault set iff the set is a closed fault set.

Proof: The sufficiency is given by applying Definition 6.3.7. If a set of faults is undetected, then each fault is masked in turn and the definition is satisfied. The necessity may be shown by contradiction. If a set is not a closed set, some fault is not masked and is detected, which is a contradiction. ∎

COROLLARY 6.3.1

A set of faults is a UFS only if its fault set graph has a directed cycle.

Proof: A set of faults can be a closed set only if its fault graph has a directed cycle by Theorem 6.3.4. Theorem 6.3.5 gives the result that a closed set is an undetected set. ∎

COROLLARY 6.3.2

If a fault graph contains only directed cycles, the fault set is an undetected fault set.

Proof: Every element of a directed cycle is a member of a closed fault set since each fault is masked by another. Since all nodes are members of a cycle, they are all members of a closed fault set, and by Theorem 6.3.5, the fault set is an undetected fault set. ∎

C. Multiple Faults in Irredundant Treelike Circuits

Now we consider multiple fault detection in a treelike circuit which is defined as a circuit in which the fan-out of every line is 1. It can be easily shown that any general combinational circuit containing gate output fan-out (no input fan-out) may be transformed into a treelike circuit. An important property of the treelike circuit is that it is functionally equivalent to the original circuit, and each fault in the original circuit is represented.

LEMMA 6.3.1

Any two faults in a treelike circuit will be detected by a T set.

Proof: The faults will have only a single convergence point. At this convergence point the faults look like faults in a single gate and, by Theorem 6.3.1, the dominant fault will always be detected. ∎

LEMMA 6.3.2

Any three faults in a treelike circuit will be detected by a T set.

Proof: The first case is that of three faults and a single convergence point. For this situation, Theorem 6.3.1 is again used and at least one fault will be detected. The second case is that of two convergence points. The dominant fault from the first convergence point may be masked if the third fault is a spanning fault at the second convergence point. If the third fault is a spanning fault, it cannot be masked and is detected; otherwise, the dominant fault of the first pair is detected. ∎

THEOREM 6.3.6

A T set applied to a treelike circuit will detect all fault sets of three or fewer elements.

Proof: This follows directly from the definition of a T set and Lemmas 6.3.1 and 6.3.2. ∎

Example 6.3.1

Consider the irredundant treelike circuit of Fig. 6.3.3(a). A T set fór the circuit can be most conveniently obtained by the path-sensitizing method, which is described as follows:

Step 1 Choose a path from the faulty line to one of the primary outputs.
Step 2 Assign the faulty line a value 0 if the fault is a s-a-0 fault.
Step 3 Along the chosen path, except the lines of the path, assign a value 0 to each input to the OR and NOR gates in the path and a value 1 to each input to the AND and NAND gates in the path.
Step 4 Trace back from gates along the sensitized path toward the circuit inputs. If a consistent input combination (a test) exists, the procedure is terminated. If, on the other hand, a contradiction is encountered, choose another path which starts at the faulty line and repeat the above procedure.

It can be easily shown that for any treelike circuit, a complete test set can always be obtained by the path-sensitizing method. For example, a T set obtained by it is {01101011, 01111110, 10110110, 11010111, 10011011, 10111001}. The fault sets {a s-a-1, d s-a-0, e s-a-1, h s-a-0} and {b s-a-0, c s-a-1, f s-a-0, g s-a-1} (there are many such sets) are undetected fault sets, because they form single directed cycles in their fault graphs, as shown in Figs. 6.3.3(b) and (c). By Corollary 6.3.2, these fault sets are undetected fault sets and cannot be detected by any test in the T set given.

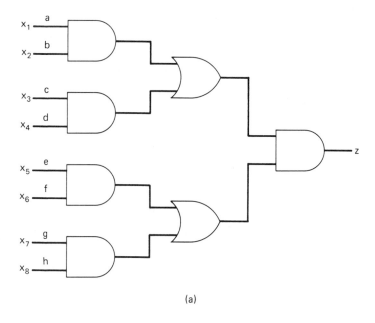

(a)

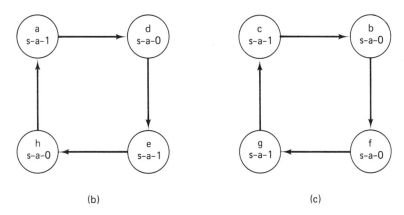

(b)   (c)

Fig. 6.3.3 Circuit of Example 6.3.3.

D. Multiple Faults in Circuits Containing Internal Fan-outs

In a circuit containing internal fan-outs, any fault that occurs prior to a fan-out point may act as a multiple fault. But in order for two faults A and B to be mutually masking they must satisfy some rather restricted conditions.

Condition 1   Fault A must be a predecessor of a fan-out node and fault B must be a successor of the same fan-out node.

Condition 2   Fault B must be a spanning fault and fault A a sensitizing fault at the convergence point, which is prior to the reconvergence node.

Condition 3   Fault A must be a spanning fault and fault B a sensitizing fault at the convergence point, which is also a reconvergent node.

It should be noted that if two faults exist on the same side of the fan-out node (assuming there is only one), then the portion of the circuit including the faults and the convergence point is a treelike circuit and, by Theorem 6.3.6, they may not be masking. Fault B must mask fault A as it is being tested and hence must be the dominant fault at the convergence point prior to the reconvergence node. Fault B must, in turn, be masked by fault A at the reconvergence node, which is the next convergence point, and hence fault A must be dominant at the convergence point. An example follows.

Example 6.3.2

Faults a s-a-0 and b s-a-1 of the circuit in Fig. 6.3.4(a) form an undetected fault set. The

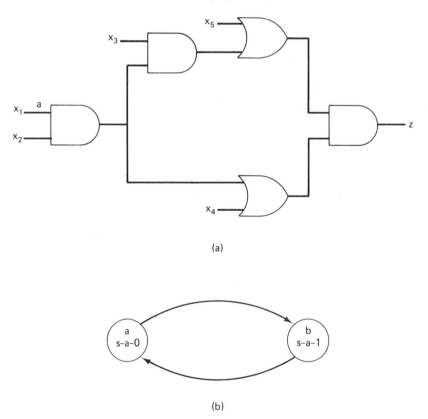

(a)

(b)

Fig. 6.3.4 Circuit of Example 6.3.4.

reason for them being undetected is that the a s-a-0 when under test is sensitized along the upper path and masked by b s-a-1. When b s-a-1 is under test a s-a-0 is sensitized along the lower path and blocks b s-a-1 at the output AND gate. The faults form a single directed cycle as shown in Fig. 6.3.4(b). A T set for this circuit is {11110, 11000, 01011, 10011}. By Corollary 6.3.2 and Definition 6.3.9, these two faults will pass the T-set test.

Exercise 6.3

1. (a) Show that the fault a s-a-1 indicated in the accompanying figure is a detectable fault and the fault b s-a-0 is an undetectable one.

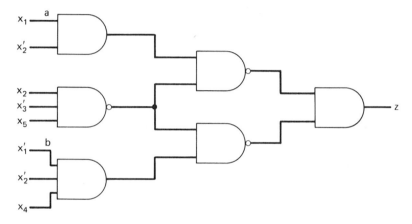

   (b) Show that when both faults a s-a-1 and b s-a-0 are present, the latter will make the former undetectable.
   (c) Can you find a remedy for it?

2. Prove that all s-a-1 and s-a-0 faults in a circuit are detectable iff the circuit is irredundant iff the function that the circuit realizes is a minimized function.

3. An advantage of considering multiple faults is that the number of checkpoints in the circuit can be greatly reduced when compared to the single-fault case. The following are the bases for reducing the number of checkpoints when multiple faults are considered.
   (a) Any multiple faults of a logic gate are equivalent to multiple faults among the input lines only.
   (b) A stuck fan-out stem line is equivalent to the stuck branches.
   (c) The set of inputs to a gate is the minimal set of checkpoints for the gate. The set of branches of a fan-out is the minimal set of checkpoints for the fan-out stem and all the branches.
   Freed from the single-fault assumption, it becomes possible to represent all the distinguishable faults by specifying a minimal number of points in a circuit such that any multiple faults in the circuit become equivalent to multiple faults among these specified checkpoints. A checkpoint labeling procedure is described below.
   (1) All the primary inputs that do not fan out are checkpoints.
   (2) All the fan-out branches are checkpoints.
   (3) NOT gates are considered as lines.
   Prove that this procedure yields the necessary and sufficient checkpoints to represent all the multiple faults in the circuit.

6.4   The Literal Proposition Method

After having discussed multiple faults in various types of combinational circuits, we now present a general method for deriving a complete test set for any fault: single or multiple. The method derives functional descriptions of each output of the circuit, which also represent the structure of the circuit. Each functional expression represents not only the relationship of the output variables to the signals on the input and internal lines, but also the effect of faults on the output. In order to include the effect of faults in the function realized by a circuit, three binary variables are used to specify the state of each line α in the circuit. These binary line variables are defined as follows:

$$\alpha_n = 1 \quad \text{iff line } \alpha \text{ is normal}$$
$$\alpha_1 = 1 \quad \text{iff line } \alpha \text{ is s-a-1}$$
$$\alpha_0 = 1 \quad \text{iff line } \alpha \text{ is s-a-0}$$

In terms of these variables, the proposition P_α, which represents the totality conditions under which the line can have a signal value 1, is

$$P_\alpha = x\alpha_n + \alpha_1$$

where x is the input proposition for the line α. What this says is that the line α will have a signal value 1 if $x = 1$ and line α is fault-free or if line α is s-a-1. Similarly, the totality condition under which the line can have a signal value 0 is represented by the proposition

$$P'_\alpha = x'\alpha_n + \alpha_0$$

which means that line α will have a value 0 if $x = 0$ and line α is fault-free or if line α is s-a-0.

Now consider an AND gate having two inputs x and y connected to the input of the gate by lines α and β, and its output z is led by a line γ. Based on the discussion above, the proposition P_γ and P'_γ of the totality of conditions under which line γ can have a signal 1 and 0 are represented by

$$P_\gamma = (P_\alpha \cdot P_\beta)\gamma_n + \gamma_1$$
$$P'_\gamma = (P'_\alpha + P'_\beta)\gamma_n + \gamma_0$$

respectively, where

$$P_\alpha = x\alpha_n + \alpha_1$$
$$P'_\alpha = x'\alpha_n + \alpha_0$$

$$P_\beta = y\beta_n + \beta_1$$
$$P'_\beta = y'\beta_n + \beta_0$$

The propositions of the output of the six gate circuits are shown in Table 6.4.1. The generation of these propositions for the gate circuits with more than two inputs can be easily obtained. For any single-output function, the output propositions representing the conditions under which the output assumes the values of 1 and 0 can be expressed in terms of the input propositions of the circuit and the variables associated with each line. If we let $P_z(n)$ and $P'_z(n)$ be the output propositions of the normal

TABLE 6.4.1 Propositions of the Outputs of the Six Gate Networks

Gate	Symbol	Proposition
AND		$P_\gamma = (P_\alpha \cdot P_\beta)\gamma_n + \gamma_1$ $P'_\gamma = (P'_\alpha + P'_\beta)\gamma_n + \gamma_0$
OR		$P_\gamma = (P_\alpha + P_\beta)\gamma_n + \gamma_1$ $P'_\gamma = (P'_\alpha \cdot P'_\beta)\gamma_n + \gamma_0$
NOT		$P_\gamma = P_\alpha \gamma_n + \gamma_1$ $P'_\gamma = P'_\alpha \gamma_n + \gamma_0$
NAND		$P_\gamma = (P'_\alpha + P'_\beta)\gamma_n + \gamma_1$ $P'_\gamma = (P_\alpha \cdot P_\beta)\gamma_n + \gamma_0$
NOR		$P_\gamma = (P_\alpha + P_\beta)\gamma_n + \gamma_1$ $P'_\gamma = (P'_\alpha \cdot P'_\beta)\gamma_n + \gamma_0$
EXCLUSIVE -OR		$P_\gamma = (P_\alpha P'_\beta + P'_\alpha P_\beta)\gamma_n + \gamma_1$ $P'_\gamma = (P'_\alpha P'_\beta + P_\alpha P_\beta)\gamma_n + \gamma_0$

Note: $P_\alpha = x\alpha_n + \alpha_1$ $P_\beta = y\beta_n + \beta_1$

$P'_\alpha = x'\alpha_n + \alpha_0$ $P'_\beta = y'\beta_n + \beta_0$

circuit and $P_z(\epsilon)$ be the output propositions of the circuit with a fault ϵ, then the set $T(\epsilon)$ of tests for detecting the fault ϵ is

$$T(\epsilon) = P_z(n) \oplus P_z(\epsilon)$$
$$= P_z(n)P_z'(\epsilon) + P_z'(n)P_z(\epsilon)$$

Example 6.4.1

Consider the circuit of Fig. 6.4.1. The complete set of tests for detecting the s-a-0 on line a may be obtained using the literal proposition method as follows. From the discussions

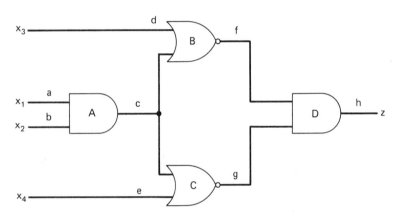

$$T(\epsilon) = P_z(n) + P_z(\epsilon)$$
$$= P_z(n)P_z'(\epsilon) + P_z'(n)P_z(\epsilon)$$

Fig. 6.4.1 Circuit of Example 6.4.1.

above, the propositions of all the input and internal lines are

$$P_a = x_1 a_n + a_1$$
$$P_a' = x_1' a_n + a_0$$
$$P_b = x_2 b_n + b_1$$
$$P_b' = x_2' b_n + b_0$$
$$P_c = (P_a \cdot P_b)c_n + c_1 = (x_1 a_n + a_1)(x_2 b_n + b_1)c_n + c_1$$
$$P_c' = (P_a' + P_b')c_n + c_0 = (x_1' a_n + a_0)(x_2' b_n + b_0)c_n + c_0$$
$$P_d = x_3 d_n + d_1$$
$$P_d' = x_3' d_n + d_0$$
$$P_e = x_4 e_n + e_1$$
$$P_e' = x_4' e_n + e_0$$
$$P_f = (P_d' P_c')f_n + f_1 = (x_3' d_n + d_0)[(x_1' a_n + a_0) + (x_2' b_n + b_0)c_n + c_0]f_n + f_1$$

$$P_f' = (P_d + P_c)f_n + f_0 = [(x_3d_n + d_1 + (x_1a_n + a_1)(x_2b_n + b_1)c_n + c_1]f_n + f_0$$
$$P_g = (P_e'P_c')g_n + g_1 = (x_4'e_n + e_0)[(x_1'a_n + a_0) + (x_2'b_n + b_0)c_n + c_0]g_n + g_1$$
$$P_g' = (P_e + P_c)g_n + g_0 = [x_4e_n + e_1 + (x_1a_n + a_1)(x_2b_n + b_1)c_n + c_1]g_n + g_0$$

and the propositions of the output line are

$$P_h = (P_fP_g)h_n + h_1$$
$$P_h' = (P_f' + P_g')h_n + h_0$$

The output propositions $P_h(n)$ and $P_h'(n)$ of the normal circuit are obtained by letting all the binary-line variables with subscript n equal to 1 and all the binary-line variables with subscripts 1 and 0 equal to 0 in the equations above,

$$P_h(n) = (x_1' + x_2')x_3'x_4'$$
$$P_h'(n) = x_1x_2 + x_3 + x_4$$

The output propositions of the circuit with a s-a-0 on line a are obtained by letting all the binary-line variables with subscripts n equal to 1 except $a_n = 0$ and all the binary-line variables with subscripts 0 and 1 equal to 0 except $a_0 = 1$:

$$P_h(a_0) = x_3'x_4'$$
$$P_h'(a_0) = x_3 + x_4$$

The set $T(a_0)$ of all the tests for testing the fault s-a-0 on line a is obtained from

$$
\begin{aligned}
T(a_0) &= P_h(n)P_h'(a_0) + P_h'(n)P_h(a_0) \\
&= (x_1' + x_2')x_3'x_4'(x_3 + x_4) + (x_1x_2 + x_3 + x_4)x_3'x_4' \qquad (6.4.1) \\
&= x_1x_2x_3'x_4'
\end{aligned}
$$

Thus, there is only one test for detecting the fault a_0, namely, 1100.

Suppose we want to detect the multiple fault: line d s-a-0 and line e s-a-1. Then the complete test set for detecting this fault is

$$
\begin{aligned}
T(d_0 \text{ and } e_1) &= P_h(n)P_h'(d_0 \text{ and } e_1) + P_h'(n)P_h(d_0 \text{ and } e_1) \\
&= (x_1' + x_2')x_3'x_4' \cdot 1 + (x_1x_2 + x_3 + x_4) \cdot 0 \qquad (6.4.2) \\
&= x_1'x_2x_3'x_4' + x_1'x_2'x_3'x_4' + x_1x_2'x_3'x_4'
\end{aligned}
$$

Thus, $T(d_0 \text{ and } e_1) = \{0000, 0100, 1000\}$.

As another example, consider the multiple fault: line a s-a-1, c s-a-0, and f s-a-1; the complete test set for detecting this fault is

$$
\begin{aligned}
T(a_1 \text{ and } c_0 \text{ and } f_1) &= P_h(n)P_h'(a_1 \text{ and } c_0 \text{ and } f_1) + P_h'(n)P_h(a_1 \text{ and } c_0 \text{ and } f_1) \\
&= (x_1' + x_2')x_3'x_4' \cdot x_4 + (x_1x_2 + x_3 + x_4) \cdot x_4' \qquad (6.4.3) \\
&= x_1x_2x_4' + x_3x_4'
\end{aligned}
$$

Thus, $T(a_1 \text{ and } c_0 \text{ and } f_1) = \{1100, 1110, 0010, 0110, 1010, 1110\}$.

A more convenient representation for the propositions for any line in a circuit is in terms of literal propositions, which are defined as follows:

DEFINITION 6.4.1

A *literal proposition* is an expression containing a single input literal and having a nested structure as $(\ldots((((xa_n + a_0)b_n + b_1)c_n + c_1)d_n + d_0)\ldots)$, which will be denoted by $x(a_0, b_1, c_1, d_0, \ldots)$.

Example 6.4.2

The propositions of Example 6.4.1 may be represented in terms of literal propositions as:

$$P_a = x_1(a_1)$$

$$P'_a = x'_1(a_0)$$

$$P_b = x_2(b_1)$$

$$P'_b = x'_2(b_0)$$

$$P_c = (P_a \cdot P_b)c_n + c_1 = (P_a c_n + c_1)(P_b c_n + c_1) = x_1(a_1, c_1)x_2(b_1, c_1)$$

$$P'_c = (P'_a + P'_b)c_n + c_0 = (P'_a c_n + c_0) + (P'_b c_n + c_0) = x'_1(a_0, c_0) + x'_2(b_1, c_0)$$

$$P_d = x_3(d_1)$$

$$P'_d = x'_3(d_0)$$

$$P_e = x_4(e_1)$$

$$P'_e = x'_4(e_0)$$

$$P_f = (P'_d P'_c)f_n + f_1 = (P'_d f_n + f_1)\cdot(P'_c f_n + f_1) = x'_3(d_0, f_1)\cdot[x'_1(a_0, c_0, f_1) + x'_2(b_1, c_0, f_1)]$$

$$P'_f = (P_d + P_c)f_n + f_0 = (P_d f_n + f_0) + (P_c f_n + f_0)$$
$$= x_3(d_1, f_0) + x_1(a_1, c_1, f_0)x_2(b_1, c_1, f_0)$$

$$P_g = (P'_e P'_c)g_n + g_1 = x'_4(e_0, g_1)[x'_1(a_0, c_0, g_1) + x'_2(b_1, c_0, g_1)]$$

$$P'_g = (P_e + P_c)g_n + g_0 = x_4(e_1, g_0) + x_1(a_1, c_1, g_0)x_2(b_1, c_1, g_0)$$

and, finally,

$$P_h = (P_f, P_g)h_n + h_1 = x'_3(d_0, f_1, h_1)\cdot[x'_1(a_0, c_0, f_1, h_1) + x'_2(b_1, c_0, f_1, h_1)]x'_4(e_0, g_1, h_1)$$
$$\cdot[x'_1(a_0, c_0, g_1, h_1) + x'_2(b_1, c_0, g_1, h_1)] \tag{6.4.4}$$

$$P'_h = (P'_f + P'_g)h_n + h_0 = x_3(d_1, f_0, h_0) + x_1(a_1, c_1, f_0, h_0)x_2(b_1, c_1, f_0, h_0)$$
$$+ [x_4(e_1, g_0, h_0) + x_1(a_1, c_1, g_0, h_0)x_2(b_1, c_1, g_0, h_0)] \tag{6.4.5}$$

It is important to note that a literal proposition represents a path from a primary input to a primary output. The following theorem is concerned with the value of a literal proposition under various line conditions along the path.

THEOREM 6.4.1

Let $x(a_{i_1}, a_{i_2}, \ldots, a_{i_m})$ be a literal proposition where i_j's are binary numbers. Then

$$x(a_{i_1}, a_{i_2}, \ldots, a_{i_m}) = x \qquad \text{if all } a_{n_j} = 1 \text{ and } a_{i_j} = 0, j = 1, 2, \ldots, m$$

$$= 0 \qquad \begin{array}{l} \text{if the subscript } i_k \text{ of the binary-line} \\ \text{variable } a_{i_k} \text{ that equals 1 appearing last} \\ \text{in the literal proposition is 0} \end{array}$$

$$= 1 \qquad \begin{array}{l} \text{if the subscript } i_k \text{ of the binary line} \\ \text{variable } a_{i_k} \text{ that equals 1 appearing last} \\ \text{in the literal proposition is 1} \end{array}$$

In words, a literal proposition will have the same value as the literal if the elements contained in it are normal. If one or more elements in the literal proposition are faulty, the value of the proposition will be 1 if the variable corresponding to the faulty wire appearing last in the literal proposition is 1, and 0 otherwise.

Proof: From the definition of literal proposition,

$$x(a_{i_1}, a_{i_2}, \ldots, a_{i_k}, \ldots, a_{i_m}) = ((\ldots((\ldots((xa_{n_1} + a_{i_1}) \cdot a_{n_2} + a_{i_2}) \ldots) a_{n_k} + a_{i_k}) \ldots) a_{n_m} + a_{i_m})$$

If all $a_{n_i} = 1$ and all $a_{i_j} = 0$, it is obvious that $x(0, 0, \ldots, 0) = x$. If one or more a_{i_j} is equal to 1, that is, one or more a_{n_j} is equal to 0, let the one that appears last in the literal proposition be denoted by a_{n_k}; then,

$$x(a_{i_1}, a_{i_2}, \ldots, a_{i_k}, \ldots, a_{i_m}) = ((\ldots(x(a_{i_1}, a_{i_2}, \ldots, a_{i_{k-1}}) \cdot a_{n_k} + a_{i_k}) \ldots) a_{n_m} + a_{i_m})$$

$$= ((\ldots(x(a_{i_1}, a_{i_2}, \ldots, a_{i_{k-1}})0 + 0) \ldots)1 + 0) = 0 \qquad \text{if } a_{i_k} = 0$$

$$= ((\ldots(x(a_{i_1}, a_{i_2}, \ldots, a_{i_{k-1}})0 + 1) \ldots)1 + 0) = 1 \qquad \text{if } a_{i_k} = 1$$

Hence, the theorem is proved. ∎

The convenience offered by the use of literal proposition in obtaining complete test sets is demonstrated by the following example.

Example 6.4.3

By using literal propositions and applying Theorem 6.4.1, the $P_h(n)$, $P'_h(n)$, $P_h(a_0)$, and $P'_h(a_0)$ of Example 6.4.1 may be easily found from the propositions of Example 6.4.2:

$$P_h(n) = (x'_1 + x'_2)x'_3x'_4$$

$$P'_h(n) = x_1x_2 + x_3 + x_4$$

$$P_h(a_0) = x'_3(0, 0, 0) \cdot [x'_1(1, 0, 0, 0) + x'_2(0, 0, 0, 0)]x'_4(0, 0, 0)$$

$$\cdot [x'_1(1, 0, 0, 0) + x'_2(0, 0, 0, 0)]$$

$$= x'_3[1 + x'_2]x'_4[1 + x'_2]$$

$$= x'_3x'_4$$

$$P'_h(a_0) = x_3(0, 0, 0) + x_1(0, 0, 0, 0)|_{a_n = 0}x_2(0, 0, 0, 0) + x_4(0, 0, 0)$$

$$+ x_1(0, 0, 0, 0)|_{a_n = 0}x_2(0, 0, 0, 0)$$

$$= x_3 + x_4$$

Similarly, the $P_h(d_0$ and $e_1)$ and $P_h'(d_0$ and $e_1)$ may be conveniently found by

$$P_h(d_0 \text{ and } e_1) = x_3'(1,0,0) \cdot [x_1'(0,0,0,0) + x_2'(0,0,0,0)]x_4'(0,0,0)|_{e_n=0}$$
$$\cdot [x_1'(0,0,0,0) + x_2'(0,0,0,0)]$$
$$= 1[x_1' + x_2] \cdot 0 \cdot [x_1' + x_2'] = 0$$
$$P_h'(d_0 \text{ and } e_1) = x_3(0,0,0)|_{d_n=0} + x_1(0,0,0,0)x_2(0,0,0,0) + x_4(1,0,0)$$
$$+ x_1(0,0,0,0)x_2(0,0,0,0)$$
$$= 0 + x_1x_2 + 1 + x_1x_2 = 1$$

The verification of $P_h(a_1$ and c_0 and $f_1)$ and $P_h'(a_1$ and c_0 and $f_1)$ is left to the reader as an exercise. It is seen that the propositions obtained by using and without using literal propositions are the same, but the procedure for obtaining them by using literal propositions is systematic and easy to compute, especially by use of a digital computer.

Another advantage of using literal propositions is the convenience offered by them in determining the complete *set of faults* that can be detected by a given input combination. This may be seen from the following example.

Example 6.4.4

Consider the same circuit of Fig. 6.4.1 with $x_1 = x_2 = 1$ and $x_3 = x_4 = 0$. Then we see that

$$P_h(n) = 0$$
$$P_h'(n) = 1$$

Since

$$T(a_{i_1}, a_{i_2}, \ldots, a_{i_m}) = P_h(n)P_h'(a_{i_1}, a_{i_2}, \ldots, a_{i_m})$$
$$+ P_h'(n)P_h(a_{i_1}, a_{i_2}, \ldots, a_{i_m})$$

then, for $x_1 = x_2 = 1$ and $x_3 = x_4 = 0$,

$$T(a_{i_1}, a_{i_2}, \ldots, a_{i_m})\Big|_{\substack{x_1=x_2=1\\x_3=x_4=0}} = P_h(a_{i_1}, a_{i_2}, \ldots, a_{i_m})\Big|_{\substack{x_1=x_2=1\\x_3=x_4=0}}$$
$$= x_3'(d_0,f_1,h_1) \cdot [x_1'(a_0,c_0,f_1,h_1) + x_2'(b_1,c_1,f_1,h_1)]$$
$$\cdot x_4'(e_0,g_1,h_1) \cdot [x_1'(a_0,c_0,f_1,h_1) + x_2'(b_1,c_0,f_1,h_1)]\Big|_{\substack{x_1=x_2=1\\x_3=x_4=0}}$$

$$(6.4.6)$$

The set of faults that can be detected by the test 1100 is the set of all possible combinations of the values of the variables in Eq. (6.4.6) which make the equation equal to 1. One such combination, for instance, is $a_0 = 1$ and $b_1 = 1$.

$$T(a_{i_1}, a_{i_2}, \ldots, a_{i_m})\Big|_{\substack{x_1=x_2=1\\x_3=x_4=0}} = x_3' \cdot [0+1] \cdot x_4'[0+1]\Big|_{\substack{x_1=x_2=1\\x_3=x_4=0}}$$

Other combinations that will make Eq. (6.4.6) equal to 1 are ($f_1 = 1$ and $g_1 = 1$), ($h_1 = 1$), and ($f_1 = 1$, $g_1 = 1$, and $h_1 = 1$). Hence, the literal proposition can be used not only for deriving a complete test set for any given multiple fault, but also for determining the set of faults that can be detected by a specified test.

Exercise 6.4

1. For the circuit in Fig. P6.4.1, find the complete set of tests using the literal proposition method.

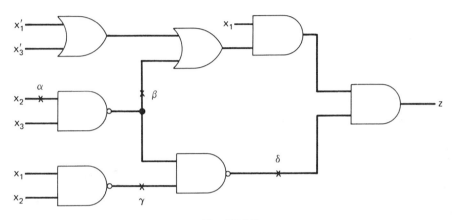

Fig. P6.4.1

   (a) The single fault α s-a-0.
   (b) The multiple fault α s-a-0 and β s-a-1.
   (c) The multiple fault α s-a-0, β s-a-1, γ s-a-0, and δ s-a-1.

2. For the following combinational circuit, find the complete test set for detecting the multiple fault: s-a-0 occurring at lines α, β, γ, δ, and ϵ, using the literal proposition method.

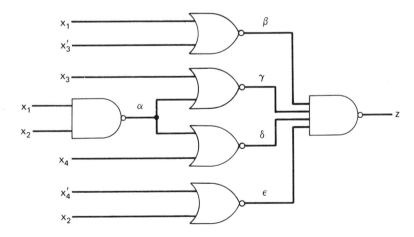

6.5 *D*-Calculus

Among all the algorithms that so far have been developed in the literature for handling the large-circuit fault-detection problem by computer, the *D*-algorithm based on the *D*-calculus is most practical. The main difference between this approach and the others (the main advantage of this approach over the others) is that it does not require representing and storing the complete circuit to be tested in terms of tables, formulas, or equations; this makes it possible to apply it to large multilevel circuits, without encountering severe storage problems.

Before presenting the *D*-calculus, first make the following observation. The values of the signals along any sensitized paths are fault-dependent and thus may be denoted by a binary variable D which represents a signal that is 1 in the normal circuit and 0 in the faulty circuit, or the complement of D, that is, D', which represents the signal that is normally 0 but becomes 1 when the fault is present. Note that the definitions of D and D' could be interchanged, but should be consistent throughout the circuit. Thus, all D in a circuit implies the same value whether 0 or 1 and all D' will have the opposite value. With this meaning associated with D, the OR, AND, and NOT operations operating on 0, 1, D, and D' are given below.

$+$	0	1	D	D'		$\cdot$	0	1	D	D'		$'$	
0	0	1	D	D'		0	0	0	0	0		0	1
1	1	1	1	1		1	0	1	D	D'		1	0
D	D	1	D	1		D	0	D	D	0		D	D'
D'	D'	1	1	D'		D'	0	D'	0	D'		D'	D

Now, *we try to propagate the D from the place where the fault occurs to the circuit output.* Suppose that we choose to use the convention: all the lines with a s-a-0 fault are assigned with a D and all the lines with a s-a-1 fault are assigned with a D'. For example, the fault b s-a-1 in the circuit of Fig. 6.5.1(a) will be represented by a D'. Suppose that it is desired to detect this fault. We attempt to propagate the D' on line b to the output z along the path *beprs*. Recall step 3 of the path-sensitizing method; we set $j = 0, q = m = 1$, and $n = 1$. The $n = 1$ and $q = 1$ requires $k = 1$ and $l = 0$, respectively, which leads to contradiction. In other words, the chosen path *beprs* is *not* sensitizable. But if we try to propagate the D' on line b to the output along the two paths simultaneously, as indicated in dark lines in the circuit of Fig. 6.5.1(b), the D' on line b can propagate to the output as shown in Fig. 6.5.1(b) and a test for detecting this fault is obtained which is $x = 1, x_2 = 0$, and one of the inputs x_3 and x_4 is 0.

From this example, we see that for small circuits such as the one in Fig. 6.5.1, it is generally not too difficult to try to propagate a D or a D' signal to the primary output terminals, even considering all possible multiple paths. Since a stuck-at fault may occur anywhere in a circuit, say it occurs on an internal line in a circuit, then we must try to propagate the D or D' signal that represents it to both the primary

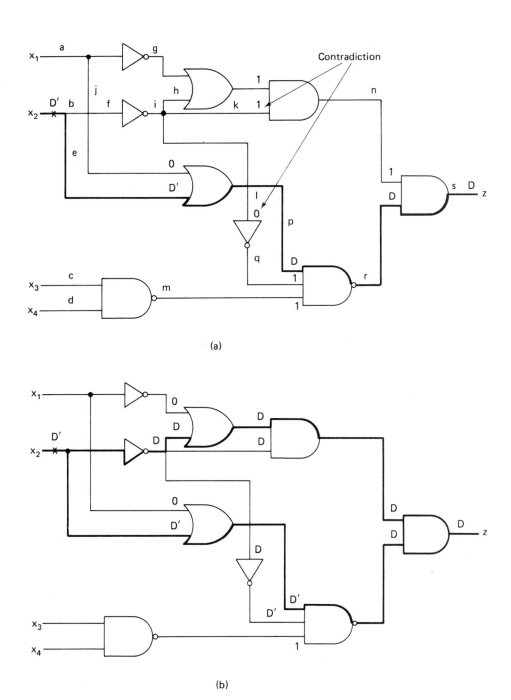

(a)

(b)

Fig. 6.5.1

outputs and the primary inputs. For large-circuit fault detection (which is our main objective), a means for effectively performing the necessary computations involving D signals in any part of a circuit and a systematic way to try to propagate a D or D' signal to the primary inputs and outputs and to find a consistent input variable assignment, i.e., a test to detect and/or to locate the fault, are needed. The former is the D-calculus and the latter is the D-algorithm.

From the above example, it is seen that a five-valued logic $(0, 1, d, D, D')$ will be used to describe the behavior of a circuit with failures. The symbol $D(D')$ denotes a logic value which is $1(0)$ in the "good" circuit C, while it is $0(1)$ in the same circuit with failure F, denoted C_F. In order to propagate or to drive a D or D' signal from a faulty line to the primary inputs and outputs of a circuit, we need to describe each logic block (gate) by a *singular cover*. Let the functional characteristics of a logic block (gate) be $f(x_1, \ldots, x_n)$, which is represented as a function of $(n+1)$ variables $g(x_1, \ldots, x_n, x_{n+1})$, such that $g(a_1, \ldots, a_n, a_{n+1}) = 1$ iff $a_{n+1} = f(a_1, \ldots, a_n)$. For example, consider an n-input AND-gate whose output function $f(x_1, \ldots, x_n) = 1$ iff all $x_1 = \ldots = x_n = 1$. Therefore, the function $f(x_1, \ldots, x_n, f) = 1$ iff

$$f(1, 1, \ldots, 1) = 1$$
$$f(0, d, \ldots, d) = 0$$
$$\vdots$$
$$f(d, d, \ldots, 0) = 0$$

where d denotes "don't care," which is one of the five logic values in the D-calculus. The condensed truth table of the function $g(x_1, \ldots, x_n, f)$ may be represented by

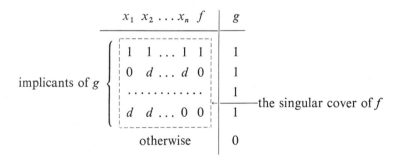

DEFINITION 6.5.1

The set of all prime implicants of the function g is called the *singular cover* of the function f.

For example, the singular cover of an n-input AND-gate is

x_1	x_2	...	x_n	f
1	1	...	1	1
0	d	...	d	0
...	...	...	...	...
d	d	...	0	0

The singular covers of the six gate elements are shown in Table 6.5.1. The function g may also be represented by

$$x_1 x_2 \ldots x_n$$

	00...0	00...1			11...10	11...1
f 0	1	1	1	...	1	0
1	0	0	0	...	0	1

$$g(x_1, \ldots, x_n, f)$$

and each row in the singular cover which corresponds to a cube marked with 1 in the above map is called a *singular cube*. Such cubes are called *prime*, when no larger singular cube of the function can exist which contains the given cube (prime singular cubes correspond to prime implicants).

DEFINITION 6.5.2

A *primitive D-cube of failure* (*pdcf*) of the fault of a block is a test (a set of completely specified inputs) of the block, which causes the output of the block (assuming single output blocks) to be different from its normal value, if a given fault is present in the block.

For example, if a s-a-0 fault is present at any one or more of the inputs $x_1, \ldots, x_n$ of an AND gate, a test $x_1 = \ldots = x_n = 1$ will detect the fault because the output will be 1 if no fault is present and 0 if the s-a-0 fault is present. This is represented by the following primitive *D*-cube of the fault

$$x_1 \ldots x_k \ldots x_n \quad f$$
$$1 \ldots 1 \ldots 1 \quad D$$

where D represents the condition under which the normal output is 1 and the faulty output is 0. For example, the primitive *D*-cubes of the faults s-a-0 and s-a-1 of an OR

TABLE 6.5.1 The Singular D-cubes, of Failure and Propagation D-cubes of the Six Gate-Elements

Gate	Symbol	Singular cover (condensed truth table) (SC)	Primitive D-cubes of failure (pdcf)	Propagation D-cubes (pdc)
AND	x1, x2 → f	x1 x2 f 1　1　1 0　–　0 –　0　0	x1 x2 f 1　1　D 0　–　D' –　0　D'	x1 x2 f D　1　D 1　D　D
OR	x1, x2 → f	x1 x2 f 1　–　1 –　1　1 0　0　0	x1 x2 f 1　–　D –　1　D 0　0　D'	x1 x2 f D　0　D 0　D　D
NOT	x1 → f	x1 f 0　1 1　0	x1 f 0　D 1　D'	x1 f D　D'
NAND	x1, x2 → f	x1 x2 f 0　–　1 –　0　1 1　1　0	x1 x2 f 0　–　D –　0　D 1　1　D'	x1 x2 f D　1　D' 1　D　D'
NOR	x1, x2 → f	x1 x2 f 0　0　1 –　1　0 1　–　0	x1 x2 f 0　0　D –　1　D' 1　–　D'	x1 x2 f D　0　D' 0　D　D'
	x1, x2 → f	x1 x2 f 1　0　1 0　1　1 0　0　0 1　1　0	x1 x2 f 1　0　D 0　1　D 0　0　D' 1　1　D'	x1 x2 f D　0　D D　1　D' 0　D　D 1　D　D'

gate are

$$
\begin{array}{cc} x_1 & f \\ 1 & D' \end{array} \quad \text{and} \quad \begin{array}{cc} x_1 & f \\ 0 & D \end{array}
$$

respectively.

A.   Procedure for Constructing pdcf
of a Gate Element

A systematic procedure for obtaining *pdcf* of any prescribed single fault of a gate (logic block) is described as follows:

1.   Let sc_N and sc_F denote the singular covers of the normal and fault blocks. Let N_0 and N_1 be sets of cubes in sc_N whose output coordinates are 0 and 1, respectively, and let F_0 and F_1 be sets of cubes in sc_F whose output coordinates are 0 and 1, respectively. For example, the singular covers T_N and T_F of the normal and faulty EXCLUSIVE-OR gate is

sc_N:	x_1	x_2	f
	1	0	1 } N_1
	0	1	1 }
	0	0	0 } N_0
	1	1	0 }

	x_1	x_2	f		sc_F:	x_1	x_2	f
x_1 s-a-0	1	d	d			0	0	1 } F_1
x_1 s-a-0	d	1	d	$\Rightarrow$		1	1	1 }
x_2 s-a-1	0	d	d'			0	1	0 } F_0
x_2 s-a-1	d	0	d'			1	0	0 }

2.   Primitive *D*-cubes of the fault are obtained by intersecting members of sets *N* and *F* with different subscripts, ignoring the output coordinates, using the rules given in Table 6.5.2.

TABLE 6.5.2   Intersection Rules for Obtaining the
Primitive *D*-Cubes of Failure and the Propagation
D-Cubes

$\sqcap$	0	1	d
0	0	$\varnothing$	0
1	$\varnothing$	1	1
d	0	1	d

The output coordinates in the intersection of members of N_1 and F_0 are assigned the value D, and the output coordinates in the intersection of members of N_0 and F_1 are assigned the value D'. If any coordinate of the intersection is $\varnothing$ (meaning that it is empty) the intersection is empty and is denoted by $\varnothing$. The nonempty primitive D-cubes of the EXCLUSIVE-OR gate are shown in Table 6.5.3. The nonempty primi-

TABLE 6.5.3   The $N_1 \sqcap F_0$ and $N_0 \sqcap F_1$

tive D-cubes obtained from $N_1 \sqcap F_0$ correspond to those with inputs that produce a 1 output from the normal circuit and a 0 output from the faulty circuit, and the nonempty primitive D-cubes obtained from $N_0 \sqcap F_1$ correspond to those with inputs that produce a 0 output from the normal circuit and a 1 output from the faulty circuit. For example, the primitive D-cube $01D$ indicates that for inputs $x_1 = 0$ and $x_2 = 1$, the circuit will output a 1 if there is no fault and a 0 if the fault is present. The primitive D-cube $00D'$ tells us that for inputs $x_1 = x_2 = 0$, the circuit will output a 0 if no fault is present, and a 1 if the circuit is faulty.

The *pdcf* for single-fault detection of the six gate elements are shown in Table 6.5.1. Another building block for effectively performing the necessary computations for propagating D signals in a faulty circuit is the propagation D-cubes of each gate of the circuit.

B.   Procedure for Constructing the Propagation
D-cubes (*pdc*) of a Gate Element

1.   Construct a set F_1' of cubes from N_0 by complementing the output and the faulty input entity if it is not "don't care" and a set F_0' of cubes from N_1 by complementing the output and the faulty input entity if it is not "don't care."

2.   The propagation *D*-cubes of a block are obtained by intersecting the *D*-cubes of N_1 and F_1' (or N_0 and F_0') using the rules given in Table 6.5.3.

3.   Assign *D* to the faulty input and *D* or *D′* to the output coordinate depending on whether the values of the output coordinate and the faulty input coordinate are the same or opposite to each other.

For example, for changes in input x_1, the F_1' and F_0' of the EXCLUSIVE-OR gate are

x_1	x_2	f	
1	0	1	F_1'
0	1	1	
0	0	0	F_0'
1	1	0	

The propagation *D*-cubes of the gate for these two faults at x_1 are shown in Table 6.5.4. It is seen that the same set of propagation *D*-cubes are obtained by $N_1 \sqcap F_1'$ and $N_0 \sqcap F_0'$. They are $D0D$ and $D1D'$. Here, the interpretation of the symbol is slightly different. *D* may be 0 or 1, but all *D′*s in a *D*-cube always have the same value. *D′* always has the value complementary to *D*. The two propagation *D*-cubes of the EXCLUSIVE-OR gate merely state that the output and the x_1 input have the same value if $x_2 = 0$ and have the opposite value if $x_2 = 1$.

The propagation *D*-cubes for single-fault two-input gate elements are given in Table 6.5.1.

TABLE 6.5.4   Propagation *D*-Cubes of the EXCLUSIVE-OR Gate

N_1			F_1'			$N_1 \sqcap F_1'$					
x_1	x_2	f	x_1	x_2	F	x_1	x_2	f		x_1	x_2	f
1	0	1	1	0	1	1	0	1	$\Rightarrow$ D	0	D
0	1	1	1	0	1	[$\varnothing$	$\varnothing$	1]	= Φ		
1	0	1	0	1	1	[$\varnothing$	$\varnothing$	1]	= Φ		
0	1	1	0	1	1	0	1	1	$\Rightarrow$ D	1	D'

N_1			F_0'			$N_0 \sqcap F_0'$					
x_1	x_2	f	x_1	x_2	f	x_1	x_2	f		x_1	x_2	f
0	0	0	0	0	0	0	0	0	$\Rightarrow$ D	0	D
1	1	0	0	0	0	[$\varnothing$	$\varnothing$	0]	= Φ		
0	0	0	1	1	0	[$\varnothing$	$\varnothing$	0]	= Φ		
1	1	0	1	1	0	1	1	0	$\Rightarrow$ D	1	D'

After having defined the primitive D-cubes of failure and the propagation D-cubes of gate elements and having shown the procedures for constructing them, now we want to show how a D signal can be propagated along a path from where the fault occurs to a primary output using the primitive D-cubes of failure and the propagation D-cubes. This is best illustrated through use of an example. Consider the circuit of Fig. 6.5.2(a). Suppose that there is a s-a-0 fault on line 3 and we want to derive a test to detect it. Assign a D signal to this line. The process of propagating a D signal from this line to the primary output using the primitive and propagation D-cubes given in Table 6.5.1 is described as follows:

1.   To test line 3 for being s-a-0 it is necessary to force a 1 on line 3 in the *good* circuit. So search for a primitive D-cube of failure of a NAND gate whose output signal is a D. From Table 6.5.5, we find that it must be 1 1 D, which is shown in the first row of the table of Fig. 6.5.2(b).

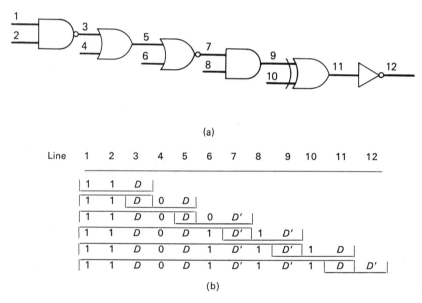

(a)

Line	1	2	3	4	5	6	7	8	9	10	11	12
	1	1	D									
	1	1	D	0	D							
	1	1	D	0	D	0	D'					
	1	1	D	0	D	1	D'	1	D'			
	1	1	D	0	D	1	D'	1	D'	1	D	
	1	1	D	0	D	1	D'	1	D'	1	D	D'

(b)

Fig. 6.5.2 Example showing how to propagate a D signal along a path from where the fault occurs to a primary output of a circuit using the propagation D-cubes.

2.   Since the signal on line 3 is a D signal, and the gate that it connects to is an OR-gate, search for a propagation D-cube of an OR-gate which is driven by a D input. We find that the propagation D-cube is D 0 D (see Table 6.5.5). Cascade the D-cube with the previous one; we obtain

Line	1	2	3	4	5		Line	1	2	3	4	5
	1	1	D					1	1	D	0	D
			D	0	D	$\Rightarrow$						

which is shown in the second row of the table of Fig. 6.5.2(b). Now, the D signal has been driven to line 5.

3.   The propagation of this D signal from line 5 to the output (line 12) is done similarly.

DEFINITION 6.5.3

Let α and β be two D-cubes. Then their D-*intersection* $\alpha \sqcap \beta$ is defined using the coordinate D intersection in Table 6.5.5 and the following rules:

1. $\alpha \sqcap \beta = \varnothing$ (empty) if any coordinate intersection is $\varnothing$.
2. $\alpha \sqcap \beta = \Psi$ (undefined) if any coordinate intersection is Ψ.
3. $\alpha \sqcap \beta =$ the cube formed from the respective coordinate intersections if neither 1 nor 2 holds.

Thus,

$$01xDx \sqcap 0x1DD' = 011DD' \quad \text{and} \quad 01xDx \sqcap 0x1D'D = \Psi$$

TABLE 6.5.5   Coordinate *D*-Intersection

$\sqcap$	0	1	x	D	D'
0	0	$\varnothing$	0	Ψ	Ψ
1	$\varnothing$	1	1	Ψ	Ψ
x	0	1	x	D	D'
D	Ψ	Ψ	D	D	Ψ
D'	Ψ	Ψ	D'	Ψ	D'

$\varnothing$, empty; Ψ, undefined.

The D-cube that describes the propagation of a D signal from an internal line to a primary output can be constructed rather conveniently by using the D-intersection. For example, the complete table of Fig. 6.5.2(b) may be replaced by the following equation.

$$11Dxxxxxxxxx \sqcap xxD0Dxxxxxxx \sqcap xxxxD0D'xxxxx$$
$$\sqcap xxxxxxD'1D'xxx \sqcap xxxxxxxxD'1Dx \sqcap xxxxxxxxxxDD'$$
$$= 11D0D1D'1D'1DD'$$

which is the last row of the table of Fig. 6.5.2(b).

The above discussion indicates that the propagation of a D signal along a path which corresponds to sensitizing the path may be obtained analytically by means of the primitive and propagation D-cubes. A formal procedure for propagating a D signal from anywhere in the circuit to the output will be described in the next section.

It should be noted that the D-calculus described above can be extended and used to propagate D signals in a circuit with a multiple fault. In detecting multiple faults, it is quite possible that there are two or more faults present along a path, and the effect of a fault may propagate to some other faulty line. It is conceivable that some additional rules must be established before we can propagate the D for such cases.

DEFINITION 6.5.4

Let $x = 0, 1, D,$ or D' and $y = D$ or D'. Define an operation $x \longrightarrow y$ as shown in Table 6.5.6.

TABLE 6.5.6 The $\longrightarrow$ Operation for Propagation of Multiple Faults

$x \to y$ x	D	D'
0	0	0
1	D	1
D	D	1
D'	0	D'

The meaning of $x \longrightarrow y$ is that "a signal x reaches a faulty line whose state is represented by y." These rules represent the consistency operation in the presence of multiple faults. The validity of the rules given in Table 6.5.6 may be seen from the following interpretation. Take $D \longrightarrow D = D$, for instance. It means that a signal that is normally 1 but changes to 0 propagates through a wire which is s-a-0 (represented by the second D). The meaning of $D \longrightarrow D = D$ is the signal on this wire is 1 if neither of the faults is present and is 0 if both of the faults are present. For the cases where one is present and the other is absent, the operation $D \longrightarrow D$ is undefined. As

N_1			F_0			$N_1 \sqcap F_0$		
x_1	x_2	f	x_1	x_2	f	x_1	x_2	f
1	0	1	0	1	0	$[\phi$	ϕ	$D] = \phi$
0	1	1	0	1	0	0	1	D
1	0	1	1	0	0	1	0	D
0	1	1	1	0	0	$[\phi$	ϕ	$D] = \phi$

N_0			F_1			$N_0 \sqcap F_1$		
x_1	x_2	f	x_1	x_2	f	x_1	x_2	f
0	0	0	0	0	1	0	0	D'
1	1	0	0	0	1	$[\phi$	ϕ	$D'] = \phi$
0	0	0	1	1	1	$[\phi$	ϕ	$D'] = \phi$
1	1	0	1	1	1	1	1	D'

Fig. 6.5.3 A circuit for illustrating D signals in the multiple fault case.

another example, $D \longrightarrow D' = D'$ means that the value on a wire having a s-a-1 fault, propagated through by a signal of truth value 0, is 0 if the fault is not present and is 1 if the fault is present. The rest of the rules can be explained similarly.

For example, consider the circuit in Fig. 6.5.3. Assume that there is a multiple fault to be detected which is composed of α s-a-0, β s-a-0, γ s-a-1, and δ s-a-0. The paths chosen to be sensitized are indicated. At the point γ, a signal D is propagated through it which results in $D \longrightarrow D' = 1$. A similar situation occurs at the point δ, which results in $D' \longrightarrow D = 0$. The test for detecting this multiple fault is then found to be $x_1 = 1$, $x_2 = 1$, and one of the inputs x_3 and x_4 is 0.

The definitions of singular cover, primitive D-cubes of failure, propagation of D-cubes, and so on, and the procedures for obtaining them for the single-fault case described above may be modified for the multiple case in a rather straightforward manner.

Exercise 6.5

1. Propagate each of the D signals indicated in the accompanying figures to one of the primary outputs using the primitive and propagation D-cubes.

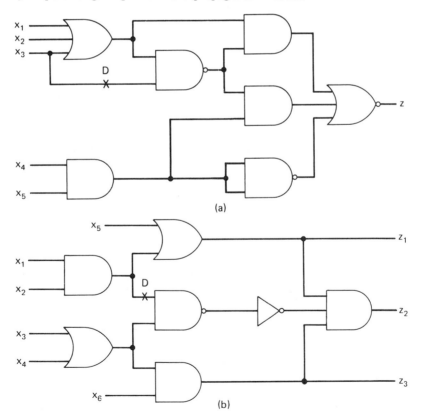

(a)

(b)

6.6 The *D*-Algorithm

The use of *pdcf* and *pdc* to propagate a D signal along a path was demonstrated in the previous section. Now we present a systematic procedure for obtaining a test for a given single fault. The procedure is known as the *D-algorithm*, which consists of the following two parts:

1. *The D-drive.* The first part of this algorithm is designed for driving a D signal from where the fault occurs to a primary output, hence the term *D-drive*. We start with a chosen primitive D-cube of the fault to successively intersect with propagation D-cubes of the blocks of the circuit in order to form a connected chain of D-coordinates to an output.

Any D-cube that represents a partially formed test during the D-drive is called a *test cube* and is represented by *tc* and a superscript denoting the step at which it is obtained. Associated with each test cube is an *activity vector*, consisting of the circuit lines to which D or D' has been propagated at this stage in the test generation. The activity vector will only consist of a single line number if the path under sensitization is a single path.

In carrying out the D-drive operation, a path and one of the primitive D-cubes of the fault under consideration are first chosen. The chosen primitive D-cube is the initial test cube, denoted by tc^0. Let r^0 be the activity vector associated with the initial test cube tc^0. The tc^0 is then intersected with one or a set of propagation D-cubes (this number depends on the number of paths, which is also the number of components of r^0), which has a D or D' at line r_i^0 and one or more D's or $\bar{D}$'s at the lines along the path that are closer to the output than line r_i^0. Let this intersection be denoted by tc^1. Treat tc^1 as tc^0 and repeat the process until a D or D' reaches the output.

2. *The consistency operation.* This operation corresponds to the tracing back from the gates in the path toward the inputs in order to specify a sufficient number of inputs to produce the desired internal signal. This operation is similar to that of step 1, except (1) the test cubes are intersected with the singular cover of each block whose output coordinate is already specified, and (2) it terminates whenever a $\varnothing$ appears in any line of the circuit. This means that the test under generation does not exist. Furthermore, the logic block to intersect with the test cubes in the consistency operation is moving from the output toward the inputs. If all the lines of the circuit are specified by the consistency operation without encountering any contradiction (i.e., a $\varnothing$ at any line), the test is thus found. Note that no consistency operation is needed if all the values of the lines of the circuit in the last test cube of the D-drive are already completed.

Thus, the D-algorithm may be described by the following five steps.

Step 1   Obtain the singular cover of each logic block of the circuit.

Step 2   Construct the primitive D-cubes of the fault under detection of each block from the singular covers of the normal and faulty blocks.

Step 3   Construct the propagation D-cubes.

Step 4   Construct the D-drive.

Step 5   The consistency operation.

The following example illustrates this algorithm.

Example 6.6.1

Construct a test to detect the fault on line 1 s-a-0 in the circuit of Fig. 6.6.1(a) using the *D*-algorithm.

The singular cover of each gate of the circuit is shown in Fig. 6.6.1(b). From the singular cover of each logic block, and following the procedure described in Step 3, the propagation *D*-cubes are found and shown in Fig. 6.6.1(c).

Now suppose the path *ABE* is chosen to be sensitized. The primitive *D*-cube of the fault is $11D$. The activity vector is 3. In the propagation *D*-cubes table of Fig. 6.6.1(c), we search for a propagation *D*-cube which has a *D* or *D'* on column 3 and *D*'s or $\bar{D}$'s on some other lines closer to the output than line 3. Such a propagation *D*-cube found is c_d. The intersection of tc^0 and c_d yields $tc^1 = 11D0D'$. The activity vector of tc^1 is 6. The propagation *D*-cube g_d having a *D* on line 6 and a *D* on line 8 is next chosen to intersect with tc^1, which is shown in the table. Since the activity vector of tc^2 is line 8, which is the output, the *D*-drive is thus completed.

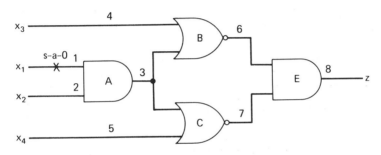

(a) A circuit with a single fault

		1	2	3	4	5	6	7	8
A	a	1	1	1					
	b	0	d	0					
	c	d	0	0					
B	d				0	0		1	
	e				1	d		0	
	f				d	1		0	
C	g				0	0		1	
	h				1	d		0	
	i				d	1		0	
E	j						1	1	1
	k						0	d	0
	l						d	0	0

(b) Singular cover

Fig. 6.6.1   Circuit and tables of Example 6.6.1.

		1	2	3	4	5	6	7	8
A	a_d	D	1	D					
	b_d	1	D	D					
B	c_d			D	0		D'		
	d_d			0	D		D'		
C	e_d			D		0		D'	
	f_d			0		D		D'	
E	g_d						D	1	D
	h_d						1	D	D

(c) Propagation D-cubes

		1	2	3	4	5	6	7	8	Activity vector
D-drive	tc^0	1	1	D						3
	$tc^1 = tc^0 \sqcap c_d$	1	1	D	0		D'			6
	$tc^2 = tc^1 \sqcap g_d$	1	1	D	0		D'	1	D'	8
Consistency operation	$tc^3 = tc^2 \sqcap g_d$	1	1	ϕ	0	0	D'	1	D'	A contradiction is present at line 3

(d) D-drive along path ABE and consistency operation

		1	2	3	4	5	6	7	8	Activity vector
D-drive	tc^0	1	1	D						3
	$tc^1 = tc^0 \sqcap c_d \sqcap e_d$	1	1	D	0	0	D'	D'		6, 7
	$tc^2 = tc^1 \sqcap g_d \sqcap h_d$	1	1	D	0	0	D'	D'	D'	8

(e) Simultaneous D-drive along paths ABE and ACE

Fig. 6.6.1 (Continued)

The final step is to examine whether the tc^2 will lead to any contradiction in assigning line values. In doing this we look for a singular cover which has a 1 on line 7, a 0 or 1 on line 5, but does not have either a 0 or a 1 on lines 3, 6, and 8. We find that such a singular cover does not exist in the table of Fig. 6.6.1(b). The closest one to such a specification is the singular cover g, which leads to an empty intersection at line 3, Fig. 6.6.1(d). Hence, the path *ABE* is not sensitizable.

The other single path of the circuit is path *ACE*. Since paths *ABE* and *ACE* are symmetrical with respect to the inputs and the output and since path *ABE* is not sensitizable, path *ACE* is also not sensitizable. We therefore next try to sensitize paths *ABE* and *ACE* simultaneously. This is shown in Fig. 6.6.1(e). The simultaneous *D*-drive along paths *ABE* and

ACE terminates at tc^2 with a completely specified test cube $11D00D'D'D'$; and thus no consistency operation is needed. The test for detecting the s-a-0 at input x_1 is $x_1 = x_2 = 1$ and $x_3 = x_4 = 0$.

As a final remark, the *D*-algorithm presented above can be modified and made applicable to circuits with multiple faults and multiple outputs.

Exercise 6.6

1. Find a test for detecting each of the following single faults in the circuit of Fig. P6.6.1 using the *D*-algorithm.

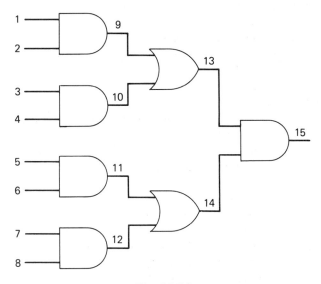

Fig. P6.6.1

(a) line 1 s-a-0
(b) line 10 s-a-1
(c) line 14 s-a-1

2. Detect the faults (a) line 7 s-a-0 and (b) line 8 s-a-1 in the circuit of Fig. P6.6.2 using the *D*-algorithm.

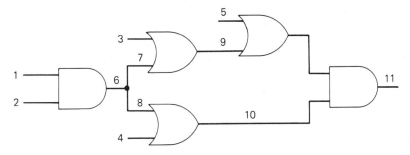

Fig. P6.6.2

6.7 DALG-II and TEST-DETECT

Based on the D-algorithm, two computer-oriented fault-detection algorithms, DALG-II (D-Algorithm Version II) and TEST-DETECT, are described. DALG-II computes a test to detect a failure in a large combinational circuit, and TEST-DETECT ascertains all failures detected by a given test. Only single faults are assumed. Both algorithms are based upon the utilization of the D-calculus that was introduced in the previous section. They are implemented on the APL (A Programming Language) interpreter at the IBM Watson Research Center since 1966, and have proved to be very efficient for detecting single faults in large combinational circuits without encountering severe storage problems. This is because in DALG-II, the sets of $pdcf$'s, pdc's, and sc's for each logic block are not actually stored; rather, these quantities are computed on an as-needed basis, since for the function types used, AND, OR, NOR, NAND, and XOR (EXCLUSIVE-OR), the computation is simple. Moreover, DALG-II does not require representing the circuit to be tested in terms of equations (such as Boolean equations, ENF, SPOOF, literal propositions) and tables (fault tables), as required by many other methods. Another relevant feature is its capability of simultaneously driving many D-paths, or sensitized paths, to the primary outputs; this is needed to guarantee that a test for a failure is always found if one exists. On the other hand, this capability may increase the computation time in practical application.

The DALG-II algorithm defines the *activity vector A* to consist of a listing of all those blocks which, at the given stage of the computation, have D's (or $\bar{D}$'s) on their inputs, an x on their output, and whose pdc intersection into the test cube has not been found to lead to an inconsistent line assignment. Thus the active blocks are on the *D-frontier*, or leading edges of the connected *D-chain* that the algorithm is driving to the primary outputs. Decisions must be made about what to do regarding driving the D's through these blocks. As soon as a decision to drive a D through a block is made (in other words, to D-intersect with a primitive D-cube for that block), the full implications of the resulting line assignments from x to 0, 1, D, or $\bar{D}$ are carried throughout the circuit. Here, "implications" means all other line assignments that are a forced consequence of the totality of assignments already made up to this point. For example, if an AND block with an output equal to x receives a 0 input, its x output must be set to 0; or if an AND block with all inputs x receives a 1 output, all its inputs must be set to 1; or whenever an AND block has all its inputs set to 1 or D, the output must be set to a D, and so on. The use of the implication concept at each step makes DALG-II very efficient, since inconsistent decisions will be discovered much sooner in the processing. It should be noted that in DALG-II, the lowest numbered block in A, skipping over those it has already found to lead to an inconsistency, is picked and processed in the drive to the primary outputs (PO's).

DEFINITION 6.7.1

Define ∂C to be the D-intersection of all the cubes in the set

$$C = \{c_1, c_2, \ldots\}, \qquad \text{that is, } \partial C = c_1 \sqcap c_2 \sqcap \ldots$$

DALG-II works like this. A primitive D-cube of failure ($pdcf$) will be selected, and all singular cubes or D-cubes whose choice is implied by $pdcf$ will be placed in a set $I^0(pdcf)$. Then an initial test cube $tc^0 = pdcf \sqcap [\partial I^0(pdcf)]$ is formed and the activity vector A^0 defined. At the kth step in the algorithm, a primitive D-cube pdc^k is developed for some block in the activity vector A^{k-1}. Then the set $I^k(tc^{k-1}, pdc^k)$ of all those cubes that are new implications of what has been done to date is generated. Finally, the test cube $tc^k = tc^{k-1} \sqcap pdc^k \sqcap [\partial I^k(tc^{k-1}, pdc^k)]$ is formed and A^k defined. The iterations cease when a primary output has been reached with the D-chain of tc^k. At this point, blocks that were assigned output values 0 and 1 by the D-drive without having appropriate inputs to account for this output, have their signals "driven" back toward the primary inputs. This is done by generating intersections with the appropriate singular cubes in an attempt to find a consistent assignment of the primary inputs that will provide the desired signals. Because the implication process was used during the D-drive, this consistency operation will always involve a choice of one of several possible singular cubes for each block it encounters; that is, if there is no choice, the implication process would have already used the singular cube in forming some tc^k.

When either the D-chain "dies" (A becomes empty) or an inconsistency is discovered, the process must back up to the last arbitrary choice of pdc^k or singular cube and make an alternative selection.

Example 6.7.1

Construct a test for the failure "line 6 stuck-at-0" for the circuit in Fig. 6.7.1 .To test a line for being stuck-at-0 it is necessary to force a 1 on that line in *the good circuit*, so *pdcf* must be $x00xxDxxxxxx$. Since there are no implications for these three line assignments, $I^0 = \varnothing$, $tc^0 = pdcf$, and $A^0 = \{9, 10\}$. In picking the block in A^0, for which pdc^1 will be gen-

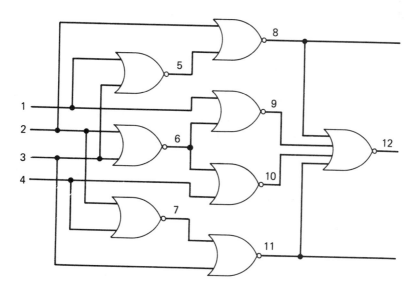

Fig. 6.7.1 Example for illustrating the D-algorithm, version II.

erated, the first arbitrary choice arises. DALG-II would pick the lowest numbered entry in A, line 9, and form $pdc^1 = 0xxxxDxx\bar{D}xxx$. The output of block 5 is forced to be a 1 since both inputs are now 0, and this assignment in turn forces the output of 8 to 0. Thus, I^1 $(tc^0, pdc^1) = \{0x0x1xxxxxxx, xxxx1xx0xxxx\}$. The new test cube tc^1 is constructed:

$$
\begin{array}{llll}
x\,0\,0\,x\,x\,D\,x\,x\,x\ \ x\,x\,x & tc^0 \\
0\,x\,x\,x\,x\,D\,x\,x\,D'\,x\,x\,x & pdc^1 \\
\left.\begin{array}{l}
0\,x\,0\,x\,1\,x\ x\,x\,x\ \ x\,x\,x \\
\Box\,x\,x\,x\,x\,1\,x\ x\,0\,x\ \ x\,x\,x
\end{array}\right\} I^1 \\
\hline
\overline{tc^1\ \ \ 0\,0\,0\,x\,1\,D\,x\,0\,D'\,x\,x\,x}
\end{array}
\tag{6.7.1}
$$

The activity vector A^1 becomes $\{10, 12\}$.

Again, picking the lowest numbered active block, in this case 10, $pdc^2 = xxx0xDxxxD'xx$. The 0's at coordinate 2 of tc^1 and coordinate 4 of pdc^2 imply a 1 on line 7, which in turn implies a 0 on line 11. At this point, line 9 and 10 are D', and line 8 and 11 are 0, thereby forcing a D on the PO line 12. Thus $I^2(tc^1, pdc^2) = \{x0x0xx1xxxxx, xxxxxx1xxx0x; xxxxxxx0D'D'0D\}$ and tc^2 is

$$
\begin{array}{llll}
0\,0\,0\,x\,1\,D\,x\,0\,D'\,x\ \ x\,x & tc^1 \\
x\,x\,x\,0\,x\,D\,x\,x\,x\ \ D'\,x\,x & pdc^2 \\
\left.\begin{array}{l}
x\,0\,x\,0\,x\,x\ 1\,x\,x\ \ x\ \ x\,x \\
x\,x\,x\,x\,x\,x\ 1\,x\,x\ \ x\ \ 0\,x \\
\Box\,x\,x\,x\,x\,x\,x\ x\,0\,D'\,D'\,0\,D
\end{array}\right\} I^2 \\
\hline
\overline{tc^2 = 0\,0\,0\,0\,1\,D\,1\,0\,D'\,D'\,0\,D}
\end{array}
\tag{6.7.2}
$$

The D-chain has reached the primary output, so the consistency operation would normally be started. However, in this example all blocks with outputs 0 or 1 have their signals already accounted for by their inputs. Thus, tc^2 is $c(T, F)$, with input test pattern $T = 0000$.

It is informative to examine what happens if block 12 in A^1, instead of block 10, is selected for forming pdc^2. Then $pdc^2 = xxxxxxx0D'00D$, and the following initial implications may be noted: the 0 on line 11 forces a 1 on line 7 (line 3 is fixed at 0), and the 0 on line 10 forces a 1 on line 4 (line 6 is currently a D). When block 7 is examined for implications, it is found to be inconsistent: the NOR-gate cannot have an input value 1 and an output value 1.

The lack of success in the second part of the example serves to illustrate that, for the given failure, the only test that exists depends on having the failure signal propagate through a chain which reconverges at block 12. "Sensitizing" any single path will *not* produce a test!

Although the D-chain in the successful test passes through three blocks (9, 10, and 12), only two decisions had to be made. Implication automatically extended the D-chain through block 12. Because DALG-II always selects the lowest numbered block in A (which it has not found to lead to an inconsistency), and because the block numbers are ordered as previously mentioned, the extension of the D-chain is an orderly level progression to the outputs. Further, DALG-II stops as soon as any segment of the D-chain reaches a PO, since leaving any other segments of the D-chain uncompleted cannot result in any contradictions to the test as currently developed.

By leaving such segments uncompleted we give them don't-care status (they may or may not reach other PO's, depending upon what is done in consistency).

The APL version of DALG-II is published in reference 10.

TEST-DETECT is a kind of converse of the D-algorithm. Given a test T TEST-DETECT computes the set of *all* faults detected by T.

DEFINITION 6.7.2

Let F be a logic fault in a circuit and T be a test. Define $c(T, F)$ as the D-cube defined by T and F in the following manner. If, for any line or coordinate i, the good and the failing circuit have the same value (either 0, 1, or x); this value is placed in $c(T, F)$ at coordinate i; if the good circuit has a 1 (0) and the failing circuit a 0 (1) on line i, then $c(T, F)$ receives a $D(D')$ in coordinate i. This choice of how the D and D' are assigned is, as with the *pdcf*, by convention.

For example, if $F =$ line 6 s-a-0 in the circuit of Fig. 6.7.1 and $T = 0000$, then $c(T, F) = 00001D10D'D'0D$, which is the same as tc^2 in the example above. The importance of the D-cube $c(T, F)$ is that it allows one to describe completely the operation of the good and failing circuit under the input T: set $D = 0$ (thus $D' = 1$) and one obtains the good circuit's response; set $D = 0$ ($D' = 1$) and one obtains the failing circuit's response. If T is indeed a test for the existence of F, then $c(T, F)$ must have a D or D' on at least one primary output so that these two responses would differ. It has been shown [8] that $c(T, F)$ must contain a connected chain of D's and/or $\bar{D}$'s linking the site of the fault to a primary output, and that $c(T, F)$ could be constructed by a D-intersection of primitive D-cubes and singular cubes of the blocks in the circuit as seen in Eq. (6.7.2). Here, *connected D-chain* is construed to mean that a block (other than the site of fault) has a D or D' on its output if and only if at least one of its inputs has value D or D'. A method for ascertaining all faults D detected by T would be to construct $c(T, F)$ for each F and assay whether or not any PO-coordinate had value D or D'. TEST-DETECT is a substantially refined version of this procedure.

If TC denotes the vector of signals assigned by the test T to each line of the logic circuit, and if TC_i has the value $\alpha = 1$ or 0, T cannot test for the failure "line i stuck-at-α," since the good and failing circuits would appear identical. Thus, if F_i denotes the failure line i stuck-at-$\bar{\alpha}$, we need only concern ourselves with investigating such F_i. Further, the set of failures detected by T can now be uniquely specified by a vector FD of lines on which T detects a failure: if $i \in FD$ and $TC_i = \alpha$, then T detects the failure F_i.

A method will now be described for effectively computing $c(T, F_i)$, and hence for computing whether or not i can be placed in FD. We do this by constructing, in the manner of the D-algorithm, the D-chain emanating from line i. The activity vector A and its use in producing an orderly level-by-level march to the PO's is as before. However, all line-signal values are fixed so that we are only concerned with whether or not the D-chain propagates through the given block, that is, whether or not the inputs correspond to the input section of a *pdc* that has a D or $\bar{D}$ at its output coordinate. For the sake of storage efficiency, $c(T, F)$ is never explicitly generated. Instead, it is recorded that a line in TC has a D or $\bar{D}$ on it by placing the line's number in a list DL. Thus, if $j \in DL$ and $TC_j = 1$ (or 0), then coordinate j in $c(T, F_i)$ is D (or $\bar{D}$); otherwise, TC and $c(T, F_i)$ agree. Since we are only interested in those lines for which active computation is being carried out, the definition of DL (and its use in the APL program) will be modified so as to exclude those lines which have already been processed and do not have successors in A.

There are three termination rules for stopping the computation of $c(T, F_i)$.

RULE T1

When the D-chain reaches a PO, $DL \sqcap PO \neq \varnothing$, terminate the generation of $c(T, F_i)$, since it can be concluded that T detects F_i.

RULE T2

When the D-chain dies, $A = \varnothing$, terminate the generation of $c(T, F_i)$, since it can be concluded that T does not detect F_i.

The third rule is based on the following theorem.

THEOREM 6.7.1

If at any stage in the computation of $c(T, F_i)$ there is only one line j in DL, then i is in FD if and only if $j \in FD$.

Proof: For the portion of the circuit corresponding to those blocks with number greater than j, the inputs (0, 1, D, or $\bar{D}$) are exactly the same as when j was originally examined since, by hypothesis, j is the only line in DL. Hence, $c(T, F_i)$ must agree with $c(T, F_j)$ in these positions, and any conclusions made regarding j being in FD will also hold for i. ∎

From Theorem 6.7.1, we have the following termination rule, which truncates the formation of $c(T, F_i)$ and greatly increases the efficiency of TEST-DETECT.

RULE T3

When there is only one line j in DL and j has been computed to be in FD, terminate the generation of $c(T, F_i)$; by Theorem 6.7.1, i is in FD.

The computation of $c(T, F_i)$ is illustrated by the following example.

Example 6.7.2

Consider the same circuit of Fig. 6.7.1, which is repeated in Fig. 6.7.2. Suppose that we would like to compute the set of all faults detected by the test $T = 0000$. For this test TC is found to be

Line	1	2	3	4	5	6	7	8	9	10	11	12
Line value	0	0	0	0	1	1	1	0	0	0	0	1

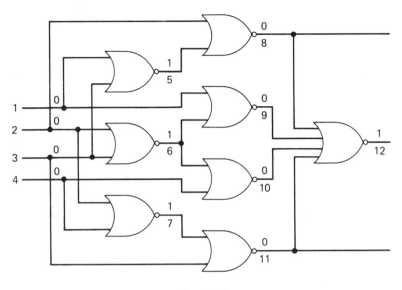

Fig. 6.7.2

which are shown as the numbers above the lines. Let us examine each line individually from the largest number to the smallest.

Lines 12, 11, and 8 If a line is a *PO*, by Rule T1, it can immediately be placed in *FD*. Thus, lines 12, 11, and 8 are in *FD*.

Lines 10 and 9 Line 10 has only one successor, 12, and when this entry in *A* is examined it is found that a *pdc* with required values (a *D'* on line 9; 0's on lines 8, 10, and 11; and a *D* on line 12) exists for block 12. Thus, $DL = 12$ and $A = 12$. Since *DL* has been reduced to a single entry that is in *FD*, by Rule T3, 10 must be in *FD*. For the similar reason, line 9 is also in *FD*.

Line 6 Investigating line 6 means that initially $D1 = 6$ and $A = 9, 10$. Since appropriate *pdc*'s exist for both 9 and 10, two iterations in the *D*-chain extension will produce $DL = 9$, 10, and $A = 12$. The cube $0D'D'0D$ is a *pdc* (involving coordinates 8 through 12, respectively) for block 12, so the *D*-chain can be extended through 12. Now $DL = 12$ and Rule *T*3 can again be applied to show line 6 is in *FD*.

Lines 5 and 7 To examine line 5, initially let $DL = 5$ and $A = 8$. There exists a *pdc* for block 8, which is

$$2 \quad 5 \quad 8$$
$$0 \quad D \quad D'$$

and by Rule *T*3, 5 is in *FD*. Similarly, line 7 is in *FD*.

Lines 4 and 1   Examining line 4, we let initially $DL = 4$ and $A = 7, 10$. There does not exist a *pdc* for block 10 with 1 on line 6, D on line 4, and D' on line 10, but there does exist a *pdc* for block 7, which is

$$
\begin{array}{ccc}
2 & 4 & 7 \\
0 & D' & D
\end{array}
$$

so the D-chain can be extended through 7. Now $DL = 7$. There is only one line in DL; Rule $T3$ can again be applied to show that line 4 is in FD. Similarly, line 1 is in FD.

Lines 3 and 2   Finally, we examine lines 3 and 2. Since both lines 3 and 2 have a value of 0 and line 6 has a value of 1, *pdc* cubes

$$
\begin{array}{cccccc}
2 & 3 & 6 & 2 & 3 & 6 \\
0 & D' & D & D' & 0 & D
\end{array}
$$

exist for block 6, so the D-chains can be extended through 6. Now $DL = 6$, which is a singleton. By Rule T3, both lines 3 and 2 are in DF.

In summary, we find that every line of this circuit is in DF, which means that any single logic fault in the circuit is detectable by the test 0000.

The APL version of TEST-DETECT is published in reference 19.

Exercise 6.7

1. The circuit in Fig. P6.7.1 is a part of the Control Matrix Network (C.M.N.) of the GE-115 Central Processing Unit. The whole C.M.N. has about 100 outputs; 60 independent inputs, and it is implemented with about 400 NOR gates. The network represented in Fig. P6.7.1 realizes the 31st output of the C.M.N. It contains 12 NOR gates and has 15 independent input lines.
   (a) Use DALG-II to derive a test that will detect the fault line 45 s-a-0.
   (b) Use TEST-DETECT to show that a test
   ```
   1 2 3 4 5 6 7 8 9 10 11 12 13 14 15
   d d d d 0 0 0 0 1 0  d  d  1  1  d
   ```
   where *d* may be either 0 or 1, will detect the fault line 45 s-a-0.
   (c) Use TEST-DETECT to find all the single faults that can be detected by the test
   ```
   1 2 3 4 5 6 7 8 9 10 11 12 13 14 15
   1 1 1 1 0 0 0 0 1 0  0  0  1  1  1
   ```

Bibliographical Remarks

Four books [1–4] on fault detection and diagnosis in digital circuits and systems and two *IEEE Transactions on Computers* special issues on fault-tolerant computing [5,6] are excellent references for materials covered in this chapter. In particular, Chapter 7 of reference 4 contains many tabular, graphical, and analytical methods for deriving fixed- and adaptive-sched-

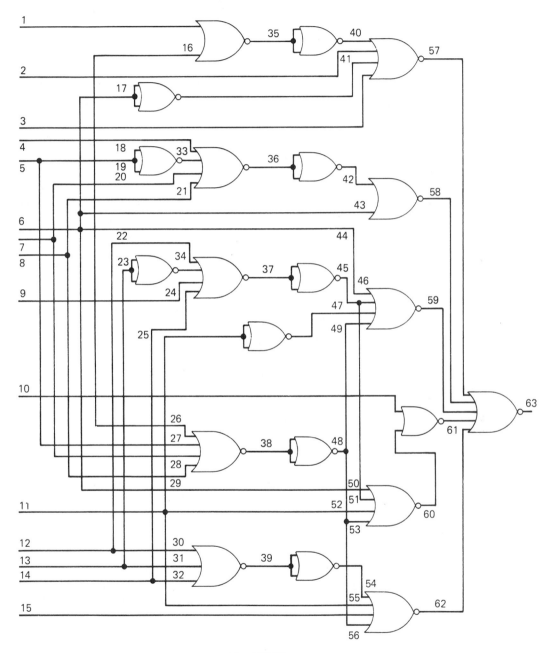

Fig. P6.7.1

uled single- and multiple-fault detection and location experiments for combinational circuits that are not included in this chapter.

The first paper on the application of Boolean differences to the derivation of fault-detection tests was published by Sellers et al. [7]. An efficient algorithm for generating a complete test set for detection of a single fault was given by Yau and Tang [8]. Multiple-fault detection in combinational circuits using the Boolean difference is described in reference 9. Other relevant references on this subject are references 10 and 11. The derivation of minimum test sets for unate logic circuits presented in Section 6.2 is based in part on the paper of Bearnson and Carroll [12]. Three excellent papers on the analysis of multiple faults in combinational circuits were given by Gault et al. [13], Bossen and Hong [14], and Du and Weiss [15]. Reference 16 is the original paper of the literal proposition method.

References 17–20 contain the original and revised versions of the D-algorithm. DALG-II and TEST-DETECT, the two practical computer programs for test generation and fault analysis, have been published by Roth et al. [19]. The algorithms for generating fault location tests can be found in Su and Cho [21].

References

Books and Journals

1. CHANG, H. Y., E. G. MAMING, and G. METZE, *Fault Diagnosis Digital Systems*, Wiley, New York, 1970.

2. FRIEDMAN, A. D., and P. R. MENON, *Fault Detection in Digital Circuits*, Prentice-Hall, Englewood Cliffs, N.J., 1970.

3. SELLERS, F. F., M. Y. HSIAO, and L. W. BEARNSON, *Error Detecting Logic*, McGraw-Hill, New York, 1968.

4. LEE, S. C. *Digital Circuits and Logic Design*, Prentice-Hall, Englewood Cliffs, N.J., 1976.

5. Special issue on Fault-Tolerant Computing, *IEEE Trans. Computers*, Vol. C-20, November 1971.

6. Special issue on Fault-Tolerant Computing, *IEEE Trans. Computers*, Vol. C-22, March 1973.

Papers

7. SELLERS, F. F., JR., M. Y. HSIAO, and L. W. BEARNSON, "Analyzing Errors with the Boolean Difference," *IEEE Trans. Computers*, Vol. C-17, July 1968, pp. 676–683.

8. YAU. S. S., and Y. S. TANG, "An Efficient Algorithm for Generating Complete Test Sets for Combinational Logic Circuits," *IEEE Trans. Computers*, Vol. C-20, November 1971, pp. 1245–1251.

9. KU, C.-T., and G. M. MASSON, "The Boolean Difference and Multiple Fault Analysis," *IEEE Trans. Computers*, Vol. C-24, January 1975, pp. 62–71.

10. SMITH, G. R., and S. S. YAU, "A Programmed Fault-Detection Algorithm for Combinational Switching Networks," in *Proc. Natl. Electronic Conf.*, Vol. 25, December 1969, pp. 668–673.

11. MARINOS, P. N., "Derivation of Minimal Complete Sets of Test-Input Sequences Using Boolean Differences," *IEEE Trans. Computers*, Vol. C-20, January 1971, pp. 25–32.

12. BEARNSON, L. W., and C. C. CARROLL, "On the Design of Minimum Length Fault Tests for Combinational Circuits," *IEEE Trans. Computers*, Vol. C-20, November 1971, pp. 1353–1356.

13. GAULT, J. W., J. P. ROBINSON, and S. M. REDDY, "Multiple Fault Detection in Combinational Networks," *IEEE Trans. Computers*, Vol. C-21, January 1972, pp. 31–36.

14. BOSSEN, D. C., and S. J. HONG, "Cause–Effect Analysis for Multiple Fault Detection in Combinational Circuits," *IEEE Trans. Computers*, Vol. C-20, November 1971.

15. DU, M. W., and C. D. WEISS, "Multiple Fault Detection in Combinational Circuits: Algorithms and Computational Results," *IEEE Trans. Computers*, Vol. C-22, March 1973, pp. 235–240.

16. POAGE, J. F., "Derivation of Optimum Tests To Detect Faults in Combinational Circuits," *Proc. Symp. Mathematical Theory of Automata*, New York, Polytechnic Press of Polytechnic Institute of Brooklyn, N.Y., 1963, pp. 483–528.

17. ROTH, J. P., "Diagnosis of Automata Failures: A Calculus and a Method," *IBM J. Res. Develop.*, Vol. 10, July 1966, pp. 278–291.

18. SCHNEIDER, P. R., "The Necessity To Examine D-Chains in Diagnostic Test Generation—An Example," *IBM J. Res. Develop.*, Vol. 11, January 1967, p. 114.

19. ROTH, J. P., W. G. BOURICIUS, and P. R. SCHNEIDER, "Programmed Algorithms To Compute Tests To Detect and Distinguish Between Failures in Logic Circuits," *IEEE Trans. Electron. Comput.*, Vol. EC-16, October 1967, pp. 567–580.

20. BOURICIUS, W. G., E. P. HSIEH, G. R. PUTZOLU, J. P. ROTH, P. R. SCHNEIDER, and C.-J. TAN, "Algorithms for Detection of Faults in Logic Circuits," *IEEE Trans. Computers*, C-20, November 1971, pp. 1258–1264.

21. SU, S. Y. H., and Y.-C. CHO, "A New Approach to the Fault Location of Combinational Circuits," *IEEE Trans. Computers*, Vol. C-21, January 1972, pp. 21–36.

7

Sequential Machines

In the previous three chapters, we considered only combinational logic. The outputs of a combinational system or circuit are dependent only on its present inputs; whereas the outputs of a sequential system or circuit depend not only on its present inputs but also on the past history of inputs. Comparing these two types of systems from the constructional point of view, the former does not contain any memory element or unit, but the latter has at least one.

The operational behavior of a combinational logic system is described by a logic truth table or a set of output switching functions which are functions of the input variables of the system. The operational behavior of a sequential logic system, on the other hand, is described by a mathematical model called a *sequential machine* which may be presented in a form of a table or a diagram. The table and the diagram in question are called the *transition table* and the *transition diagram* of the sequential machine, respectively. The operational behavior of a sequential logic system can also be described by two sets of switching functions, one is called the *next-state function* and the other, the *output function*. Both are functions of the system's inputs and state. Two basic models of deterministic finite-state automata or machines—Mealy machine and Moore machine—are defined. The terms sequential machine, finite-state machine, finite-state automaton, and automaton are synonyms.

In this chapter, three important problems of sequential machines are discussed. The first one discussed is the *state minimization problem*. This problem consists of two subproblems: one is determining all the equivalent states of a sequential machine, and the second is eliminating all the redundant states from the transition table or diagram. The state minimization of completely and incompletely specified machines is studied. The minimized machine obtained is identical in behavior to the given machine.

In the realization of a sequential machine using a sequential circuit, say a flip-flop circuit, if the states of the machine are symbols or numbers other than binary, we must first code the states into binary codes. The coding of the states into binary codes is by no means unique: different coding assignments to the state variables yield different next-state and output equations, thus different realizations. The problem of how to find coding assignments to the state variables that will yield the minimum amount of circuitry is known as the *state assignment* problem. The approach presented in this chapter is the reduced dependency method, which yields a good, though not necessarily optimum, assignment of the state variables.

For machines having a large number of state variables, even with a good choice of state assignment, the design may still be quite complex. An alternative, however, is suggested. Instead of synthesizing the given large machine, we decompose it into interconnected sets of smaller machines, realize each component machine first, and then interconnect them. The composite machine obtained is again identical in behavior to the given machine.

The determining of equivalent states, a good choice of state assignment, and the decomposition of a machine are made possible by the substitution property (S.P.) partition. It is shown that the set of all S.P. partitions on the set of states of a completely specified machine M form a lattice. Because of this property, the finding of all S.P. partitions is made easy.

The realization of minimized and decomposed sequential machines using reduced dependency state assignments by flip-flop sequential circuits will be presented in Chapter 9.

7.1 Basic Models

The theory of deterministic sequential machines is concerned with mathematical models of discrete, deterministic information processing devices and systems, such as digital computers, digital control units, and electronic circuits with synchronized delay elements. All these devices and systems have the following common properties, which are abstracted in the definition of deterministic sequential machines.

DEFINITION 7.1.1(a)

A *deterministic* sequential machine or *Mealy machine* is a system that can be characterized by a quintuple

$$M = (\Sigma, Q, Z, f, g)$$

where Σ = finite nonempty set of input symbols $\sigma_1, \sigma_2, \ldots, \sigma_l$
Q = finite nonempty set of states, $q_1, q_2, \ldots, q_n$
Z = finite nonempty set of output symbols, $z_1, z_2, \ldots, z_m$
f = next-state function, which maps

$$Q \times \Sigma \xrightarrow{f} Q$$

g = output function, which maps

$$Q \times \Sigma \xrightarrow{g} Z$$

DEFINITION 7.1.1(b)

A deterministic sequential machine is said to be of the *Moore type* (*Moore machine*) if its output function is a function of its states only, that is, $g: Q \rightarrow Z$.

Let $\sigma(t)$, $q(t)$, and $z(t)$ denote the input, states, and output variables, where $t = 0, 1, 2, \ldots$. The next-state and output functions are expressed as

$$f(q(t), \sigma(t)) = q(t + 1) \tag{7.1.1}$$

and

$$g(q(t), \sigma(t)) = z(t) \tag{7.1.2}$$

where $\sigma(t)$, $q(t)$, and $z(t)$ are the *present input, present state,* and *present output* of the machine, respectively, and $q(t + 1)$ is the *next state* of the machine. An example of a deterministic sequential machine is given below.

Example 7.1.1

Consider the 1-bit serial binary adder. It has two binary inputs x_1 and x_2 and one binary output S. In addition, it must have at least two internal states for describing its operation:

    (1) A no-carry state, designated by q_0.
    (2) A carry state, designated by q_1.

Since it has two binary inputs, there are four possible values of inputs $x_1 x_2 = 00, 01, 10,$ and 11. The next-state function and the output function of the 1-bit serial binary adder in terms of its present inputs and its present state are found to be

$$\begin{array}{lll} f(q_0, 11) = q_1, & f(q_0, x_1 x_2) = q_0, & \text{if } x_1 x_2 = 00, 01, \text{ or } 10 \\ f(q_1, 00) = q_0, & f(q_1, x_1 x_2) = q_1, & \text{if } x_1 x_2 = 10, 10, \text{ or } 11 \end{array} \tag{7.1.3}$$

$$\begin{array}{ll} g(q_0, 00 \text{ or } 11) = 0, & g(q_0, 01 \text{ or } 10) = 1 \\ g(q_1, 00 \text{ or } 11) = 1, & g(q_1, 01 \text{ or } 10) = 0 \end{array} \tag{7.1.4}$$

Thus, the serial binary adder can be modeled by a Mealy machine as

$$M_1 = (\Sigma_1, Q_1, Z_1, f_1, g_1) \tag{7.1.5}$$

where $\Sigma_1 = \{00, 01, 10, 11\}$
           $Q_1 = \{q_0, q_1\}$
           $Z_1 = \{0, 1\}$
           f_1 and g_1 as described in Eqs. (7.1.3) and (7.1.4)

It is interesting to point out that the binary adder can also be modeled by a Moore machine M_2, which is equivalent to the Mealy machine M_1, described above. Two machines with the same sets of input symbols and output symbols are indistinguishability equivalent if for *any* sequence of input symbols, they produce the identical sequence of output symbols. A detailed discussion of state and machine equivalences is given in Section 7.3.

Define the four states as

q_{00} = no-carry state (indicated by the first subscript) with an output 0 (indicated by the second subscript)

q_{01} = no-carry state with an output 1

q_{10} = carry state with an output 0

q_{11} = carry state with an output 1

Then

$$M_2 = (\Sigma_2, Q_2, Z_2, f_2, g_2) \qquad (7.1.6)$$

where

$\Sigma_2 = \Sigma_1$

$Q_2 = \{q_{00}, q_{01}, q_{10}, q_{11}\}$

$Z_2 = Z_1$

$f(q_{00} \text{ or } q_{01}, 00) = q_{00}, \quad f(q_{00} \text{ or } q_{01}, 11) = q_{10}, \quad f(q_{00} \text{ or } q_{01}, 01 \text{ or } 10) = q_{01}$

$f(q_{10} \text{ or } q_{11}, 00) = q_{01}, \quad f(q_{10} \text{ or } q_{11}, 11) = q_{11}, \quad f(q_{10} \text{ or } q_{11}, 01 \text{ or } 10) = q_{10}$

and

$$g(q_{00}) = g(q_{10}) = 0$$
$$g(q_{01}) = g(q_{11}) = 1$$

Two remarks about the models of a sequential machine are in order:

1. From the basic model of a sequential machine, for any sequence of input symbols $\tilde{x} = \sigma(0)\, \sigma(1) \dots, \sigma(K)$, the next states and the sequence of output symbols of a machine can be computed by

$$f(q(0), \sigma(0) \dots \sigma(K)) = f(\dots f(f(q(0), \sigma(0)), \sigma(1)), \dots, \sigma(K)) \qquad (7.1.7)$$

$$\underbrace{\qquad\qquad\qquad\qquad}_{= q(1)}$$

$$= q(2)$$

$$\cdot$$
$$\cdot$$

$$\underbrace{\qquad\qquad\qquad\qquad}_{= q(K+1)}$$

$$g(q(k), \sigma(k)) = z(k), \qquad k = 0, 1, \dots, K \qquad (7.1.8)$$

2. A "machine" does not need to be physical, such as computers or computerlike machines; any discrete-time system, physical or abstract, as long as it can be described by the mathematical model of Definition 7.1.1(a) or 7.1.1(b), is a sequential machine.

The Mealy machine and the Moore machine can be described by a transition table, which are shown in Tables 7.1.1(a) and (b), respectively. For example, the machines

**TABLE 7.1.1   Transition-Table Description
of Sequential Machines**

(a)   Transition-table description of the Mealy machine

Input

x q	σ_1	σ_2	$\cdots$	σ_l
q_1	$f(q_1,\sigma_1), g(q_1,\sigma_1)$	$f(q_1,\sigma_2), g(q_1,\sigma_2)$	$\cdots$	$f(q_1,\sigma_l), g(q_1,\sigma_l)$
q_2	$f(q_2,\sigma_1), g(q_2,\sigma_1)$	$f(q_2,\sigma_2), g(q_2,\sigma_2)$	$\cdots$	$f(q_2,\sigma_l), g(q_2,\sigma_l)$
$\vdots$	$\vdots$	$\vdots$		$\vdots$
q_n	$f(q_n,\sigma_1), g(q_n,\sigma_1)$	$f(q_n,\sigma_2), g(q_n,\sigma_2)$	$\cdots$	$f(q_n,\sigma_l), g(q_n,\sigma_l)$

Present state

Next-state, present output

(b)   Transition-table description of the Moore machine

Input

x q	σ_1	σ_2	$\cdots$	σ_l	z
q_1	$f(q_1,\sigma_1)$	$f(q_1,\sigma_2)$	$\cdots$	$f(q_1,\sigma_l)$	$g(q_1)$
q_2	$f(q_2,\sigma_1)$	$f(q_2,\sigma_2)$	$\cdots$	$f(q_2,\sigma_l)$	$g(q_2)$
$\vdots$	$\vdots$	$\vdots$	$\vdots$	$\vdots$	$\vdots$
q_n	$f(q_n,\sigma_1)$	$f(q_n,\sigma_2)$	$\cdots$	$f(q_n,\sigma_l)$	$g(q_n)$

Present state

Next-state   Present
output

of Example 7.1.1 can be conveniently described by the transition tables of Tables
7.1.2(a) and (b).

TABLE 7.1.2   Transition table description of M_1 and M_2

x_1x_2 q	00	01	11	10
q_0	$q_0, 0$	$q_0, 1$	$q_1, 0$	$q_0, 1$
q_1	$q_0, 1$	$q_1, 0$	$q_1, 1$	$q_1, 0$

(a)   Transition table of M_1

x_1x_2 q	00	01	11	10	S
q_{00}	q_{00}	q_{01}	q_{10}	q_{01}	0
q_{01}	q_{00}	q_{01}	q_{10}	q_{01}	1
q_{10}	q_{01}	q_{10}	q_{11}	q_{10}	0
q_{11}	q_{01}	q_{10}	q_{11}	q_{10}	1

(b)   Transition table of M_2

An alternative way of describing a sequential machine is to use a transition diagram. For example, the transition diagrams of machines M_1 and M_2 are given in Figs. 7.1.1(a) and (b), respectively.

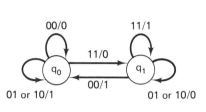

(a) Transition diagram of M_1.

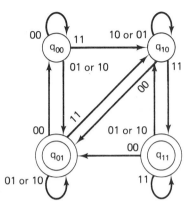

(b) Transition diagram of M_2.

Note: The conventional input symbol/output symbol is used. For example, $q_0 \xrightarrow{00/1} q_1$ means that upon application of 00 the machine will go from q_0 to q_1 with an output 1.

Note: Single-circled states denote output-0 states, and double-circled states denote output-1 states.

Fig. 7.1.1   Transition diagrams of M_2 and M_1.

It should be mentioned here that every Mealy machine can be converted into a Moore machine, and vice versa (problem 3).

Exercise 7.1

1. Give three examples of Mealy machines. Describe them using a transition table or a transition diagram.

2. Give three examples of Moore machines. Describe them using a transition table or a transition diagram.

3. Prove that every Mealy machine can be converted into a Moore machine, and vice versa.

7.2   The S. P. Partition and the Lattice L_M

The most important concept that will be used throughout this chapter is that of the substitution property (S.P.) partition.

DEFINITION 7.2.1

Let B be a subset of Q. A *partition* on the set of states is a collection of disjoint subsets of Q whose union is Q. Thus, a partition contains each state once and only once. B is called a *block* of a partition.

Since a partition is a set, it is enclosed within braces with blocks denoted by lines. Thus, $\{\overline{1}; \overline{2, 3, 4}; \overline{5, 6}\}$ and $\{\overline{1, 2, 3}; \overline{4}; \overline{5}; \overline{6}\}$ are valid partitions for a six-state machine. Because blocks that contain only one state may be implied by their absence, they will not be shown explicitly in partitions; this will simplify writing and using partitions. The two partition examples above thus become $\{\overline{2, 3, 4}; \overline{5, 6}\}$ and $\{\overline{1, 2, 3}\}$ it must be remembered, nevertheless, that all states not shown actually exist as one-state blocks.

DEFINITION 7.2.2

Two trivial partitions are contained in every machine; these are the *identity* partition *I*, which has all states in Q in the single block, and the *zero* partition *O*, which has each state in Q in a separate block.

DEFINITION 7.2.3

The block of a partition π which contains state q is denoted by $B_\pi(q)$. In a similar way, $1 \equiv 2(\pi)$ means that states 1 and 2 are in the same block of partition π:

$$\pi = \{\overline{1, 2}; \overline{3, 5, 7}\} = \{B_1; B_2\}$$

Additionally, $B_\pi(3) = B_2$; that is, state 3 is in block B_2.

DEFINITION 7.2.4

A partition π has the *substitution property* (S.P.) on the states of a machine iff for each input the next-state function maps all blocks of π into blocks of π. Thus, π has S.P. iff

$$q_1 \equiv q_2(\pi) \longrightarrow f(q_1, x) \equiv f(q_2, x)(\pi)$$

for all $q_1, q_2 \in Q$ and all $x \in \Sigma$. A partition that possesses the substitution property is called an *S.P. partition*.

Example 7.2.1

For machine M_1 in Fig. 7.2.1(a), consider the partition $= \{\overline{1, 2}; \overline{4, 5}\}$. For states 1 and 2 under inputs 0 and 1, the next-states are (2, 1) and (3, 3), respectively; and the next-states of rows 4 and 5 are (1, 1) and (5, 4), respectively. Since these next-state sets are contained in some block of the partition, the partition has S.P. The partition $\{\overline{1, 2, 3}\}$ does not have S.P., since the next-states of these states under input 1 are states 3 and 4, which are not contained in a single block of the partition. Note that the missing single-state blocks do not need to be checked, because the next-state set of a state of a single-state block has only one state and is therefore always contained in a block. This machine has 10 nontrivial S.P. partitions, which are shown in Fig. 7.2.1(b).

Now let us define three binary operations on partitions.

x q	0	1	z
1	2	3	0
2	1	3	0
3	1	4	0
4	1	5	1
5	1	4	1

(a) Machine M_1

$\pi_1 = \{\overline{1, 2}\}$
$\pi_2 = \{\overline{3, 5}\}$
$\pi_3 = \{\overline{4, 5}\}$
$\pi_4 = \{\overline{1, 2}; \overline{3, 5}\}$
$\pi_5 = \{2, 4; \overline{3, 5}\}$

$\pi_6 = \{\overline{1, 2}; \overline{4, 5}\}$
$\pi_7 = \{\overline{3, 4, 5}\}$
$\pi_8 = \{\overline{1, 2, 4}; \overline{3, 5}\}$
$\pi_9 = \{\overline{1, 2}; \overline{3, 4, 5}\}$
$\pi_{10} = \{\overline{2, 3, 4, 5}\}$

(b) S. P. partitions

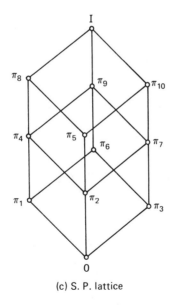

(c) S. P. lattice

Fig. 7.2.1 Machine M_1 together with the S.P. partitions and lattice.

DEFINITION 7.2.5

Let π_1 and π_2 be partitions on Q. The partition π_1 is *included in* or *contained in* partition π_2 $(\pi_1 \leq \pi_2)$ iff all states in a block of π_1 are also together in a block of π_2.

Stated another way, $\pi_1 \leq \pi_2$ iff

$$q_1 \equiv q_2(\pi_1) \longrightarrow q_1 \equiv q_2(\pi_2)$$

for all $q_1, q_2 \in Q$. If $\pi_1 \leq \pi_2$, π_1 is also said to be *smaller than or equal to* π_2; obviously if $\pi_1 < \pi_2$, π_1 is just *smaller than* π_2.

DEFINITION 7.2.6

If π_1 and π_2 are partitions on Q, the "*product*" $\pi_1 \cdot \pi_2$ is a partition on Q obtained by intersecting the individual blocks of π_1 with those of π_2.

DEFINITION 7.2.7

If π_1 and π_2 are partitions on Q, the "*sum*" $\pi_1 + \pi_2$ is a partition on Q obtained by merging those blocks of π_1 and π_2 which have at least one common state.

To illustrate the binary operations, let

$$\pi_1 = \{\overline{1, 2}; \overline{3, 4, 5, 6, 7, 8}\}$$
$$\pi_2 = \{\overline{1, 2, 3, 4}; \overline{5, 6, 7}\}$$
$$\pi_3 = \{\overline{1, 2}; \overline{3, 6, 8}\}$$
$$\pi_4 = \{\overline{1}; \overline{2}; \overline{3}; \overline{4}; \overline{5}; \overline{6}; \overline{7}; \overline{8}\} = O$$
$$\pi_5 = \{\overline{1, 2, 3, 4, 5, 6, 7, 8}\} = I$$

Then, $\pi_3 < \pi_1$, while π_4 is contained in all other partitions and π_5 contains all other partitions. There is no relationship between the other pairs. The "sum" and "product" of the possible pairs are:

$$\pi_1 + \pi_2 = I \qquad\qquad\qquad \pi_1 \cdot \pi_2 = \{\overline{1, 2}; \overline{3, 4}; \overline{5, 6, 7}\}$$
$$\pi_1 + \pi_3 = \{\overline{1, 2}; \overline{3, 4, 5, 6, 7, 8}\} \qquad \pi_1 \cdot \pi_3 = \{\overline{1, 2}; \overline{3, 6, 8}\} = \pi_3$$
$$\qquad\quad = \pi_1$$
$$\pi_2 + \pi_3 = I \qquad\qquad\qquad \pi_2 \cdot \pi_3 = \{\overline{1, 2}\}$$
$$\pi_i + O = \pi_i \qquad\qquad\qquad \pi_i \cdot O = O$$
$$\pi_i + I = I \qquad\qquad\qquad\quad \pi_i \cdot I = \pi_i$$

where π_i is any of the five partitions.

In Fig. 7.2.1(b), π_1 and π_2 are both contained in π_4. Note that there is no relationship between π_1 and π_2.

THEOREM 7.2.1

If π_1 and π_2 are S.P. partitions on Q, then $\pi_1 + \pi_2$ and $\pi_1 \cdot \pi_2$ are also S.P. partitions on Q.

Proof: Assume that $q \equiv p(\pi_1)$ and $q \equiv p(\pi_2)$, where q and p are two states of the machine. Then, by definition,

$$q \equiv p(\pi_1 \cdot \pi_2)$$

For any input x in Σ,

$$f(q, x) \equiv f(p, x)(\pi_1)$$

and

$$f(q, x) \equiv f(p, x)(\pi_2)$$

by the hypotheses of the theorem, but then clearly

$$f(q, x) \equiv f(p, x)(\pi_1 \cdot \pi_2)$$

by the definition of $\pi_1 \cdot \pi_2$.

To show that $\pi_1 + \pi_2$ has S.P. we recall that

$$q \equiv p(\pi_1 + \pi_2)$$

implies that there exists a chain

$$q = q_0, \quad q_1, \quad q_2, \quad \ldots, \quad q_m = p$$

such that

$$q_j = q_{j+1}(\pi_1) \quad \text{or} \quad q_j = q_{j+1}(\pi_2), \qquad j = 0, 1, 2, \ldots, m - 1$$

We now use this to show that for every input x in Σ,

$$f(q, x) \equiv f(p, x)(\pi_1 + \pi_2)$$

Since π_1 and π_2 have S.P.,

$$f(q, x) \equiv f(q_1, x)(\pi_1) \quad \text{or} \quad f(q_1, x) \equiv f(q, x)(\pi_2)$$

and since π_1 and π_2 are both smaller than $\pi_1 + \pi_2$, we conclude that

$$f(q, x) \equiv f(q_1, x)(\pi_1 + \pi_2)$$

Similarly,

$$f(q_1, x) \equiv f(q_2, x)(\pi_1 + \pi_2)$$
$$f(q_2, x) \equiv f(q_3, x)(\pi_1 + \pi_2)$$
$$\cdots\cdots\cdots\cdots\cdots\cdots\cdots\cdots$$
$$f(q_{m-1}, x) \equiv f(p, x)(\pi_1 + \pi_2)$$

Thus, since the equivalence relation is transitive, we have

$$f(q, x) \equiv f(p, x)(\pi_1 + \pi_2)$$

as was to be shown.   ∎

There are basically two ways of obtaining S.P. partitions. The first way is to work with the machine description until a partition which satisfies Definition 7.2.4 is found.

The second way is to combine previously obtained S.P. partitions using the sum and product operations. Since the second way is much easier than the first, the most efficient procedures are those which rely most heavily on the second way. Of course, some use must be made of the first way since no nontrivial partitions are known a priori to have S.P. Before giving the procedure for generating all the S.P. partitions of a given machine, the following definition is first presented.

DEFINITION 7.2.8

A pair of states is said to be *identified* when both states appear in the same block of a partition.

The S.P. partitions of a sequential machine can be obtained by carrying out the following two steps:

Step 1   For every pair of states q and p, compute the smallest S.P. partition, $\pi^m_{q,p}$, which identifies the pair.

Step 2   Find all possible sums of the $\pi^m_{q,p}$. These sums constitute all the S.P. partitions.

Step 1 is performed by first placing the states q and p in a block; that is, the partition is just $\overline{\{q,p\}}$. Then for each input, find the next-states for states q and p; these next-states must also be identified and the transitive law applied where possible. Finally, states are added as necessary to make the partition have S.P. An example to illustrate this procedure is given below.

Example 7.2.2

Consider machine M_1 of Fig. 7.2.1(a). We start by computing the S.P. partitions obtained by identifying a pair of states. The $\pi^m_{q,p}$ are found first.
$\pi^m_{1,2}$ is obtained by first placing states 1 and 2 in a block; $\overline{\{1,2\}}$. Then we find that

$$f(q,x) = f(1,0) = 2, \quad f(2,0) = 1$$

and   (7.2.1)

$$f(1,1) = 3, \quad f(2,1) = 3$$

Since any state can be identified with itself,

$$\pi^m_{1,2} = \overline{\{1,2\}}$$

Now consider the construction of another $\pi^m_{q,p}$, say $\pi^m_{1,3}$. From Fig. 7.2.1(a),

$$f(1,0) = 2, \quad f(3,0) = 1$$

and

$$f(1,1) = 3, \quad f(3,1) = 4$$

which says simply that states (1, 2) and (3, 4) must also be identified by $\pi_{1,3}^m$. Therefore,

$$\pi_{1,3}^m \geq \{\overline{1, 2, 3, 4}\}.$$

Since

$$f(3, 0) = 1, \qquad f(4, 0) = 1$$
$$f(3, 1) = 4, \qquad f(4, 1) = 5$$

etc., we find that states 4 and 5 must also be identified by $\pi_{1,3}^m$. Thus,

$$\pi_{1,3}^m = \{\overline{1, 2, 3, 4, 5}\} = I$$

We conclude that states 1 and 3 *cannot* be identified in a block of any *nontrivial* S.P. partition. In a similar manner, the other $\pi_{q,p}^m$ partitions are found:

(1) $\pi_{1,2}^m = \{\overline{1, 2}\}$ (4) $\pi_{2,4}^m = \{\overline{2, 4}; \overline{3, 5}\}$

  $\pi_{1,3}^m = I$ $\pi_{2,5}^m = \pi_{2,3}^m$

(2) $\pi_{1,4}^m = \{\overline{1, 2, 4}; \overline{3, 5}\}$ (5) $\pi_{3,4}^m = \{\overline{3, 4, 5}\}$

  $\pi_{1,5}^m = I$ (6) $\pi_{3,5}^m = \{\overline{3, 5}\}$

(3) $\pi_{2,3}^m = \{\overline{2, 3, 4, 5}\}$ (7) $\pi_{4,5}^m = \{\overline{4, 5}\}$

It is seen that not only state pair (1, 3), but also state pair (1, 5), cannot be in one block of any nontrivial S.P. partition. By finding all possible sums of the seven unique, nontrivial S.P. partitions that have been discovered for machine M_1, we obtain the following three additional partitions:

(8) $\pi_{1,2}^m + \pi_{4,5}^m = \{\overline{1, 2}; \overline{4, 5}\}$

(9) $\pi_{1,2}^m + \pi_{3,5}^m = \{\overline{1, 2}; \overline{3, 5}\}$

and

(10) $\pi_{1,2}^m + \pi_{3,4}^m = \{\overline{1, 2}; \overline{3, 4, 5}\}$

Because of Theorem 7.2.1, these partitions must have S.P.

THEOREM 7.2.2

The set of all S.P. partitions on the set of states of a sequential machine M forms a lattice L_M with both O and I, under the natural partition ordering.

Proof: From Theorem 7.2.1, we know that the set of S.P. partitions on Q of M is closed under the "$\cdot$" and "$+$" operations. Thus, the set of all S.P. partitions forms a sublattice of the lattice of all partitions on Q, and therefore a lattice in the natural ordering of partitions. The fact that the smallest and largest elements of this lattice are the O partition and I partition, respectively, is obvious. ∎

For example, the lattice L_{M_1} of machine M_1 of Fig. 7.2.1(a) is depicted in Fig. 7.2.1(c), in which there are a total of 12 partitions: two trivial partitions O and I and

10 nontrivial partitions which were found in Example 7.2.2 and are shown in Fig. 7.2.1(b).

It should be noted that in step 1 of the above procedure for finding all S.P. partitions, there are $n(n-1)/2$ pairs of $\pi_{q,p}^m$ to examine for an n-state machine. For example, in Example 7.2.2, n was 5, so we examine $5 \cdot 4/2 = 10$ state pairs and found that there were seven distinct nontrivial S.P. partitions. Then in step 2, we needed to examine $7 \cdot (7-1)/2$ possible sums to obtain all S.P. partitions. Now with the a priori knowledge that the set of all S.P. partitions form a lattice, given by Theorem 7.2.2, the computation required by step 2 may be greatly reduced. This is because only those partitions belonging to one level of the lattice need be summed to find the partitions at the next level. This means that all partitions not belonging to the present level may be eliminated, with sums of partitions taken among only those that belong to the present level. This method markedly reduces the number of sums that must be performed.

Example 7.2.3

In Example 7.2.2, we found that there were seven distinct, nontrivial S.P. partitions. However, note that $\pi_{1,4}^m = \pi_{1,2}^m + \pi_{2,4}^m$, $\pi_{2,3}^m = \pi_{2,4}^m + \pi_{4,5}^m$, and $\pi_{3,4}^m = \pi_{3,5}^m + \pi_{4,5}^m$. These three partitions, $\pi_{1,4}^m$, $\pi_{2,3}^m$, and $\pi_{3,4}^m$, will be generated later and so may be eliminated here. Also, consider $\pi_{2,4}^m$; it contains $\pi_{3,5}^m$ but no other partition. $\pi_{2,4}^m$ does not therefore belong on the same level as $\pi_{3,5}^m$, and it will not be generated later. $\pi_{2,4}^m$ will thus not be summed with first-level partitions, but neither may it be eliminated; it must be saved for later use. The three remaining partitions are partitions in the first level or row of the S.P. lattice [see Fig. 7.2.1(c)]. Let $\pi_1 = \{\overline{1,2}\}$, $\pi_2 = \{\overline{3,5}\}$, and $\pi_3 = \{\overline{4,5}\}$. The three sums of these partitions yield:

$$\pi_1 + \pi_2 = \{\overline{1,2}; \overline{3,5}\} = \pi_4$$

$$\pi_1 + \pi_3 = \{\overline{1,2}; \overline{4,5}\} = \pi_6$$

$$\pi_2 + \pi_3 = \{\overline{3,4,5}\} = \pi_7$$

Because of Theorem 7.2.1, these three partitions must have S.P. These new partitions are candidates to belong to row 2 of the S.P. lattice. Because there is no relationship between these partitions, it is concluded that they do belong to row 2.

At this time, $\pi_{2,4}^m$ must be reconsidered. Noting that none of the row 2 partitions is contained in $\pi_{2,4}^m$, it is added to row 2. Row 2 now has four partitions—π_4, π_6, π_7, and $\pi_{2,4}^m = \pi_5$. This ends the calculations for row 2; the previous steps are repeated to obtain the next level.

Sums of row 2 partitions yield three different nontrivial partitions:

$$\pi_8 = \{\overline{1,2,4}; \overline{3,5}\}$$

$$\pi_9 = \{\overline{1,2}; \overline{3,4,5}\}$$

$$\pi_{10} = \{\overline{2,3,4,5}\}$$

Since no inclusion relationship exists between these three partitions, they all belong to row 3; and since there are no other partitions being saved, row 3 has only these three partitions.

Taking sums of the partitions in row 3, only the identity partition I is obtained, and so the lattice is complete [Fig. 7.2.1(c)].

A formal algorithm for generating the lattice L_M and hence all S.P. partitions will be presented after introducing the following definitions, lemmas, and theorems.

DEFINITION 7.2.9

Partitions are called *generators* if the entire lattice can be obtained by taking sums of these partitions.

DEFINITION 7.2.10

A partition is a *basic generator* if it is a generator which cannot itself be obtained by the summing of other partitions.

In Fig. 7.2.1(b), the partitions π_1, π_2, π_3, and π_5 are basic generators. Basic generators are readily discerned in a lattice diagram by the fact that they have only one line leading up to them.

LEMMA 7.2.1

Partitions in the first row of a lattice are basic generators.

LEMMA 7.2.2

The $\pi_{q,p}^m$ partitions are generators for the S.P. lattice.

Because the $\pi_{q,p}^m$ partitions are generators of a lattice, they must include the set of basic generators, including the first row. This means that the first row of partitions, plus any which will not be generated later, may be used to generate the S.P. lattice. The following two theorems are obvious.

THEOREM 7.2.3

The set of basic generators is the minimum set necessary to construct the lattice by taking sums of partitions.

In addition, once the sums of partitions in a row are obtained, that row need not be used further.

THEOREM 7.2.4

In any row r, a partition that is not a basic generator is the sum of two other partitions in the row just below row r.

Thus, not only can the lattice be generated with the $\pi_{q,p}^m$ partitions, but some of these partitions may be ignored.

DEFINITION 7.2.11

A basic generator that does not belong to the first row of a partition diagram is called an *unassigned basic generator* (UBG).

ALGORITHM 7.2.1

The algorithm for generating the lattice of S.P. partitions is as follows:

1.   Obtain all $\pi_{q,p}^m$ S.P. partitions, eliminating any which are duplicates or are the identity partition.

2.   Examine the remaining partitions. Eliminate those that are sums of other partitions. The remaining partitions are basic generators.

3.   If a basic generator contains another basic generator, it does not belong to the first row and must be saved until a later row. This is an unassigned basic generator (UBG). The remaining basic generators comprise the first row.

4.   Call the row just completed row n. If there is only one partition in row n, go to step 7. Otherwise, take the sum of all row n partitions. These sums are candidates for row $n + 1$. Row n may now be ignored.

5.   If there are any UBG, compare each one to all row $n + 1$ candidates, and all other UBG. If an UBG does not contain any of these partitions, the UBG belongs to row $n + 1$; if an UBG does contain a candidate or another UBG, it remains unassigned.

6.   Now compare all row $n + 1$ candidates to each other and to any basic generators assigned to row $n + 1$ in step 5. In these comparisons, a candidate is eliminated if it contains another partition. The remaining partitions comprise row $n + 1$. Return to step 4.

7.   If there are no UBG, the S.P. lattice is complete and the partition in row n is the maximal partition with S.P.; go to step 9. Otherwise, the unassigned basic generators will form row $n + 1$.

8.   Compare the UBG among themselves: an UBG belongs to row $n + 1$ if it contains no other UBG; otherwise, it remains an UBG. Return to step 4.

9.   Stop.

Exercise 7.2

1.   Compute all S.P. partitions for machines M_1, M_2, M_3, and M_4 described in Table P7.2.1.

TABLE P7.2.1

q \ x	a	b
A	C, 0	B, 1
B	A, 1	D, 0
C	A, 1	A, 0
D	B, 0	B, 1

(a) M_1

q \ x	a	b	c
A	D, 0	F, 1	B, 0
B	C, 0	E, 1	A, 0
C	E, 0	A, 1	D, 0
D	F, 1	B, 1	C, 0
E	A, 1	D, 1	F, 0
F	B, 1	C, 1	E, 0

(b) M_2

TABLE P7.2.1   (Continued)

q \ x	a	b	z
A	C	G	1
B	E	F	0
C	G	D	0
D	B	B	1
E	B	H	0
F	E	B	0
G	C	A	0
H	G	G	0

(c) M_3

q \ x	a	b	c	d	e	z
A	E	E	B	F	A	0
B	H	D	A	G	B	0
C	F	G	H	C	D	1
D	A	B	E	D	C	0
E	F	A	D	E	H	0
F	H	H	G	A	F	0
G	E	C	F	B	G	0
H	A	F	C	H	E	0

(d) M_4

2. (a) For each of the four machines described in Table P7.2.1, find the partitions that are basic generators.
   (b) Construct the S.P. lattice for each of these four machines using Algorithm 7.2.1.

7.3 State Minimization Using the O.C.S.P. Partition

In this section, two important problems of sequential machines, equivalence and minimization, will be discussed. For deterministic sequential machines there are two types of equivalence: indistinguishability equivalence and tape equivalence. Only indistinguishability equivalence will be discussed here; tape equivalence will be discussed in Chapter 8. For indistinguishability equivalence, two equivalence relations, state equivalence and machine equivalence, will be considered. They are defined as follows:

DEFINITION 7.3.1

Let q_a and q_b be two states of machines M_a and M_b (M_a and M_b may be the same machine). q_a and q_b are said to be *equivalent* if for any sequence of input symbols applied to them, the output sequences are identical. If q_a and q_b are not equivalent, we say they are *distinguishable*.

DEFINITION 7.3.2

Let M_a and M_b be two machines. M_a and M_b are said to be *equivalent* if for every state of M_a there exists at least one equivalent state in M_b and vice versa. Similarly, if M_a and M_b are not equivalent, we say that they are *distinguishable*.

We shall first be concerned with state equivalence. In the preceding definition of state equivalence, instead of considering two machines, we could view these two machines as two submachines of a machine. In doing so, the discussion of state equivalence may be simplified considerably. This can be done as follows: Given two machines $M_a = (\Sigma_a, Q_a, Z_a, f_a, g_a)$ and $M_b = (\Sigma_b, Q_b, Z_b, f_b, g_b)$, where $\Sigma_a = \Sigma_b = \Sigma$, $Z_a = Z_b = Z$; that is, M_a and M_b have the same set Σ of input symbols and the same set Z of output symbols, construct a machine $M = (\Sigma, Q, Z, f, g)$ with

$$Q = Q_a \cup Q_b$$

$$f(q_i, \sigma_j) = \begin{cases} f_a(q_i, \sigma_j) & \text{for } q_i \in Q_a \\ f_b(q_i, \sigma_j) & \text{for } q_i \in Q_b \end{cases}$$

$$g(q_i, \sigma_j) = \begin{cases} g_a(q_i, \sigma_j) & \text{for } q_i \in Q_a \\ g_b(q_i, \sigma_j) & \text{for } q_i \in Q_b \end{cases}$$

From now on, instead of considering the equivalence of two states in two machines M_a and M_b, we can simply consider the two states in the machine M defined above.

In this section a state minimization method using the output consistent substitution property (O.C.S.P) partition, known as the *O.C.S.P partition minimization method*, is presented. We first present the state minimization of completely specified machines using this method, and then apply it to incompletely specified machines.

A. Completely Specified Machines

Before introducing the method, the following definition is needed.

DEFINITION 7.3.1

A partition π on the states of a machine is said to be *output consistent (O.C.)* iff for any input the outputs for all states in a block B of π are the same. In other words,

$$q_1 \equiv q_2(\pi) \longrightarrow g(q_1, x) = g(q_2, x)$$

for all $q_1, q_2 \in Q$ and all $x \in \Sigma$. In Fig. 7.2.1(a), the partition $\{\overline{1, 2, 3}; \overline{4, 5}\}$ is output consistent.

THEOREM 7.3.1

If π is an S.P. partition which is also O.C., then the blocks of π are sets of states which are equivalent to each other.

Proof: The proof follows directly from the definitions of S.P., O.C., and equivalence. Since π has S.P., applying an input starting from any state q_i in a block B_i will cause the machine to enter a block B_j which contains the next-state q_j; that is, $f(q_i, x) = q_j$. It is not necessary to know what state the machine is in, only what block contains the present state. An input sequence $\tilde{x}$ will move the machine from block to block, but since the partition also is output consistent, the output from every state in a particular block is the same. Then for any two states in a block and any finite input sequence, the next-state will always be in the same block, and hence their outputs will always be the same. Thus, blocks of an O.C.S.P. partition are sets of equivalent states. ∎

Following directly from Theorem 7.3.1, we have

LEMMA 7.3.1

The product and the sum of two S.P. partitions are O.C.S.P. partitions iff the two partitions are O.C.S.P. partitions.

THEOREM 7.3.2

The set of all O.C.S.P. partitions on the states of a completely specified machine M forms a sublattice L_M^* of the lattice of all S.P. partitions, L_M.

Proof: The partition O is an O.C.S.P. partition. By Lemma 7.3.1, the O.C. basic generators of the lattice L_M must generate a sublattice of L_M.   ∎

For example, the two O.C. generators of the sublattice $L_{M_1}^*$ of machine M_1 of Fig. 7.2.1(a) are π_1 and π_3. In Fig. 7.2.1(c), the O.C. sublattice generated by these two O.C.S.P. partitions is denoted by dashed lines.

The O.C.S.P. lattice L_M^* of a completely specified machine M can be obtained from Algorithm 7.2.1 with slight modification.

ALGORITHM 7.3.1

The O.C.S.P. lattice can be generated from Algorithm 7.2.1 if all S.P. partition statements in the algorithm are replaced by O.C.S.P. partition statements.

The higher in an O.C.S.P. lattice diagram a partition is located, the larger the partition; by larger is meant more states in a block. One would expect that the highest level would contain the largest partition—the partition giving the most state reduction.

THEOREM 7.3.3

For completely specified sequential machines, there is a unique partition which yields the minimized machine. Furthermore, this partition is located at the highest level of the O.C.S.P. lattice and is the only partition at that level.

In Fig. 7.3.1(b), π_3 is the highest partition on the O.C.S.P. lattice and yields the minimized machine of Fig. 7.3.1(c).

A method of minimizing completely specified machines by the use of output consistent S.P. partitions is now obvious and simple: obtain the lattice of O.C.S.P. partitions, and use the highest partition on the lattice to minimize the machine. Each block of this *maximal* partition (including any implied one-state blocks) becomes one state in the minimized machine. The only significant computation thus required to minimize a machine is that of generating the O.C.S.P. lattice.

Example 7.3.1

Machine M_1, repeated in Fig. 7.3.1(a), illustrates how partitions can minimize a transition table. Figure 7.3.1(b) displays the O.C.S.P. lattice diagram for machine M_1. This diagram is constructed by drawing a line between two partitions (represented by nodes) whenever one

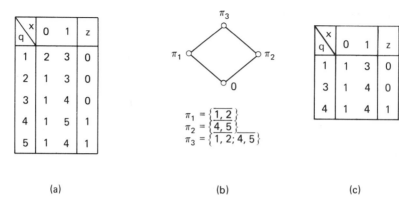

x q	0	1	z
1	2	3	0
2	1	3	0
3	1	4	0
4	1	5	1
5	1	4	1

$\pi_1 = \{\overline{1,\ 2}\}$
$\pi_2 = \{\overline{4,\ 5}\}$
$\pi_3 = \{\overline{1,\ 2;\ 4,\ 5}\}$

x q	0	1	z
1	1	3	0
3	1	4	0
4	1	4	1

(a)   (b)   (c)

Fig. 7.3.1 (a) Transition table, (b) lattice of O.C.S.P. partitions, and (c) minimized transition table for the completely specified sequential machine M_1.

partition is contained in the other and there is no partition intermediate between the first two. Stated another way, nodes π_1 and π_2 are connected by a direct line if $\pi_1 < \pi_2$ and there is no other partition π_3 for which $\pi_1 < \pi_3 < \pi_2$. Also, if $\pi_1 < \pi_2$, then π_2 is placed on a higher level or row than π_1. The levels of a partition diagram are numbered starting with the first level at the bottom (ignoring the O partition) and counting consecutively up to the highest level.

The l.u.b. and g.l.b. of π_1 and π_2 on Fig. 7.3.1(b) are π_3 and O, respectively, since $\pi_1 + \pi_2 = \pi_3$ and $\pi_1 \cdot \pi_2 = O$. The l.u.b. of two partitions may be found on a lattice diagram by taking the shortest upward route from each node to the closest common node. Similarly, the g.l.b. of two partitions is found by going downward on the lattice diagram.

There are three O.C.S.P. partitions: π_1, π_2, and π_3. Consider π_3. Since states 1 and 2 are equivalent and states 4 and 5 are equivalent, one state in each pair may be eliminated to give a minimized machine. The table of Fig. 7.3.1(c) results from Fig. 7.3.1(a) when states 2 and 5 are eliminated by the following procedure. The rows for the present states 2 and 5 are crossed off; whenever state 2 or 5 is encountered in the table, state 1 or 4, respectively, is substituted.

It should be noted that the maximal O.C.S.P. partition can be directly obtained from Algorithm 7.2.1 by changing each mention of S.P. to read O.C.S.P.

B. Incompletely Specified Machines

If a designer does not specify a next-state or output entry when it normally would be specified, the reason is usually because the machine is not expected to enter that next-state condition. Since the designer does not care what the next-state or output is, it could be specified as any valid next-state or output. In fact, it could be specified differently under different machine conditions. Thus, it is reasonable to let an unspecified state-table entry assume as many different values as desired.

DEFINITION 7.3.2

If $M_1 = (Q_1, \Sigma, Z, f_1, g_1)$ and $M_2 = (Q_2, \Sigma, Z, f_2, g_2)$ are two machines with the same input and output sets, and if $\tilde{x}$ is any input sequence, then state $q_1 \in Q_1$ and state $q_2 \in Q_2$ are *compatible* ($q_1 \sim q_2$) iff for $x \in \tilde{x}$,

$$g_1(q_1, \tilde{x}) = g_2(q_2, \tilde{x})$$

whenever $g_1(q_1, x)$ and $g_2(q_2, x)$ are both specified.

From now on, the term *compatible* is used instead of *equivalent* as a reminder of the different definitions. For incompletely specified machines, the transitive law is lost and cannot be repaired. Figure 7.3.2 illustrates the transitive law breakdown.

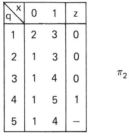

q \ x	0	1	z
1	2	3	0
2	1	3	0
3	1	4	0
4	1	5	1
5	1	4	—

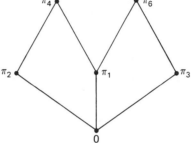

(a) Machine M'_1 (b) O.C.S.P. partition diagram for machine M'_1

Fig. 7.3.2 Machine M'_1, showing how an unspecified table entry destroys the lattice.

In this machine, since states 3 and 5 are equivalent and states 5 and 4 are equivalent, the transitive law says that state 3 should be equivalent to state 4—but it is not, because they have different outputs. Figure 7.3.2(b) shows all O.C.S.P. partitions. It is not a lattice since there are partition pairs which do not have a l.u.b.

Because the method of finding the maximal O.C.S.P. partition for completely specified machines depends upon a partition lattice, it fails when unspecified entries occur. Lemma 7.3.1 and Theorems 7.3.1, 7.3.2, and 7.3.3 do not hold for incompletely specified machines.

In the following, we would like to show that the transitive law usually holds in practical machines even though unspecified entries are present, and also show how the maximal O.C.S.P. partition may be found.

Note that practical machines do not have unspecified entries scattered at random in a transition table but instead follow a rational order. To understand this ordering, consider a table entry pair (q, z) consisting of a next-state and an output. If q is specified while z is not, then upon entering state q the output could be any valid code. It is hard to see any reason for specifying a state without also specifying a definite

output. Likewise, it does not make sense to specify the output but not the corresponding next-state. In effect, the designer would be saying that a definite output is desired
but that the machine can now go anywhere it wants to go.

Consider table entries for which both the next-state and the output are unspecified. This happens when the machine is not expected to enter this cell of the table
—this is the most common explanation. If such a cell occurs in row r and input
column i, then input i can never follow any input that causes the machine to enter r
as a state.

The descriptions of a practical machine given above are now formalized in a
definition.

DEFINITION 7.3.3

A machine that satisfies the following condition is called an *input-restricted
machine:* Any unspecified next-state entry in an input column i of the transition table
occurs because this cell of the table cannot be entered. That is, input i cannot follow
some input j.

This means that if input a cannot follow input b for any row of a transition table,
the next-state entry in input column a for all possible sequences ending with b, a (b
followed by a) must be unspecified.

In Fig. 7.3.3(a), the entry for row 2, input a, is unspecified. If the machine is in
state 1 and the input sequence b, a is applied, the machine is directed to this unspecified

x q	a	b	c
1	4, 0	2, 0	4, 0
2	—	3, 0	4, 0
3	1, 1	2, 0	4, 0
4	4, 1	—	1, 1

(a) Machine M_2

x q	a	b	c
1	4, 0	—	4, 0
2	—	3, 0	4, 0
3	—	2, 0	4, 0
4	4, 1	—	1, 1

(b) Machine M_2'

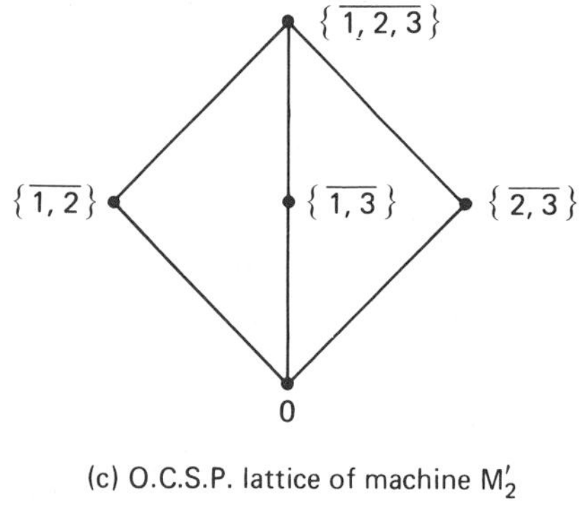

(c) O.C.S.P. lattice of machine M_2'

Fig. 7.3.3 Illustration of input restriction in a Mealy—type
machine.

entry—therefore, the input sequence b, a is not allowed. If the machine is in state 2 and the input sequence is b, a, the machine first goes to state 3 (due to input b), and then goes to state 1 under input a. But since the input sequence b, a is not allowed, specifying an entry for row 3 under input a is immaterial. This machine is thus not input-restricted. To be input-restricted, $g(3, a)$ and $f(3, a)$ must be unspecified. By similar reasoning, the unspecified entry in row 4, column b, requires the entry in row 1, column b, to be unspecified. Machine M'_2 in Fig. 7.3.3(b) has these two changes needed to make machine M_2 input-restricted.

ALGORITHM 7.3.2

An algorithm to check next-states for input restriction is as follows:

1. Locate an unspecified next-state entry; call its row and column r and c, respectively.

2. Find any column k (except the null input column in a Moore machine) that contains r as a next-state entry.

3. For all next-states q listed in column k, there must be an unspecified entry for $f(q, c)$ and $g(q, c)$.

Most, but not all, input-restricted machines are transitive. To exclude those few input-restricted machines which are not transitive, a further restriction on unspecified table entries is required.

DEFINITION 7.3.4

A sequential machine $M = (Q, \Sigma, Z, f, g)$ is said to be *transitive-restricted* (*T.R.*) if it is input-restricted and satisfies the following conditions: Let $a,b,c \in Q$, $x \in \Sigma$, and U denote unspecified; then for each unspecified output $g(b, x) = U$, such that $a \sim b$ and $b \sim c$, either

Condition 1. $f(a, x) = U$ or $f(c, x) = U$; or

Condition 2. $g(a, x) = g(c, x)$ (when both are specified) and both $f(a, x)$ and $f(c, x)$ are compatible to either a or c,

is true.

The machines in Table 7.3.1 are input-restricted, but are not T.R. machines.

TABLE 7.3.1   Input-Restricted Machines Which Are Not Transitive

q \ x	a	b	c
1	1, 0	2, 0	3, 1
2	–	2, 0	3, 1
3	4, 0	2, 0	3, 1
4	4, 1	2, 1	1, 0

(a)   Machine M_3

q \ x	a	b	c
1	1, 0	2 –	2 –
2	4 –	1 –	2, 1
3	3, 1	2 –	2 –
4	2 –	4, 1	3 –

(b)   Machine M_4

Machine M_3 fails to be T.R. because of the second condition: $f(3, a) = 4$ is not compatible with either state 1 or 3. Machine M_4, which is in fundamental mode, fails to have the T.R. property because of the other part of the second condition: $g(1, a) \neq g(3, a)$. Changing $f(3, a)$ of machine M_3 to a 3 will make that machine transitive-restricted.

THEOREM 7.3.4

A transitive-restricted sequential machine is transitive.

Proof: Let $M = (Q, \Sigma, Z, f, g)$ be a transitive-restricted machine. Since a completely specified machine is transitive, it is the unspecified entry which causes intransitivity; and to be intransitive, there must be some $a, b, c \in Q$ such that $a \nsim c$ for some input even though $a \sim b$ and $b \sim c$.

Definition 7.3.4 of a transitive-restricted machine requires that either condition 1 or condition 2 is true. If condition 1 is true, then either $f(a, x)$ or $f(c, x)$ is unspecified. Since M is input-restricted, the corresponding output, $g(a, x)$ or $g(c, x)$, respectively, must also be unspecified. This makes state a compatible with state c as far as input x is concerned.

If condition 1 is not true, condition 2 must be true and $g(a, x) = g(c, x)$ (when both are specified). If either output is unspecified or if they are equal, these two outputs cannot cause state a to be incompatible with state c. Additionally, condition 2 requires that $f(a, x) = q_1$ and $f(c, x) = q_2$ be compatible with either state a or state c. As far as compatibility is concerned, q_1 and q_2 could be replaced by states a and c, giving possible next-state pairs of a, a; a, c; c, a; and c, c. None of these pairs can cause state a to be incompatible with state c. Thus, condition 2, if true, makes state a compatible with state c as far as input x is concerned.

Since either condition 1 or condition 2 must be true for all unspecified entries and all inputs, there are no incompatible states in M; and thus M is transitive. ∎

THEOREM 7.3.5

For transitive-restricted machines, the minimization method used for completely specified machines is valid and yields a reduced machine.

The definition of transitive-restricted machines is somewhat too restrictive, as there are machines which are not input-restrictive but can be reduced to the minimal configuration. However, machines which cannot be minimized by partitions to a minimal machine may still be partially minimized. Every minimization by O.C.S.P. partitions, whether to a minimal machine or not, results in a valid machine whose input–output response is identical to the original machine for all specified outputs. Theorem 7.3.5 assures this to be true.

Example 7.3.2

Machine M_5 in Fig. 7.3.4 is an incompletely specified Mealy-type machine. The partition structure is not a lattice, as every pair of partitions does not have a l.u.b. Consider $\pi_1 + \pi_3$: the sum includes the block $\overline{0, 1, 2}$, which is not output-consistent. The sum $\pi_2 + \pi_3$ is neither output-consistent nor does it have the substitution property. Finally, note that while π_3 is the largest O.C.S.P. partition, it is not on the highest row—the maximal partition π_4 is higher. π_3 will yield the six-state minimized machine.

q \ x	0	1	2
0	0, 1	5, 0	2, 0
1	—	6, 0	2, 0
2	0, 0	6, 0	1, 0
3	0, 0	7, 1	4, 1
4	1, 0	7, 1	3, 1
5	2, 1	8, 0	3, 0
6	2, 1	9, 0	4, 0
7	3, 0	7, 1	0, 1
8	3, 0	8, 1	—
9	4, 0	9, 1	2, 0

(a) Machine M_5

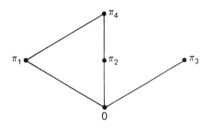

(b) O.C.S.P. partition structure of machine M_5

Fig. 7.3.4 Machine M_5 and its partition structure.

With further modification of Algorithm 7.2.1, we obtain the following algorithm, which can be used to minimize an incompletely specified machine.

ALGORITHM 7.3.3

Algorithm 7.2.1 can be modified for use on non-input-restricted machines and is rewritten here with modified steps in italics for emphasis.

1. Obtain all O.C.S.P. π_{qp} partitions, eliminating any that are duplicates or are the identity partition.

2. Examine the remaining partitions. Eliminate those that are sums of other partitions. The remaining partitions are basic generators.

3. If a basic generator contains another basic generator, it does not belong to the first row and must be saved until a later row. This is an unassigned basic generator (UBG). The remaining basic generators comprise the first row.

4. *Call the row just completed row n. If there are no partitions or if there is only one partition in row n, go to step 7. Otherwise, take the sum of all row n partitions, eliminating any sum that is not an O.C.S.P. partition. The remaining sums are candidates for row n + 1. Row n may now be ignored for the moment.*

5. If there are any UBG, compare each one to all row $n + 1$ candidates, and all other UBG. If an UBG does not contain any of these partitions, the UBG belongs to row $n + 1$; if an UBG does contain a candidate or another UBG, it remains unassigned.

6. Now compare all row $n + 1$ candidates to each other and to any basic generators assigned to row $n + 1$ in step 5. In these comparisons, a candidate is eliminated if it contains another partition. The remaining partitions comprise row $n + 1$. Return to step 4.

7.   *If there are no UBG, the O.C.S.P diagram is complete; search the diagram for the largest O.C.S.P. partition, then go to step 9. Otherwise, the unassigned basic generators will form row $n + 1$.*

8.   Compare the UBG among themselves: an UBG belongs to row $n + 1$ if it contains no other UBG; otherwise, it remains an UBG. Return to step 4.

9.   Stop.

Modification of step 4 is necessary because the machine may not be transitive, and so the sum of two O.C.S.P. partitions may not itself be O.C.S.P. Step 7 is modified to take into consideration the fact that a nonlattice structure may have more than one partition in the highest level—in fact, there may be more than one maximal partition; moreover, the largest partition may not even be in the highest level.

Exercise 7.3

1.   (a) Find all O.C.S.P. partitions of the completely specified machines M_a, M_b, M_c, and M_d in Table P7.3.1.
(b) Find the basic generators of the O.C.S.P. lattices.
(c) Construct the O.C.S.P. lattices of these machines.
(d) Minimize the four machines using Algorithm 7.3.1.

TABLE P7.3.1

q \ x	0	1
A	A, 0	B, 1
B	B, 1	D, 0
C	C. 0	B, 0
D	B, 1	C, 1

(a) M_a

q \ x	0	1
A	B, 0	E, 1
B	C, 1	D, 0
C	A, 1	D, 1
D	D, 1	A, 1
E	E, 0	C, 0

(b) M_b

q \ x	0	1
A	C, 1	D, 1
B	B, 0	C, 1
C	C, 1	A, 0
D	D, 0	C, 0
E	E, 0	C, 0
F	F, 0	C, 1

(c) M_c

q \ x	0	1
A	A, 0	G, 1
B	B, 0	D, 0
C	D, 1	E, 0
D	G, 1	E, 1
E	E, 0	G, 1
F	F, 0	D, 0
G	C, 0	F, 1

(d) M_d

2.   Using the state-minimization criterion described in this section, show that no state reduction of the machine described in Table P7.3.2 is possible.

TABLE P7.3.2

q\x	a	b	c	z
A	E	F	D	γ
B	F	E	D	α
C	D	D	B	β
D	C	C	F	γ
E	B	A	C	γ
F	A	B	C	α

M_e

3. Minimize the incompletely specified machines M_f, M_g, M_h, and M_i described in Table P7.3.3 using Algorithm 7.3.2.

TABLE P7.3.3

q\x	0	1	z
A	E	C	1
B	-	C	1
C	D	-	1
D	-	A	-
E	F	B	0
F	F	B	-

(a) M_f

q\x	a	b	c	d	z
A	A	J	G	-	1
B	B	I	D	-	0
C	-	J	F	C	0
D	B	-	D	H	0
E	-	I	F	E	0
F	B	-	F	E	0
G	-	-	G	C	0
H	--	I	D	H	0
I	A	I	-	H	1
J	A	J	-	-	1

(b) M_g

q\x	a	b	c
A	A, 1	F, 0	C, 0
B	-	G, 0	C, 0
C	A, 0	G, 0	B, 0
D	A, 0	H, 1	E, 1
E	B, 0	H, 1	D, 1
F	C, 1	I, 0	D, 0
G	C, 1	J, 0	E, 0
H	D, 0	H, 1	A, 1
I	D, 0	I, 1	-
J	E, 0	J, 1	C, 0

(c) M_h

q\x	a	b	c	d	e	f	g
A	-	E, 1	A, -	A, 0	D, 0	B, 0	-
B	D, 1	-	A, 1	B, 0	A, -	A, -	-
C	D, 1	-	-	B, 0	A, 1	-	G, 0
D	E, -	B, -	-	-	-	B, 0	A, -
E	E, -	-	E, -	B, -	A, -	B, -	A, 1
F	C, -	H, 1	G, 0	B, 0	-, 1	F, 1	-
G	C, 1	E, 1	G, 0	-	-	-	F, 0
H	E, 0	B, 0	E, -	A, 1	D, 1	B, -	A, 1

(d) M_i

7.4 State Assignment Using S.P. Partitions

State assignment is the association of a valuation of the state variables with each machine state. The state variable valuations associated with any pair of nonequivalent states must be distinct for proper functioning of the sequential circuit realization.

Example 7.4.1

Let us consider three different state assignments for machine M_6 described in Table 7.4.1.

State q	State assignment 1 $y_1 y_2 y_3$	State assignment 2 $y_1 y_2 y_3$	State assignment 3 $y_1 y_2 y_3$
1	001	000	000
2	010	001	001
3	011	010	011
4	100	011	010
5	101	100	110
6	110	101	111
7	111	110	101
8	000	111	100

TABLE 7.4.1   Machine M_6

q \ $x_1 x_2$	00	01	11	10
1	1	4	3	1
2	1	4	4	1
3	1	4	5	8
4	1	4	6	8
5	8	8	7	8
6	8	8	8	8
7	1	1	1	1
8	1	1	2	1

Three state variables y_1, y_2, and y_3 are used, since machine M_6 has eight states. The transition tables in coded form using these three state assignments are shown in Tables 7.4.2, 7.4.3, and 7.4.4, respectively. The next-state functions Y_1, Y_2, and Y_3 after minimization (see Figs. 7.4.1–7.4.3) are found to be as follows:

State Assignment 1:

$$Y_1 = y_1' y_2 x_2 + y_1 y_2' y_3' x_2 + y_1' y_3 x_1' x_2 + y_1 y_2' x_1 x_2$$

$$Y_2 = y_2' x_1 x_2$$

$$Y_3 = y_1' y_3' x_2' + y_1 y_2 y_3 + y_2' y_3' x_1' x_2' + y_3 x_1 x_2 + y_1' x_1' x_2'$$
$$\quad + y_1' y_2' y_3' x_1' + y_1' y_2' y_3 x_1$$

State Assignment 2:

$$Y_1 = y_1 y_2' + y_1' y_2 x_1$$

$$Y_2 = y_1 y_2' + y_2' x_2 + y_1' x_1' x_2 + y_1' y_2 x_1 x_2' + y_1' y_2 x_1 x_2'$$

$$Y_3 = y_1 y_2' x_2' + y_3 x_1 x_2 + y_1' x_1' x_2 + y_1' y_2 x_1 x_2' + y_2' x_1' x_2$$

State Assignment 3:

$$Y_1 = y_1 y_2 + y_2 x_1$$

$$Y_2 = y_1' x_2$$

$$Y_3 = y_3' x_1 x_2$$

**TABLE 7.4.2 Transition Table in Coded Form
Obtained by State Assignment 1**

$y_1 y_2 y_3$ \\ $x_1 x_2$	00	01	11	10
001	001	100	011	001
010	001	100	100	001
011	001	100	101	000
100	001	100	110	000
101	000	000	111	000
110	000	000	000	000
111	001	001	001	001
000	001	001	010	001

$$Y_1 Y_2 Y_3$$

From the three sets of next-state functions, it is seen that the next-state functions of the third state assignment is much simpler than the other two. A realization of the next-state functions corresponding to the third state assignment is shown in Fig. 7.4.5, in contrast to the much more complicated circuit realization of the same machine shown in Fig. 7.4.4 when a poor state assignment (state assignment 1) is employed. A comparison of the number of gate elements needed to realize these three equivalent sets of next-state functions and the total number of gate inputs upon assuming a two-level AND–OR gate circuit are tabulated in Table 7.4.5.

TABLE 7.4.3   Transition Table in Coded Form Obtained by State Assignment 2

$y_1y_2y_3$ \ x_1x_2	00	01	11	10
000	000	011	010	000
001	000	011	011	000
010	000	011	100	111
011	000	011	101	111
100	111	111	110	111
101	111	111	111	111
110	000	000	000	000
111	000	000	001	000

$$Y_1Y_2Y_3$$

TABLE 7.4.4   Transition Table in Coded Form Obtained by State Assignment 3

$y_1y_2y_3$ \ x_1x_2	00	01	11	10
000	000	010	011	000
001	000	010	010	000
011	000	010	110	100
010	000	010	111	100
110	100	100	101	100
111	100	100	100	100
101	000	000	000	000
100	000	000	001	000

$$Y_1Y_2Y_3$$

TABLE 7.4.5   Comparison of the Number of Gates Needed to Realize the Three Equivalent Sets of Next-State Functions

Gate	Realization using state assignment 1	Realization using state assignment 2	Realization using state assignment 3
AND	12	11	4
OR	2	3	1
NOT	5	4	2
Total no. of gates	19	18	7
Total no. of gate inputs	47	36	11

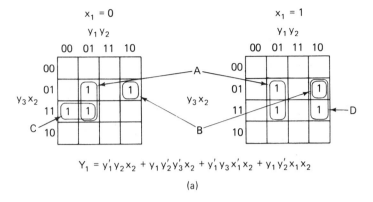

$$Y_1 = y_1' y_2 x_2 + y_1 y_2' y_3' x_2 + y_1' y_3 x_1' x_2 + y_1 y_2' x_1 x_2$$

(a)

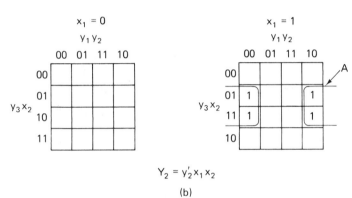

$$Y_2 = y_2' x_1 x_2$$

(b)

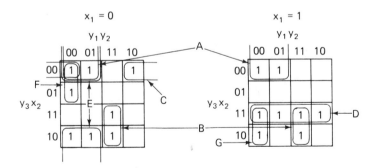

$$Y_3 = y_1' y_3' x_2' + y_1 y_2 y_3 + y_2' y_3' x_1' x_2' + y_3 x_1 x_2 + y_1' x_1' x_2' + y_1' y_2' y_3' x_1' + y_1' y_2' y_3 x_1$$

(c)

Fig. 7.4.1 Minimization of the next-state functions Y_1, Y_2, and Y_3 obtained from state assignment 1.

265

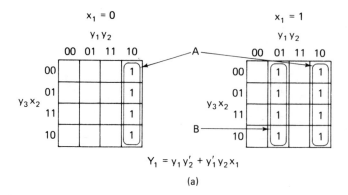

$$Y_1 = y_1 y_2' + y_1' y_2 x_1$$

(a)

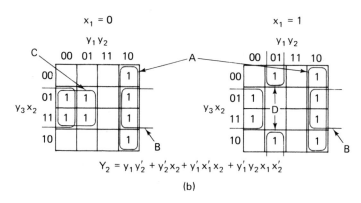

$$Y_2 = y_1 y_2' + y_2' x_2 + y_1' x_1' x_2 + y_1' y_2 x_1 x_2'$$

(b)

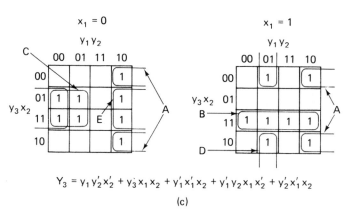

$$Y_3 = y_1 y_2' x_2' + y_3 x_1 x_2 + y_1' x_1' x_2 + y_1' y_2 x_1 x_2' + y_2' x_1' x_2$$

(c)

Fig. 7.4.2 Minimization of the next-state functions Y_1, Y_2, and Y_3 obtained from state assignment 2.

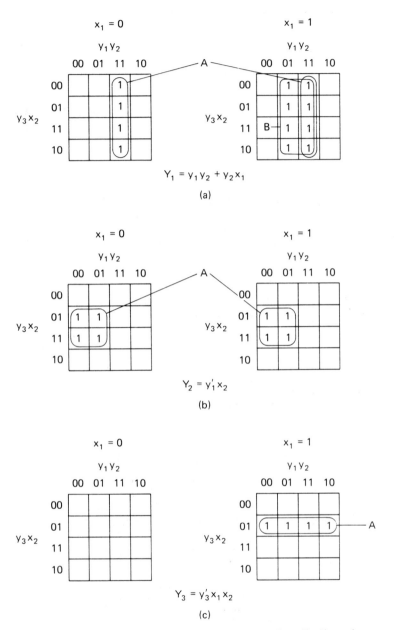

Fig. 7.4.3 Minimization of the next-state functions Y_1, Y_2, and Y_3 obtained from state assignment 3.

267

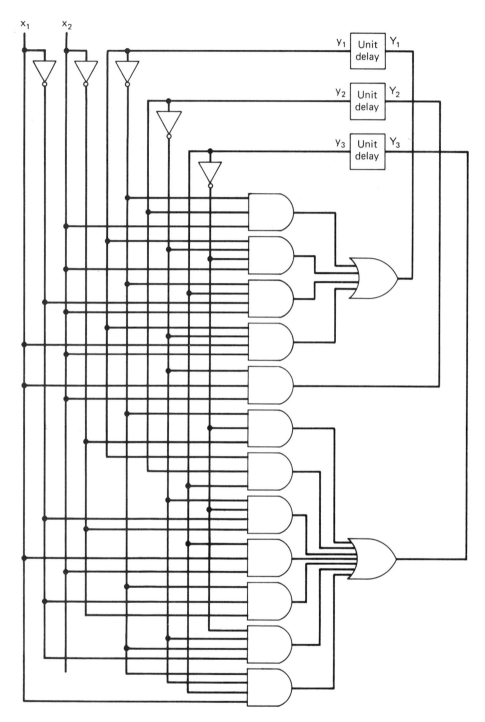

Fig. 7.4.4 Realization of machine M_6 when state assignment 1 (a poor state assignment) is used.

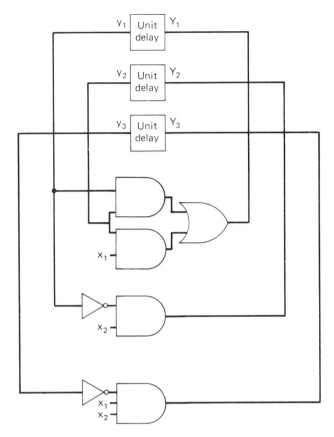

Fig. 7.4.5 Realization of machine M_6 when state assignment 3 (a good state assignment) is used.

Example 7.4.2

As another example of showing that different state assignments result in circuit realizations with different structural complexity, consider the realization of machine M_7 described in Table 7.4.6. This machine has eight states and is reduced. Again, three state variables are

TABLE 7.4.6   Machine M_7

q \ x	0	1
A	C, 1	B, 1
B	G, 0	H, 0
C	F, 0	A, 0
D	D, 0	E, 0
E	C, 0	B, 0
F	D, 1	E, 1
G	F, 1	A, 1
H	G, 1	H, 1

sufficient to code these eight states. Consider a randomly chosen state assignment 1 and a judicious state assignment 2, which are shown below.

State q	State assignment 1 $y_1 y_2 y_3$	State assignment 2 $y_1 y_2 y_3$
A	000	101
B	001	110
C	010	010
D	011	000
E	100	100
F	101	001
G	110	011
H	111	111

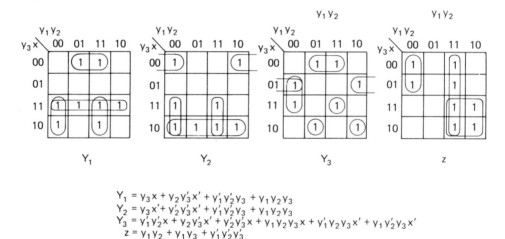

$$Y_1 = y_3 x + y_2 y_3' x' + y_1' y_2' y_3 + y_1 y_2 y_3$$
$$Y_2 = y_3 x' + y_2' y_3 x' + y_1' y_2' y_3 + y_1 y_2 y_3$$
$$Y_3 = y_1' y_2' x + y_2 y_3' x' + y_2' y_3 x + y_1 y_2 y_3 x + y_1' y_2 y_3 x' + y_1 y_2' y_3 x'$$
$$z = y_1 y_2 + y_1 y_3 + y_1' y_2' y_3'$$

(a) State assignment 1 is used

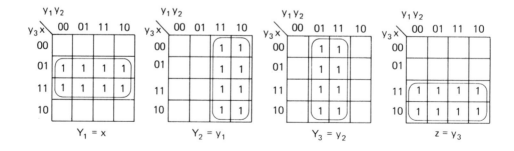

$$Y_1 = x \qquad Y_2 = y_1 \qquad Y_3 = y_2 \qquad z = y_3$$

(b) State assignment 2 is used

Fig. 7.4.6 The minimization of next-state functions using Karnaugh maps.

The next-state and output functions obtained from these two assignments after minimization (see Fig. 7.4.6) are found to be as follows:

State Assignment 1:

$$Y_1 = y_3 x + y_2 y_3' x' + y_1' y_2' y_3 + y_1 y_2 y_3$$

$$Y_2 = y_3 x' + y_2' y_3' x' + y_1' y_2' y_3 + y_1 y_2 y_3$$

$$Y_3 = y_1' y_2' x + y_2 y_3' x' + y_2' y_3' x + y_1 y_2 y_3 x + y_1' y_2 y_3 x' + y_1 y_2' y_3 x'$$

$$z = y_1 y_2 + y_1 y_3 + y_1' y_2' y_3'$$

State Assignment 2:

$$Y_1 = x$$
$$Y_2 = y_1$$
$$Y_3 = y_2$$
$$z = y_3$$

The transition tables in coded form are shown in Table 7.4.7. Their realizations using delay memory devices are depicted in Fig. 7.4.7. It is seen that even with the technique of sharing

TABLE 7.4.7 Transition Tables in Coded Form

$y_1 y_2 y_3$ \ x	0	1
000	010, 1	001, 1
001	110, 0	111, 0
010	101, 0	000, 0
011	011, 0	100, 0
100	010, 0	001, 0
101	011, 1	100, 1
110	101, 1	000, 1
111	110, 1	111, 1

$y_1 y_2 y_3$ \ x	0	1
101	010, 1	110, 1
110	011, 0	111, 0
010	001, 0	101, 0
000	000, 0	100, 0
100	010, 0	110, 0
001	000, 1	100, 1
011	001, 1	101, 1
111	011, 1	111, 1

(a) Transition table corresponding to state assignment 1

(b) Transition table corresponding to state assignment 2

components in the realizations of Y_1, Y_2, and Y_3, the realization obtained by using state assignment 2 is much simpler than that obtained by using state assignment 1. From the circuit of Fig. 7.4.7(b), it is easy to recognize that machine M_7 is a three-moment delay machine.

From the examples above we observe that different state assignments to code the states of a sequential machine will result in quite different sequential circuit realizations of the machine with respect to their structural complexity. In other words, a judicious choice of state assignment will give simple next-state functions, thus a simpler realization.

The problem of state assignment is to find a state assignment which will give, if not the simplest, at least a much simpler sequential circuit realization compared to that obtained by using a randomly selected state assignment.

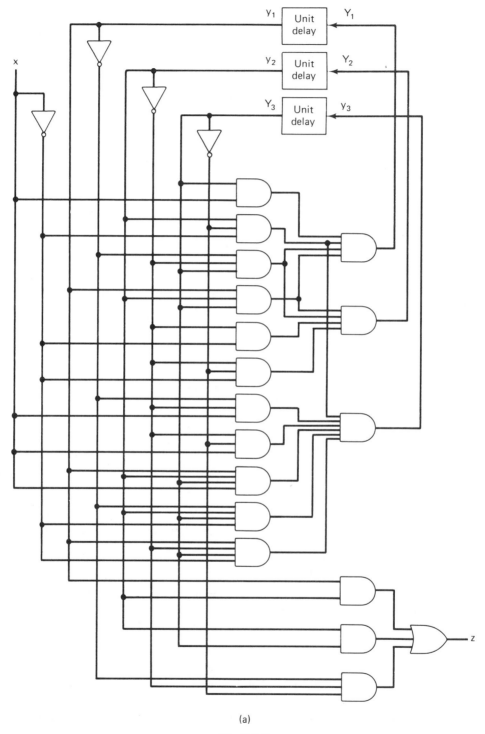

(a)

Fig. 7.4.7

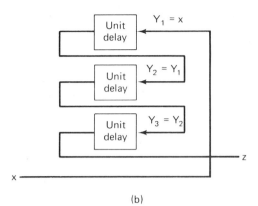

Fig. 7.4.7 (Continued) (a) Realization of machine M_7 when state assignment 1 is used. (b) Realization of machine M_7 when state assignment 2 is used.

(b)

A. Criterion of Reduction of State Variable Dependency

If, for a given state assignment, the state variable Y_j is an essential variable in the next-state variable Y_i, then Y_i is said to depend on Y_j for the given assignment. Although a general rule cannot be stated for all machines, usually the fewer state variables on which Y_i depends, the simpler is the realization of the next-state functions Y_i's.

In general, Y, the set of next-state variables, depends upon the input set X and the set of present-state variables V;

$$Y = f(V, X)$$

The criterion of reduction of state-variable dependency may be stated as follows: If a state variable is coded according to the blocks of an S.P. partition, the next-state equation for this variable depends upon X and only a reduced subset of variables R, $R \subset V$. Thus, for a random state assignment, the ith next-state variable is

$$Y_i = f_i(V, X)$$

But if Y_i is assigned over the blocks of an S.P. partition, then (with $y_i \in R$)

$$Y_i = f_i(R, X)$$

which normally will be an expression with fewer terms than for an arbitrary assignment.

DEFINITION 7.4.1

A set of v variables is *assigned* over the b blocks of a partition ($2^v \geq b$), by choosing a different v-bit binary code for each block of the partition. If $\pi = \{\overline{1, 2}; \overline{3, 4}; \overline{5, 6}\}$, then at least two variables (y_1, y_2) are needed and could be coded $(0, 0)$ for the first block, $(1, 1)$ for the second block, and $(0, 1)$ for the last block.

The following theorem expresses the concept of a reduced dependency assignment.

THEOREM 7.4.1

Let M be an n-state sequential machine with $s > \lceil \log_2 n \rceil$ state variables. Then there is an assignment of s binary variables to the states of M such that k of the next-state variables $(0 < k < s)$ can be computed without the knowledge of the other $s - k$ present-state variables iff there is an S.P. partition which identifies all the states that have identical coding in these k bits.

The proof is obvious and is omitted.
For convenience, we define

DEFINITION 7.4.2

A *self-dependent set* $y_1, \ldots, y_s$ is a set of state variables such that the next-state functions Y_i $(1 \le i \le s)$ depends only on state variables of the set $y_1, \ldots, y_s$ and perhaps also on input variables.

For example, if $\pi = \{\overline{1, 2}; \overline{3, 4, 5, 6}; \overline{7, 8, 9}\}$ for a nine-state machine, and if y_1 and y_2 are two of the four variables needed, then y_1 and y_2 can be coded over the blocks of π. Every state in a particular block of π will then have the identical $y_1 y_2$ code, and Y_1 and Y_2 could be computed without the knowledge of the other variables; specifically,

$$Y_1 = f_1(y_1, y_2, X)$$

and

$$Y_2 = f_2(y_1, y_2, X)$$

while Y_3 and Y_4 remain functions of all four present-state variables. The reverse statement is also true; if Y_1 and Y_2 can be calculated without knowledge of other variables, there must be a three- or four-block partition having S.P. The variables y_1 and y_2 form a *self-dependent subset* of the set of all state variables.

B. Conditions for State Assignment
    with Minimum Variables Using One
    S.P. Partition

However, a sufficient condition in terms of S.P. partitions for the existence of a state assignment using a *minimum* number of state variables which produces a self-dependent set, has been found [10].

THEOREM 7.4.2

If there exists a nontrivial S.P. partition π of machine M such that

$$\lceil \log_2 \#(\pi) \rceil + \lceil \log_2 m(\pi) \rceil = \lceil \log_2 \#(Q) \rceil$$

where $\#(\pi)$, $m(\pi)$, and $\#(Q)$ denote the number of blocks in π, the number of states in the largest block of π, and the number of distinct states of M, respectively, then there exists a state assignment using the minimum number of state variables which produces a self-dependent set.

Proof: Since there are $\#(\pi)$ blocks in the S.P. partition π, the first $\lceil \log_2 \#(\pi) \rceil$ state variables $y_1, y_2, \ldots, y_r$ can be used to distinguish among the blocks. That is, the state assignment for states q_i and q_j agree in $y_1, \ldots, y_r$, iff q_i and q_j are members of the same block of π. Because π is a S.P. partition, the input x and the block of π which contains q_i are sufficient to determine the set of P which contains $g(q_i, x)$. Hence, x and $y_1, y_2, \ldots, y_r$ determine $Y_1, Y_2, \ldots, Y_r$, and $y_1, y_2, \ldots, y_r$ is a self-dependent set. The remaining $\lceil \log_2 m(\pi) \rceil$ state variables $y_{r+1}, \ldots, y_s$ are used to distinguish among states within a block of π. Hence, every state is associated with a unique state assignment. ∎

To illustrate this theorem, let us go back to the first two state assignment examples presented above.

Example 7.4.1 *(continued)*

By using Algorithm 7.2.1, we find that there are only two nontrivial S.P. partitions of machine M_6:

$$\pi_1 = \{\overline{1, 4, 5, 8}; \overline{2, 3, 6, 7}\}$$
$$\pi_2 = \{\overline{1, 2}; \overline{3, 4}; \overline{5, 6}; \overline{7, 8}\}$$

and the S.P. lattice of machine M_6 is depicted in Fig. 7.4.8. Both partitions π_1 and π_2 satisfy

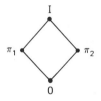

$$\pi_1 = \{\overline{1, 4, 5, 8}; \overline{2, 3, 6, 7}\}$$

Fig. 7.4.8 The S.P. lattice of machine M_6. $\pi_2 = \{\overline{1, 2}; \overline{3, 4}; \overline{5, 6}; \overline{7, 8}\}$

the conditions of Theorem 7.4.2, for

$$\lceil \log_2 \#(Q) \rceil = \lceil \log_2 8 \rceil = 3$$
$$\lceil \log_2 \#(\pi_2) \rceil + \lceil \log_2 m(\pi_2) \rceil = \lceil \log_2 4 \rceil + \lceil \log_2 2 \rceil = 3$$
$$\lceil \log_2 \#(\pi_1) \rceil + \lceil \log_2 m(\pi_1) \rceil = \lceil \log_2 2 \rceil + \lceil \log_2 4 \rceil = 3$$

Hence, both these two S.P. partitions can yield a state assignment using the minimum number of state variables which produces a self-dependent set. Each of the following assignments using partitions π_1 and π_2 is one of many which satisfy the procedure of Theorem 7.4.2.

A state assignment using π_2:

y_1	y_2	$y_3 = 0$	$y_3 = 1$
0	0	1	2
0	1	3	4
1	0	5	6
1	1	7	8

(This is state assignment 2.)

A state assignment using π_1:

y_1	y_2	$y_3 = 0$	$y_3 = 1$
0	0	1	2
0	1	4	3
1	1	5	6
1	0	8	7

(This is state assignment 3.)

These two assignments, which produce self-dependent sets, are, in fact, the state assignments 2 and 3 described earlier. It has been shown that the realizations obtained from these two state assignments were simpler than those obtained from a randomly selected state assignment such as state assignment 1. It is conceivable that the reduced dependency state assignment based on the largest S.P. partition available at each level of the S.P. lattice should in general give simpler next-state functions.

Example 7.4.2 (*continued*)

The S.P. lattice of machine M_7 obtained by Algorithm 7.2.1 is depicted in Fig. 7.4.9. It is seen from Fig. 7.4.9 that this machine has an S.P. partition

$$\pi_5 = \{\overline{A, B, E, H}; \overline{C, D, F, G}\}$$

which again satisfies Theorem 7.4.2; thus, a reduced dependency state assignment with a minimum number of state variables exists. One such assignment was state assignment 2 given above, since it satisfies the procedure of Theorem 7.4.2 as seen below.

y_2	y_3	$y_1 = 0$	$y_1 = 1$
0	0	D	E
0	1	F	A
1	0	C	B
1	1	G	H

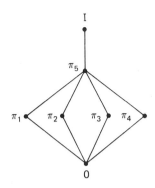

$$\pi_1 = \overline{A, E}, \quad \pi_2 = \overline{B, H}, \quad \pi_3 = \overline{C, G}, \quad \pi_4 = \overline{D, F}, \quad \pi_5 = \overline{ABEH}; \overline{CDFG}$$

Fig. 7.4.9 The S.P. lattice of machine M_7.

Hence, it also yields simple next-state functions Y_i's; as a matter of fact, this is the best assignment among all possible assignments to the states of this machine.

C. Conditions for State Assignment with Minimum Variables Using a Set of S.P. Partitions

A better approach to the state assignment problem is to use a set of S.P. partitions. The theoretical concept of this approach is straightforward; applying it has been difficult. Some of the general difficulties encountered in using a set of S.P. partitions to obtain a reduced dependency state assignment are illustrated by the following examples.

Example 7.4.3

Let $\pi = \{\overline{1, 2, 3}; \overline{4, 5}; \overline{6, 7}\}$ and $\tau = \{\overline{1, 2, 3, 4}; \overline{5, 6, 7}\}$ be two S.P. state partitions on a seven-state machine; a minimum of three state variables are needed. Now since these partitions have S.P., any variables assigned over the blocks of the partitions will form a self-dependent subset. If y_1 and y_2 are assigned over the blocks of π, and y_3 is assigned over the blocks of τ, then Y_1 and Y_2 are functions only of themselves (and the inputs), while Y_3 is a function only of itself (and the inputs). Assume that $y_1 = y_2 = 0$ is the code that is assigned to the block $\overline{1, 2, 3}$, and $y_3 = 0$ is assigned to the block $\overline{1, 2, 3, 4}$. Then states 1, 2, and 3 will not have unique codes; all three states have the code 000. In fact, there is no way to use π and τ to code the states of the machine without using two extra variables (five variables in all) to obtain unique codes for the states. Additionally, π cannot be used by itself unless one extra variable is used.

It is apparent from Example 7.4.1 that using S.P. partitions to obtain reduced dependency is not as straightforward as would appear at first; more than the minimum number of variables may be required unless partitions are chosen carefully. Example 7.4.2 points out another difficulty in choosing partitions.

Example 7.4.4

For a nine-state machine, let the following be the only partitions with S.P.:

$$\pi = \{\overline{1, 2, 3, 4}; \overline{5, 6, 7, 8, 9}\}$$
$$\beta = \{\overline{1, 2}; \overline{3, 6}; \overline{5, 7, 9}; \overline{4, 8}\}$$
$$\phi = \{\overline{1, 3, 4}; \overline{2, 5}; \overline{6, 7}; \overline{8, 9}\}$$

Four state variables are the minimum needed. The obvious first choice of a partition is π; only one state variable is needed to code its blocks, making that variable a function only of itself. However, attempts to also use one of the other two partitions results in five variables required, instead of four. If π is used alone, the other three state variables must be assigned at random to provide unique codes for the states; these three variables will then depend upon all state variables.

If partitions β and ϕ are chosen—with two variables assigned over each partition—the minimum of four variables is possible, with each variable a function of itself and the other variable assigned over the same partition. This assignment will yield simpler next-state equations, when considering all the state variables, than any assignment which uses partition π—even though the one variable assigned over π will have a very simple next-state equation.

The problem of deciding which valid set of partitions to use has no general solution, since each machine design has its own definition of what is best. Even when "best" has been defined, choosing a set of partitions can involve a tree structure, where each branch represents a valid choice of partitions; the number of such branches is often too large to calculate each one. However, in the following, we shall be confined to the search for a set of S.P. partitions which will provide a reduced dependency state assignment with a minimum number of state variables. The necessary and sufficient conditions for selecting a set of S.P. partititions which require only the minimum number of state variables has been found [17]. These conditions naturally include those stated in Theorem 7.4.2 as a special case.

THEOREM 7.4.3

Let M be a minimized n-state sequential machine, and A be a nonempty set of m S.P. partitions in machine M. Then there is a state assignment of s state variables such that the subsets of variables assigned over each partition in A are each self-dependent iff

1. $k_{\pi_i} + \mu_{\pi_i} = s$, for all $\pi_i \in A$.

2. $\displaystyle\sum_{i=1}^{m} k_{\pi_i} = k_{p^m}$.

3. $k_{p^m} + \mu_{p^m} = s$.

where $s = \lceil \log_2 n \rceil$, $k_{\pi_i} = \lceil \log_2 \#(\pi_i) \rceil$, $\mu_{\pi_i} = \lceil \log_2 m(\pi_i) \rceil$, and $p^m = \displaystyle\prod_{i=1}^{m} \pi_i$.

Proof: Theorem 7.4.1 states that the variables assigned over a partition form a self-dependent subset of the state variables iff the partition has S.P. Obviously, this is also true for a set of partitions. What remains to be proved for the set A is that the minimum number (s) of variables is assured by the three conditions of this theorem.

Condition 1 states that each partition in A uses just s variables for a unique code, while condition 2 states that the number of variables assigned over the partitions in A is exactly the correct minimum number to assign over partition p^m. Thus, no extra variables are assigned over the set A. This leaves $s - k_{p^m}$ variables to assign over the states in the largest block of p^m. Condition 3 demands that no extra variables be used to uniquely code the states in the largest block of p^m; so exactly s variables are used. Therefore, since exactly s variables are used, and an S.P. partition yields a self-dependent subset, each partition in A has a self-dependent subset such that exactly s variables are used to uniquely code the states of M.

To prove the necessity of the three conditions, assume that variables are assigned over the partitions in A such that exactly s variables are used to yield unique state codes; furthermore, assume that the subset of variables assigned over each partition in A is a self-dependent subset. Since the partitions yield a self-dependent subset, each partition has S.P. Because exactly s variables are needed for unique codes, each partition that has one or more variables assigned over its blocks must have $k + \mu = s$; otherwise, $k + \mu > s$ and more than s variables would be needed. Thus, conditions 1 and 3 are true. It can be shown that if the sum of the k's is not equal to the k of the product, more than s variables are needed. Since M uses only s variables, condition 2 must be true. Therefore, conditions 1, 2, and 3 are true for any machine M with s variables having self-dependent subsets; thus, these conditions are necessary and sufficient. ∎

This theorem is illustrated by the following example.

Example 7.4.5

Consider the minimized six-state machine M_8 whose transition table is described in Table 7.4.11. Using Algorithm 7.2.1, we find there are only two nontrivial partitions, which are

TABLE 7.4.11   Machine M_8

q ╲ x	0	1	z
A	D	C	0
B	F	C	0
C	E	B	0
D	B	E	1
E	A	D	0
F	C	D	0

$$\pi_1 = \{\overline{A, B, C};\ \overline{D, E, F}\}$$

and

$$\pi_2 = \{\overline{A, F};\ \overline{B, E};\ \overline{C, D}\}$$

Let $A = \{\pi_1, \pi_2\}$, so $n = 6$ and $m = 2$:

$$s = \lceil \log_2 6 \rceil = 3$$
$$k_{\pi_1} = \lceil \log_2 \#(\pi_1) \rceil = \lceil \log_2 2 \rceil = 1$$
$$\dot{k}_{\pi_2} = \lceil \log_2 \#(\pi_2) \rceil = \lceil \log_2 3 \rceil = 2$$
$$\mu_{\pi_1} = \lceil \log_2 m(\pi_1) \rceil = \lceil \log_2 3 \rceil = 2$$
$$\mu_{\pi_2} = \lceil \log_2 m(\pi_2) \rceil = \lceil \log_2 2 \rceil = 1$$
$$p^2 = \prod_{i=1}^{2} \pi_i = \{A, B, C; \overline{D, E, F}\} \cdot \{A, F; \overline{B, E}; \overline{C, D}\}$$
$$= \{\bar{A}; \bar{B}; \bar{C}; \bar{D}; \bar{E}; \bar{F}\}$$
$$k_{p^2} = \lceil \log_2 \#(p^2) \rceil = 3$$
$$\mu_{p^2} = \lceil \log_2 m(p^2) \rceil = \lceil \log_2 1 \rceil = 0$$

It is found that
1. $k_{\pi_1} + \mu_{\pi_1} = 3 = s$
    $k_{\pi_2} + \mu_{\pi_2} = 3 = s$
2. $k_{\pi_1} + k_{\pi_2} = 1 + 2 = 3 = k_{p^2}$
3. $k_{p^2} + \mu_{p^2} = 3 + 0 = 3 = s$

All three conditions in Theorem 7.4.3 are satisfied; and, therefore, there is a reduced dependency state assignment of three state variables. We may assign codes to the blocks of the two partitions using three state variables as follows:

$$\begin{array}{cc} y_1 = 0 & y_1 = 1 \\ \{A, B, C; & D, E, F\} \end{array}$$

$$\begin{array}{ccc} y_2 y_3 = 00 & y_2 y_3 = 01 & y_2 y_3 = 10 \\ \{ \quad A, F \quad ; & B, E \quad ; & C, D \quad \} \end{array}$$

which yield the following state assignments:

State	Assignment
q	$y_1 y_2 y_3$
A	000
B	001
C	010
D	110
E	101
F	100

Notice that these six codes are distinct. The finding of minimized next-state functions corresponding to this state assignment, and the comparison of those obtained by a randomly selected state assignment, are left to the reader.

D. Selection of an S.P. Partition Using the S.P. Lattice

The algorithm for selecting the set of partitions meeting the conditions of Theorem 7.4.3 is presented below. This algorithm will not necessarily yield that set of partitions that will ensure the simplest set of next-state assignments, since finding such a set requires an exhaustive search of all possible lattice structures—a lengthy procedure, even with a computer. Instead, an attempt is made to use the largest partition or partitions available at each level of the lattice. With machines that have a large number of S.P. partitions, this procedure will yield a partition set which—if not the optimum set—at least approaches the optimum set. Unfortunately, the variance from this optimum set becomes larger as the number of S.P. partitions becomes smaller.

DEFINITION 7.4.3

An *ordered list* of partitions is a listing of partitions according to their k's, from smallest k at top, down to the largest k. If two or more partitions have the same k, there is no relative order among them.

ALGORITHM 7.4.1

The largest partition or partitions at each level of the S.P. lattice of a machine M can be obtained as follows:

1. Generate the S.P. partition structure using Algorithm 7.2.1. If there are no S.P. partitions go to step 14. Let s be the minimum number of variables needed. Let m and n be given levels of the state assignment structure (to be generated) and the S.P. structure respectively.

2. For each S.P. partition calculate k and μ; delete any partition from the structure if its sum of k and μ is not equal to s.

3. Start the partition search with the highest level in the partition structure; this is level n. Set m equal to n. If the number of partitions in this level is zero, go to step 10; if not, arrange the partitions in row n in an ordered list.

4. Add to this list any partitions from the next lower row $(n - 1)$ which are not contained in any partition in row n.

5. Set q equal to s; if q is greater than the number of partitions in the ordered list, then set q equal to the number of partitions in the list.

6. Examine groups of q partitions, testing them for conditions 2 and 3 of Theorem 7.4.3. If no group of q partitions satisfies these conditions, go to step 9.

7. If the state-assignment structure is empty, go to step 8. If not, compare the sum of all q partitions to the partitions in the next-highest $(m + 1)$ level. If the sum is not contained in any of these partitions, this group of q partitions is not suitable; return to step 6 and continue the search for suitable groups of q partitions.

8. This group of q partitions becomes level m of the state assignment structure. Now decrease m by one. If the product of all q partitions is the zero partition, the state

assignment structure is finished—go to step 11. If the product is not zero, go to step 10.

9. No group of q partitions satisfies all the conditions. If q is one, go to step 10; otherwise, reduce q by one and go to step 6.

10. If n equals 1, the state assignment structure is finished; go to step 11. If not, reduce n by one. If the number of partitions in level n is zero, return to the beginning of this step. Otherwise, start a new ordered list using the partitions of this new level n; go to step 4.

11. Assign different variables over each partition in the highest level of the state-assignment structure; each such partition π is assigned k_π variables.

12. Go to the next lower level in the state assignment structure. Let π be a partition in this level and τ be a partition (in the next higher level) which contains π. If $k_\pi = k_\tau$, ignore π; otherwise, assign additional v variables over $\pi(k_\pi = k_\tau + v)$. (The k_τ variables assigned over τ are also assigned over groups of blocks of π, since $\pi < \tau$.)

13. If s variables have been assigned, or if all levels of the state assignment structure have been used, go to step 14. Otherwise, go to step 12.

14. Stop.

This algorithm will select, at each level, the largest set of S.P. partitions with the smallest k's. Therefore, each self-dependent next-state variable will be dependent upon the smallest possible subset of present state variables.

Exercise 7.4

1. Show that there exists an S.P. partition of the machine of Table P7.4.1 which satisfies the

TABLE P7.4.1

q \ x	a	b	c
A	E	F	D
B	F	E	D
C	D	D	B
D	C	C	F
E	B	A	C
F	A	B	C

conditions of Theorem 7.4.2. Give a reduced dependency state assignment using this partition.

2. (a) Show that there also exists a set of S.P. partitions of the machine of Table P7.4.1 which satisfies the conditions of Theorem 7.4.3. Give a state assignment using such a set of S.P. partitions.

(b) Construct the reduced next-state functions of this machine for the two assignments obtained in problem 1 and in problem 2(a). Assume that these functions are realized by a three-level AND–OR circuit plus delay memory devices. Compare the structure complexity of the two realizations by comparing the total number of gates and the total number of gate inputs.

3. (a) Minimize the machine described in Table P7.4.3 by the method described in Section 7.3.

TABLE P7.4.3

q \ x	a	b	c	a	b	c
A	A	F	C	1	0	0
B	—	G	C	—	0	0
C	A	G	B	0	0	0
D	A	H	E	0	1	1
E	B	H	D	0	1	1
F	C	I	D	1	0	0
G	C	J	E	1	0	0
H	D	H	A	0	1	1
I	D	I	—	0	1	—
J	E	J	C	0	1	0

(b) Give an optimum (or nearly optimum) state assignment for the reduced machine obtained in part (a).

7.5 Serial Decomposition of Sequential Machine

When a sequential machine has a large number of states, even if a "good" state assignment is used, it may still require a large amount of computation to realize it using a synchronous sequential circuit. In this section an alternative is suggested. Instead of synthesizing the given machine, we can first decompose it into smaller machines, realize each component machine individually, and then interconnect them together. The composite machine obtained is identical in behavior to the given machine.

Two basic types of decomposition of a sequential machine, serial decomposition and parallel decomposition, will be investigated. Both are derived from S.P. partitions. The necessary and sufficient conditions for a machine to be decomposed into a serial and parallel decomposition are presented. Procedures for obtaining these two types of decomposition are given and illustrated by examples. First, let us investigate the serial decomposition. The parallel decomposition will be discussed in the next section.

We now make precise what we mean by a serial connection of two machines.

DEFINITION 7.5.1

The *serial connection* of two machines

$$M_1 = (Q_1, \Sigma_1, Z_1, f_1, g_1) \quad \text{and} \quad M_2 = (Q_2, \Sigma_2, Z_2, f_2, g_2)$$

for which

$$Z_1 = \Sigma_2$$

is the machine

$$M = M_1 \ominus M_2 = (Q_1 \times Q_2, \Sigma_1, Z_2, f, g)$$

where

$$f[(q, p), x] = (f_1(q, x), f_2[p, g_1(q, x)])$$

and

$$g[(q, p), x] = g_2[p, g_1(q, x)]$$

A schematic representation of a serial connection is shown in Fig. 7.5.1.

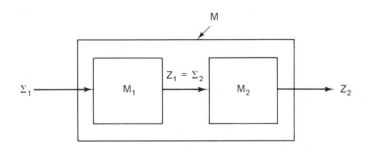

$$M_1 = (Q_1, \Sigma_1, Z_1, f_1, g_1)$$
$$M_2 = (Q_2, \Sigma_2, Z_2, f_2, g_2)$$
and
$$M = (Q_1 \times Q_2, \Sigma_1, Z_2, f, g)$$
where
$$f[(q, p), x] = (f_1(q, x), f_2[p, g_1(q, x)])$$
and
$$g[(q, p), x] = g_2[p, g_1(q, x)]$$

Fig. 7.5.1   Serial connection of two machines.

DEFINITION 7.5.2

The machine $M_1 \ominus M_2$ is a *serial decomposition* of M iff $M_1 \ominus M_2$ realizes M. The decomposition is nontrivial iff M_1 and M_2 have fewer states than M.

The necessary and sufficient conditions for a sequential machine having a serial decomposition is stated in the following theorem.

THEOREM 7.5.1

The sequential machine M has a nontrivial serial decomposition iff there exists a nontrivial S.P. partition π on the set of states of M.

Proof: Assume that M is realized by a nontrivial $M_1 \ominus M_2$. Let ξ be a one-to-one mapping

$$\xi: \quad Q \longrightarrow Q_1 \times Q_2$$

The mapping ξ induces an equivalence relation or partition π on Q if we consider all those states equivalent whose first components under ξ are identical; that is,

$$q \equiv p(\pi) \qquad \text{iff } q_1 = p_1$$

where

$$\xi(q) = (q_1, q_2) \quad \text{and} \quad \xi(p) = (p_1, p_2)$$

To show that π is an S.P. partition, note that by definition

$$\xi[f(q, a)] = (f_1(q_1, \psi(a)), f_2[q_2, g_1(q_1, \psi(a))])$$
$$\xi[f(p, a)] = (f_1(p_1, \psi(a)), f_2[p_2, g_1(p_1, \psi(a))])$$

and so if $q \equiv t(\pi)$, then $q_1 = t$ and thus $f_1(q_1, \psi(a)) = f_1(p_1, \psi(a))$. This means that $f(q, a) \equiv f(p, a)(\pi)$ and so π has S.P. By assumption, M_1 has fewer states than M, which implies that $\pi \neq O$; and M_2 has fewer states than M, which implies that $\pi \neq I$. Thus, π must be a nontrivial partition on Q.

Now we show if there exists a nontrivial S.P. partition π on Q of M, then M has a nontrivial serial realization. We assume that π has l blocks and that the largest block has k states. Since π is nontrivial, $n > k$ and $n > l$. Let τ be a k block partition on Q such that $\pi \cdot \tau = O$. The basic idea of the proof is to design two machines, M_1 and M_2, which, when connected in series, operate so that M_1 computes the block of π, which contains the state of M, and M_2 computes the corresponding block of τ. Since $\pi \cdot \tau = O$, the serial connection of M_1 and M_2 computes the state of M and we can obtain a realization of M by defining the proper output function. To construct M_1, we let $M_1 = M_\pi$ and let its output be the present state and input. Thus,

$$M_1 = (\{B_\pi\}, \Sigma, \{B_\pi\} \times X, f_1 = f_\pi, e)$$

where e stands for the identity function. Let

$$M_2 = (\{B_\tau\}, \{B_\pi\} \times \Sigma, Z, f_2, g_2 = g)$$

where

$$f_2[B_\tau, (B_\pi, x)] = B_\tau$$

B_τ' is such that

$$f(B_\tau \cap B_\pi, x) \in B_\tau'$$

The output

$$g_2[B_\tau, (B_\pi, x)] = g(B_\tau \cap B_\pi, x)$$

Note that if $B_\tau \cap B_\pi = \bigcirc$, then we have a don't-care condition, but this pair of states is never entered by $M_1 \ominus M_2$ in the simulation of M. Thus, $M_1 \ominus M_2$ is a realization of M, which proves that M has a nontrivial serial decomposition of M. This completes the proof. ∎

The procedure for obtaining a serial decomposition of a given machine may be explicitly outlined as follows:

1. Find a nontrivial S.P. partition π.
2. Find a nontrivial partition τ such that $\pi \cdot \tau$ is the O partition.
3. Construct two machines M_1 and M_2 using the π and τ partitions as described above.

The following example illustrates this procedure.

Example 7.5.1

Find a serial decomposition of machine M_6 in Fig. 7.5.2(a).
From the $\pi^m_{q,\,p}$, which are as follows:

$q,\,p$	$\pi^m_{q,\,p}$
$\overline{1,2}$	$\overline{1,2} = \pi_1$
$\overline{1,3} \longrightarrow \overline{2,3} \longrightarrow$	$\overline{1,2,3} = \pi_4$
$\overline{1,4} \longrightarrow \overline{1,3} \longrightarrow \overline{1,3,4} \longrightarrow$	$\overline{1,2,3,4} = \pi_7$
$\overline{1,5} \longrightarrow \overline{3,6}; \overline{1,4} \longrightarrow \overline{1,4,5}; \overline{3,6} \longrightarrow$	$/$
$\overline{1,6} \longrightarrow \overline{3,5}; \overline{1,4} \longrightarrow \overline{1,4,6}; \overline{3,6} \longrightarrow$	$/$
$\overline{2,3} \longrightarrow$	$\overline{2,3} = \pi_2$
$\overline{2,4} \longrightarrow \overline{1,3} \longrightarrow \overline{2,4}; \overline{1,3}$	$\overline{1,2,3,4}$
$\overline{2,5} \longrightarrow \overline{3,6}; \overline{1,4} \longrightarrow \overline{2,5}; \overline{3,6}; \overline{1,4} \longrightarrow$	$/$
$\overline{2,6} \longrightarrow \overline{3,5}; \overline{1,4} \longrightarrow \overline{2,6}; \overline{3,5}; \overline{1,4} \longrightarrow$	$/$
$\overline{3,4} \longrightarrow \overline{1,2}; \overline{1,3} \longrightarrow$	$\overline{1,2,3,4}$
$\overline{3,5} \longrightarrow \overline{2,6}; \overline{1,4} \longrightarrow \overline{3,5}; \overline{2,6}; \overline{1,4} \longrightarrow$	$/$
$\overline{3,6} \longrightarrow \overline{2,5}; \overline{1,4} \longrightarrow \overline{3,6}; \overline{2,5}; \overline{1,4} \longrightarrow$	$/$
$\overline{4,5} \longrightarrow \overline{1,6}; \overline{3,4} \longrightarrow \overline{3,4,5}; \overline{1,6} \longrightarrow$	$/$
$\overline{4,6} \longrightarrow \overline{1,5}; \overline{3,4} \longrightarrow \overline{3,4,6}; \overline{1,5} \longrightarrow$	$/$
$\overline{5,6} \longrightarrow$	$\overline{5,6} = \pi_3$

we find that in addition to these five smallest nontrivial S.P. partitions, there are four other nontrivial S.P. partitions:

$$\pi_1 + \pi_3 = \{\overline{1,2}; \overline{5,6}\} = \pi_5$$
$$\pi_2 + \pi_3 = \{\overline{2,3}; \overline{5,6}\} = \pi_6$$
$$\pi_4 + \pi_6 = \{\overline{1,2,3}; \overline{5,6}\} = \pi_8$$

and

$$\pi_7 + \pi_8 = \{\overline{1,2,3,4}; \overline{5,6}\} = \pi_9$$

The S.P. lattice L_{M_6} is depicted in Fig. 7.5.2(b). There are a total of nine nontrivial S.P. partitions. So we may use any one of them to obtain a serial decomposition of M_6.

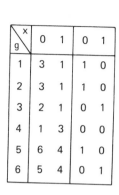

x g	0	1	0	1
1	3	1	1	0
2	3	1	1	0
3	2	1	0	1
4	1	3	0	0
5	6	4	1	0
6	5	4	0	1

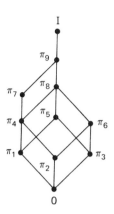

(a) Machine M_6 (b) The S. P. lattice L_{M_6}

Fig. 7.5.2   Machine M_6 and its S.P. lattice L_{M_6}.

Suppose that π_9 is used.

$$\pi = \pi_9 = \{\overline{1, 2, 3, 4}; \overline{5, 6}\} = \{A, B\} = \{B_\pi\}$$

Since π_9 has a four-state block and a two-state block, the partition τ must have four (or more) blocks. It is seen that there are $2 \times C_2^4 = 12$ such four-block partitions. We choose

$$\tau = \{\overline{1, 5}; \overline{2, 6}; \overline{3}; \overline{4}\} = \{a, b, c, d\} = \{B_\tau\}$$

The component machine $M_{6\,(1)}$ obtained from π_9 is then

$$M_{6\,(1)} = (\{A, B\}, \{0, 1\}, \{(A, 0), (A, 1), (B, 0), (B, 1)\}, f_{\pi_9}, g)$$

where

$$f_{\pi_9}(A, 0) = A, \qquad f_{\pi_9}(B, 0) = B, \qquad f(A, 1) = f(B, 1) = A$$
$$g(A, x) = (A, x) \quad \text{and} \quad g(B, x) = (B, x), x = 0, 1$$

The reason $f_{\pi_9}(A, 0) = A$ is because upon an application of input symbol 0, the block A is mapped to block A under π_9. The other three mappings are obtained in the similar manner. Note that there are four possible outputs of $M_{6\,(1)}$ corresponding to the four pairs of present state and present external input: $A0$, $A1$, $B0$ and $B1$, which for convenience are denoted by $\alpha, \beta, \gamma,$ and δ, respectively. The component machine $M_{6\,(1)}$ is shown in Table 7.5.1(a).

Next, let us construct the second component machine $M_{6\,(2)}$ from partitions π and τ. As a first step, replace each present- and next-state in M_6 by its corresponding block designator. This is shown in Table 7.5.1(b-1). Since the output of $M_{6\,(1)}$ is the input of $M_{6\,(2)}$, we reconstruct the transition table of Table 7.5.1(b-1), which is shown in Table 7.5.1(b-2), in which the set of input symbols are expanded from $\{0, 1\}$ to $\{A0, A1, B0, B1\}$ to "match" the set of output symbols of $M_{6\,(1)}$, and a dash in the table indicates that the entry is undefined. After

TABLE 7.5.1   Serial Decomposition of Machine M_6
of Example 7.5.1

q \ x	0	1	0	1
A	A	A	A, 0	A, 1
B	B	A	B, 0	B, 1

(a-1)

$\longrightarrow$

q \ x	0	1	0	1
A	A	A	α	β
B	B	A	γ	δ

(a-2)

(a)   Component machine $M_{6(1)}$

	q \ x	0	1	0	1
A	a	c	a	0	1
	b	c	a	1	0
	c	b	a	0	1
	d	a	c	0	0
B	a	b	d	1	0
	b	a	d	0	1

(b-1)

$\longrightarrow$

q \ x	A0	A1	B0	B1	A0	A1	B0	B1
a	c	a	b	d	0	1	1	0
b	c	a	a	d	1	0	0	1
c	c	b	—	—	0	1	—	—
d	d	a	—	—	0	0	—	—

(b-2)

q \ x	α	β	γ	δ	α	β	γ	δ
a	a	c	b	d	0	1	1	0
b	c	a	a	d	1	0	0	1
c	c	b	—	—	0	1	—	—
d	d	a	—	—	0	0	—	—

(b-3)

(b)   Component machine $M_{6(2)}$

replacing the four input symbols by α, β, γ, and δ, we obtain $M_{6(2)}$, which is depicted in Table 7.5.1(b-3).

7.6   Parallel Decomposition of Sequential Machine

In this section, we study the decomposition of a sequential machine into two machines which operate in parallel. We shall follow the same sequence of development as we did in the previous section. First, we define the parallel connection of two machines.

DEFINITION 7.6.1

The *parallel connection* of the two machines

$$M_1 = (Q_1, \Sigma_1, Z_1, f_1, g_1) \quad \text{and} \quad M_2 = (Q_2, \Sigma_2, Z_2, f_2, g_2)$$

is the machine

$$M_1 \| M_2 = (Q_1 \times Q_2, \Sigma_1 \times \Sigma_2, Z_1 \times Z_2, f, g)$$

with

$$f[(q_1, q_2), (x_1, x_2)] = (f_1(q_1, x_1), f_2(q_2, x_2))$$

and

$$g[(q_1, q_2), (x_1, x_2)] = (g_1(q_1, x_1), g_2(q_2, x_2))$$

A schematic representation of a parallel connection is shown in Fig. 7.6.1.

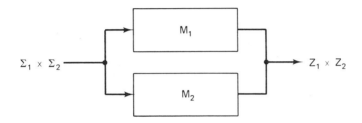

$$M_1 = (Q_1, \Sigma_1, Z_1, f_1, g_1)$$
$$M_2 = (Q_2, \Sigma_2, Z_2, f_2, g_2)$$

and

$$M = M_1 \| M_2 = (Q_1 \times Q_2, \Sigma_2 \times X_2, Z_1 \times Z_2, f, g)$$

with

$$f[(q_1, q_2), (x_1, x_2)] = (f_1(q_1, x_1), f_2(q_2, x_2))$$

and

$$g[(q_1, q_2), (x_1, x_2)] = (g_1(q_1, x_1), g_2(q_2, x_2))$$

Fig. 7.6.1 Parallel connection of two machines.

DEFINITION 7.6.2

The machine $M_1 \| M_2$ is a *parallel decomposition* of M iff $M_1 \| M_2$ realizes M. The decomposition is nontrivial iff M_1 and M_2 have fewer states than M.

The following theorem describes the necessary and sufficient condition for a given machine having a nontrivial parallel decomposition.

THEOREM 7.6.1

The sequential machine M has a nontrivial parallel decomposition iff there exist two nontrivial S.P. partitions π_1 and π_2 on M such that $\pi_1 \cdot \pi_2 = 0$.

Proof: Suppose that M has a nontrivial parallel decomposition. We want to prove the following:

(a) There exist two partitions π_1 and π_2 on the set Q of states of M such that $\pi_1 \cdot \pi_2 = O$.

(b) π_1 and π_2 have S.P.

(c) π_1 and π_2 are nontrivial.

To show the existence of two partitions π_1 and π_2 such that $\pi_1 \cdot \pi_2 = O$, we let (ξ, ψ, ζ) be the assignment map such that $\xi: Q \longrightarrow Q_1 \times Q_2$ is one-to-one, where Q_1 and Q_2 denote the sets of states of M_1 and M_2. The mapping ξ define two equivalence relations π_1 and π_2 on Q as follows:

$$q \equiv p(\pi_1) \text{ iff } q_1 = p_1 \quad \text{where } \xi(q) = (q_1, q_2) \quad \text{and} \quad \xi(p) = (p_1, p_2)$$
$$q \equiv p(\pi_2) \text{ iff } q_2 = p_2 \quad \text{where } \xi(p) = (p_1, p_2) \quad \text{and} \quad \xi(p) = (p_1, p_2)$$

Since the mapping ξ is one-to-one, we know that $q \equiv p(\pi_1 \cdot \pi_2)$ implies that $q = p$, and thus $\pi_1 \cdot \pi_2 = O$.

To see that π_1 and π_2 have S.P., note that if

$$q \equiv p(\pi_1)$$

then

$$\xi(q) = (q_1, q_2) \quad \text{and} \quad \xi(p) = (q_1, p_2)$$

But then

$$\xi[f(q, x)] = [f_1(q_1, \psi(x)), f_2(q_2, \psi(x))]$$
$$\xi[f(p, x)] = [f_1(q_1, \psi(x)), f_2(p_2, \psi(x))]$$

and therefore, the first components of the next states are again identical under ξ and we have $f(q, x) \equiv f(p, x)(\pi_1)$. The same argument shows that π_2 also has S.P.

Finally, we want to show that π_1 and π_2 are nontrivial. Since, by assumption, $M_1 \| M_2$ is a nontrivial realization, then $|Q| > |Q_1|$, $|Q| > |Q_2|$, and $|Q_1| \cdot |Q_2| > |Q|$. Thus, π_1 and π_2 have less than $|Q|$ blocks and more than one block and are therefore nontrivial partition.

To show the converse, assume that there exist two nontrivial S.P. partitions, π_1 and π_2, on M such that $\pi_1 \cdot \pi_2 = O$. To construct M_1 and M_2, we take their image machines M_{π_1} and M_{π_2} and add outputs:

$$M_1 = (\{B_{\pi_1}\}, \Sigma, \{B_{\pi_2}\} \times X, f_1 = f_{\pi_1}, e)$$

and

$$M_2 = (\{B_{\pi_1}\}, \Sigma, \{B_{\pi_2}\} \times X, f_2 = f_{\pi_2}, e)$$

Obviously, since $\pi_1 \cdot \pi_2 = O$, the output of $M_1 \| M_2$ determines a unique pair in $(B_{\pi_1} \times B_{\pi_2}) \times \Sigma$, and thus there is a mapping λ which maps $Z_1 \times Z_2$ onto Z so that $M_1 \| M_2$ realizes M. This completes the proof. ∎

Example 7.6.1

To illustrate the construction of a nontrivial parallel decomposition of a sequential machine, consider the machine M_7 in Fig. 7.6.2(a) whose S.P. partition lattice is shown in Fig. 7.6.2(b). It is seen that this machine has only two nontrivial S.P. partitions, which are

$$\pi_1 = \{\overline{1, 2, 3}; \overline{4, 5, 6}\} = \{A, B\} = \{B_{\pi_1}\}$$

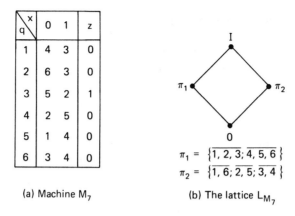

x q	0	1	z
1	4	3	0
2	6	3	0
3	5	2	1
4	2	5	0
5	1	4	0
6	3	4	0

(a) Machine M_7

$\pi_1 = \{\overline{1, 2, 3}; \overline{4, 5, 6}\}$
$\pi_2 = \{\overline{1, 6}; \overline{2, 5}; \overline{3, 4}\}$

(b) The lattice L_{M_7}

Fig. 7.6.2 Machine M_7 and its S.P. lattice L_{M_7}.

and

$$\pi_2 = \{\overline{1, 6}; \overline{2, 5}; \overline{3, 4}\} = \{C, D, E\} = \{B_{\pi_2}\}$$

The image machines of π_1 and π_2 are shown in Figs. 7.6.3(a) and (b), in which we have chosen outputs so that $g(q) = g_1[B_{\pi_1}(q)] \cdot g_2[B_{\pi_2}(q)]$. The parallel decomposition is depicted in Fig. 7.6.3(c). The output of state 3, for example, is calculated by

$$g(3) = g_1[B_{\pi_1}(3)] \cdot g_2[B_{\pi_2}(3)]$$
$$= g_1[A] \cdot g_2[E]$$
$$= 1 \cdot 1$$
$$= 1$$

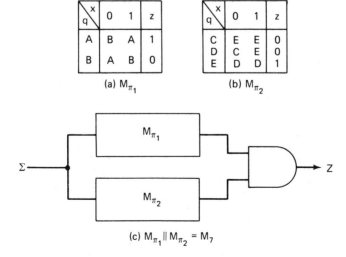

x q	0	1	z
A	B	A	1
B	A	B	0

(a) M_{π_1}

x q	0	1	z
C	E	E	0
D	C	E	0
E	D	D	1

(b) M_{π_2}

(c) $M_{\pi_1} \| M_{\pi_2} = M_7$

Fig. 7.6.3 Parallel decomposition of M_7.

and the output of state 4 is computed by

$$g(4) = g_1[B_{\pi_1}(4)] \cdot g_2[B_{\pi_2}(4)]$$
$$= g_1(B) \cdot g_2(E)$$
$$= 0 \cdot 1$$
$$= 0$$

In the following, two interesting examples are given to demonstrate some close relationships between machine decomposition theory and lattice theory introduced in Section 1.2.

Example 7.6.2

Consider machine M_8 of Fig. 7.6.4(a), which has the following three nontrivial partitions with S.P.:

$$\pi_1 = \{\overline{1, 2}; \overline{3, 4}; \overline{5, 6}; \overline{7, 8}\} = \{a, b, c, d\}$$
$$\pi_2 = \{\overline{1, 2, 3, 4}; \overline{5, 6, 7, 8}\} = \{A, B\}$$
$$\pi_3 = \{\overline{1, 8}; \overline{2, 6}; \overline{3, 7}; \overline{4, 5}\} = \{\text{I, II, III, IV}\}$$

Since $\pi_1 \cdot \pi_3 = O$, we know that it has a corresponding parallel decomposition into component machines M_{π_1} and M_{π_3} (using appropriate output logic). The same logic can be used about $\{\pi_2, \pi_3\}$, and so there is also a parallel decomposition using M_{π_2} and M_{π_3}. Therefore, we see that either M_{π_1} or M_{π_2} can be used in parallel with M_{π_3}. Now the question is: Which of the two should generally be chosen? Before answering this question, let us first examine the lattice L_{M_8}, which is depicted in Fig. 7.6.4(b). We find that both π_1 and π_2 have identical

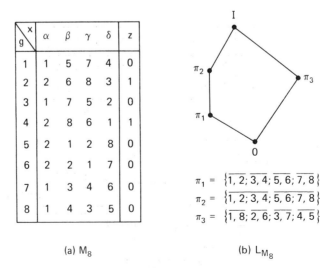

x g	α	β	γ	δ	z
1	1	5	7	4	0
2	2	6	8	3	1
3	1	7	5	2	0
4	2	8	6	1	1
5	2	1	2	8	0
6	2	2	1	7	0
7	1	3	4	6	0
8	1	4	3	5	0

$$\pi_1 = \{\overline{1, 2}; \overline{3, 4}; \overline{5, 6}; \overline{7, 8}\}$$
$$\pi_2 = \{\overline{1, 2; 3, 4}; \overline{5, 6; 7, 8}\}$$
$$\pi_3 = \{\overline{1, 8}; \overline{2, 6}; \overline{3, 7}; \overline{4, 5}\}$$

(a) M_8 (b) L_{M_8}

Fig. 7.6.4   Machine M_8 and lattice L_{M_8}.

interactions with π_3. Since π_2 is greater than π_1, we see that M_{π_1}, when used with M_{π_3}, compute more information than is really necessary. Thus, as a rule of thumb, when confronted with such a choice, the larger partition should generally be chosen. The three image machines, M_{π_1}, M_{π_2}, and M_{π_3}, are shown in Table 7.6.1. It is seen that the four-state machine M_{π_1} may

TABLE 7.6.1   Transition Tables of M_{π_1}, M_{π_2}, and M_{π_3}

$q \backslash x$	α	β	γ	δ	z
a	a	c	d	b	1
b	a	d	c	a	1
c	a	a	a	d	0
d	a	b	b	c	0

(a) M_{π_1}

$q \backslash x$	α	β	γ	δ	z
A	A	B	B	A	1
B	A	A	A	B	0

(b) M_{π_2}

$q \backslash x$	α	β	γ	δ	z
I	I	IV	III	IV	0
II	II	II	I	III	1
III	I	III	IV	II	0
IV	II	I	II	I	1

(c) M_{π_3}

be reduced to the two-state machine M_{π_2} using the method described in Section 7.3. Hence, the parallel decomposition of M_8 using M_{π_2} and M_{π_3} is simpler than that using M_{π_1} and M_{π_3}.

Our second example illustrates that even when a component machine cannot be replaced by a reduced version, we may still replace it by some other machine with fewer states.

Example 7.6.3

Consider the parallel decomposition of machine M_9 of Fig. 7.6.5(a). The nontrivial S.P. partitions for machine M_9 are found to be

$$\pi_1 = \{\overline{1,2}; \overline{3,4}; \overline{5,6}\} = \{a, b, c\}$$
$$\pi_2 = \{\overline{1,4}; \overline{2,5}; \overline{3,6}\} = \{A, B, C\}$$
$$\pi_3 = \{\overline{1,3,5}; \overline{2,4,6}\} = \{\text{I, II}\}$$
$$\pi_4 = \{\overline{1,6}; \overline{2,3}; \overline{4,5}\}$$

It is interesting to note that

$$\pi_i \cdot \pi_j = O$$

and

$$\pi_i + \pi_j = I$$

for $i, j = 1, 2, 3$, and 4, and $i \neq j$. Since $\pi_1 \cdot \pi_2 = O$, we can obtain a parallel decomposition from them, which is shown in Table 7.6.2. Although neither M_{π_1} nor M_{π_2} can be replaced by a reduced version, yet either M_{π_1} or M_{π_2} can be replaced by M_{π_3} [see Table 7.6.2(c)], which has one state less than M_{π_1} and M_{π_2}.

TABLE 7.6.2   Transition Tables of M_{π_1}, M_{π_2}, and M_{π_3}

q \ x	0	1	z
a	c	a	0
b	a	c	0
c	b	b	1

(a) M_{π_1}

q \ x	0	1	z
A	B	B	0
B	C	A	0
C	A	C	1

(b) M_{π_2}

q \ x	0	1	z
I	I	II	0
II	II	I	1

(c) M_{π_3}

It is known (problem 18 of Exercise 1.2) that lattices which do not contain sublattices isomorphic to the lattice of Fig. 7.6.4(b) and the sublattices in the dotted frame of Fig. 7.6.5(b) are distributive lattices.

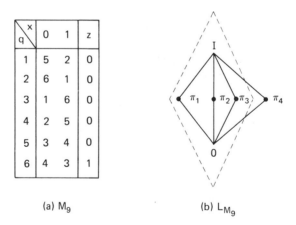

q \ x	0	1	z
1	5	2	0
2	6	1	0
3	1	6	0
4	2	5	0
5	3	4	0
6	4	3	1

(a) M_9

(b) L_{M_9}

Fig. 7.6.5 Machine M_9 and lattice L_{M_9}.

If the S.P. lattice is distributive, the last two kinds of component machine substitution cannot occur. Thus, we discover the following interesting relationship between machine decomposition theory and lattice theory.

DEFINITION 7.6.3

A parallel realization for a machine M based on S.P. partition π_1 and π_2 is called *prime* iff $\pi_1 + \pi_2 = I$.

THEOREM 7.6.2

If the S.P. lattice L_M for M is distributive and π is an S.P. partition, there is *at most* one partition π_1 such that $\{\pi, \pi_1\}$ gives a prime parallel decomposition.

Proof: Suppose that there are two S.P. partitions π_1 and π_2 such that

$$\pi \cdot \pi_i = O \quad \text{and} \quad \pi + \pi_i = I$$

for $i = 1, 2$. Then

$$\pi_1 = \pi_1 \cdot I = \pi_1 \cdot (\pi + \pi_2)$$
$$= \pi_1 \cdot \pi + \pi_1 \cdot \pi_2 \qquad \text{by distributive law}$$
$$= O + \pi_1 \cdot \pi_2 = \pi_1 \cdot \pi_2$$

Similarly, we obtain $\pi_2 = \pi_2 \cdot \pi_1$. Hence, $\pi_1 = \pi_2$. ∎

Bibliographical Remarks

The notion of a sequential machine is usually attributed to McCulloch and Pitts [5]. The formalism and the initial description of the properties of completely specified sequential machines were first introduced by Mealy [6], Moore [7], and Rabin and Scott [8]. The minimization of completely specified sequential machines appeared originally in Huffman [4] and Moore [7], and the minimization of incompletely specified sequential machines was published by Paull and Unger [22] and Grasselli and Luccio [14].

Reference 1 has an extensive discussion of most of the topics covered in this chapter, hence is strongly recommended. The state assignment problem is treated extensively in references 2 and 3. A considerable amount of research work is still being carried out concerning the state-assignment problem. Some of the recent development on this subject can be found in references 9–27.

References

Books

1. HARTMANIS, J., and R. E. STEARNS, *Algebraic Structure Theory of Sequential Machines*, Prentice-Hall, Englewood Cliffs, N.J., 1966.

2. MILLER, R. E., *Switching Theory*, Vol. II, Wiley, New York, 1965.

3. WOOD, P. E., JR., *Switching Theory*, McGraw-Hill, New York, 1968.

Papers

4. HUFFMAN, D. A., "The Synthesis of Sequential Switching Networks," *J. Franklin Inst.*, Vol. 257, 1954, pp. 161–190.

5. McCULLOCH, W. S., and W. PITTS, "A Logical Calculus of the Ideas Immanent in Nervous Activity," *Bull. Math. Biophys.*, Vol. 5, 1953, pp. 115–133.

6. MEALY, G. E., "Method for Synthesizing Sequential Circuits," *Bell System Tech. J.*, Vol. 34, 1955, pp. 1054–1079.

7. MOORE, E. F., "Gedanken-Experiments on Sequential Machines," in *Automata Studies*, C. E. Shannon and J. McCarthy, eds., Princeton University Press, Princeton, N.J., 1956, pp. 129–156.

8. RABIN, M. O., and SCOTT, D., "Finite Automata and Their Decision Problems," *IBM J. Res. Develop.*, Vol. 3, 1959, pp. 114–125.

9. CURTIS, H. A., "Multiple Reduction of Variable Dependency of Sequential Machines," *ACM*, Vol. 9, July 1962, pp. 324–344.

10. DOLOTTA, T. A., and E. J. MCCLUSKEY, "The Coding of Internal States of Sequential Circuits," *IEEE Trans. Electronic Computers*, Vol. EC-13, No. 5, October 1964, pp. 549–562.

11. EPLEY, D. L., and P. T. WANG, "On State Assignments and Sequential Machine Decomposition from S.P. Partitions," *Proc. 5th Ann. Symp. Switching Theory and Logical Design*, November 1964, pp. 228–233.

12. FARR, E. H., "Lattice Properties of Sequential Machines," *J. ACM*, Vol. 10, No. 3, July 1963, pp. 365–385.

13. GRASSELLI, A., "Minimal Closed Partitions for Incompletely Specified Flow Tables," *IEEE Trans. Electronic Computers*, Vol. EC-15, No. 2, April 1966, pp. 245–249.

14. GRASSELLI, A., and F. LUCCIO, "A Method for Minimizing the Number of Internal States in Incompletely Specified Sequential Networks," *IEEE Trans. Electronic Computers*, Vol. EC-14, No. 3, June 1965, pp. 350–359.

15. HARTMANIS, J., "On the State Assignment Problem for Sequential Machines I," *IRE Trans. Electronic Computers*, Vol. EC-10, June 1961, pp. 157–165.

16. HARTMANIS, J., and R. E. STEARNS, "Some Dangers in State Reduction of Sequential Machines," *Information and Control*, Vol. 5, No. 3, September 1962, pp. 252–260.

17. HARTMANIS, J., and R. E. STEARNS, "Pair Algebra and Its Application to Automata Theory," *Information and Control*, Vol. 7, No. 4, December 1964, pp. 485–507.

18. KARP, R. M., "Some Techniques of State Assignment for Synchronous Sequential Machines," *IEEE Trans. Electronic Computers*, Vol. EC-13, No. 5, October 1964, pp. 507–518.

19. MCCLUSKEY, E. J., "Minimum-State Sequential Circuits for a Restricted Class of Incompletely Specified Flow Tables," *Bell System Tech. J.*, Vol. 41, No. 6, November 1962, pp. 1759–1768.

20. MCCLUSKEY, E. J., and S. H. UNGER, "A Note on the Number of Internal Variable Assignments for Sequential Switching Circuits," *IRE Trans. Electronic Computers*, Vol. EC-8, No. 4, December 1959, pp. 439–440.

21. MEISEL, W. S., "A Note on Internal State Minimization in Incompletely Specified Sequential Networks," *IEEE Trans. Electronic Computers*, Vol. EC-16, No. 4, August 1967, pp. 508–509.

22. PAULL, M. C., and S. H. UNGER, "Minimizing the Number of States in Incompletely Specified Sequential Switching Functions," *IEEE Trans. Electronic Computers*, Vol. EC-8, September 1959, pp. 356–367.

23. STEARNS, R. E., and J. HARTMANIS, "On the State Assignment Problem for Sequential

Machines II," *IRE Trans. Electronic Computers*, Vol. EC-10, No. 4, Dec., 1961, pp. 593–603.

24. TORNG, H. C., "An Algorithm for Finding Secondary Assignments of Synchronous Sequential Circuits," *IEEE Trans. Electronic Computers*, Vol. C-17, No. 5, May 1968, pp. 461–469.

25. UNGER, S. H., "Flow Table Simplification—Some Useful Aids," *IEEE Trans. Electronic Computers*, Vol. EC-14, No. 3, June 1965, pp. 472–475.

26. WEINER, P., and E. J. SMITH, "Optimization of Reduced Dependencies for Synchronous Sequential Machines," *IEEE Trans. Electronic Computers*, Vol. EC-16, No. 6, December 1967, pp. 835–847.

27. ZAHLE, T. U., "On Coding the States of Sequential Machines with the Use of Partition Pairs," *IEEE Trans. Electronic Computers*, Vol. EC-15, No. 2, April 1966, pp. 249–253.

8

Regular Expressions

In Chapter 7, the indistinguishability (state and machine) equivalence denoted by $\equiv_I$ was discussed. In this chapter, another type of equivalence is introduced, tape equivalence, which is derived from the concept of "machine recognition of input tapes," defined on a class of binary-input/binary-output machine known as the *Rabin–Scott machine*. Two Rabin–Scott machines are tape-equivalent iff they recognize the exact same set of tapes. It is shown that for any two deterministic Rabin–Scott machines M_1 and M_2, $M_1 \equiv_I M_2$ iff $M_1 \equiv_T M_2$.

A set of tapes or sequences of input symbols is *regular* iff there exists a sequential machine which recognizes it. A compact, systematic, and convenient way to represent regular sets is the *regular expression*. A systematic procedure for obtaining a regular expression from a transition diagram and a procedure for constructing a transition diagram from a regular expression are presented.

The nondeterministic sequential machine, a generalization of a deterministic sequential machine, is then introduced. The method for converting a nondeterministic machine into a tape-equivalent deterministic one is known as the *subset construction*. This conversion technique, together with the method for constructing a regular expression from a deterministic machine, shows that every Rabin–Scott nondeterministic machine may also be represented by a regular expression.

There are two types of state equivalence and minimization in nondeterministic sequential machines. One is state equivalence and minimization based on indistinguishability equivalence, which is a straightforward generalization of that of deterministic machines. The second one is state minimization based on tape equivalence. For nondeterministic sequential machines, indistinguishability equivalence implies tape equivalence, but the converse is not true.

Besides its application to the state minimization of nondeterministic sequential machines, subset construction also offers a convenient way to construct transition tables of long and complicated regular expressions that can be broken down into several short regular expressions connected by the union, intersection, concatenation, or other operations.

8.1   Regular Sets and Regular Expressions

After having discussed the indistinguishability equivalence, we would like to introduce another type of equivalence, tape equivalence. This type of equivalence is defined on a special type of machine, known as the *Rabin–Scott machine*.

DEFINITION 8.1.1

A *Rabin–Scott machine* over the binary alphabet $\Sigma = \{0, 1\}$ is a Moore-type machine with a specified initial state described by $M = (Q, \delta, q_0, F)$, where Q is the set of internal states, $q_0 \in Q$ is the initial state of the machine, and $F \in Q$ is a set of designated final states. The present outputs of a nonfinal state and a final state are 0 and 1, respectively.

Tape equivalence is defined based on the concept of "machine recognition of a tape." For notational simplicity, we shall use $x = \sigma_0\sigma_1 \cdots \sigma_{n-1}$ (instead of $\tilde{x}$) to denote a sequence of input symbols $\sigma_0, \sigma_1, \ldots, \sigma_{n-1}$, or a tape, where each σ_i is either 0 or 1.

DEFINITION 8.1.2

Let $A = (Q, \delta, q_0, F)$ be a Rabin–Scott machine. A sequence of input symbols or a tape x is to be *recognized* (or accepted, or remembered) by M if $\delta(q_0, x) \in F$; otherwise, it is *not recognized* (or not accepted, or not remembered) by M.

For example, machine M_1 described in Fig. 8.1.1(a) recognizes any tape with 0's, if any, appearing before 1's.

DEFINITION 8.1.3

The *length* of a sequence or tape is the number of symbols (zeros and ones) in it.

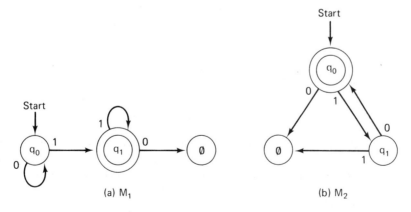

(a) M_1 (b) M_2

Fig. 8.1.1 M_1 and M_2.

DEFINITION 8.1.4

The *empty sequence*, denoted by Λ, is the sequence of zero length. If x is a sequence, $\Lambda x = x\Lambda = x$.

DEFINITION 8.1.5

A set of sequences or tapes is called *regular* if there exists a finite-state Rabin–Scott machine which recognizes *exactly* the sequences in this set, and no others; otherwise, we call it *nonregular*.

The following are examples of regular sets.

1. $\{0^m 1^n \mid m \geq 0, n \geq 1\}$. The machine that recognizes this set is M_1.
2. $\{\Lambda, 10, 1010, 101010, \ldots\}$. The machine which recognizes this set is M_2.

The state $\varnothing$ denotes the "dead state," which indicates that whenever the machine lands at this state, it will then never be able to leave this state. This example has also illustrated that the initial state of a Rabin–Scott machine can also be a designated final state of the machine. The sets

(1) $\{0^n 1^n \mid n \geq 0,$ where $0^0 \triangleq \Lambda$ and $1^0 \triangleq \Lambda\}$

(2) $\{1^{2^n} \mid n = 0, 1, 2, \ldots\} = \{1, 11, 1111, \ldots\}$

are examples of nonregular sets. To show that they are nonregular is to show that there do not exist machines which recognize them. The following example illustrates a general procedure for proving a set to be nonregular.

Example 8.1.1

The set $\{0^n 1^n \mid n \geq 0,$ where $0^0 = \Lambda$ and $1^0 = \Lambda\}$ is nonregular.

Proof: Suppose that there exists a sequential machine which recognizes this set. Since a sequential machine can only have a finite number of states and since there are infinitely many sequences in this set, there must exist n_1 and n_2, where $n_1 \neq n_2$, such that

$$f(q_0, 0^{n_1}) = q_i$$
$$f(q_0, 0^{n_2}) = q_i$$

which implies that

$$g(q_0, 0^{n_1} 1^{n_1}) = (q_0, 0^{n_2} 1^{n_1}) = 1$$

This equation tells us that if there exists a sequential machine which recognizes the set $\{0^n 1^n \mid n \geq 0\}$, it will recognize some input sequences of $0^{n_2} 1^{n_1}$ with $n_2 \neq n_1$ which are *not* in the given set. This is to say that there *does not* exist a sequential machine which *only* recognizes the sequences of the set $\{0^n 1^n \mid n \geq 0\}$ and no other input sequences. Hence, the set $\{0^n 1^n \mid n \geq 0\}$ is nonregular. ∎

A compact, systematic, and convenient way to represent a regular set is the regular expression which consists of the following:

1.   A set of primitive objects: 0 (zero), 1 (one), Λ (lambda), and $\varnothing$, where Λ is the empty sequence, the sequence of zero length, and $\varnothing$ is the empty set.

2.   A set of operations: * (star), $\cdot$ (dot), $\cup$ (union), $\cap$ (intersection), and $\sim$ (complementation). If A_1 and A_2 are sets, $A_1 \cdot A_2$, or simply $A_1 A_2 = \{x_1 x_2$ (it reads "the concatenation of x_1, x_2") $\mid x_1 \in A_1$ and $x_2 \in A_2\}$. Clearly, in general $A_1 A_2 \neq A_2 A_1$.

DEFINITION 8.1.6

A *regular expression* is defined recursively as follows:

(a)   A sequence consisting of a single 0, a single 1, a single Λ, or a single $\varnothing$ is a regular expression.

(b)   If R is a regular expression, so are $R*$ and $\sim R$.

(c)   If R_1 and R_2 are regular expressions, so are $R_1 R_2$, $R_1 \cup R_2$, and $R_1 \cap R_2$.

(d)   No sequence is a regular expression unless its being so follows from (a), (b), and (c).

However, it should be noted that a regular expression denotes both sequences and a set of sequences.

DEFINITION 8.1.7

Two regular expressions R_1 and R_2 are said to be *equal*, denoted by $R_1 = R_2$, if they denote the same set.

For example, the two expressions $(0 \cup 1)*$ and $(0*1*)*$ are equal, since they both denote the same set:

$$(0 \cup 1)* = (0*1*)* = \{\Lambda, 0, 1, 00, 01, 10, 11, 000, 001, \ldots\}$$

A.   Properties of Regular Expressions

First, the properties of union, intersection, and complementation on sets all hold for regular expressions, since regular expressions are sets. Besides, the empty sequence Λ and the empty set $\varnothing$ have the following properties:

$$\Lambda* = \Lambda \qquad (8.1.1)$$

$$\varnothing* = \Lambda \qquad (8.1.2)$$

$$\varnothing \cup R = R \qquad (8.1.3)$$

$$\varnothing \cap R = \varnothing \qquad (8.1.4)$$

$$\Lambda R = R\Lambda = R \qquad (8.1.5)$$

The concatenation is both right- and left-distributive over the union

$$R_1(R_2 \cup R_3) = R_1 R_2 \cup R_1 R_3 \tag{8.1.6}$$

$$(R_2 \cup R_3)R_1 = R_2 R_1 \cup R_3 R_1 \tag{8.1.7}$$

$$R_1^*(R_2 \cup R_3) = R_1^* R_2 \cup R_1^* R_3 \tag{8.1.8}$$

$$(R_2 \cup R_3)R_1^* = R_2 R_1^* \cup R_3 R_1^* \tag{8.1.9}$$

Additionally, the star operation has the following properties:

$$RR^* = R^*R \tag{8.1.10}$$

$$(R_1 R_2)^* R_1 = R_1 (R_2 R_1)^* \tag{8.1.11}$$

$$\Lambda \cup RR^* = R^* \tag{8.1.12}$$

$$\Lambda \cup (R_1 \cup R_2)^* R_2 = (R_1^* R_2)^* \tag{8.1.13}$$

$$R^* R^* = R^* = (R^*)^* \tag{8.1.14}$$

$$(R_1 \cup R_2)^* = (R_1^* \cup R_2^*)^* = R_1^*(R_2 R_1^*)^* = (R_1^* R_2)^* R_1^* \tag{8.1.15}$$

$$R_1^* R_2 = R_2 \cup R_1^* R_1 R_2 = R_2 \cup R_1 R_1^* R_2 \tag{8.1.16}$$

The properties of Eqs. (8.1.1)–(8.1.5) follow directly from the definitions of Λ, $\varnothing$, and the concatenation operation. The properties of Eqs. (8.1.6)–(8.1.9) may be proved by using the ordinary way of proving a set equality, namely by proving that every element in the set of the left-hand side of the equality is also in the set of the right-hand side of the equality, and vice versa. For example, Eq. (8.1.6) may be proved as follows. If one, two, or all three sets R_1, R_2, and R_3 are Λ or $\varnothing$, it will be trivially held by virtue of Eqs. (8.1.3)–(8.1.5). If, on the other hand, all three sets are neither Λ nor $\varnothing$, let a, b, and c be typical elements of R_1, R_2, and R_3, respectively. Then a typical element of the set on the left-hand side is $a(b \cup c)$ and a typical element of the set on the right-hand side is $ab \cup ac$. By the definitions of the $\cdot$ and $\cup$ operations, $a(b \cup c) = ab \cup ac$. Hence $R_1(R_2 \cup R_3) = R_1 R_2 \cup R_1 R_3$.

Equations (8.1.10)–(8.1.16) may be proved from the definition of the $*$ operation, which is left to the reader as an exercise (problem 5).

The reader should be cautious about some pairs of regular expressions which look equal but actually are not. For example, the following are unequal:

$$R_1(R_2 \cap R_3) \neq R_1 R_2 \cap R_1 R_3 \tag{8.1.17}$$

$$(R_2 \cap R_3)R_1 \neq R_2 R_1 \cap R_3 R_1 \tag{8.1.18}$$

$$(R_1 \cup R_2)^* \neq R_1^* \cup R_2^* \tag{8.1.19}$$

$$(R_1 \cap R_2)^* \neq R_1^* \cap R_2^* \tag{8.1.20}$$

$$\Lambda \cup R \neq R \tag{8.1.21}$$

$$RR^* \neq R^* \tag{8.1.22}$$

One way to prove these relations is by constructing counterexamples. This is left to the reader as an exercise (Problem 6).

If x is any tape, it can be turned end for end and written backward.

DEFINITION 8.1.8

The *reverse* of a tape x, denoted by $\overleftarrow{x}$, is the tape resulting from writing x backward.

For example, if $x = \sigma_0 \sigma_1 \ldots \sigma_{n-1}$, then $\overleftarrow{x} = \sigma_{n-1}\sigma_{n-2} \ldots \sigma_0$. Following directly from the definition, we find that the reverse operation has the following properties:

$$\overleftarrow{\sigma} = \sigma$$

$$\overleftarrow{\Lambda} = \Lambda$$

$$\overleftarrow{\overleftarrow{x}} = x$$

$$\overleftarrow{xy} = \overleftarrow{y}\,\overleftarrow{x}$$

DEFINITION 8.1.9

The *reverse* of a regular expression R, denoted by $\overleftarrow{R}$, is defined as the set consisting of the members of R, each reversed in time.

We can prove that

THEOREM 8.1.1

The reverse of any regular expression is always a regular expression.

Proof: For regular expressions with length 1,

$$\overleftarrow{\Lambda} = \Lambda, \quad \overleftarrow{\varnothing} = \varnothing, \quad \overleftarrow{0} = 0, \quad \text{and} \quad \overleftarrow{1} = 1 \tag{8.1.23}$$

they are regular expressions. By induction, if $\overleftarrow{R_1}$ and $\overleftarrow{R_2}$ are the reverses of two regular expressions R_1 and R_2, respectively, then

$$\overleftarrow{(R_1 R_2)} = \overleftarrow{R_2}\,\overleftarrow{R_1} \tag{8.1.24}$$

$$\overleftarrow{(R_1 \cup R_2)} = \overleftarrow{R_1} \cup \overleftarrow{R_2} \tag{8.1.25}$$

$$\overleftarrow{(R_1 \cap R_2)} = \overleftarrow{R_1} \cap \overleftarrow{R_2} \tag{8.1.26}$$

$$(\sim \overleftarrow{R_1}) = \overleftarrow{(U - R_1)} = \overleftarrow{U} - \overleftarrow{R_1} \tag{8.1.27}$$

$$\overleftarrow{(R_1^*)} = (\overleftarrow{R_1})^* \tag{8.1.28}$$

where U denotes the set of all sequences composed of 0 and 1. Hence, every reverse of a regular expression is a regular expression. ∎

Example 8.1.2

Let $R = ((01*0 \cup 1)(11*0)*0)*$. What is $\overleftarrow{R}$?

$$\overleftarrow{R} = \overleftarrow{[((01*0 \cup 1)(11*0)*0)*]}$$

$$= [\overleftarrow{((01*0 \cup 1)(11*0)*0)}]* \qquad \text{by (8.1.28)}$$

$$= [0[\overleftarrow{(11*0)*}]\overleftarrow{(01*0 \cup 1)}]* \qquad \text{by (8.1.24)}$$

$$= [0(01*1)*(01*0 \cup 1)]* \qquad\qquad\qquad (8.1.29)$$

B. Construction of a Regular Expression from a Transition Diagram

The regular expressions of simple machines such as M_1 and M_2 may be obtained by inspection. For example, the regular expressions of M_1 and M_2 are $0*11*$ and $(10)*$, respectively. But for more complicated machines, the obtaining of their regular expressions by inspection becomes difficult, sometimes even impossible. A systematic procedure for obtaining a regular expression of a sequential machine is described below.

1. Eliminate the dead state, if it exists, and all the transition lines going to and coming out from it.

2. Construct a state equation for each state. In constructing state equations, each state q_i is considered as a variable and each input symbol is considered as a constant. The left-hand side of the state equation for state q_i is q_i and its right-hand side is the union of $q_k \sigma_{ki}$, where σ_{ki} is a single input symbol, upon application of which, the machine goes from state q_k to state q_i.

3. From the state equations, solve one final state variable in terms of the initial state variable at a time by the following procedure.

(3-1) Express every noninitial and nonfinal state variable in terms of the initial state variable and the final state variable under investigation. In obtaining these expressions, the following formula is needed when the right-hand side of the state equation for state q_i contains the state variable q_i, which is to be eliminated.

Let q_i and q_j be two variables and a and b be two constants. Then the solution of q_i in terms of q_j of the equation

$$q_i = q_i a \cup q_j b \qquad\qquad\qquad (8.1.30)$$

is

$$q_i = q_j ba* \qquad\qquad\qquad (8.1.31)$$

The proof may be seen from repeatedly substituting q_i of Eq. (8.1.30) into itself:

$$q_i = (q_i a \cup q_j b)a \cup q_j b$$
$$= (((q_i a \cup q_j b)a \cup q_j b)a \cup q_j b)a \cup q_j b$$

.
.
.

$$= (q_i \underbrace{aa \ldots a}_{n+1}) \cup (q_j \underbrace{baa \ldots a}_{n}) \cup (q_j \underbrace{baa \ldots a}_{n-1}) \ldots (q_j ba) \cup (q_j b)$$

.
.
.

Thus,

$$q_i = q_i a^* \cup q_j ba^* \qquad\qquad\qquad (8.1.32)$$

Since we solve q_i in terms of q_j, we drop the first term; this gives the result of Eq. (8.1.31). Since whenever an equation is of the form $q_i = q_j \sigma_{ji} \cup q_k \sigma_{ki}$, where $i \neq j \neq k$, then q_i can be substituted into all other equations to yield a system with fewer equations and unknowns. Also, whenever an equation has the form $q_i = q_i \sigma_{ii} \cup q_j \sigma_{ji}$. Eq. (8.1.31) can be applied to yield $q_i = q_j \sigma_{ji} \sigma_{ii}^*$, which can now be substituted for q_i in the other equations. Thus, by applying this procedure, every noninitial and nonfinal state variable can be expressed in terms of the initial state variables and a final state variable.

(3-2)   Replace all the noninitial and nonfinal state variables on the right-hand sides of the state equations of the initial and the final states by their expressions obtained in (3-1).

(3-3)   Finally, substituting the initial state variable on the right-hand side of the final state equation by the initial state equation, whose right-hand side now contains at most two variables, the initial and final state variables, and applying Eq. (8.1.31) to the resulting equation to eliminate the final state variable on the right-hand side of the equation, we obtain the regular expression recognized by this final state.

(3-4)   If there are more than one final state, repeat the above procedure for each one. The total regular expression for the machine is the union of the regular expressions recognized by each of the final states of the machine.

This procedure is illustrated by the following example.

Example 8.1.3

Consider machine M_3 in Fig. 8.1.2. The state equations are

$$q_0 = q_0 \Lambda \cup q_2 1 \qquad\qquad\qquad (8.1.33)$$
$$q_1 = q_1 1 \cup q_0 1 \Longrightarrow q_1 = q_0 11^* \qquad \text{[by Eq. (8.1.31)]} \qquad (8.1.34)$$
$$q_2 = q_1 0 \cup q_3 1 \qquad\qquad\qquad (8.1.35)$$
$$q_3 = q_2 0 \cup q_3 0 \Longrightarrow q_3 = q_2 00^* \qquad \text{[by Eq. (8.1.31)]} \qquad (8.1.36)$$

Substituting Eq. (8.1.33) into Eq. (8.1.34) and then substituting the resulting equation and

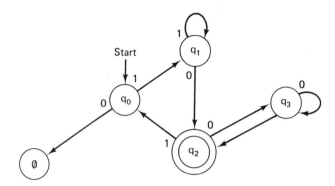

Fig. 8.1.2 M_3.

Eq. (8.1.36) into Eq. (8.1.35), we obtain

$$q_2 = q_0(\Lambda 11^*0) \cup q_2(111^*0 \cup 00^*1) \qquad (8.1.37)$$

Applying Eq. (8.1.31) to Eq. (8.1.37), we obtain

$$q_2 = q_0(\Lambda 11^*0)(111^*0 \cup 00^*1)^* \qquad (8.1.38)$$

Hence, the regular expression of M_3 is $11^*0(111^*0 \cup 00^*1)^*$.

THEOREM 8.1.2

For any given sequential machine, we can always obtain a regular set in regular expression form recognized by it.

The advantages offered by the regular-expression representation of a sequential machine are that

1. It provides a means of constructing a regular set from a given sequential machine, and
2. It provides a means of constructing a transition diagram from a given regular set.

The former was clearly indicated in the above illustrations; the latter is shown in the following subsection.

C. Construction of a Transition Diagram
from a Regular Expression

Now we want to show that every regular expression uniquely represents a sequential machine. To show this, we present the following general procedure for constructing a transition diagram from a given regular expression, which is best illustrated by examples.

Example 8.1.4

Let $\Sigma = \{0, 1\}$ be an alphabet and L be a regular set over Σ having the property that every sequence of the set may have arbitrary 0's but can only have exactly two 1's. Find a sequential machine to represent this set.

The regular expression of this set is

$$R_L = 0^*10^*10^* \tag{8.1.39}$$

The construction of a sequential machine from this regular expression consists of the following steps:

Step 1   Name the 0's and 1's as

$$0_1^*1_10_2^*1_20_3^*$$

We shall use five states denoted by 0_1, 1_1, 0_2, 1_2, and 0_3, plus an initial state I and the dead state $\varnothing$ to construct this machine.

Step 2   Find all the possible transitions among 0_i's and 1_i's. From the regular expression above we can obtain the following transitions:

Initial state	Next- state	State- transition representation	Initial state	Next- state	State- transition representation
0_1 $\xrightarrow{\text{input a } 0}$	0_1	$(0_1, 0_1)$	0_1 $\xrightarrow{\text{input a } 1}$	1_1	$(0_1, 1_1)$
1_1 $\xrightarrow{\text{input a } 0}$	0_2	$(1_1, 0_2)$	1_1 $\xrightarrow{\text{input a } 1}$	1_2	$(1_1, 1_2)$
0_2 $\xrightarrow{\text{input a } 0}$	0_2	$(0_2, 0_2)$	0_2 $\xrightarrow{\text{input a } 1}$	1_2	$(0_2, 1_2)$
1_2 $\xrightarrow{\text{input a } 0}$	0_3	$(1_2, 0_3)$			
0_3 $\xrightarrow{\text{input a } 0}$	0_3	$(0_3, 0_3)$			

Note that this state-transition representation is unambiguous. The general form is (a_i, b_j). The state transition that it represents is that the machine is initially at state a_i; upon an input b, it goes to state b_j.

Step 3   Start with the initial state I. An input of 0 applied to the initial state I lands us in state 0_1, and an input 1 applied to the initial state I lands us in state 1_1. This is because the leftmost two symbols of the regular expression are $0_1^*1_1$. Any sequence of this set may start with 0 or 1. An input 0 applied to state 0_1 remains in state 0_1, since $(0_1, 0_1)$ and an input 1 applied to state 0_1 results in state 1_1, since $(0_1, 1_1)$. On the other hand, when an input 0 is applied to state 1_1, because of the transition $(1_1, 0_2)$, it results in state 0_2; and when an input 1 is applied to state 1_1, because of the transition $(1_1, 1_2)$, it results in state 1_2. The rest of the state transitions are constructed in the same way. It is clear from the regular expression $0_1^*1_10_2^*1_20_3^*$ that when the machine has been either at state 1_2 or state 0_3 and an input 1 is applied, regardless of the input symbols, the machine will never recognize this sequence, for it becomes

$$0*10*10*1z_1$$

or

$$0*10*11z_2$$

where z_1 and z_2 are two arbitrary sequences, which are not sequences of the regular expression of Eq. (8.1.39). Hence, in either of these situations, the machine goes to the dead state. Since the machine which realizes the expression $0_1^*1_10_2^*1_20_3^*$ counts *exactly* 2 in unary representation, the two final states of this machine are 1_2 and 0_3. The transition diagram of this machine is given in Fig. 8.1.3(a). From the transition diagram, it is easy to recognize that states I and 0_1, 1_1 and 0_2, and 1_2 and 0_3 are equivalent. The minimized machine is shown in Fig. 8.1.3(b). From a recognition point of view, this machine represents the given set in the sense that it recognizes exactly the sequences in the set and no others.

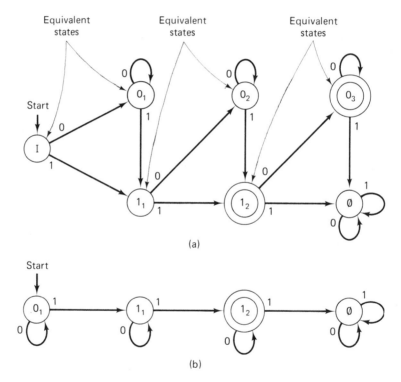

Fig. 8.1.3   (a) The transition diagram obtained from the regular expression of Eq. (8.1.9) ; (b) the minimal form of the machine.

Example 8.1.5

As another example, consider a set S over the symbols $\Sigma = \{0, 1\}$ having the following property: All sequences with all their 0's at the beginning and the first symbol must be 0. The closed-form set description of this set is

$$S = \{0^n 1^m \mid n \geq 1 \quad \text{and} \quad m \geq 0\}$$

The regular expression R of it is

$$R = 00^*1^* \qquad (8.1.40)$$

With the 0's and 1's being so named: $0_1 0_2^* 1_1^*$, the possible transitions among the 0_i's and 1_i's, are

Input a 0	Input a 1
$(0_1, 0_2)$	$(0_1, 1_1)$
$(0_2, 0_2)$	$(0_2, 1_1)$
	$(1_1, 1_1)$

Using a total of five states—states 0_1, 0_2, and 1_1 plus the initial state and the dead state—and constructing a sequential machine as described in the previous example, the result is shown in Fig. 8.1.4. A close examination of the states of this machine indicates that no equivalent

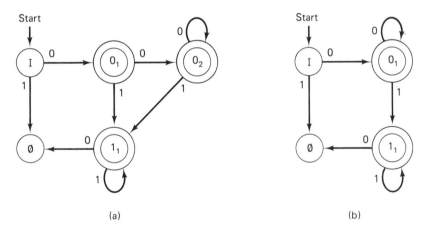

(a)   (b)

Fig. 8.1.4 The transition diagram of the sequential machine which recognizes the regular expression 00*1*

states exist among them; hence, this is the minimal form of the sequential machine that represents the given set.

The procedure illustrated by the examples above is completely general; therefore, we have

LEMMA 8.1.1

Every regular expression can be realized by a sequential machine.

LEMMA 8.1.2

Let A be a set over the alphabet $\Sigma = \{0, 1\}$. A is regular iff A can be described by a regular expression.

Proof: Suppose that A is regular. Then, by definition, there exists a sequential machine which recognizes exactly the sequences in it and no others. By Theorem 8.1.2, A can be described by a regular expression. Now suppose that A can be described by a regular expression. Then A is regular, which follows directly from Lemma 8.1.1.   ∎

From Lemmas 8.1.1 and 8.1.2, we have

THEOREM 8.1.3

Every regular set can be represented by a sequential machine.

Following from this theorem, we have

COROLLARY 8.1.1

No nonregular sets can be described by a regular expression.

COROLLARY 8.1.2

No nonregular sets can be represented by a sequential machine.

D.   Relation Between Indistinguishability Equivalence
     and Tape Equivalence

From the discussion above we arrive at the following important theorem.

THEOREM 8.1.4

For any deterministic Rabin–Scott machines M_1 and M_2, $M_1 \equiv_I M_2$ iff $M_1 \equiv_T M_2$.

Proof: Suppose that M_1 and M_2 are indistinguishability equivalent. From the definition of indistinguishability equivalence given in Definition 7.2.2, we find that by applying the same input symbol to M_1 and M_2 in their initial states, the two machines should produce the same output, which is either a 0 or a 1, depending on whether the states on which the machines land are nonfinal or final states. Because of the assurance of identical output sequence to identical input sequence, by the definition of indistinguishability equivalence, any tape recognized (rejected) by one machine will also be recognized (rejected) by the other. Hence, they are also tape-equivalent.

Conversely, suppose that $M_1 \equiv_T M_2$; we want to show that M_1 and M_2 are also indistinguishability equivalent. We shall prove this by the method of contradiction. Suppose that $M_1 \not\equiv_I M_2$. Then there must exist at least one state in one of the two machines which is not equivalent to any state in the other. Without loss of generality, assume that state q_1 in machine M_1 is not equivalent to any state in machine M_2. Let x_1 be the input sequence that will bring M_1 from its initial state to state q_1. Suppose that the outputs of M_1 and M_2 to x_1 are the same. Since state q_1 of M_1 is not equivalent to any state in M_2, there must exist at least one input sequence x_2 such that the output sequences of the two machines to the input sequence x_1x_2 (the concatenation of x_1 and x_2) would be different. This implies that there exists a tape x_1x_2 which is recognized by one machine but rejected by the other—a contradiction. Thus, M_1 and M_2 must also be indistinguishability equivalent. ∎

Exercise 8.1

1. Determine which of the following sets are regular and which are nonregular.
   (a) The set of all sequences 0, 1, 00, 01, 10, 11, 000, ...
   (b) Any finite set of sequences over a finite set.
   (c) $\{0, 101, 11011, \ldots, 1^n01^n, \ldots\}$, or $\{1^n01^n \mid n \geq 0\}$.
   (d) $\{0^n1^m \mid n \geq 5 \text{ and } m \geq 10\}$
   (e) $\{0^n1^m \mid n \leq m\}$
   (f) $\{0^n1^m \mid m > 0 \text{ and } n > m\}$

2. Define the ideal numbers $I = 2^n - 1$, $n \geq 1$.
   (a) Show that the set of ideal numbers in binary representation is regular.
   (b) Show that the set of ideal numbers in unary representation is nonregular.

3. Find the regular expressions of the regular sets of problem 1.

4. Give the regular expression of the set of ideal numbers in binary representation.

5. Prove Eqs. (8. 1. 10)–(8.1.16).

6. Prove, by constructing a counterexample, that the expressions of Eqs. (8.1.17)–(8.1.22) are not equalities.

7. Show that
   (a) $\sim((0 \cup 1)^*) = \emptyset$
   (b) $\sim((0 \cup 1)(0 \cup 1)^*) = \Lambda$

8. Which of the following are true?
   (a) $0^*1 = (0 \cup 0^*)1$
   (b) $(10 \cup 1100)^* = (1(10)^*0)^*$
   (c) $(0 \cup 1)^* = (0^* \cup 1^*)(1^* \cup 0^*)^*$
   (d) $0^*1^* = 0^*01^* \cup 0^*(11)^*$

9. Prove that the following equivalences hold for the regular expressions.
   (a) $0^* = \Lambda \cup 00^*$
   (b) $(0 \cup 01)(\Lambda \cup 1) = (0 \cup 01 \cup 011)$
   (c) $(0 \cup 1 \cup 110)^* = (0 \cup 1 \cup 01)^*$
   (d) $(0 \cup 11^*0)^* = (0 \cup 1)^*$

10. Replace the following regular expressions by regular expressions that do not use the union operation.
    (a) $(0 \cup 1)^*$

(b) $(0 \cup 01 \cup 11)*$

(c) $(0 \cup (01 \cup 11)*)*$

(d) $(0 \cup \overleftarrow{(01 \cup 11)})*$

11. Find the reverse of each of the following regular expressions.

(a) $0 \cup 1(10)*(11 \cup 0)*$

(b) $10*(11 \cup 0*10*1) \cup (10*11 \cup 01*0*1)$

(c) $[(0 \cup 1)*00*0(0 \cup 1)*] \cup [(0 \cup 1) \cup 0*1*]$

(d) $\sim[(0 \cup 1)*000(0 \cup 1)*] \cup (0 \cup 1)*111 \cup [(0 \cup 1)*000(0 \cup 1*)*]$

12. Let R_1 and R_2 be two regular expressions. Does the equality $R_1 R_2 = R_2 R_1$ imply that $R_1 = R_2$? Justify your answer.

13. Find a regular expression for each of the machines of Fig. P8.1.13.

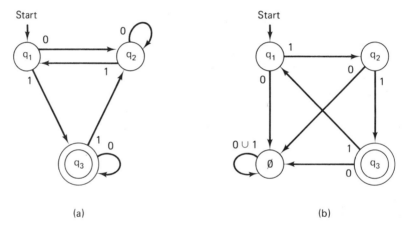

(a)   (b)

Fig. P8.1.13

14. Construct a transition diagram to recognize (a) the empty sequence Λ and (b) the empty set $\varnothing$.

15. Show that it is always possible to construct a sequential machine that recognizes a sequence iff that sequence is divisible by b without remainder, where b is a fixed positive integer.

16. Construct a transition diagram for each of the following regular expressions.

(a) $01*0(10* \cup 0*1)$

(b) $(101)* \cup (1*0 \cup 0*1) \cup \overleftarrow{(11*0 \cup 10*1)}$

(c) $10*(0110 \cup 10*1*0)\overleftarrow{(11*0 \cup 10*0)}$

17. Let M_1 and M_2 of Fig. P8.1.17 be binary recognizers. The recognized set of M_1 is given by $(1 \cup 0)(1 \cup 01)*$. It is shown that $z(k) = 0$, for all $k \geq 0$ (i.e., from the time the starting signal is applied, z is never 1). Draw the transition diagram of M_2.

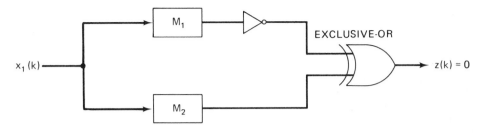

Fig. P8.1.17

18. Let M_s be a given sequential machine. Let a regular expression $R = 0^*10110^*0$. Construct a sequential machine M which recognizes R iff M_s recognizes 0110^*0 or 10^*0.

8.2 Nondeterministic Sequential Machines: Equivalence and Minimization

Nondeterministic sequential machines are generalizations of deterministic sequential machines.

DEFINITION 8.2.1

A Mealy- or Moore-type sequential machine $M = (\Sigma, Q, Z, \delta, \lambda)$ is said to be *nondeterministic* if there exists at least a state $q_i \in Q$ and an input $\sigma_j \in X$ such that either $\delta(q_i, \sigma_j)$ or $\lambda(q_i, \sigma_j)$ or both are multivalued. By a multivalued $\delta(q_i, \sigma_j)$ and a multivalued $\lambda(q_i, \sigma_j)$, we mean that

$$\delta(q_i, \sigma_j) = \begin{cases} q_k \\ \cdot \\ \cdot \\ \cdot \\ q_s \end{cases} \qquad \lambda(q_i, \sigma_j) = \begin{cases} z_k \\ \cdot \\ \cdot \\ \cdot \\ z_s \end{cases}$$

where $q_k, \ldots, q_s \in Q$, and $z_k, \ldots, z_s \in Z$.

The machine recognition of a tape by a nondeterministic machine is only defined on the class of Rabin–Scott nondeterministic machines, which is defined below.

DEFINITION 8.2.2

A *Rabin–Scott nondeterministic machine* over the binary alphabet $\Sigma = \{0, 1\}$ is a nondeterministic machine denoted by $A = (Q, \delta, Q_0, F)$, where Q and $F \in Q$ are the set of internal states and the set of designated final states, respectively; $Q_0 \in Q$ is a set of initial states; and δ is a multivalued next-state function. Machine M_4 in Fig. 8.2.1 is an example of a Rabin–Scott nondeterministic machine.

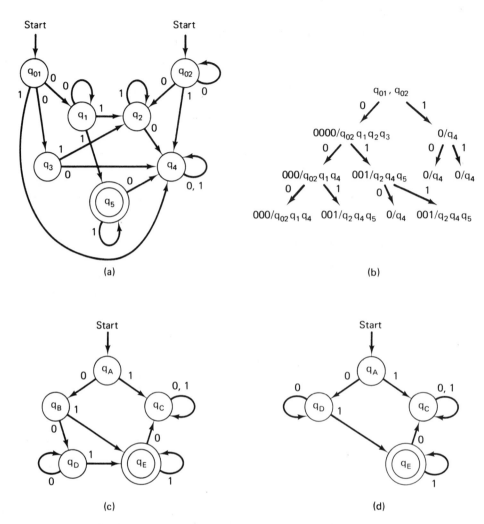

Fig. 8.2.1   (a) A nondeterministic machine M_4; (b) its transition tree; (c) its deterministic machine equivalent $D(M_4)$; (d) the $D(M_4)$ after minimization, denoted by $D_m(M_4)$.

For nondeterministic machines, there may be more than one initial state [such as q_{01} and q_{02} of the machine of Fig. 8.2.1(a)], and responding to each input σ_i, there may be several choices in its move. Consequently, for a tape $x = \sigma_0 \ldots \sigma_{n-1}$, we may have a number of sequences of choices, which result in many possible paths from the initial state(s) to the final state(s) via the tape x. Of course, some sequence of choices will lead to either situations from which no moves are possible or to final states that are not in F. Now let us agree to *disregard all such failures* and define the tape recognition of a nondeterministic machine as follows:

DEFINITION 8.2.3

A tape x is said to be *recognized* or *accepted* by a nondeterministic machine if there is (at least) one winning combination of choices of states leading to a designated final state.

For example, consider the machine M_4 of Fig. 8.2.1(a) for an input tape $x = 011$; there will be five possible paths:

$$q_{01} \quad q_1 \quad q_2 \quad q_2 \qquad \text{(failure)}$$
$$q_{01} \quad q_3 \quad q_2 \quad q_2 \qquad \text{(failure)}$$
$$q_{01} \quad q_1 \quad q_5 \quad q_5 \qquad \text{(success)}$$
$$q_{02} \quad q_2 \quad q_2 \quad q_2 \qquad \text{(failure)}$$
$$q_{02} \quad q_{02} \quad q_4 \quad q_4 \qquad \text{(failure)}$$

Among them, we see that there is one combination which leads to the state q_5, the designated final state. By definition, 011 is recognized by the machine. On the other hand, consider another tape, $y = 0110$. Then we find that all possible paths are:

$$q_{01} \quad q_1 \quad q_2 \quad q_2 \quad q_4$$
$$q_{01} \quad q_3 \quad q_2 \quad q_2 \quad q_4$$
$$q_{01} \quad q_1 \quad q_5 \quad q_5 \quad q_4$$
$$q_{02} \quad q_2 \quad q_2 \quad q_2 \quad q_4$$
$$q_{02} \quad q_{02} \quad q_4 \quad q_4 \quad q_4$$

The last state of all of them is q_4, a nonfinal state. Hence, the machine does not recognize the tape 0110.

Recall that in Section 8.1 it is shown that every deterministic sequential machine can be represented by a regular expression, and vice versa. We now develop a similar theory for nondeterministic machines.

DEFINITION 8.2.4

Let M be a nondeterministic machine. *The set of all tapes recognized by M*, or the *event recognized by M* denoted by $T(M)$, is a collection of all tapes $x = \sigma_0 \ldots \sigma_{n-1}$, for which there exists a sequence $q_0 q_1 \ldots q_n$ of internal states of M such that q_0 is in the set of initial states Q_0, q_i is in $\delta(q_{i-1}, \sigma_{i-1})$ for $i = 1, 2, \ldots, n$, and q_n is in F.

DEFINITION 8.2.5

For a given nondeterministic machine $M = (Q, \delta, Q_0, F)$, we can always construct a deterministic machine $D(M) = (P, \delta', p_0, F')$ as follows:

1.   Construct the transition tree of the given nondeterministic machine M.
2.   The set of states in each leaf of the tree is considered as a state of the new machine.
3.   The output of a state of the new machine is 1 if there is (at least) a final state of M in the set and is 0 otherwise.
4.   After the states and outputs are determined, construct a machine using the same transition tree that was obtained in 1.

In mathematical language, the deterministic machine $D(M)$ so constructed may be described, in terms of the states and transition functions of M, by

$$p_0 = Q_0$$
$$P = \{f(s, S_0) \,|\, x \in \Sigma^*\} = \{p_1, p_2, \ldots, p_m\}$$
$$F' = \{p \in P \,|\, p \cap F \neq \varnothing\}$$
$$\delta'(p_i, \sigma) = p_j \quad \text{iff} \quad \delta(p_i, \sigma) = p_j \quad \text{for all } i, j \text{ and all } \sigma \in \Sigma$$

This procedure of obtaining $D(M)$ from any M is called the *subset construction*.

Example 8.2.1

Consider the nondeterministic machine M_4 of Fig. 8.2.1(a). The transition tree of this machine is shown in Fig. 8.2.1(b). Define the sets of states in the leaves of the transition tree to be

$$q_A = \{q_{01}, q_{02}\}, \qquad q_B = \{q_{02}, q_1, q_2, q_3\}, \qquad q_C = \{q_4\}$$
$$q_D = \{q_{02}, q_1, q_4\}, \qquad q_E = \{q_2, q_4, q_5\}$$

From Definition 8.2.5, q_A is an initial state, q_A, q_B, q_C, and q_D are nonfinal states of $D(M_4)$, and q_E is a final state of $D(M_4)$. Hence,

$$p_0 = q_A$$
$$P = \{q_A, q_B, q_C, q_D, q_E\}$$
$$F' = \{q_E\}$$

and the δ' is described by the transition tree of Fig. 8.2.1(b). The machine $D(M_4)$ is shown in Fig. 8.2.1(c).

DEFINITION 8.2.6

Two nondeterministic machines M_a and M_b are said to be *tape-equivalent* if $T(M_a) = T(M_b)$. Symbolically, $M_a \equiv_T M_b$.

Then we have

THEOREM 8.2.1

Let M be a nondeterministic machine. Then $D(M) \equiv_T M$.

Proof: To prove that $D(M) \equiv_T M$ is to prove that $T(M) \equiv T(D(M))$. Let $x = \sigma_0 \sigma_1 \ldots$ σ_{n-1} be a tape of $T(M)$, and $q_0 q_1 \ldots q_n$ be a sequence of internal states of M satisfying the conditions of Definition 8.2.4. Let $p_0 p_1 \ldots p_n$ be a sequence of $n + 1$ states of $D(M)$, where p_0 is the starting state and p_n is a designated final state of $D(M)$. It is clear that state q_0 is in the starting state p_0 of $D(M)$, because it contains all the initial states of M. Since $q_1 = \delta(q_0, \sigma_0)$, q_1 is in the state of $D(M)$ that consists of all the states in $\delta(q_0, \sigma_0)$; denote this state of $D(M)$ by p_1. Likewise, every q_i is in the state p_i of $D(M)$ composed of all the states in $\delta(q_{i-1}, \sigma_{i-1})$. Since q_n is a designated final state and in p_n, p_n is a designated final state of $D(M)$. Hence, x is also recognized by $D(M)$. Since x is an arbitrary tape in $T(M)$,

$$T(M) \subseteq T(D(M)) \tag{8.2.1}$$

Now assume that $x = \sigma_0 \sigma_1 \ldots \sigma_{n-1}$ is a tape of $T(D(M))$. Let $p_0 p_1 \ldots p_n$ be a sequence of internal states of $D(M)$ satisfying the conditions of Definition 8.2.5. From the way we construct the machine $D(M)$, in each state p_{i-1} there exists a state q_{i-1} such that the next-state $q_i = \delta(q_{i-1}, \sigma_{i-1})$ is an element of p_i. Thus, there exists a sequence of states $q_0 q_1 \ldots q_n$ of M which satisfy the following conditions:

(1) q_0 is an initial state of M since $q_0 \in p_0$.
(2) q_i is in $\delta(q_{i-1}, \sigma_{i-1})$ for $i = 1, 2, \ldots, n$.
(3) q_n is in F since $q_n \in p_n$.

By Definition 8.2.5, the tape $x = \sigma_0 \sigma_1 \ldots \sigma_{n-1}$. is also in $T(M)$. Therefore,

$$T(D(M)) \subseteq T(M) \tag{8.2.2}$$

Combining Eqs. (8.2.1) and (8.2.2),

$$T(D(M)) = T(M)$$

Hence, the theorem is proved. ∎

It should be noted that the machine $D(M_4)$, obtained from M_4 above, may have redundant states. For example, by inspection, we see that the states q_B and q_D are equivalent since they both go to state q_D if they are excited by an input 0 and state q_E if they are excited by an input 1. Thus, one of them can be removed from the machine. Suppose that we remove the state q_B; the machine $D_m(M_4)$ after minimization is shown in Fig. 8.2.1(d). It is easy to see that the set of all tapes $T(D_m(M_4))$ recognized by $D_m(M_4)$ is

$$T(D_m(M_4)) = \{0^n 1^m \mid n \geq 1 \quad \text{and} \quad m \geq 1\} \tag{8.2.3}$$

which by Theorem 8.2.1 is also the set of all tapes $T(M_4)$ recognized by M_4.

Using the method introduced in Section 8.1, we find that

$$q_E = 00^*11^* \tag{8.2.4}$$

which is the regular expression of the machine $D_m(M_4)$. It is, of course, also a regular expression of machines M_4 and $D(M_4)$, because it is merely a formal way of representing the set of tapes $T(D_m(M_4))$.

Now we would like to explore another interesting property about nondeterministic machines.

DEFINITION 8.2.7

Let $M = (Q, \delta, Q_0, F)$ be a Rabin–Scott nondeterministic machine. The *dual* of M, denoted by $\overleftarrow{M} = (Q, \overleftarrow{\delta}, F, Q_0)$, is defined as follows:

   (a)   The set of internal states of $\overleftarrow{M}$ is the same as that of M.

   (b)   The relation between $\overleftarrow{\delta}$ and δ is that

$$q' \text{ is in } \overleftarrow{\delta}(q, \sigma) \qquad \text{iff } q \text{ is in } \delta(q', \sigma)$$

   (c)   The set of final states of $\overleftarrow{M}$ are the set of initial states of M.

   (d)   The set of initial states of $\overleftarrow{M}$ are the set of final states of M.

For example, the duals of M_4, $D(M_4)$, and $D_m(M_4)$ of Fig. 8.2.1 are shown in Fig. 8.2.2. Note that the dual of M in general does not have a complete state-transition specification. For instance, the state transitions of q_{01} of $\overleftarrow{M_4}$ for either input 0 or input 1 are unspecified. Clearly, for any finite-state machine M, $\overleftarrow{\overleftarrow{M}} = M$.

Then we can prove the following.

THEOREM 8.2.2

For any Rabin–Scott nondeterministic machine M, $\overleftarrow{T(M)} = T(\overleftarrow{M})$.

Proof: (a) Proof of $\overleftarrow{T(M)} \subseteq T(\overleftarrow{M})$: Let $x = \sigma_0\sigma_1 \ldots \sigma_{n-1}$ be a tape of $T(M)$. By Definition 8.2.4, there exists a sequence of states of M, $q_0q_1 \ldots q_n$, such that q_0 is an initial state, q_i is in $f(\sigma_{i-1}, q_{i-1})$, and q_n is in F. By the definition of the reverse of a set of tapes, the tape $\overleftarrow{x} = \sigma_{n-1}\sigma_{n-2} \ldots \sigma_0$ is in $\overleftarrow{T(M)}$. By the definition of $\overleftarrow{M}$, q_n is an initial state of M^d, q_0 is a final state of $\overleftarrow{M}$, and q_{i-1} is in $f(\sigma_{i-1}, q_i)$ for $i = 1, 2, \ldots, n$. Hence, there exists a sequence of states $q_nq_{n-1} \ldots q_0$ satisfying the conditions of Definition 8.1.8. So $\overleftarrow{x}$ is also in $T(\overleftarrow{M})$. This means that

$$\overleftarrow{T(M)} \subseteq T(\overleftarrow{M}) \tag{8.2.5}$$

   (b) Proof of $\overleftarrow{T(M)} \supseteq T(\overleftarrow{M})$: Letting M in Eq. (8.2.5) be replaced by $\overleftarrow{M}$,

$$\overleftarrow{T(\overleftarrow{M})} \subseteq T((\overleftarrow{\overleftarrow{M}}))$$

or

$$\overleftarrow{T(\overleftarrow{M})} \subseteq T(M)$$

Applying the reverse operation on both sides of this inequality, we obtain

$$\overleftarrow{(T(\overleftarrow{M}))} \subseteq \overleftarrow{T(M)}$$

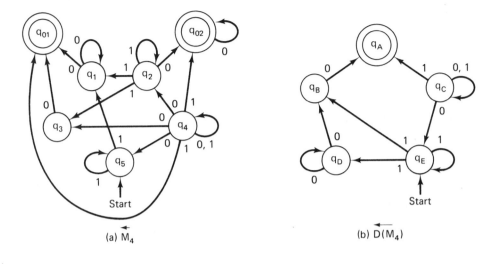

(a) $\overleftarrow{M}_4$

(b) $\overleftarrow{D(M_4)}$

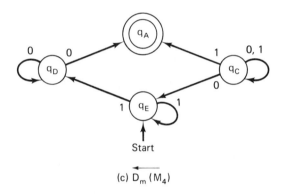

(c) $\overleftarrow{D_m(M_4)}$

Fig. 8.2.2 The duals of M_4, $D(M_4)$, and $D_m(M_4)$ of Fig. 8.2.1.

or

$$T(\overleftarrow{M}) \subseteq \overleftarrow{T(M)} \tag{8.2.6}$$

Combining Eqs. (8.2.5) and (8.2.6), we have $\overleftarrow{T(M)} = T(\overleftarrow{M})$.
Hence, the theorem is proved. ∎

COROLLARY 8.2.1

If $M_a \equiv_T M_b$, then $\overleftarrow{M}_a \equiv_T \overleftarrow{M}_b$.

Proof: From Theorem 8.2.2,

$$\overleftarrow{T(M_a)} = T(\overleftarrow{M}_a) \tag{8.2.7}$$

$$\overleftarrow{T(M_b)} = T(\overleftarrow{M}_b) \tag{8.2.8}$$

By definition, $M_1 \equiv_T M_2$ iff

$$T(M_a) = T(M_b) \tag{8.2.9}$$

which implies that

$$\overleftarrow{T(M_a)} = \overleftarrow{T(M_b)} \tag{8.2.10}$$

From Eqs. (8.2.7), (8.2.8), and (8.2.10), we obtain

$$T(\overleftarrow{M_a}) = T(\overleftarrow{M_b}) \tag{8.2.11}$$

Hence, $\overleftarrow{M_a} \equiv_T \overleftarrow{M_b}$. ∎

DEFINITION 8.2.8

Given a machine $M = (Q, \delta, Q_0, F)$, the *succeeding event* and *preceding event* of a state q_i of M, $sc_M(q_i)$ and $pr_M(q_i)$, respectively, are defined by

$$sc_M(q_i) = T((Q, \delta, q_i, F)) = \{x \in \Sigma^* \mid \delta(x, q_i) \cap F \neq \varnothing\}$$
$$pr_M(q_i) = T((Q, \delta, q_0, q_i)) = \{x \in \Sigma^* \mid q_i \in \delta(x, Q_0)\}$$

Example 8.2.2

The succeeding events and preceding events of the states of M_4 of Fig. 8.2.1 are

$$sc_{M_4}(q_{01}) = 00^*11^* \qquad pr_{M_4}(q_{01}) = \Lambda$$
$$sc_{M_4}(q_{02}) = \varnothing \qquad pr_{M_4}(q_{02}) = 0^*$$
$$sc_{M_4}(q_1) = 0^*11^* \qquad pr_{M_4}(q_1) = 00^*$$
$$sc_{M_4}(q_2) = \varnothing \qquad pr_{M_4}(q_2) = (0^*0 \cup 00^*1 \cup 01)1^*$$
$$sc_{M_4}(q_3) = \varnothing \qquad pr_{M_4}(q_3) = 0$$
$$sc_{M_4}(q_4) = \varnothing \qquad pr_{M_4}(q_4) = (1 \cup 00 \cup 011^*0 \cup 00^*11^*0$$
$$\cup\ 0^*1 \cup 0^* \cup 01^*0)$$
$$= (1 \cup 00^*11^*0)(0^* \cup 1^*)^*$$
$$sc_{M_4}(q_5) = 1^* \qquad pr_{M_4}(q_5) = 00^*11^*$$

The following theorem shows some relations of the succeeding events and preceding events between the states of a nondeterministic machine M and the states of $D(M)$.

THEOREM 8.2.3

Let $M = (Q, \delta, Q_0, F)$ and $D(M) = (P, \delta', p_0, F')$.
(a) $sc_{D(M)}(p) = \bigcup_{q \in p} sc_M(q)$.
(b) $pr_{D(M)}(p) = \bigcap_{q \in p} pr_M(q)$.
(c) $pr_{D(M)}(p) \cap pr_M(q) = \varnothing$ for all $q \in Q - p$.

Proof:

(a) $sc_{D(M)}(p) = \{x \mid f'(x, p) \cap F \neq \varnothing\}$

$= \{x \mid \underset{q \in p}{\bigcup} f(x, q) \cap F \neq \varnothing\}$

$= \underset{q \in p}{\bigcup} \{x \mid f(x, q) \cap F \neq \varnothing\}$

$= \underset{q \in p}{\bigcup} sc_M(q)$

(b) $pr_{D(M)}(p) = \{x \mid p = f'(x, p_0)\}$

$= \{x \mid p = f(x, Q_0)\}$

$\subseteq \{x \mid q \subseteq f(x, Q_0)\}$

$= \{x \mid \underset{q \in p}{\bigcap} [q \in f(x, Q_0)]\}$

$= \underset{q \in q}{\bigcap} \{x \mid q \in f(x, Q_0)\}$

$= \underset{q \in p}{\bigcap} pr_M(q)$

(c) For all $q \notin p$, that $x \in pr_{D(M)}(p)$ implies that $q \notin f(x, Q_0)$, which implies that $q \notin pr_M(q)$. Hence, for all $q \in Q - p$, $pr_{D(M)}(p) \cap pr_M(q) = \varnothing$. ∎

Example 8.2.3

The succeeding events and preceding events of the states of $D(M_4)$ are

$sc_{D(M_4)}(q_A) = 0(00^* \cup \Lambda)11^*$ $pr_{D(M_4)}(q_A) = \Lambda$

$sc_{D(M_4)}(q_B) = (00^* \cup \Lambda)11^*$ $pr_{D(M_4)}(q_B) = 0$

          $= 0^*11^*$

$sc_{D(M_4)}(q_C) = \varnothing$ $pr_{D(M_4)}(q_C) = (1 \cup 00^*11^*0)(0^* \cup 1^*)^*$

$sc_{D(M_4)}(q_D) = 0^*11^*$ $pr_{D(M_4)}(q_D) = 000^*$

$sc_{D(M_4)}(q_E) = 1^*$ $pr_{D(M_4)}(q_E) = 0(00^* \cup \Lambda)11^*$

          $= 00^*11^*$

They can also be obtained from the succeeding and preceding events of the states of M_4 found in Example 8.2.1 as

$sc_{D(M_4)}(q_A) = sc_{M_4}(q_{01}) \cup sc_{M_4}(q_{02})$

         $= 00^*11^* \cup \varnothing = 00^*11^*$

$sc_{D(M_4)}(q_B) = sc_{M_4}(q_{02}) \cup sc_{M_4}(q_1) \cup sc_{M_4}(q_2) \cup sc_{M_4}(q_3)$

         $= \varnothing \cup 0^*11^* \cup \varnothing = 0^*11^*$

$sc_{D(M_4)}(q_C) = sc_{M_4}(q_4) = \varnothing$

$sc_{D(M_4)}(q_D) = sc_{M_4}(q_{02}) \cup sc_{M_4}(q_1) \cup sc_{M_4}(q_4)$

         $= \varnothing \cup 0^*11^* \cup \varnothing = 0^*11^*$

$sc_{D(M_4)}(q_E) = sc_{M_4}(q_2) \cup sc_{M_4}(q_4) \cup sc_{M_4}(q_5)$

         $= \varnothing \cup \varnothing \cup 1^* = 1^*$

and

$$pr_{D(M_4)}(q_A) = pr_{M_4}(q_{01}) \cap pr_{M_4}(q_{02})$$
$$= \Lambda \cap 0^* = \Lambda$$
$$pr_{D(M_4)}(q_B) = pr_{M_4}(q_{02}) \cap pr_{M_4}(q_1) \cap pr_{M_4}(q_2) \cap pr_{M_4}(q_3)$$
$$= 0^* \cap 00^* \cap [(0^*0 \cup 00^*1 \cup 00^*1)1^*] \cap 0 = 0$$
$$pr_{D(M_4)}(q_C) = pr_{M_4}(q_4) = (1 \cup 00^*11^*0)(0^* \cup 1^*)^*$$
$$pr_{D(M_4)}(q_D) = pr_{M_4}(q_{02}) \cap pr_{M_4}(q_1) \cap pr_{M_4}(q_4)$$
$$= 0^* \cap 00^* \cap [1 \cup 00^*11^*0)(0^* \cup 1^*)^*] = 000^*$$
$$pr_{D(M_4)}(q_E) = pr_{M_4}(q_2) \cap pr_{M_4}(q_4) \cap pr_{M_4}(q_5)$$
$$= [(0^*0 \cup 00^*1 \cup 01)1^*] \cap [1 \cup 00^*11^*0)(0^* \cup 1^*)^*] \cap [00^*11^*] = 00^*11^*$$

Consider a state of $D(M_4)$, say, $q_E = \{q_2, q_4, q_5\}$ which does not contain $q_{01}, q_{02}, q_1,$ and q_3. We find that

$$pr_{D(M_4)}(q_E) \cap pr_{M_4}(q_{01}) = pr_{D(M_4)}(q_E) \cap pr_{M_4}(q_{02}) = \Lambda$$
$$pr_{D(M_4)}(q_E) \cap pr_{M_4}(q_1) = pr_{D(M_4)}(q_E) \cap pr_{M_4}(q_3) = \emptyset$$

THEOREM 8.2.4

(a) $sc_M(q) = \overleftarrow{pr_{\overline{M}}}(q).$

(b) $pr_M(q) = \overleftarrow{sc_{\overline{M}}}(q).$

Proof:

(a) $\overleftarrow{pr_{\overline{M}}}(q) = \{x \mid q \in \overleftarrow{\delta}(F, x)\}$
$$= \{x \mid \text{for at least one state } q_F \text{ in } F \text{ such that } q_F \in \overleftrightarrow{\delta}(q, x)\}$$
$$= \{x \mid q_F \in \delta(q, x)\}$$
$$= \{x \mid \delta(q, x) \cap F \neq \emptyset\}$$
$$= sc_M(q)$$

(b) $\overleftarrow{sc_{\overline{M}}}(q) = \{x \mid \overleftarrow{\delta}(q, x)Q_0 \neq \emptyset\}$
$$= \{x \mid \text{for at least one } q_0 \text{ in } Q_0 \text{ such that } q_0 \in \overleftrightarrow{\delta}(q, x)\}$$
$$= \{x \mid q \in \overleftrightarrow{\delta}(q_0, x)\}$$
$$= \{x \mid q \in \delta(q_0, x)\}$$
$$= \{x \mid q \in \delta(Q_0, x)\}$$
$$= pr_M(q) \quad \blacksquare$$

Example 8.2.4

The succeeding events and preceding events of the states of $\overleftarrow{M_4}$ of Fig. 8.2.2(a) are

$$sc_{\overleftarrow{M_4}}(q_{01}) = \Lambda \qquad\qquad pr_{\overleftarrow{M_4}}(q_{01}) = 1^*10^*0$$

$$sc_{\overleftarrow{M_4}}(q_{02}) = \varnothing 0^* \qquad\qquad pr_{\overleftarrow{M_4}}(q_{02}) = \varnothing$$

$$sc_{\overleftarrow{M_4}}(q_1) = 0^*0 \qquad\qquad pr_{\overleftarrow{M_4}}(q_1) = 1^*0^*$$

$$sc_{\overleftarrow{M_4}}(q_2) = 1^*(00^* \cup 10^*0 \cup 10) \qquad pr_{\overleftarrow{M_4}}(q_2) = \varnothing$$

$$sc_{\overleftarrow{M_4}}(q_3) = 0 \qquad\qquad pr_{\overrightarrow{M_4}}(q_3) = \varnothing$$

$$sc_{\overleftarrow{M_4}}(q_4) = (0^* \cup 1^*)^* \qquad\qquad pr_{\overleftarrow{M_4}}(q_4) = \varnothing$$

$$(1 \cup 00 \cup 01^*10 \cup 01^*10^*0 \cup 10^* \cup 01^*00^*$$

$$= (0^* \cup 1^*)^*(1 \cup 01^*10^*0)$$

$$sc_{\overleftarrow{M_4}}(q_5) = 1^*10^*0 \qquad\qquad pr_{\overleftarrow{M_4}}(q_5) = 1^*$$

From the succeeding events and preceding events of the states of M_4 given in Example 8.2.1 and those of $\overleftarrow{M_4}$ shown above, we see that $sc_{M_4}(q) = pr_{\overleftarrow{M_4}}(q)$ and $pr_{M_4}(q) = sc_{\overleftarrow{M_4}}(q)$ for every state q of M_4.

Note that Theorems 8.2.3(a) and 8.2.4 offer another proof of Theorems 8.2.1 and 8.2.2, respectively. If we let $p = p_0 = Q_0$ in Theorem 8.2.3(a), then Theorem 8.2.1 follows immediately from it. From Theorem 8.2.4,

$$sc_M(Q_0) = \bigcup_{q_0 \in Q_0} sc_M(q_0)$$

$$= \bigcup_{q_0 \in Q_0} pr_{\overleftarrow{M}}(q_0)$$

$$= \bigcup_{q_F \in F} sc_{\overleftarrow{M}}(q_F)$$

$$= sc_{\overleftarrow{M}}(F)$$

Hence, $T(M) = \overleftarrow{T(\overleftarrow{M})}$. Since $\overleftarrow{T(M)} = T(\overleftarrow{M})$, $T(M) = T(\overleftarrow{\overleftarrow{M}})$.

THEOREM 8.2.5 (Brzozowski)

Let $M = (Q, \delta, q_0, F)$ be a deterministic machine (i.e., f is single-valued), not necessarily minimized. $D(\overleftarrow{M})$ is a minimized deterministic machine tape-equivalent to $\overleftarrow{M}$.

Proof: From Theorem 8.2.1, $D(\overleftarrow{M}) \equiv_T \overleftarrow{M}$. Now we want to show that $D(\overleftarrow{M})$ is minimized. We shall prove it by the method of contradiction.

Suppose that two distinct states s_i and s_j of $D(\overleftarrow{M})$ were equivalent. By the hypothesis,

$$sc_{D(\overleftarrow{M})}(s_i) = sc_{D(\overleftarrow{M})}(s_j)$$

which implies [by Theorem 8.2.3(a)] that

$$\bigcup_{q \in s_i} sc_{\overleftarrow{M}}(q) = \bigcup_{q \in s_j} sc_{\overleftarrow{M}}(q)$$

Using Theorem 8.2.4, we get

$$\bigcup_{q \in s_i} \overleftarrow{pr_M}(q) = \bigcup_{q \in s_j} \overleftarrow{pr_M}(q)$$

Hence,

$$\bigcup_{q \in s_i} pr_M(q) = \bigcup_{q \in s_j} pr_M(q)$$

This means that the union of the (nonempty) events in a subset of the set $\{pr_M(q) | q \in Q\}$ is identical to the union of the events in another subset distinct from the first subset. This is a contradiction to the fact that the preceding events of a deterministic machine are mutually disjoint. Hence, no two states of $D(\overleftarrow{M})$ are equivalent. In other words, $D(\overleftarrow{M})$ is a minimized deterministic machine. ∎

Example 8.2.5

The transition tree and the state diagram of $D(\overleftarrow{M_4})$ are shown in Fig. 8.2.3. It is easy to verify that $D(\overleftarrow{M_4})$ is minimal.

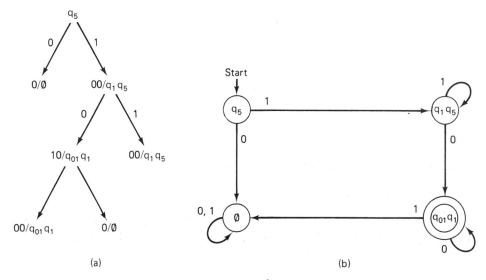

Fig. 8.2.3   (a) Transition tree of $D(\overleftarrow{M_4})$; (b) transition diagram of $D(\overleftarrow{M_4})$.

Exercise 8.2

1. Give an example of a system encountered in everyday life that may be represented by a nondeterministic sequential machine.

2. Determine the indistinguishable equivalent states of each of the nondeterministic machines in Fig. P8.2.2 and find their minimal forms.

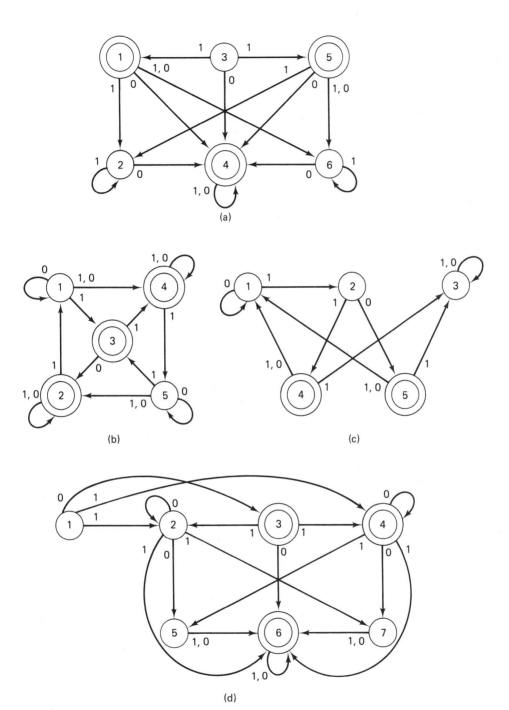

(a)

(b)

(c)

(d)

Fig. P8.2.2

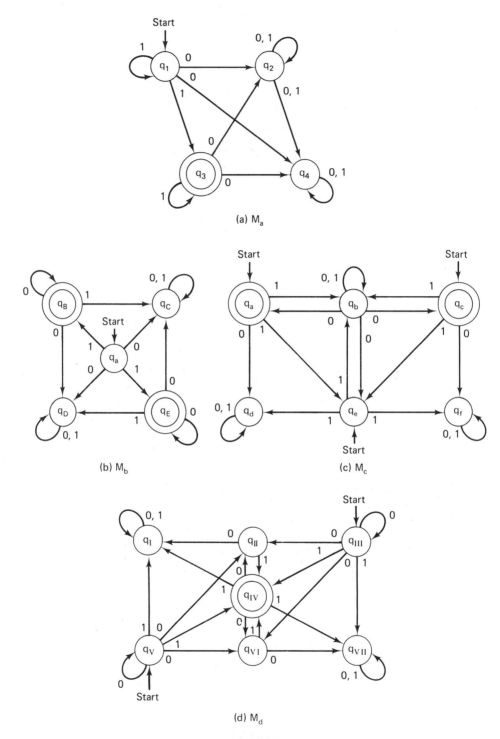

(a) M_a

(b) M_b

(c) M_c

(d) M_d

Fig. P8.2.3

3. Find the deterministic equivalent machines of the Rabin–Scott nondeterministic machines of Fig. P8.2.3.

4. (a) Find the set of tapes recognized by each of the machines of Fig. P8.2.2.
    (b) Find the duals of the machines of Fig. P8.2.2.
    (c) Determine the set of tapes recognized by each of the duals obtained in (b).
    (d) Verify the relation $\overleftarrow{T(M)} = T(\overleftarrow{M})$ using the machines of Fig. P8.2.2.

5. Find a physical circuit using AND's, OR's, and NOT's and unit delays which realizes the set of tapes recognized by M_e of Fig. P8.2.5.

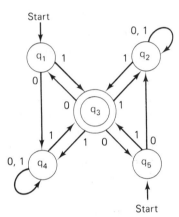

Fig. P8.2.5 M_e.

6. (a) Find the regular expression R_{M_e} of M_e of Fig. P8.2.5.
    (b) Construct the dual of M_e from $\overleftarrow{R}_{M_e}$ using the method given in Section 8.1.
    (c) Show that the dual of M_e constructed by the procedure described in Definition 8.2.7 and the dual of M_e obtained in part (b) are equivalent.
    (d) Verify Theorem 8.2.3 with M_e and $D(M_e)$.

7. Define the operation $\mathfrak{M}(M)$ as the machine-minimization operation operating on M based on the indistinguishability equivalence. For the class of machines with a single designated final state, we define the operation $\mathfrak{I}(M)$ as a direct machine-minimization operation operating on M based on the tape equivalence. For the case where there is more than one transition line excited by the same input symbol from a state to other states, including one to the designated final state, the operation is simply to delete all the transition lines from the machine except the one that goes to the designated final state.

    Consider machine M_d of Fig. P8.2.3.
    (a) Find $\mathfrak{M}(M_d)$ and then $\mathfrak{I}[\mathfrak{M}(M_d)]$.
    (b) Find $D(M_d)$ and then $\mathfrak{M}[D(M_d)]$.
    (c) Show that $\mathfrak{I}[\mathfrak{M}(M_d)] \equiv \mathfrak{M}[D(M_d)]$, where the symbol $M_a \equiv M_b$ denotes that M_a and M_b are identical.

8. Consider machine M_4 of Example 8.2.1. The $D(M_4)$ and $\mathfrak{M}[D(M_4)]$ are shown in Figs. 8.2.1(c) and (d), respectively. For this machine, again show that $\mathfrak{I}[\mathfrak{M}(M_4)] \equiv \mathfrak{M}[D(M_4)]$, where the symbol $\equiv$ is the same as defined in problem 7.

9. From the results of problems 7 and 8, can we conclude that $\mathfrak{J}[\mathfrak{M}(M)] \equiv \mathfrak{M}[D(M)]$ for any Rabin–Scott nondeterministic machine with a single designated final state? Give reasons to support your answer.

10. Show that $D(M_4)$ and $D_m(M_4)$ of Figs. P8.2.3(c) and (d) recognize exactly the same set of tapes as do $\overleftarrow{D(M_4)}$ and $\overleftarrow{D_m(M_4)}$.

11. (a) Find $\overleftarrow{M_e}$ and $D(\overleftarrow{M_e})$ for M_e of Figure P8.2.5.

    (b) Verify Theorem 8.2.3 with M_e and $\overleftarrow{M_e}$.

    (c) Show that $D(\overleftarrow{M_e})$ is a minimal machine.

8.3 Construction of Transition Diagrams of $R_1 R_2$, $R_1 \cup R_2$, $R_1 \cap R_2$, $R_1 \triangle R_2$, etc.

The construction of a transition diagram from a regular expression, presented in Section 8.1, was obtained from the entire expression. In some cases, especially when the regular expression is a long and complicated one, it is much more convenient to break the expression into several short, regular expressions connected by the union, intersection, concatenation, or other operation. Perform the rather simple task of constructing transition diagrams of the partial regular expressions first, and then convert these transition diagrams to a transition diagram that recognizes the given regular expression.

Let R_1 and R_2 be two regular expressions and M_1 and M_2 be the transition diagrams obtained from R_1 and R_2. For example, $R_1 = 01^*$ and $R_2 = (01)^*$. Their realizations are shown in Fig. 8.3.1. We shall use these two regular expressions as an example to illustrate various procedures for constructing the transition diagrams presented below.

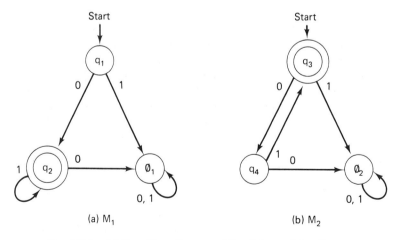

(a) M_1 (b) M_2

Fig. 8.3.1 (a) Transition diagram of R_1; (b) transition diagram of R_2.

1. Transition Diagram of ~R Constructed from M

Let M denote a transition diagram of R. By definition, $\sim R = \Sigma^* - R$, where $\Sigma = \{0, 1\}$ and Σ^* is the set of all possible sequences of 0 and 1.

ALGORITHM 8.3.1

The transition diagram of $\sim R$ can be obtained by changing all the final states (double-circled states, or state-output-1 states) of M into nonfinal states (single-circled states, or state-output-0 states) and all the nonfinal states into final states.

Example 8.3.1

The transition diagram of $\sim R$, for $R_1 = 01^*$, whose transition diagram M_1 is shown in Fig. 8.3.1(a), is given in Fig. 8.3.2.

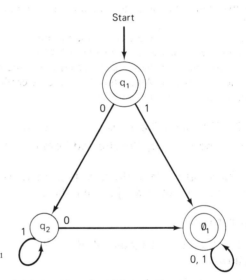

Fig. 8.3.2 Transition diagram of $\sim R_1$ constructed from M_1.

2. Transition Diagram of $R_1 R_2$ Constructed from M_1 and M_2

ALGORITHM 8.3.2

The transition diagram of $R_1 R_2$ can be obtained as follows:

1. Superimpose the final states of M_1 with the starting state of M_2, resulting in a nondeterministic machine.

2. Convert the nondeterministic machine into a deterministic one, and denote it by $M_{R_1 R_2}$.

3. Suppose that $q_{\alpha_1 \alpha_2 \ldots \alpha_n}$ is a state of $M_{R_1 R_2}$, which is composed of states q_{α_1}, $q_{\alpha_2}, \ldots, q_{\alpha_n}$ of M_1 and M_2. Determine the type of $q_{\alpha_1 \alpha_2 \ldots \alpha_n}$ of $M_{R_1 R_2}$ using the following rules:

(3-1) $q_{\alpha_1\alpha_2\ldots\alpha_n}$ is a nonfinal state if all q_{α_i} are nonfinal states.

(3-2) $q_{\alpha_1\alpha_2\ldots\alpha_n}$ is the dead state of the deterministic machine if *all* q_{α_i} are the dead states of M_1 and M_2.

(3-3) If any q_{α_i} are dead states of M_1 and M_2, delete those states from $q_{\alpha_1\alpha_2\ldots\alpha_n}$.

(3-4) $q_{\alpha_1\alpha_2\ldots\alpha_n}$ is a final state of $M_{R_1R_2}$ if one of the states $q_{\alpha_1}, q_{\alpha_2}, \ldots, q_{\alpha_n}$ is a final state of M_2.

This algorithm is illustrated by the following example.

Example 8.3.2

Construct the transition diagram of R_1R_2 from M_1 and M_2 of Fig. 8.3.1 by Algorithm 8.3.2.

1. The final state of M_1 is q_2 and the starting state of M_2 is q_3. After superimposing these two states, the transition diagram is shown in Fig. 8.3.3(a), which is a nondeterministic machine.

2. Convert the nondeterministic machine into a deterministic one by constructing the transition tree which is given in Fig. 8.3.3(b).

3. By applying the rules for determining the types of the states of the deterministic machine described in Algorithm 8.3.2, the deterministic machine that recognizes R_1R_2 is obtained and shown in Fig. 8.3.3(c).

3. Transition Diagrams of $R_1 \cup R_2$, $R_1 \cap R_2$, and $R_1 \triangle R_2$ Constructed from M_1 and M_2

The construction of the transition diagrams of $R_1 \cup R_2$, $R_1 \cap R_2$, and $R_1 \triangle R_2$ is similar to that of R_1R_2, just described.

ALGORITHM 8.3.3

The transition diagrams of $R_1 \cup R_2$, $R_1 \cap R_2$, and $R_1 \triangle R_2$ can be obtained as follows:

1. Consider the two deterministic machines M_1 and M_2 as one nondeterministic machine.

2. Convert the nondeterministic machine into a deterministic one.

3. Determine whether state $q_{\alpha_1\alpha_2\ldots\alpha_n}$ of the deterministic machine obtained in 2 is a nonfinal, final, or dead state using the following rules:

(3-1) ⎫
(3-2) ⎬ The same as those of Algorithm 8.3.2.
(3-3) ⎭

(3-4) $q_{\alpha_1\alpha_2\ldots\alpha_n}$ is a final state of the deterministic machine of

$$\left.\begin{array}{c} R_1 \cup R_2 \\ R_1 \cap R_2 \\ R_1 \triangle R_2 \end{array}\right\} \text{ iff } \left\{\begin{array}{l} \text{there is a state } q_{\alpha_i} \text{ which is a final state of } M_1 \text{ or } M_2. \text{ (Rule A)} \\ \text{there are states } q_{\alpha_i} \text{ and } q_{\alpha_j} \text{ which are final states of } M_1 \text{ and } M_2, \\ \quad \text{respectively. (Rule B)} \\ q_{\alpha_1\alpha_2\ldots\alpha_n} \text{ satisfies Rule A but not Rule B.} \end{array}\right.$$

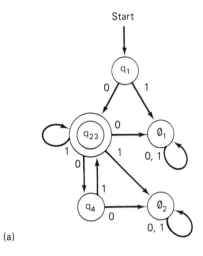

(a)

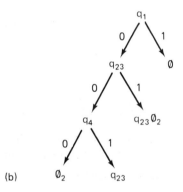

(b)

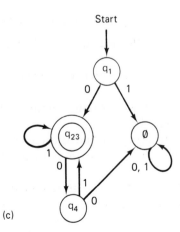

(c)

Fig. 8.3.3 (a) A nondeterministic machine of R_1R_2; (b) its transition tree; (c) the deterministic equivalent machine.

Example 8.3.3

Consider the constructions of transition diagrams of $R_1 \cup R_2$, $R_1 \cap R_2$, and $R_1 \triangle R_2$ from M_1 and M_2 of Fig. 8.3.1. First, we put the two machines M_1 and M_2 together and consider them as *one* machine, as shown in Fig. 8.3.4(a). Then we construct the transition tree of this nondeterministic machine, which is shown in Fig. 8.3.4(b). Using the rules for determining the types of the states of the deterministic machines of $R_1 \cup R_2$, $R_1 \cap R_2$, and $R_1 \triangle R_2$, the complete transition diagrams of these three regular expressions are, respectively, shown in Figs. 8.3.4(c), (d), and (e). Notice that the three machines have the same graph. The only difference among them is that they have different final states. The reader can easily verify that the three machines recognize the sets

$$R_1 \cup R_2 = \{01, 011, 0111, \ldots, 01, 0101, 010101, \ldots\}$$
$$R_1 \cap R_2 = \{01, 0101, 010101, \ldots\}$$

and

$$R_1 \triangle R_2 = \{011, 0111, \ldots, 0101, 010101, \ldots\}$$

respectively. From either these sets or the transition diagrams, it is seen that the relation

$$R_1 \cup R_2 = (R_1 \cap R_2) \cup (R_1 \triangle R_2)$$

holds.

4. Transition Diagram of $\overleftarrow{R}$ Constructed from M

Another interesting problem of constructing the transition diagram of a given regular expression from the transition diagram of its subexpression is the construction of $\overleftarrow{R}$ from M, which is described as follows.

ALGORITHM 8.3.4

The transition diagram of $\overleftarrow{R}$ can be obtained as follows:

1. Change all the final states to the starting states and the starting state to the final state.
2. Remove the transition lines that go to the dead state and reverse the remaining ones.
3. Complete the transition diagram of step 2 by filling up all the missing transitions to the dead state.
4. If the resulting machine of step 3 is deterministic, the desired transition diagram is found. If the resulting machine of step 3 is nondeterministic, convert it into a deterministic one.
5. A state $q_{\alpha_1 \alpha_2 \ldots \alpha_n}$ is a final state of the deterministic equivalent machine if it contains the starting state of M.

This procedure is illustrated by the following examples.

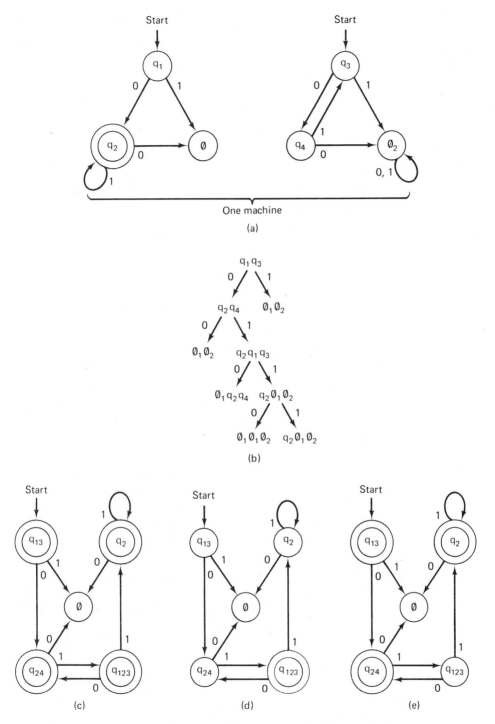

Fig. 8.3.4 (a) A nondeterministic machine; (b) its transition tree; (c) the deterministic equivalent machine.

Example 8.3.4

Consider the construction of transition diagram of $\overleftarrow{R}_1$ from M_1 of Fig. 8.3.1. For this simple example, the machine obtained at step 3 of Algorithm 8.3.4 is a deterministic one; thus, no nondeterministic-to-deterministic machine conversion is needed. Figure 8.3.5 shows each of the steps of obtaining the transition diagram of $\overleftarrow{R}_1$ from M_1.

Example 8.3.5

As another example, consider the construction of transition diagram of $\overleftarrow{R}$, where $R = 0*10*1*$, whose transition diagram is shown in Fig. 8.3.6(a). Applying Algorithm

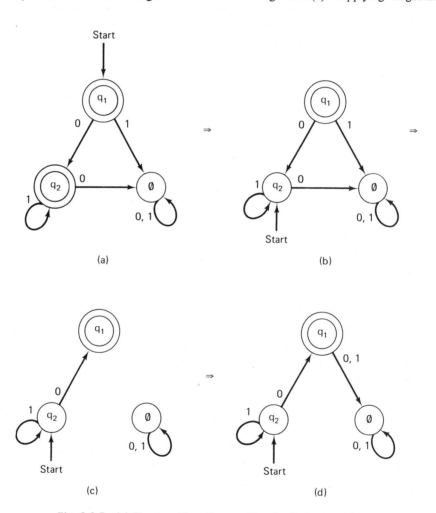

Fig. 8.3.5 (a) The transition diagram M_1; (b–d) the transition diagrams corresponding to steps 1-3 of Algorithm 8.3.4.

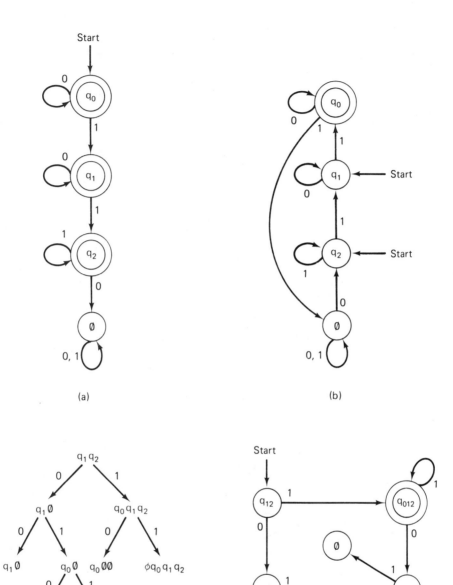

Fig. 8.3.6 (a) The transition diagram of $R = 0^*10^*1^*$; (b) the transition diagram obtained by applying Algorithm 8.3.4; (c) convert the transition diagram of (b) into a deterministic one; (d) the transition diagram of $\overleftarrow{R} = \overrightarrow{0^*10^*1^*}$.

8.3.4(1)–(3), we obtain the transition diagram of Fig. 8.3.6(b). It is seen that this machine is nondeterministic; we need to convert it into a deterministic one. The transition tree and the deterministic equivalent machine which recognizes $\overleftarrow{0^*10^*1^*}$ are shown in Figs. 8.3.6(c) and (d).

5.   Transition Diagram of R^* Constructed from M

Another important operation used in regular expressions is the star operation $*$. The construction of the transition diagram of R^* from M is described in Algorithm 8.3.5.

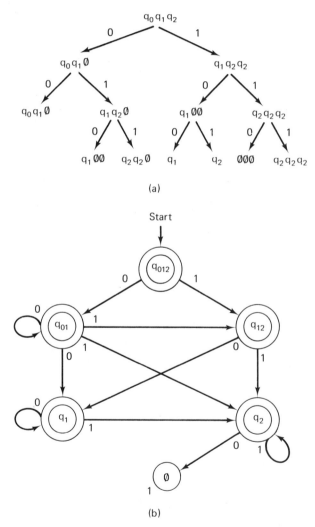

(a)

(b)

Fig. 8.3.7   (a) Conversion of a nondeterministic machine into a deterministic one; (b) the machine that recognizes $(0^*10^*1^*)^*$.

ALGORITHM 8.3.5

The transition diagram of R^* can be obtained as follows:

1. Superimpose all the final states of M with the starting state, which is the same as making all the final states be both final states and starting states.
2. Convert the nondeterministic machine of step 1 into a deterministic one.
3. A state $q_{\alpha_1\alpha_2\ldots\alpha_n}$ of the deterministic machine is a final state if one of the q_{α_i} is a final state of M.

Example 8.3.6

Suppose that $R = 0^*10^*1^*$. The transition diagram of R was given in Fig. 8.3.6(a). Applying Algorithm 8.3.5, the transition diagram of $(0^*10^*1^*)^*$ is shown in Fig. 8.3.7.

Exercise 8.3

1. Realize the following regular expressions.
    (a) $((0^*10^*1^*) \cap (0^*101^*)) \cup (11^*00^*)$
    (b) $\overleftarrow{(0^*10^*1^*)(0^*101^*)}$
    (c) $\overline{(\sim(0^*10^*1)) \cup (\sim(0^*101^*))}$
    (d) $((\sim(0^*10^*1^*)(0^*101^*))^*$
    (e) $\overleftarrow{0^*10^*1^*} \cup (0^*101^*)^*$
    (f) $\overleftarrow{0^*10^*1^*} \cap \overleftarrow{0^*101^*}$
    (g) $(0^*10^*1^*)^* \triangle (0^*101^*)^*$

2. Let R_1 and R_2 be regular expressions and their realizations be as shown in Fig. P8.3.2. Find:

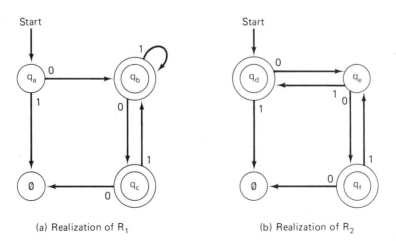

(a) Realization of R_1 (b) Realization of R_2

Fig. P8.3.2

(a) A sequential machine which realizes $R_1 \cup R_2$.
(b) A sequential machine which realizes $R_1 \cap R_2$.
(c) Sequential machines which realize R_1^* and R_2^*.
(d) A sequential machine which realizes $R_1 R_2$.
(e) Sequential machines which realize $\overleftarrow{R_1 \cup R_2}$, $\overleftarrow{R_1^*}$, $\overleftarrow{R_2^*}$, and $\overleftarrow{R_1 R_2}$.

Bibliographical Remarks

   Tape equivalence and the subset construction of nondeterministic sequential machines are discussed in Rabin and Scott [9]. The state minimization of nondeterministic sequential machines is described by Kameda and Weiner [4, 5]. The regular set and the regular expression are from Kleene [6]. Many results concerning regular expressions can be found in Brzozowski [2, 3]. The construction of a regular expression from a transition diagram was first introduced by Arden [1], and the construction of a transition diagram from a regular expression was originally published by McNaughton and Yamada [7].

References

1. ARDEN, D. N., "Delay Logic and Finite State Machines," *Proc. Second Ann. Symp. Switching Theory and Logical Design*, 1961, pp. 133–151.

2. BRZOZOWSKI, J. A., "A Survey of Regular Expressions and Their Applications," *PGEC*, Vol. 11, No. 3, 1962, pp. 324–335.

3. BRZOZOWSKI, J. A., "Derivation of Regular Expression," *J. ACM*, Vol. 5, 1964, pp. 481–494.

4. KAMEDA, T., "On the Reduction of Nondeterministic Automata," Ph.D. dissertation, Department of Electrical Engineering, Princeton University, Princeton, N.J. 1968.

5. KAMEDA, T., and P. WEINER, "On the State Minimization of Nondeterministic Finite Automata," *IEEE Trans. Computers*, Vol. C-19, No. 7, 1970.

6. KLEENE, S. C., "Realization of Events in Nerve Nets and Finite Automata," in *Automata Studies*, C. E. Shannon and J. McCarthy, eds., Princeton University Press, Princeton, N.J., 1956, pp. 3–42.

7. McNAUGHTON, R., and H. YAMADA, "Regular Expressions and State Graphs for Automata," in *Sequential Machines*, E. F. Moore, ed., Addison-Wesley, Reading, Mass., 1964, pp. 157–174.

8. PAULL, M. C., and S. H. UNGHER, "Minimizing the Number of States in Incompletely Specified Sequential Switching Functions," *IRE Trans. Electronic Computers*, Vol. EC-8, 1959, pp. 356–367.

9. RABIN, M. O., and D. SCOTT, "Finite Automata and Their Decision Problems," *IBM J. Res. Develop.*, Vol. 3, 1959, pp. 114–125.

9

Sequential-Machine
Realization

From the constructional point of view, sequential circuits can be divided into two classes: *clocked sequential circuits* and *unclocked sequential circuits*. In a clocked sequential circuit, there is a (synchronizing) clock connected to the clock inputs of *all* the flip-flops (memory elements) of the circuit; hence, the operation of the entire circuit is controlled and synchronized by the periodic pulses of the clock, whereas in an unclocked sequential circuit, such a clock is not present. In the class of unclocked circuits, there are *pulse-mode* and *fundamental-mode* (pulseless) *sequential* circuits. From the behavioral point of view, clocked sequential circuits and pulse-mode sequential circuits are *synchronous sequential circuits* in which every state is a stable state and no transient states may exist. Fundamental-mode sequential circuits are *asynchronous sequential circuits*, which have transient states (unstable states) in addition to the usual stable states.

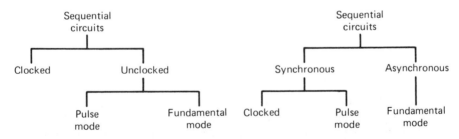

The objective of this chapter is twofold. One is to show that any sequential machine can be physically realized by a clocked sequential circuit—and how to obtain it. Various types of commonly used flip-flops are introduced and their characteristic functions are studied. Based on the application tables of the flip-flops, which are derived from their characteristic functions, a general procedure for synthesizing sequential machines by a clocked flip-flop circuit is presented and illustrated by an example. This procedure is applicable to the realization of any sequential machine, decomposed machine, and regular expression discussed in the previous two chapters.

The second objective is to present the realizations of sequential machines using the pulse-mode and fundamental-mode circuits. The analysis and design of pulse-mode sequential circuits are similar to those of clocked sequential circuits. But the analysis and design of fundamental-mode sequential circuits are different from those of clocked sequential circuits. The two undesirable transient phenomena of fundamental circuits —races and hazards—and their elimination are discussed.

9.1 Delay Flip-Flop Sequential Circuit Realization of Binary Machines

Before giving a formal procedure for realizing sequential machines using flip-flop sequential circuits, we present the realization of binary machines using the delay flip-flop (DFF). The binary machine is one whose input symbols, states, and output symbols are binary numbers; therefore, no state assignment is needed for this class of machines. The binary counter and the shift registers are examples of binary machines. The DFF is the simplest type among all the flip-flops used. It has one input (D input), two outputs (one is the complement of the other, denoted by y and y'), and a clock input [see Fig. 9.1.1(a)]. The clock input ensures synchronization between the input and the outputs; a transition occurs only at the presence of a clock pulse.

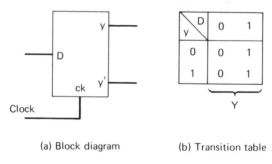

(a) Block diagram   (b) Transition table

Fig. 9.1.1   DFF.

The input at the time that the clock pulse occurs completely determines the output until the next clock pulse. Thus, if $D = 1$, the next state of the flip-flop will become $y = 1$ when the clock pulse appears, regardless of the value of y before the clock pulse. The transition table for this flip-flop is given in Fig. 9.1.1(b). Examining this table, we see that the following logical expression can be used to describe the input/output relationship of this flip-flop,

$$Y(k + 1) = D(k) \qquad\qquad (9.1.1)$$

where $Y(k + 1)$ denotes the next state of the flip-flop. In words, the next state of the DFF is the present input to the flip-flop.

The realization of a binary machine using DFF sequential circuit is best illustrated by examples.

Example 9.1.1   Realization of Binary Counters Using DFF Circuits

Counters are important building blocks in digital systems, and they can take on a variety of forms, depending on the applications for which they are used. A counter will count through a sequence of numbers, then reset itself to an initial value, and then repeat the counting process. The simplest types of counters are binary counters. These counters start at 0 and count to $2^N - 1$, where N is a positive integer, and reset to 0. For example, if $N = 3$, a counter will count through the successive binary numbers 0 through 7, and therefore has eight states. The output of this machine is the same as its state. The transition table of this counter is given in Fig. 9.1.2(a). The three next-state functions, Y_1, Y_2, and Y_3, can be represented on the Karnaugh maps in Fig. 9.1.2(b), from which minimized next-state functions are obtained. Three DFF's are needed to realize the three state variables, y_1, y_2, and y_3. From the input/output relationship of DFF described in Eq. (9.1.1), we find that the minimized flip-flop input logic functions D_1, D_2, and D_3 are

$$D_1 = y_1 y_2' + y_1 y_3' + y_1' y_2 y_3$$
$$D_2 = y_2' y_3 + y_2 y_3' \qquad\qquad (9.1.2)$$
$$D_3 = y_3'$$

The DFF realization of the modulo-8 binary counter is shown in Fig. 9.1.2(c).

Example 9.1.2   Realization of Shift Registers Using DFF Circuits

One of the major uses of flip-flops is to form registers that are used to store information during some portion of the information-processing task. One of the simplest operations that can be performed by a register is to shift information serially from one end of the register to the other. A 3-bit shift-right serial-input/serial-out shift register can be described by a transition table as shown in Fig. 9.1.3(a). This machine has one binary input x and, again, eight states, 000, 001, . . . , 111. The output of this machine is the same as its state. Again, three DFF's are needed to realize this machine. Referring to Fig. 9.1.3(b) and Eq. (9.1.1), we obtain the minimized flip-flop input logic functions, D_1, D_2, and D_3, which are

$$D_1 = x$$
$$D_2 = y_1 \qquad\qquad (9.1.3)$$
$$D_3 = y_2$$

and the circuit realization is shown in Fig. 9.1.3(b).

Exercise 9.1

1. Design a modulo-8 binary counter using a DFF sequential circuit. The input of this counter is not the clock input, but a binary input that can take on values 0 and 1. The transition table of this counter is shown in Table P9.1.1.

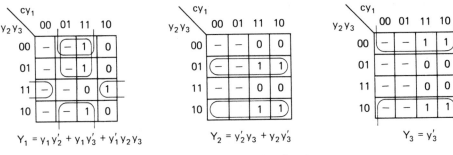

c(clock input)			1		
y_1	y_2	y_3			
0	0	0	0	0	1
0	0	1	0	1	0
0	1	0	0	1	1
0	1	1	1	0	0
1	0	0	1	0	1
1	0	1	1	1	0
1	1	0	1	1	1
1	1	1	0	0	0

$$Y_1 Y_2 Y_3$$

(a) Transition table of modulo-8 binary counter

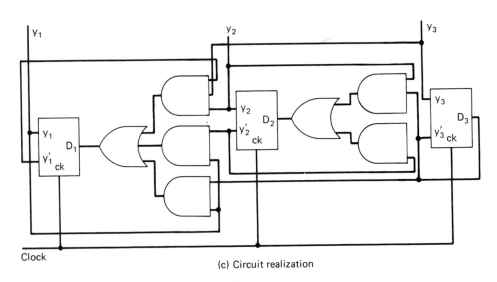

$$Y_1 = y_1 y_2' + y_1 y_3' + y_1' y_2 y_3$$

$$Y_2 = y_2' y_3 + y_2 y_3'$$

$$Y_3 = y_3'$$

(b) Minimized next-state functions

(c) Circuit realization

Fig. 9.1.2 DFF realization of modulo-8 binary counter.

342

x $y_1 y_2 y_3$	0	1
0 0 0	0 0 0	1 0 0
0 0 1	0 0 0	1 0 0
0 1 0	0 0 1	1 0 1
0 1 1	0 0 1	1 0 1
1 0 0	0 1 0	1 1 0
1 0 1	0 1 0	1 1 0
1 1 0	0 1 1	1 1 1
1 1 1	0 1 1	1 1 1

$$Y_1 Y_2 Y_3$$

(a) Transition table of 3-bit shift-right shift register

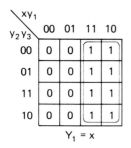
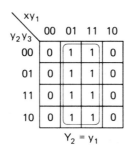
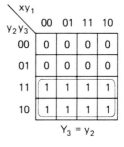

(b) Minimized next-state functions

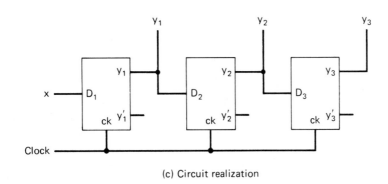

(c) Circuit realization

Fig. 9.1.3 DFF realization of 3-bit shift-right shift register.

TABLE P9.1.1

$y_1y_2y_3$ $\diagdown$ x	0	1
000	000	001
001	001	010
010	010	011
011	011	100
100	100	101
101	101	110
110	110	111
111	111	000

$$Y_1Y_2Y_3$$

2. Realize the transition table of the serial binary adder using a sequential circuit using one DFF. It is assumed that the no-carry and carry states are represented by $y = 0$ and 1, respectively, where y is the state variable or the output of the DFF to be used in this realization. The transition table of the serial binary adder was given in Table 7.1.2(a).

9.2   Analysis of Clocked Flip-Flop Circuits

In the previous section the DFF and the realization of binary machines using DFF's were discussed. Since the input/output relationship of the DFF [see Eq. (9.1.1)] is very simple, the synthesis of sequential machines using DFF's was straight-forward. In this section three other types of flip-flops, set–reset flip-flops, J-K flip-flops, and toggle flip-flops, are presented. Their input/output relationships are not as simple as that of the DFF, and neither is the synthesis procedure using those flip-flops, which will be presented in the next section.

A.   Clocked S-R Flip-Flops

The clocked S-R flip-flop (SRFF) is a two-state machine which has two inputs, set (S) input and reset (R) input, and two outputs which are complementary to each other, denoted by y and y'. The block diagram of an SRFF is shown in the second row and second column of Table 9.2.1. The clock input is used for synchronization between the inputs and outputs of the flip-flop as well as between it and other flip-flops in a circuit. The clock pulse must last for a sufficient time to allow the flip-flop to assume its new state.

The operation of the SRFF may be described as follows:

1. If $S = R = 0$, the flip-flop will remain in its present state.
2. If $S = 1$ and $R = 0$, the flip-flop is forced into the SET state ($y = 1$).

TABLE 9.2.1 Various Types of Flip-Flops and Their Characteristic Functions

Type of flip-flop	Block diagram	Transition table	Characteristic function
Delay flip-flop (DFF)	D, y, y', ck	D\y: 0 1 / 0: 0 1 / 1: 0 1 (Y)	$Y = D$
Set-reset flip-flop (SRFF)	S, R, y, y', ck	SR\y: 00 01 11 10 / 0: 0 0 — 1 / 1: 1 0 — 1 (Y)	$Y = S + R'y$
J-K flip-flop (JKFF)	J, K, y, y', ck	JK\y: 00 01 11 10 / 0: 0 0 1 1 / 1: 1 0 0 1 (Y)	$Y = Jy' + K'y$
Toggle flip-flop (TFF)	T, y, y', ck	T\y: 0 1 / 0: 0 1 / 1: 1 0 (Y)	$Y = T'y + Ty'$

345

3. If $S = 0$ and $R = 1$, the flip-flop is forced into the RESET state ($y = 0$).
4. If $S = R = 1$, then the next state of the flip-flop is indeterminant.

The transition table of the SRFF is shown in the second row and third column of Table 9.2.1. Note that the indeterminant next-states are denoted by dashes (don't cares). Also notice that the order of the inputs is so arranged that this transition table can be viewed as the Karnaugh map of the next-state function Y. This Karnaugh map can be used in obtaining the minimized next-state function, which is shown in the second row and fourth column of the table. Since the function

$$Y = S + R'y \qquad (9.2.1)$$

contains every piece of information that is contained in the transition table and the transition table describes completely the operation of the flip-flop, this function is called the *characteristic function* of the set–reset flip-flop.

B. Clocked J-K Flip-Flops

The J-K flip-flop (JKFF) is very similar to the SRFF. As a matter of fact, the operation of the JKFF is exactly the same as that of the SRFF except when both inputs (J and K) of the JKFF are 1; then its next-state is defined and is equal to the complement of its present state, as shown in the JKFF transition table in Table 9.2.1. From the transition table, the characteristic function of the JKFF can be obtained, which is shown in the third row and last column of Table 9.2.1. The ability of the J-K flip-flop to give the toggling response when both J and K are enabled will often lead to a simplification of its control logic relative to that required by the S-R flip-flop.

C. Clocked Toggle Flip-Flops

The toggle flip-flop (TFF) can be constructed by combining the J and K inputs of a JKFF into one input, designated by T. The operation of the TFF may be described as follows.

1. If $T = 0$, the flip-flop will remain in its present state.
2. If $T = 1$, the next-state of the flip-flop will be the complement of its present state.

The block diagram, transition table, and characteristic function of the TFF are shown in the fourth row of Table 9.2.1.

Two important remarks about the flip-flop are in order.

1. A clocked flip-flop is the basic building block of the modern synchronous sequential circuit. The clock, which produces a train of equally spaced pulses, is fed into the circuit in such a way that the arrival of the appropriate synchronization pulse is assured. This process ensures an orderly execution of the various operations and

logical decisions to be made by the circuit. To avoid the occurrence of situations where the next-states and outputs of a clocked sequential circuit are nondeterministic, the values of inputs must be kept constant during a pulse interval. The width of the clock pulse should satisfy the following requirements.

Let T_W denote the width of the clock pulse of a clocked sequential circuit, T_p the minimum time required for propagating input signals from the input terminals to all the flip-flops of the circuit and to change the states of the flip-flops, and T_0 the minimum time required for propagating input signals and the present values of the flip-flops to all the output terminals of the circuit. Then

$$2T_p > T_W \geq \max [T_p, T_0]$$

The reason for requiring $2T_p > T_W$ is to prevent the flip-flops from changing their values twice. In constructing a clocked flip-flop circuit, we require that

$$2T_p > T_0$$

This can always be achieved by introducing necessary delays in the input logic circuits of flip-flops and/or their feedback loops.

2. Any flip-flop can be constructed from a flip-flop of different type. An example follows.

Example 9.2.1

Construct (a) an SRFF, (b) a JKFF, and (c) a TFF from a DFF.

Referring to the last column of Table 9.2.1, we obtain the following DFF input logic functions:

For the realization of SRFF,

$$D = S + R'y$$

For the realization of JKFF,

$$D = Jy' + K'y$$

For the realization of TFF,

$$D = T'y + Ty'$$

The realizations of an SRFF, JKFF, and TFF using a DFF are shown in Table 9.2.2.

The study of synchronous or clocked digital circuits containing flip-flops can be divided into two parts: analysis and synthesis. The problem of analyzing flip-flop circuits is: Given a flip-flop circuit, find the transition table of the circuit. A general flip-flop circuit analysis procedure is presented below and is illustrated by the following example.

TABLE 9.2.2 Realization of an SRFF, JKFF, and TFF Using a DFF

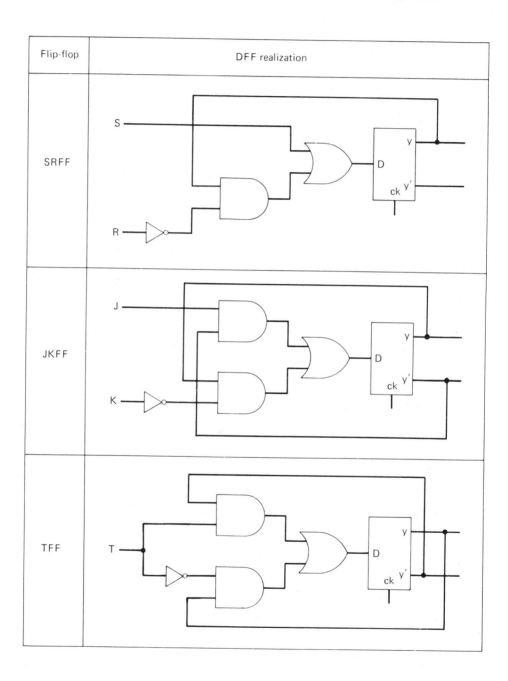

Flip-flop	DFF realization
SRFF	
JKFF	
TFF	

Example 9.2.2

Consider the SRFF circuit of Fig. 9.2.1. The procedure for obtaining the state-transition function and the output function of this circuit is given as follows.

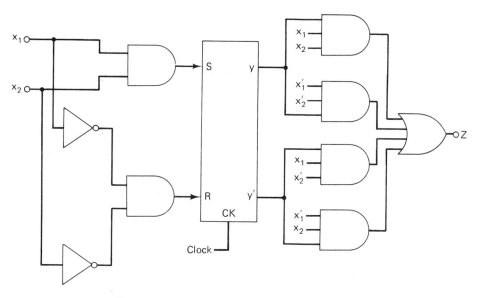

Fig. 9.2.1 A sequential circuit containing an SRFF.

Step 1 Find the excitation functions in terms of the circuit inputs and the flip-flop outputs from the given circuit diagram, namely $S = f_S(x_1, \ldots, x_n; y_1, \ldots, y_m)$ and $R = f_R(x_1, \ldots, x_n; y_1, \ldots, y_m)$ for every flip-flop. In this example,

$$S = x_1 x_2$$
$$R = x_1' x_2'$$

Step 2 Obtain the excitation table. Display the functions S and R as shown in Table 9.2.3, which is known as the *excitation table*.

TABLE 9.2.3   Excitation Table of the Circuit of Fig. 9.2.1

y \ $x_1 x_2$	00	01	11	10
0	01	00	10	00
1	01	00	10	00

$$SR$$

Step 3   Obtain the state transition table from the excitation table using the characteristic functions of the flip-flops in the circuit. The state transition table of this example is shown in Table 9.2.4. The entities of this table are obtained by using the characteristic function of Eq. (9.2.1). The first-row and third-column entry, for example, is obtained by substituting the values of S, R, and y at that position into the characteristic function $Y = S + R'y = 1 + 0'0 = 1$. The rest are obtained similarly.

**TABLE 9.2.4   State Transition Table of the
Circuit of Fig. 9.2.1**

y \ $x_1 x_2$	00	01	11	10
0	0	0	1	0
1	0	1	1	1

Step 4   Obtain the transition table. The output of the circuit is determined by substituting the values of the circuit inputs and the next-state into the output functions. For our example, the output function is

$$z = x_1 x_2 y + x_1' x_2' y + x_1 x_2' y' + x_1' x_2 y'$$

If we name the states $y = 0$ and 1 by q_0 and q_1, respectively, then the transition table of the circuit of Fig. 9.2.1 is the one shown in Table 9.2.5, which shows that this circuit is a serial binary adder.

**TABLE 9.2.5   Transition Table of the
Serial Binary Adder**

y \ $x_1 x_2$	00	01	11	10
q_0	q_0, 0	q_0, 1	q_1, 0	q_0, 1
q_1	q_0, 1	q_1, 0	q_1, 1	q_1, 0

Exercise 9.2

1.   Find the transition tables of the two sequential circuits in Fig. P9.2.1 and show that they act as J-K flip-flops.

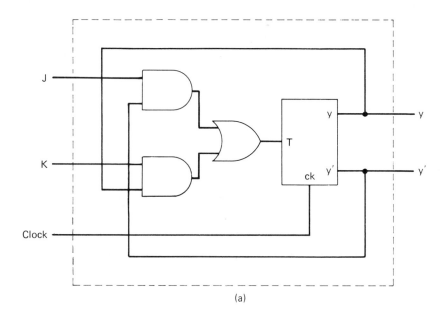

(a)

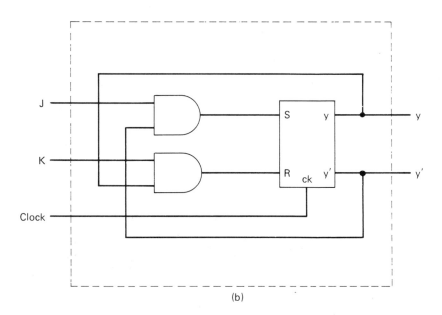

(b)

Fig. P9.2.1

2. (a) Find the transition table of the circuit in Fig. P9.2.2(a).
   (b) Plot the output z of the circuit in response to the input x shown in Fig. P9.2.2(b).
   (c) What does this circuit do?

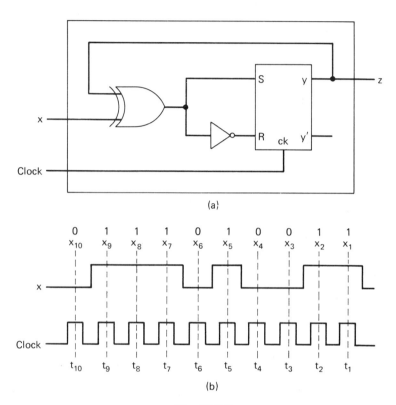

(a)

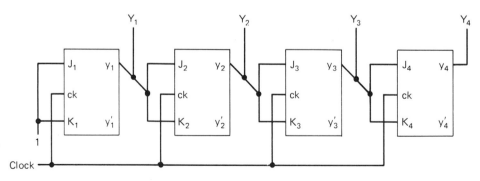

(b)

Fig. P9.2.2

3. (a) Find the transition table of the circuit in Fig. P9.2.3.
   (b) What does this circuit do?

Fig. P9.2.3

9.3 General Synthesis Procedure

Having introduced the analysis of clocked flip-flop circuits, we now turn to the synthesis problem; that is, given a machine specification (transition table or diagram), find a clocked flip-flop circuit which realizes it. The block diagram of a general clocked flip-flop circuit that can realize any synchronous sequential machine is depicted in Fig. 9.3.1.

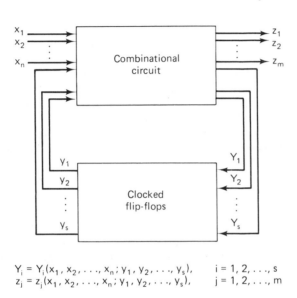

$$Y_i = Y_i(x_1, x_2, \ldots, x_n; y_1, y_2, \ldots, y_s), \qquad i = 1, 2, \ldots, s$$
$$z_j = z_j(x_1, x_2, \ldots, x_n; y_1, y_2, \ldots, y_s), \qquad j = 1, 2, \ldots, m$$

Fig. 9.3.1   Block diagram of a general clocked flip-flop circuit that can realize any sequential machine.

The approach to be followed is to reverse the steps which were used in the analysis procedure described above. From Example 9.2.2, it is seen that the crucial step in analyzing a flip-flop circuit is that of going from an excitation table to the corresponding state-transition table. This is achieved by use of the characteristic function $Y = S + R'y$. In the flip-flop circuit realization of sequential machines, this step in the reverse direction is still the crucial one. One method is to make a truth table from the characteristic function considering S and R as dependent variables and y and Y as independent variables. This table, known as the flip-flop *application table*, is shown in Table 9.3.1(b). The procedure for synthesizing sequential machines using flip-flop circuits is illustrated by the following example.

Example 9.3.1

Suppose that it is desired to realize the binary adder described in Table 9.2.5 using an *S-R* flip-flop circuit.

TABLE 9.3.1   Flip-Flop Application Table

y	Y	D		y	Y	S	R		y	Y	J	K		y	Y	T
0	0	0		0	0	0	–		0	0	0	–		0	0	0
0	1	1		0	1	1	0		0	1	1	–		0	1	1
1	0	0		1	0	0	1		1	0	–	1		1	0	1
1	1	1		1	1	–	0		1	1	–	0		1	1	0

(a) DFF   (b) SRFF   (c) JKFF   (d) TFF

Step 1   Find the state transition table in coded form. For this example, since there are only two states, one variable is needed to code them. If we code q_0 and q_1 by 0 and 1, respectively, and denote this variable by y, and denote the two input variables by x_1 and x_2, then the transition table of this machine in coded form is shown in Table 9.3.2(a).

Step 2   Obtain the excitation table. Using the flip-flop application table of Table 9.3.1, the excitation table of this example is shown in Table 9.3.2(b).

Step 3   Obtain the minimized excitation functions. The excitation functions S and R presented by Karnaugh maps are shown in Tables 9.3.2(c) and (d). After minimization, they are $S = x_1 x_2$ and $R = x_1' x_2'$.

Step 4   Realize the minimized excitation functions obtained in step 3 and the output function after minimization. The output function of the machine after minimization is

TABLE 9.3.2   (a) The State Transition Table in Coded Form; (b) The Excitation Table; (c) The Karnaugh Map of the Excitation Function S; (d) The Karnaugh Map of the Excitation Function R of Example 9.3.1.

y \ $x_1 x_2$	00	01	11	10		00	01	11	10
0	0	0	1	0		0	1	0	1
1	0	1	1	1		1	0	1	0

| | Y | | z |

(a)

y \ $x_1 x_2$	00	01	11	10
0	0–	0–	10	0–
1	01	–0	–0	–0

SR

(b)

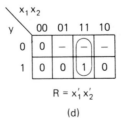

$R = x_1' x_2'$

(d)

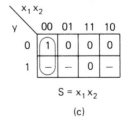

$S = x_1 x_2$

(c)

$z = x_1x_2y + x_1x_2'y' + x_1'x_2y + x_1'x_2y'$. The complete flip-flop realization of the serial binary adder was given in Fig. 9.2.1.

Example 9.3.2

As another design example, consider designing a synchronous J-K flip-flop circuit having an input x and an output z. Whenever the input sequence consists of two consecutive 0's followed by two consecutive 1's or two consecutive 1's followed by two consecutive 0's, the output will be 1. Otherwise, the output will be 0.

To construct the transition table of this circuit, eight states are needed. Let the eight states correspond to the most current input sequences which will potentially satisfy the requirements for $z = 1$. Let $q_0, q_1, q_{00}, q_{11}, q_{001}, q_{110}, q_{0011}$, and q_{1100} denote a current input sequence of one 0, one 1, two consecutive 0's, two consecutive 1's, two consecutive 0's followed by a 1, two consecutive 1's followed by a 0, two consecutive 0's followed by two consecutive 1's, and two consecutive 1's followed by two consecutive 0's, respectively. The transition table of this circuit is given in Fig. 9.3.2(a). Applying the sequential machine minimization technique described in Section 7.3 to this machine, it is easy to show that

$$\text{states } q_{11} \text{ and } q_{0011} \text{ are equivalent}$$
$$\text{states } q_{00} \text{ and } q_{1100} \text{ are equivalent}$$

If we delete states q_{0011} and q_{1100}, the resulting minimized transition table is given in Fig. 9.3.2(b).

The minimized machine has an S.P. partition,

$$\{\overline{q_0q_{00}q_{110}, q_1q_{11}q_{001}}\}$$

Hence, a good state assignment for this machine is

y_2	y_3	$y_1 = 0$	$y_1 = 1$
0	0	q_0	q_1
0	1	q_{00}	q_{11}
1	0	q_{110}	q_{001}

The coded transition table using this assignment is given in Fig. 9.3.2(c). Applying the application table of the J-K flip-flop to this table, we obtain the excitation table, which is shown in Fig. 9.3.2(d). The Karnaugh map representations of $J_1, K_1, J_2, K_2, J_3, K_3$, and z are depicted in Fig. 9.3.2(e). From these maps, we obtain the following minimized functions:

$$J_1 = x$$
$$K_1 = x'$$
$$J_2 = x'y_1y_3 + xy_1'y_3$$
$$K_2 = 1$$
$$J_3 = x'y_1' + xy_1$$
$$K_3 = x'y_1 + xy_1'$$

and

$$z = x'y_1'y_2 + xy_1y_2$$

The complete design of the circuit is shown in Fig. 9.3.2(f).

As a final remark, the procedure presented in this section is completely general and can be used to realize any sequential machine, decomposed sequential machine, or regular expression discussed in Chapters 7 and 8.

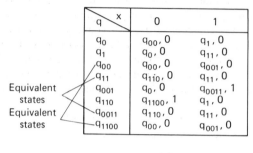

q \ x	0	1
q_0	q_{00}, 0	q_1, 0
q_1	q_0, 0	q_{11}, 0
q_{00}	q_{00}, 0	q_{001}, 0
q_{11}	q_{110}, 0	q_{11}, 0
q_{001}	q_0, 0	q_{0011}, 1
q_{110}	q_{1100}, 1	q_1, 0
q_{0011}	q_{110}, 0	q_{11}, 0
q_{1100}	q_{00}, 0	q_{001}, 0

Equivalent states (bracket to q_{11} and q_{0011})
Equivalent states (bracket to q_{001} and q_{1100})

(a)

q \ x	0	1
q_0	q_{00}, 0	q_1, 0
q_1	q_0, 0	q_{11}, 0
q_{00}	q_{00}, 0	q_{001}, 0
q_{11}	q_{110}, 0	q_{11}, 0
q_{001}	q_0, 0	q_{11}, 1
q_{110}	q_{00}, 1	q_1, 0

(b)

$y_1 y_2 y_3$		x: 0	1
0 0 0	(q_0)	001, 0	100, 0
0 0 1	(q_{00})	001, 0	110, 0
0 1 0	(q_{110})	001, 1	100, 0
1 0 0	(q_1)	000, 0	101, 0
1 0 1	(q_{11})	010, 0	101, 0
1 1 0	(q_{001})	000, 0	101, 1

(c)

Fig. 9.3.2   (a) Transition table of the machine of Example 9.3.1 ; (b) minimized transition table; (c) coded transition table; (d) excitation table ; (e) Karnaugh maps of the excitation functions and the output function and their minimization; (f) J-K flip-flop circuit realization.

$y_1 y_2 y_3$ \ x	0	1
0 0 0	0–, 0–, 1–; 0	1–, 0–, 0–; 0
0 0 1	0–, 0–, –0; 0	1–, 1–, –1; 0
0 1 0	0–, –1, 1–; 1	1–, –1, 0–; 0
1 0 0	–1, 0–, 0–; 0	–0, 0–, 1–; 0
1 0 1	–1, 1–, –1; 0	–0, 0–, –0; 0
1 1 0	–1, –1, 0–; 0	–0, –0, 0–; 1

$$\underbrace{}$$
$$J_1 K_1, J_2 K_2, J_3 K_3; z$$

(d)

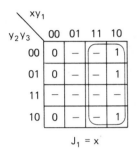

$J_1 = x$

$J_2 = x'y_1 y_3 + xy_1' y_3$

$J_3 = x'y_1' + xy_1$

$K_1 = x'$

$K_2 = 1$

$K_3 = x'y_1 + xy_1'$

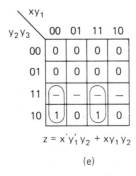

$z = x'y_1' y_2 + xy_1 y_2$

(e)

Fig. 9.3.2 (Continued)

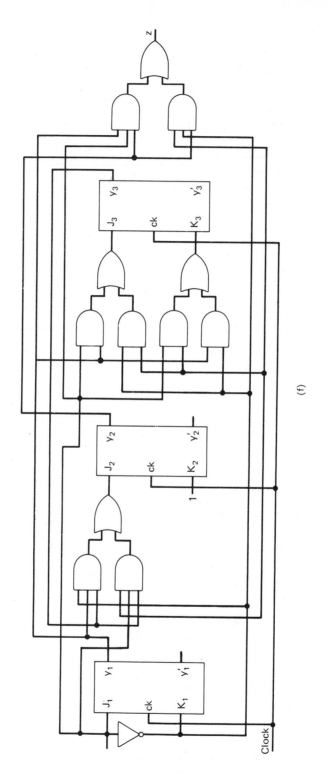

Fig. 9.3.2 (Continued)

(f)

Exercise 9.3

1. Find a JKFF sequential circuit realization of each of the four machines M_a, M_b, M_c, and M_d of problem 1 of Exercise 7.3 after all the redundant states are removed. In each realization, a reduced dependency state assignment is preferred.

2. Repeat problem 1 for machines M_f, M_g, M_h, and M_i of problem 3 of Exercise 7.3.

3. Realize machine M_6 of Fig. 7.5.2 and its decomposed version in Table 7.5.1 using SRFF's.

4. Realize machine M_7 of Fig. 7.6.2(a) and its decomposed version in Fig. 7.6.3 using TFF's.

5. A sequential machine is to receive a sequence of 1's and 0's and to produce an output of 1, only if the total number of 1's received from the starting time is odd.
   (a) Find a transition diagram for this sequential machine.
   (b) Realize it by an S-R flip-flop circuit.

6. (a) Minimize the machine whose transition table is described in Table P9.3.6.
   (b) Realize the minimal form of this machine using AND-gates, OR-gates, NOT-gates, and DFF's.
   (c) Realize the minimal form of this machine using AND-gates, OR-gates, NOT-gates, and J-K flip-flops.

TABLE P9.3.6

q \ x	a	b	c	a	b	c
1	4	1	1	1	1	0
2	1	4	2	1	1	1
3	4	4	3	1	0	0
4	2	3	4	0	1	0
		Next-state			Present output	

7. Realize the machine described in Table P9.3.7 by using JKFF's and two different state

TABLE P9.3.7

q \ x	0	1	0	1
1	4	3	1	0
2	6	3	0	1
3	5	2	0	1
4	2	5	1	1
5	1	4	1	0
6	3	4	1	1
	Next-state		Present output	

assignments. One of them is a reduced dependency state assignment and the other is arbitrary. Compare the two realizations in terms of the number of gates used in the realizations.

8. (a) The characteristic function of a flip-flop is described by

$$Y = x_1 \oplus x_2 \oplus y$$

Realize this flip-flop using (1) SRFF, (2) TFF, (3) DFF, and (4) JKFF.
   (b) Realize the transition table of Table P9.3.6 using this flip-flop.

9. Realize the following regular expressions using (1) DFF and (2) JKFF.
   (a) $(11)*(1 \cup 01*)$
   (b) $(1*0 \cup 001)*01$
   (c) $(0*001 \cup 11*010*)*$
   (d) $01*(10*01*)* \cap (0*001 \cup 11*010*)*$

9.4 Sequential-Machine Realization Using Pulse-Mode Sequential Circuits

A pulse-mode sequential circuit is an unclocked, yet synchronous, sequential circuit. It is defined as follows.

DEFINITION 9.4.1

An unclocked sequential circuit is operating in *pulse mode* if the following conditions are satisfied:
   (a)   The inputs are pulses whose width must be sufficiently wide to trigger the flip-flops used in the circuit.
   (b)   The pulse may occur on *only one* input line at a time.
   (c)   Changes in internal state occur only in response to the occurrence of a pulse at one of the pulse inputs, and each pulse input causes *only one* change in internal state.

Each occurrence of a pulse on any input will constitute an input to the circuit and will trigger a transition from one state to another. Since simultaneous pulses are forbidden, *the number of distinct inputs, and thus the number of columns in the next-state table, is equal to the number of input lines.* The output of a pulse-mode circuit may either be pulses or levels. If the outputs are to be functions of the inputs as well as the state variables [Mealy machine, see Definition 7.1.1(a)], *they must be pulses,* obtained by gating the state variables with the input pulses. In this case, a distinct output may be defined for every possible combination of states and inputs. If, on the other hand, the outputs are functions of state variables only [Moore machine, see Definition 7.1.1(b)], *they will be levels* whose value will be defined in the intervals *between* input pulses rather than at the time of the input pulses. In this case, the number of distinct outputs is, of course, no greater than the number of distinct states.

Example 9.4.1

Design a pulse-mode vending machine which sells three kinds of merchandise: gums for 15 cents, cookies for 25 cents, and rolls for 35 cents. The machine accepts nickels, dimes, and quarters.

Assume that the select pushbutton switches of this machine are mechanical switches. Let PB_{g_i}, PB_{c_j}, and PB_{r_k} denote the pushbuttons for gum i, cookies j, and roll k, respectively. Since the machine is to be operating in pulse mode, in addition to PB_{g_i}, PB_{c_j}, and PB_{r_k}, we need three pulse input lines, one for nickels, one for dimes, and one for quarters; they are denoted by x_N, x_D, and x_Q, respectively. The following states are needed for designing this machine: $q_{0¢}, q_{5¢}, q_{10¢}, q_{15¢}, q_{20¢}, q_{25¢}$, and $q_{30¢}$ denoting that 0, 5, 10, 15, 20, 25, and 30 cents has been deposited, respectively. If the Mealy type (pulse outputs) is used, three output lines, denoted by z_1, z_2, and z_3, are required. When a sufficient amount of money is deposited and the select button PB of a desired merchandise is pushed, an output pulse z will be generated to activate the appropriate relay to release the merchandise. Assume that the z_{g_i} pulse is to release a pack of gums i, the z_{c_j} pulse is to release a pack of cookies j, and the z_{r_k} pulse is to release a roll k. The transition table for this pulse-mode circuit is shown in Fig. 9.4.1. Note that this machine has two outputs. One is the output pulse required to release the merchandise and the other is the change. A 0 output pulse indicates that no output pulse is generated.

From the discussions above, it is seen that the analysis and design of pulse-mode circuits are the same as those for clocked circuits given in Sections 9.2 and 9.3, except that each input symbol is input by an individual input line. If a pulse-mode circuit has n input lines, the number of possible inputs is equal to n, not 2^n.

An example of designing pulse-mode circuits is given below.

Example 9.4.2

A pulse-mode circuit is to be designed having two pulse inputs x_1 and x_2 and one level output z. The inputs are restricted so that x_1 and x_2 are never simultaneously equal to 1. Whenever an x_1 pulse is received, the output is to become equal to 1, provided that there has been exactly one x_2 pulse after the last previous x_1 pulse. Otherwise, the output is to remain equal to 0. Once the output becomes equal to 1, it is to remain equal to 1 until the next x_2 pulse. Whenever an x_2 pulse is received, the output is to become equal to 0. Assume that the J-K flip-flops are to be used in this design.

To design this circuit, a minimum of four internal states is needed. They are defined as follows:

q_0:   no x_1 pulse is received
q_1:   one or more x_1 pulses are received
q_2:   one x_2 pulse, preceded by one (or more) x_1 pulse, is received
q_3:   one (or more) x_1 pulse, preceded by exactly one x_2 pulse and then one (or more) x_1 pulse

The transition table of this circuit is shown in Fig. 9.4.2(a).

The circuit realization of the machine of Fig. 9.4.2(a) may be carried out by the following steps:

Step 1   Find the binary-coded state transition table. Since the machine in Fig. 9.4.2(a) is minimal and it has four states, two J-K flip-flops will be sufficient to code them. Suppose the state assignment

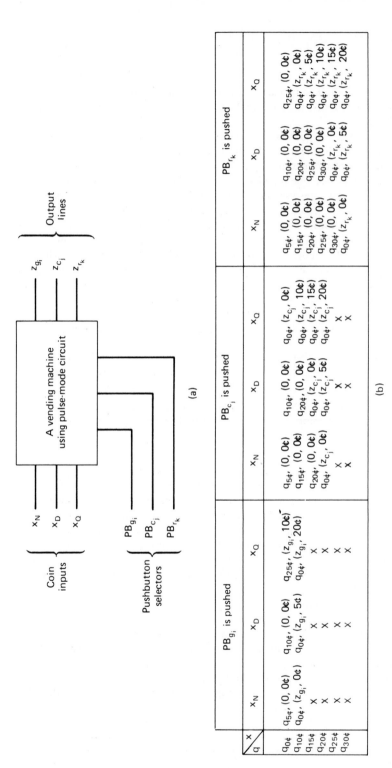

Fig. 9.4.1 Example 9.4.1.

Input State	x_1	x_2	z
q_0	q_1	q_0	0
q_1	q_1	q_2	0
q_2	q_3	q_0	0
q_3	q_3	q_2	1

(a) Transition table

x_2x_1 / y_2y_1	00	01	10	11
0 0	00	01	00	—
0 1	01	01	10	—
1 0	10	11	00	—
1 1	11	11	10	—

States do not change   Can never occur

(b) Coded transition table

x_2x_1 / y_2y_1	00	01	10	11
00	0–, 0–	0–, 1–	0–, 0–	– –
01	0–, 0–	0–, –0	1–, –1	– –
10	–0, 0–	–0, 1–	–1, 0–	– –
11	–0, –0	–0, –0	–0, –1	– –

$J_2 K_2, J_1 K_1$

(c) Excitation table

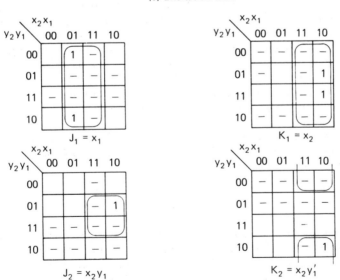

(d) Minimization of the excitation functions

Fig. 9.4.2 Example 9.4.2.

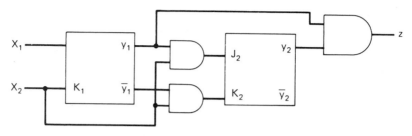

(e) Pulse-mode circuit realization of the machine of (a)

Fig. 9.4.2   (Continued)

q	$y_2 y_1$
q_0	00
q_1	01
q_2	10
q_3	11

is chosen. The binary-coded state transition table of this machine is shown in Fig. 9.4.2(b). Notice that in this table, under the input 00 column, the next states are the same as the present state, since no input is applied to the circuit. Under the input 11 column, the dashes indicate that no transition is possible, since x_1 and x_2 are never simultaneously equal to 1.

Step 2   Obtain the excitation table. Applying the J-K flip-flop application table to the binary-coded state-transition table of Fig. 9.4.2(b), we obtain the excitation table shown in Fig. 9.4.2(c).

Step 3   Obtain the minimized excitation functions. Karnaugh maps of the functions J_1, K_1, J_2, and K_2 are depicted in Fig. 9.4.2(d), and the minimized excitation functions are shown in the same figure.

Step 4   Circuit construction. From the minimized excitation functions J_1, K_1, J_2, and K_2, the pulse-mode circuit realization of the machine in Fig. 9.4.2(a) can be easily constructed, and is shown in Fig. 9.4.2(e).

Exercise 9.4

1. A pulse-mode sequential circuit using J-K flip-flops is to be designed having two pulse inputs x_1 and x_2 and one dc output z. The inputs are restricted so that x_1 and x_2 are never simultaneously equal to 1. The output is to become equal to 1 whenever three or more consecutive x_1 pulses or three or more consecutive x_2 pulses are received. Otherwise, the output is to remain equal to 0.

2. It is desired to design a pulse-mode sequential circuit having two pulse inputs x_1 and x_2 and two dc outputs, z_1 and z_2. The inputs are restricted so that x_1 and x_2 are never simultaneously equal to 1. When either x_1 or x_2 is equal to 1, the corresponding output z_1 or z_2 is to be equal to 1. When x_1 and x_2 are both equal to 0, z_1 is to be equal to 1 if x_1 was

the last input equal to 1 and z_2 is to be equal to 1 if x_2 was the last input equal to 1. z_1 and z_2 are never both equal to 1. Design this circuit using J-K flip-flops.

3. (a) Design a decade up-counter using S-R flip-flops.
   (b) Design a decade down-counter using J-K flip-flops.

4. Design a counter which will count modulo-8 if a level input $L = 1$ or will count modulo-16 if $L = 0$. The count is to be incremented each time a pulse appears on line P_1 and decremented each time a pulse appears on line P_2. Use any type of flip-flops you prefer.

5. Find a circuit realization of the vending machine whose transition table is described in Fig. 9.4.1 using J-K flip-flops.

9.5 Sequential-Machine Realization Using Fundamental-Mode (Pulseless) Sequential Circuits

Let us now consider the second type of unclocked circuit, the fundamental-mode circuit.

DEFINITION 9.5.1

An unclocked sequential circuit is operating in *fundamental mode* iff

1. The inputs are voltage levels.
2. The inputs are never changed unless the circuit is stable internally.

The circuit is stable internally if none of the internal signals is unstable or changing.

This type of circuit (asynchronous) differs from the clocked circuit and the pulse-mode circuit (synchronous) in two major respects:

1. It has stable as well as unstable or transient states. From a stable state to another stable state, it usually "goes through" several transient states before it lands at a final stable state.

2. It may have transient phenomena, which sometimes makes the circuit performance unpredictable. Whenever such situations arise, one must modify or redesign the circuit so that they can be eliminated. In addition, the asynchronous circuit is faster but more difficult to design than the synchronous circuit.

Although there are many forms that an asynchronous (sequential) circuit might take, the one shown in Fig. 9.5.1 is the most straightforward and widely used in the study of the theory of asynchronous circuits or machines. Externally, the inputs and outputs are represented by levels rather than pulses. Internally, it is characterized by the use of delay elements as memory devices. Symbolically, they are denoted by

$\longrightarrow \boxed{\quad \Delta \quad} \longrightarrow$, where Δ denotes the amount of time delay between the

input and the output. It should be noted, however, that in practice, when the inherent

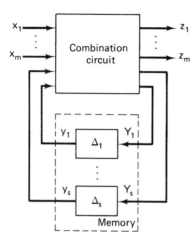

Fig. 9.5.1 Basic model of asynchronous circuit.

delay of the combinational logic is large enough, the external delay elements may not be necessary. But for clarity of presentation, we shall assume them to be present. The delay-type memory elements in Fig. 9.5.1 are characterized as follows:

$$y_i(t) = Y_i(t - \Delta), \qquad i = 1, 2, \ldots, s$$

Although the Δ's, in general, may not be equal, for illustration simplicity, it is assumed that

1. $\Delta_1 = \Delta_2 = \ldots = \Delta_s = \Delta$
2. The delay Δ is larger than any one may encounter in propagating a signal through the combinational circuit.

A. Analysis

Any unclocked combinational circuit with feedback loops can be operated as a sequential circuit in fundamental mode. Consider a simple circuit in Fig. 9.5.2(a), which may be represented by the circuit in Fig. 9.5.2(b). From the circuit, it is seen that for a given present state and input, the next-state (which may be stable or unstable) can be found by the following next-state equations:

$$Y_1 = x_1' y_2' \tag{9.5.1}$$
$$Y_2 = x_2' y_1' \tag{9.5.2}$$

From these equations, the transition table of the circuit is obtained and shown in Fig. 9.5.2(c). In this table, unlike the transition table of a clocked circuit, there are two types of next-states: stable states and unstable (transient) states. The next-state $Y_1 Y_2$ is a stable state if $Y_1 Y_2 = y_1 y_2$; otherwise, it is unstable. To differentiate the stable states from the unstable ones, the former are encircled. Upon an application of an input, the chain of transitions from any present state to a final stable state can be

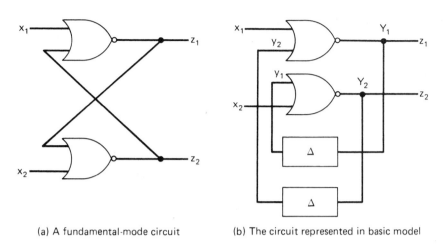

(a) A fundamental-mode circuit   (b) The circuit represented in basic model

$x_1 x_2$ / $y_1 y_2$	00	01	11	10
00	11*	10	(00)	01
01	(01)	00	00	(01)
11	00*	00*	00*	00*
10	(10)	(10)	00	00

(c) Transition table

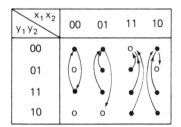

(d) Transition diagram

$x_1 x_2$ / $y_1 y_2$	00	01	11	10
00	?	10	(00)	01
01	(01)	10	00	(01)
11	?	10	00	01
10	(10)	(10)	00	01

$Y_1 Y_2$

(e) State flow table

Fig. 9.5.2 An example for illustrating the analysis of a fundamental-mode circuit.

obtained from this table. For example, consider the final states of, say, state 10, under various different inputs. If the input is $x_1 x_2 = 00$ or 01, the final stable state is state 10. This means that the state is unchanged under the input 00 or 01. If the input is

$x_1x_2 = 11$, the following transitions would occur:

$$\text{(10)} \longrightarrow 00 \longrightarrow \text{(00)}$$

present   stable
state   next-
                               state

If the input is $x_1x_2 = 10$, the transitions are

$$\text{(10)} \longrightarrow 00 \longrightarrow 01 \longrightarrow \text{(01)}$$

present   stable
state   next-
                                        state

These chains of state transitions are best displayed by use of a *transition diagram*, which is shown in Fig. 9.5.2(d). In this table, stable and unstable states are represented by ∘ and •, respectively. In the column under $x_1x_2 = 00$, there are two isolated stable next-states, 01 and 10, and two unstable states, which form a closed path. The latter is called a *cycle* or *loop*. Whenever a transition path is a cycle or contains a cycle, the final state is indeterminate. We denote such an indeterminate next-state by "?".

It is interesting to observe that under the columns of $x_1x_2 = 01$, 11, and 10 of the transition diagram, regardless of its initial state, the final (stable) states of the machine are states 10, 00, and 01, respectively. The table that indicates all the final stable states of the state transitions is shown in Fig. 9.5.2(e). This table is called the *state flow table* of the circuit.

B. Races and Their Elimination

In the analysis above it was assumed that the changes in the values of state variables associated with a state transition occur simultaneously. In practice, it is almost impossible to have several state variables change at exactly the same time instant. Whenever a state transition in the transition table involves more than one state-variable change, we say that it involves a *race condition*. It is indicated by an asterisk * at the upper right corner of the next-state. There are five race conditions in this transition table, which are shown in Fig. 9.5.3. If the final stable states of a race are the same, the race is called a *noncritical race*. If the final stable states of a race are different, the race is called a *critical race*. The races shown in Figs. 9.5.3(a) and (b) are critical races, and the races shown in Figs. 9.5.3(c), (d), and (e) are noncritical races. *Critical races are to be avoided in the design of asynchronous circuits because they would lead to erroneous operations of the circuit.* Several methods for eliminating races are described below.

Method 1: Race Elimination by Restricting Input and State Variables   One way to eliminate races is to put restrictions on the input variables and state variables such that the race conditions can never arise. For the example above, if the following restrictions are imposed on the circuit:

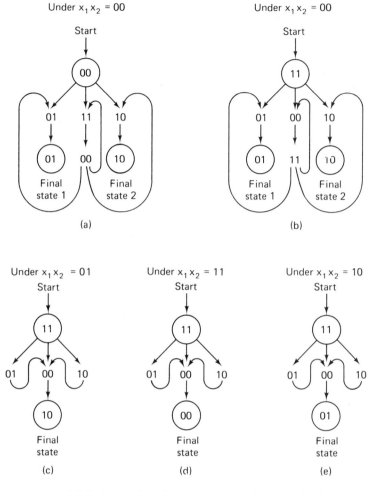

Fig. 9.5.3 Illustration of races in an asynchronous circuit.

1. The input $x_1x_2 = 11$ can never occur, and
2. y_1 and y_2 are always complementary to each other, that is, $y_2 = y_1'$,

the circuit becomes a well-known asynchronous S-R flip-flop or a NOR latch, which is shown in Fig. 9.5.4. This circuit operating under the above restrictions would have no races nor cycles.

Method 2: Race Elimination by Way of State Assignment

Race elimination by way of adjacent state assignment: In some cases, races can be completely eliminated by way of an adjacent state assignment. For example, consider the modulo-8 counter of Fig. 9.1.2(c). Suppose that the clock input is changed to a level input x. The transition table of this asynchronous circuit is shown in Fig. 9.5.5(b). It is seen that there are four race conditions in this circuit. Race conditions can,

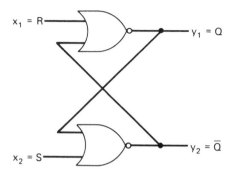

Q \ RS	00	01	11	10
0	(0)	1	–	(0)
1	(1)	(1)	–	0

(Inputs R and S cannot be simultaneously equal to 1)

Fig. 9.5.4   Asynchronous S-R flip-flop (NOR latch).

however, be (completely or partially) eliminated by use of a *state adjacent diagram*. This diagram consists of all the states of the machine. A line is connected to two states q_i and q_j if there is a state transition between them. The state adjacent diagram of the modulo-8 counter of Fig. 9.5.5(b) is depicted in Fig. 9.5.5(c). It is seen that if we use an adjacent state assignment such as

State	$y_1 y_2 y_3$
0	000
1	001
2	011
3	010
4	110
5	111
6	101
7	100

all the races will be eliminated. This is shown in Fig. 9.5.5(d). Notice that in this table every state transition only involves a change of one state variable; thus, no races exist in this circuit. The code given above is known as the *Gray code*. Also, it should be pointed out that this circuit does not have any stable (next) state, and the state transitions of the circuit form a cycle, as shown in Fig. 9.5.5(e).

Let us consider another example, say, the machine of Fig. 9.5.6(a), whose state adjacent diagram is shown in Fig. 9.5.6(b). From the adjacency relations shown in this diagram, it is seen that state q_2 is adjacent to three states, q_0, q_1, and q_3; therefore, a two-variable race-free state assignment is impossible. In this case, what we can do is to use a "nearly adjacent" state assignment such as the one shown in Fig. 9.5.6(c). The corresponding transition table is shown in Fig. 9.5.6(d). It is seen that there is a race existing in the transition from state 01 to state 10. It is easy to verify that this race is noncritical. It need not be avoided because it would not lead to erroneous operations

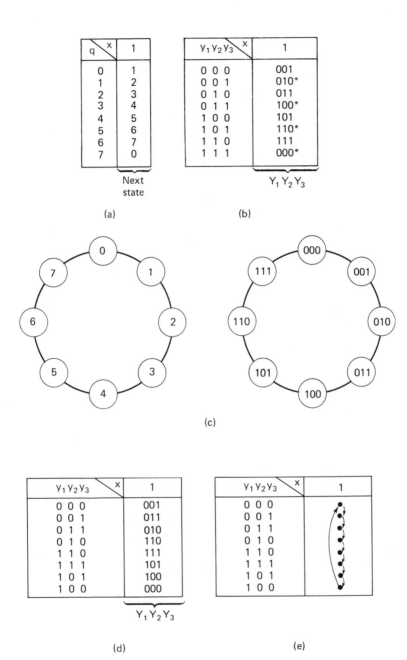

q \ x	1
0	1
1	2
2	3
3	4
4	5
5	6
6	7
7	0

Next state

(a)

$y_1 y_2 y_3$ \ x	1
0 0 0	001
0 0 1	010*
0 1 0	011
0 1 1	100*
1 0 0	101
1 0 1	110*
1 1 0	111
1 1 1	000*

$Y_1 Y_2 Y_3$

(b)

(c)

$y_1 y_2 y_3$ \ x	1
0 0 0	001
0 0 1	011
0 1 1	010
0 1 0	110
1 1 0	111
1 1 1	101
1 0 1	100
1 0 0	000

$Y_1 Y_2 Y_3$

(d)

$y_1 y_2 y_3$ \ x	1
0 0 0	
0 0 1	
0 1 1	
0 1 0	
1 1 0	
1 1 1	
1 0 1	
1 0 0	

(e)

Fig. 9.5.5 (a) Transition table of a modulo-8 counter; (b) transition table of the modulo-8 counter circuit of Fig. 9.1.2(c) operating in fundamental mode; (c) state adjacent diagram of (a); (d) Gray-coded transition table; (e) transition diagram of (d).

371

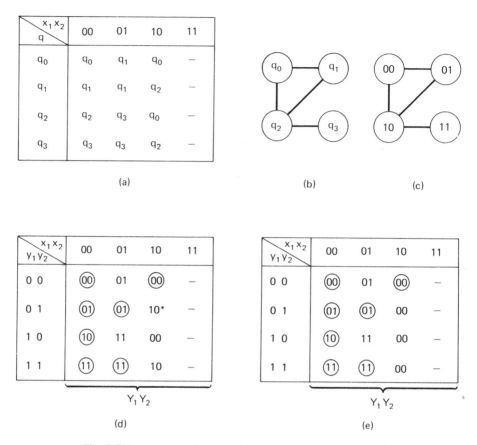

$x_1 x_2$ q	00	01	10	11
q_0	q_0	q_1	q_0	—
q_1	q_1	q_1	q_2	—
q_2	q_2	q_3	q_0	—
q_3	q_3	q_3	q_2	—

(a)

(b)

(c)

$x_1 x_2$ $y_1 y_2$	00	01	10	11
0 0	(00)	01	(00)	—
0 1	(01)	(01)	10*	—
1 0	(10)	11	00	—
1 1	(11)	(11)	10	—

$Y_1 Y_2$

(d)

$x_1 x_2$ $y_1 y_2$	00	01	10	11
0 0	(00)	01	(00)	—
0 1	(01)	(01)	00	—
1 0	(10)	11	00	—
1 1	(11)	(11)	00	—

$Y_1 Y_2$

(e)

Fig. 9.5.6 An example for illustrating the use of "nearly adjacent" state assignment to eliminate critical races.

of the machine. The state flow table of this machine is shown in Fig. 9.5.6(e).

General race-free state-assignment method: Let us return to the machine of Fig. 9.5.6(a). There are four states in the machine and we decide to use four state variables. We generate a state-assignment scheme by the following procedure:

	State	$y_1 y_2 y_3 y_4$
Original states	q_0	0001
	q_1	0010
	q_2	0100
	q_3	1000
Added transition states	$q_0 q_1$	0011
	$q_0 q_2$	0101
	$q_1 q_2$	0110
	$q_2 q_3$	1100

First, we assign to each state a 4-tuple, which consists of only one 1. Arbitrarily, we assign 0001 to q_0, and so on. Examination of row q_0 of the transition table of Fig. 9.5.6(a) reveals that there is a transition from state q_0 to state q_1. We create a transition state, $q_0 q_1$, and assign to it the 4-tuple 0011. We can envision arriving at the code 0011 by adding the codes for q_0 and q_1, component by component. We create the other transition states in the same fashion. Every transition $q_i \rightarrow q_j$ in the transition table will be replaced by $q_i \rightarrow q_i q_j \rightarrow q_j$. For example, the transition $q_0 \rightarrow q_1$ is replaced by $q_0 \rightarrow q_0 q_1 \rightarrow q_1$. Represented by their codes, this transition is

$$0001 \xrightarrow{01} 0011 \xrightarrow{01} 0010$$

It is seen that no race conditions exist in this chain of transitions. The transition table obtained by this race-free state-assignment technique is shown in Table 9.5.1. This transition table is race-free and equivalent to the original one.

TABLE 9.5.1   A Race-free Transition Table of the Machine
of Fig. 9.5.6(a)

$x_1 x_2$ / $y_4 y_3 y_2 y_1$	00	01	10	11
0 0 0 1	(0001)	0011	(0001)	—
0 0 1 0	(0010)	(0010)	0110	—
0 1 0 0	(0100)	1100	0101	—
1 0 0 0	(1000)	(1000)	1100	—
0 0 1 1	—	0010	—	—
0 1 0 1	—	—	0001	—
0 1 1 0	—	—	0100	—
1 1 0 0	—	1000	0100	—

C. Hazards and Their Elimination

Another undesirable transient phenomenon of an asynchronous machine is the *hazard*. Just as the state variables, it is also impossible to have several input variables change at exactly the same time instant. Thus, to avoid possible undesirable operations, only one input variable of an asynchronous sequential circuit is allowed to change at a time. Also, to avoid critical races, the state assignment must be made in such a manner that only one state variable is to change at a time. However, even in the change of a single variable, it is still possible to have some undesirable operating situations because a variable has two forms, the uncomplemented form x_i and the complemented form x_i', which may not change at exactly the same time instant. For example, x_i' may be obtained from x_i through an inverter. When x_i changes, say from 0 to 1, x_i' also changes, from 1 to 0, but the change of x_i' will be slightly later because of the delay in the inverter. Because of this slight difference in time of change, the circuit may momentarily give some incorrect outputs. The hazard is a transient phenomenon which happens during the change of such a single input variable. If the outputs before and after the change of the input are the same, the hazard is called a *static hazard*. If the outputs before and after the change of input are both 1, with an incorrect output 0 in between (i.e., the output sequence is 1–0–1), the hazard is a *static 1 hazard*. If the outputs before and after the change of input are both 0, with an incorrect output 1 in between (i.e., the output sequence is 0–1–0), the hazard is a *static 0 hazard*.

Now let us consider the realization of the transition table of Fig. 9.5.6(d). In order to design the necessary combinational circuit, we proceed to identify the functions for next-state variables Y_1 and Y_2. Since each next-state variable is a function of

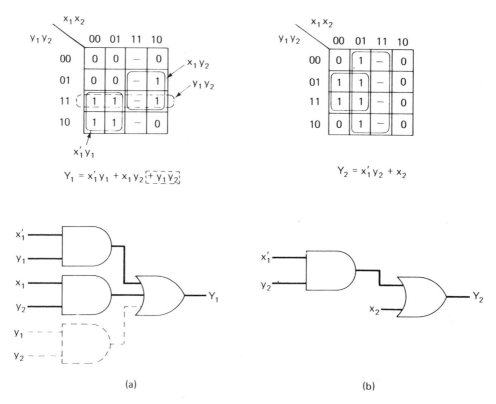

$$Y_1 = x_1' y_1 + x_1 y_2 \overline{(+ y_1 y_2)}$$

$$Y_2 = x_1' y_2 + x_2$$

(a)   (b)

Fig. 9.5.7 An example to illustrate static 1 hazards.

four variables, x_1, x_2, y_1, and y_2, from the transition table of Fig. 9.5.6(d), we obtain the two Karnaugh maps shown in Fig. 9.5.7. The realizations of Y_1 and Y_2 are given in the same figure. The circuit realization of Y_2 does not have any hazards, because it does not have input variables x_i and x_i' both present at the input. But the circuit realization of Y_1 does. In fact, there are two static 1 hazards in this circuit. One exists in the transition $y_1 y_2 = 11$, x_1 changing from 1 to 0 and x_1' from 0 to 1, and the other in the transition $y_1 y_2 = 11$, x_1 changing from 0 to 1 and x_1' from 1 to 0. Let the input combination before time t_1 be $x_1 x_2 y_1 y_2 = 1\text{-}11$, where the dash denotes that x_2 could be either 0 or 1. Let x_1 change from 1 to 0 at t_1 and let x_1' change from 0 to 1 at a slightly later time t_2. Thus, the input combination after t_2 is $x_1 x_2 y_1 y_2 = 0\text{-}11$. Consider x_1 and x_1' as two distinct variables. The values of the variables and the function Y_1 before t_1, between t_1 and t_2, and after t_2 are shown in Table 9.5.2(a). A similar analysis for the second static 1 hazard is given in Table 9.5.2(b).

These hazards can be avoided by using a sort of redundancy. Referring to the Karnaugh map in Fig. 9.5.7(a), if we add the term $y_1 y_2$ to the function Y_1, then the value of Y_1 will be 1 before t_1, between t_1 and t_2, and after t_2, because $y_1 y_2$ is independent of x_1 and x_1'. This is shown in the last columns of Tables 9.5.2(a) and (b). The addition of the product term $y_1 y_2$ to Y_1 corresponds to an addition of an AND-gate realizing $y_1 y_2$, in the circuit realization of Y_1, shown in Fig. 9.5.7(a). In a very similar manner, we can analyze and eliminate state 0 hazards in two-level OR–AND circuits by using the product-of-sums Karnaugh map and adding redundant OR-gates to the circuit, respectively.

Besides static hazards, there is another type of hazard, called *essential hazard*, which originates in a sequential circuit. This type of hazard happens when the change

TABLE 9.5.2 Values of Y_1 During Static 1 Hazards in the Circuit Realization of Y_1

	x_1	x_1'	x_2	y_1	y_2	$Y_1 = x_1' y_1 + x_1 y_2$	$Y_1 = x_1' y_1 + x_1 y_2 + y_1 y_2$
$t < t_1$	1	0	—	1	1	1	1
$t_1 < t < t_2$	0	0	—	1	1	0	1
$t > t_2$	0	1	—	1	1	1	1
	Don't care			Fixed		Static 1 hazard	No hazard

(a)

	x_1	x_1'	x_2	y_1	y_2	$Y_1 = x_1' y_1 + x_1 y_2$	$Y_1 = x_1' y_1 + x_1 y_2 + y_1 y_2$
$t < t_1$	0	1	—	1	1	1	1
$t_1 < t < t_2$	1	1	—	1	1	0	1
$t > t_2$	1	0	—	1	1	1	1
	Don't care			Fixed		Static 1 hazard	No hazard

(b)

Time (a)

Time (b)

TABLE 9.5.3 Transition Table of an
Asynchronous Sequential Circuit

$y_1 y_2$ \ x	0	1
00	01	(00)
01	(01)	11
11	10	(11)
10	(10)	00

$$Y_1 Y_2$$

of one form of an input variable x_i, say the complemented form, is slower than the change of the state variables resulting from the change of the uncomplemented form of the input variable. This can be best explained through a simple example. The transition table of a certain asynchronous sequential circuit is shown in Table 9.5.3. The next-state functions

$$Y_1 = x'y_1 + xy_2$$
$$Y_2 = x'y_1' + xy_2$$

In order to avoid static 1 hazards, redundant terms are added to the expressions of Y_1 and Y_2. Thus, we have

$$Y_1 = x'y_1 + xy_2 + y_1y_2$$
$$Y_2 = x'y_1' + xy_2 + y_1'y_2$$

Assume that initially $x = 1$, $y_1 = 1$, $y_2 = 1$. Thus, $Y_1 = 1$, $Y_2 = 1$, and the circuit is in stable state (11). Two cases are analyzed in Table 9.5.4.

CASE 1

The changes of values of x_1 and x_1' occur at exactly the same time instant.

CASE 2

The changes of values of x_1 and x_1' do not occur at the same time instant.

The sequence of changes of values of the variables for case 1 is shown in Table 9.5.4(a). From this table, we obtain the following sequence of state transitions:

$$(11) \dashrightarrow 10 \longrightarrow (10)$$

If the change of values of $x(1 \longrightarrow 0)$ occurs before that of x' does, then from Table

TABLE 9.5.4 (a) Sequence of Changes of the Values of the Variables for the Case where x_1 and x_1' Changes Values at the Same Instant; (b) Sequence of Changes of the Values of the Variables for the Case where x_1 Changes from 1 to 0 First and x_1' Changes from 0 to 1 at a Later Time

	t_0	t_1	t_2
x	1	0	0
x'	0	1	1
y_1	1	1	1
y_2	1	1	0
Y_1	1	1	1
Y_2	1	0	0
$Y_1 Y_2$	⑪→ 10 → ⑩		

(a)

	t_0	t_1	t_2	t_3	t_4
x	1	0	0	0	0
x'	0	0	0	1	1
y_1	1	1	1	0	0
y_2	1	1	0	0	1
Y_1	1	1	0	0	0
Y_2	1	0	0	1	1
$Y_1 Y_2$	⑪→ 10 → 00 → 01 → ㉛				

(b)

9.5.4(b), we see that the sequence of state transitions would then be

$$⑪ \longrightarrow 10 \longrightarrow 00 \longrightarrow 01 \longrightarrow ㉛$$

Essential hazards in an asynchronous circuit cannot be tolerated because they could cause erroneous transitions of the circuit. Fortunately, essential hazards can be avoided by inserting delay units in the feedback paths from the next-state variables, thus delaying the change of the state variables until the change of the input variable has completed.

In summary, the following remarks are made.

1. Static 1 and 0 hazards originate in combinational circuits. Static 1 hazards occur in two-level AND–OR circuits and static 0 hazards occur in two-level OR–AND circuits. Both can be eliminated by adding redundant gates to the circuits.

2. Essential hazards originate in sequential circuits. They can be eliminated by inserting delay units in the feedback paths from the next-state variables.

3. Besides static and essential hazards, there is a third type of hazard, called *dynamic hazard*, which originates in combinational circuits. If the outputs before and after the change of input are different, and the output changes three times instead of once and passes through an additional temporary sequence of 01 or 10 in going to the final output (i.e., the output sequence is either 1–0–1–0 or 0–1–0–1), the hazard is a dynamic hazard. Dynamic hazards, however, do not occur in two-level AND-OR and OR-AND circuits.

D. Synthesis

Analyses of the fundamental-mode circuit in terms of transition and flow tables were presented above. Now consider the synthesis problem: Given a verbal description of a design problem, find a fundamental-mode circuit that meets all the design specifications. Just as it was not possible to give a procedure for going from a flow table to a word statement of the circuit performance, it is impossible to develop a formal procedure for going from a word statement to a flow table. The basic reason for this difficulty is that the word statement is not a formal description of the circuit action. The task of translating the problem statements into a flow table is made easy through use of a *primitive flow table*, which is a flow table in which only one stable state appears in each row.

Example 9.5.1

   Design a fundamental-mode (asynchronous) circuit with two inputs x_1 and x_2, and one output z. The output is 1 if the two successive input symbols up to the present are in ascending order as binary numbers. Otherwise, the output is 0.
   There are four input symbols, 00, 01, 10, and 11. For simplicity, we shall denote them by the decimal numbers 0, 1, 2, and 3. Since the two input variables are not allowed to change at the same time, the following transitions do *not* occur.

$x_1 x_2$		$x_1 x_2$
00	$\longleftrightarrow$	11
01	$\longleftrightarrow$	10

All the possible sequences (in decimal) of length 2 are: 01, 02, 10, 13, 20, 23, 31, and 32. As in the case of synchronous machines (see Examples 9.2.2 and 9.4.1), we can define states according to the various input sequences. Thus, we have a machine M of eight states, the flow table of which is shown in Table 9.5.5. It is found that

$$\text{states } q_3 \text{ and } q_5 \text{ are equivalent}$$

$$\text{states } q_4 \text{ and } q_6 \text{ are equivalent}$$

The minimized flow table is shown in Table 9.5.5(b). The state adjacent diagram of this machine is given in Fig. 9.5.8(a). It is seen that state q_3 is adjacent to the four states q_1, q_2, q_7, and q_8, which are also adjacent to q_4. Under these adjacent relationships, it is quite difficult to find a race-free state assignment for this machine. But a close observation of the minimized flow table of Table 9.5.5(b) reveals that in the columns under inputs $I = 0$ and $I = 3$, critical races can never happen. The state adjacent diagram for the partial flow table of the columns under inputs $I = 1$ and $I = 2$ is depicted in Fig. 9.5.8(b). A critical-race-free state assignment for this machine is

y_2	y_3	$y_1 = 0$	$y_1 = 1$
0	0	q_3	q_4
0	1	q_1	q_7
1	0	q_2	q_8

TABLE 9.5.5 (a) Primitive Flow Table of the Machine of Example 9.5.1; (b) Minimized Flow Table

I \ q	0 ($x_1 x_2 = 00$)	1 ($x_1 x_2 = 01$)	3 ($x_1 x_2 = 11$)	2 ($x_1 x_2 = 10$)	Present output
q_1 (01)	q_3	$⑨_1$	q_4	—	1
q_2 (02)	q_5	—	q_6	$⑨_2$	1
q_3 (10)	$⑨_3$	q_1	—	q_2	0
q_4 (13)	—	q_7	$⑨_4$	q_8	1
q_5 (20)	$⑨_5$	q_1	—	q_2	0
q_6 (23)	—	q_7	$⑨_6$	q_8	1
q_7 (31)	q_3	$⑨_7$	q_4	—	0
q_8 (32)	q_5	—	q_6	$⑨_8$	0

Next state

(a)

I \ q	0 ($x_1 x_2 = 00$)	1 ($x_1 x_2 = 01$)	3 ($x_1 x_2 = 11$)	2 ($x_1 x_2 = 10$)	Present output
q_1 (01)	q_3	$⑨_1$	q_4	—	1
q_2 (02)	q_3	—	q_4	$⑨_2$	1
q_3 (10)	$⑨_3$	q_1	—	q_2	0
q_4 (13)	—	q_7	$⑨_4$	q_8	1
q_7 (31)	q_3	$⑨_7$	q_4	—	0
q_8 (32)	q_3	—	q_4	$⑨_8$	0

Next state

(b)

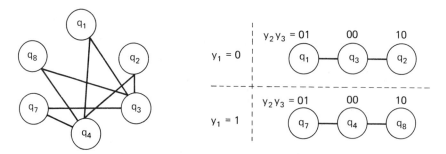

Fig. 9.5.8 (a) State adjacent diagram; (b) state adjacent diagram for the partial flow table of the columns under inputs $I = 1$ and $I = 2$.

The coded transition table is shown in Table 9.5.6. The Karnaugh maps of next-state functions Y_1, Y_2, Y_3 and the output function z are shown in Fig. 9.5.9, from which the minimized Y_1, Y_2, Y_3, and z are found:

$$Y_1 = y_1x_1 + y_1x_2 + y_2x_2 + y_3x_1$$
$$Y_2 = x_1x_2'$$
$$Y_3 = x_1'x_2$$
$$z = y_1'y_2' + y_2'y_3'$$

Since the functions Y_1, Y_2, Y_3 do not have an input variable with both uncomplemented and complemented forms, no static hazards can happen in the AND–OR gating circuit realizations of Y_1, Y_2, and Y_3. The circuit realization of this asynchronous machine is shown in Fig. 9.5.10.

TABLE 9.5.6 Coded Minimized Flow Table of the Machine of Example 9.5.1

$y_1 y_2 y_3$ \ $x_1 x_2$	00	01	11	10	z
0 0 0	(000)	001	–	010	1
0 0 1	000	(001)	100	–	1
0 1 0	000	–	100	(010)	0
1 0 0	–	101	(100)	110	1
1 0 1	000	(101)	100	–	0
1 1 0	000	–	100	(110)	0

$$Y_1 Y_2 Y_3$$

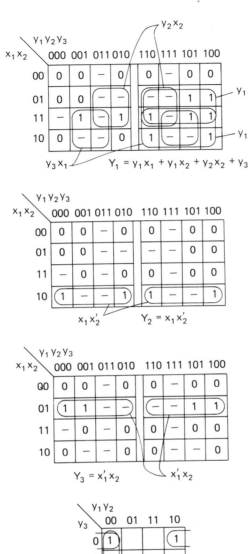

Fig. 9.5.9 Karnaugh maps of Y_1, Y_2, Y_3, and Z.

Exercise 9.5

1. Consider the flow table of Table P9.5.1.
   (a) Find a transition table that avoids critical races.

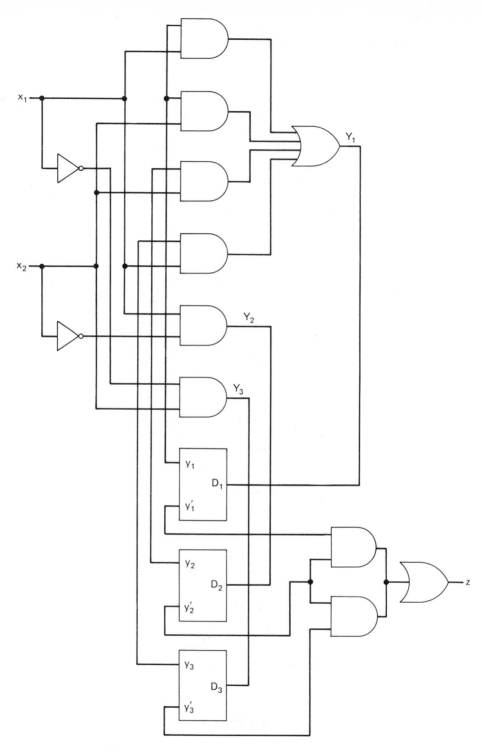

Fig. 9.5.10 Circuit realization of the asynchronous machine of
Example 9.5.1.

(b) Find a realization without static hazards.

TABLE P9.5.1

$x_1 x_2$ \ q	00	01	11	10
A	(A)	B	(A)	D
B	A	(B)	(B)	(B)
C	A	(C)	B	(C)
D	A	C	A	C

2. Find the essential hazards, if any, of the machine described by the transition table of Fig. 9.5.5(d).

3. The primitive flow table of an asynchronous machine is shown in Table P9.5.3.

TABLE P9.5.3

$x_1 x_2$ \ q	00	01	11	10	z
A	(A)	E	–	H	1
B	C	(B)	F	–	1
C	(C)	B	–	H	1
D	(D)	J	–	–	1
E	A	(E)	F	–	1
F	–	J	(F)	I	0
G	–	J	(G)	I	0
H	A	–	F	(H)	1
I	D	–	G	I	0
J	C	(J)	G	–	1

(a) Determine the equivalent states of this machine.
(b) Find the minimized flow table.
(c) Find a critical-race-free state assignment.
(d) Find the minimized next-state functions and the minimized output function.
(e) Find a static-hazard-free realization of this asynchronous machine.

4. (a) Find the transition table of the circuit of Fig. P9.5.4.
   (b) Find all the critical and noncritical races of this circuit, if any.
   (c) Find all the static hazards of this circuit, if any.

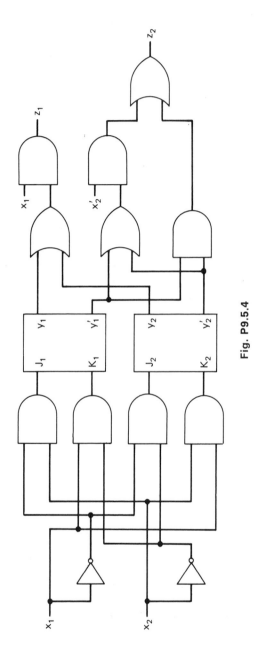

Fig. P9.5.4

5. Determine the primitive flow table of a fundamental-mode sequential circuit with two inputs x_1 and x_2 and one output z. The output is 1 iff $x_1 x_2$ changes from 00 to 10.

6. A fundamental-mode sequential circuit is to be designed having two inputs and one output z. When $x_1 = 1$, z will be identical to the x_2 input, but when $x_1 = 0$, z will remain the same as its value prior to x_1 becoming 0.
   (a) Construct the primitive flow table of the circuit.
   (b) Find the minimized flow table.
   (c) Find a critical-race-free state assignment.
   (d) Find the next-state functions and the output function.
   (e) Realize them in two-level AND–OR circuits. Eliminate all the static 1 hazards, if any.
   (f) Realize them in two-level OR–AND circuits. Eliminate all the static 0 hazards, if any.

7. A fundamental-mode circuit is to be designed having two inputs, x_1 and x_2, and two outputs, z_1 and z_2. The output is $z_1 z_2 = 01$ iff the three successive input symbols up to the present are in ascending order as binary numbers; the output is $z_1 z_2 = 10$ iff the three successive input symbols up to the present are in descending order; otherwise, the output is $z_1 z_2 = 00$.
   (a) Construct the primitive flow table of this circuit.
   (b) Minimize the primitive flow table.
   (c) Find a critical-race-free state assignment.
   (d) Find a static-hazard-free realization.

Bibliographical Remarks

The materials covered in this chapter can be found in the references listed below. The analysis and synthesis of asynchronous sequential circuits are treated extensively in references 4 and 10.

   1. TORNG, H. C., *Introduction to the Logical Design of Switching Systems*, Addison-Wesley, Reading, Mass., 1964.

   2. MCCLUSKEY, E. J., *Introduction to the Theory of Switching Circuits*, McGraw-Hill, New York, 1965.

   3. MILLER, R. E., *Switching Theory*, Vol. II, Wiley, New York, 1965.

   4. HILL, J. H., and G. R. PETERSON, *Introduction to Switching Theory and Logical Design*, Wiley, New York, 1968.

   5. WOOD, P. E., Jr., *Switching Theory*, McGraw-Hill, New York, 1968.

   6. GIVONE, D. D., *Introduction to Switching Circuit Theory*, McGraw-Hill, New York, 1970.

   7. BOOTH, T. L., *Digital Networks and Computer Systems*, Wiley, New York, 1971.

   8. RYNE, V. T., *Fundamentals of Digital System Design*. Prentice-Hall, Englewood Cliffs, N.J., 1973.

   9. KOHAVI, Z., *Switching and Finite Automata Theory*, McGraw-Hill, New York, 1970.

   10. SHENG, C. L., *Introduction to Switching Logic*, Intext Educational Publishers, New York, 1972.

   11. BRZOZOWSKI, J. A., and M. YOELI, *Digital Networks*, Prentice-Hall, Englewood Cliffs, N.J., 1976.

   12. LEE, S. C., *Digital Circuits and Logic Design*, Prentice-Hall, Englewood Cliffs, N.J., 1976.

10

Sequential Machine Fault Detection

The objective of this chapter is to introduce methods for designing a fault-detection experiment for a sequential machine. An experiment is the process of applying a sequence of input symbols to the input terminals of a machine and recording the resulting output sequence obtained at the output terminals. The problem of designing a fault detection or checking experiment is actually a restricted problem of machine identification. An experimenter is supplied with a machine and its transition table. His task is to determine from terminal experiments whether the given table accurately describes the behavior of the machine, that is, if the actual machine is isomorphic to the one described by the transition table. If at any time during the execution of the experiment the machine being tested produces a different response from that of the correctly operating machine, the test can be stopped, and we conclude that the machine definitely is faulty. The subtlest part of fault-detection experiment design is to make sure that only a fault-free machine or its isomorphisms will respond correctly to the experiment.

A fault-detection experiment for an n-state, m-input, p-output machine consists of three phases:

1. The initialization phase. During the initialization phase the machine to be tested must be brought to a specific state from which the second phase will begin.

2. The state-identification phase. During the state-identification phase, a distinguishing sequence is repeatedly applied to the machine to see whether it has n different responses to the distinguishing sequence, indicating n distinct states.

3. The transition-verification phase. During this phase the machine is made to go through all possible transitions.

The machine under test is assumed to be strongly connected, minimal, and diagnosable, with no growing states.

10.1 Homing Sequence, Synchronizing Sequence, and Experiment Initialization

The first step in designing a fault-detection experiment for a sequential machine is to bring the machine to a known state from which the actual checking experiment will begin. Generally, the initialization phase is *adaptive* in that a homing sequence is applied to the machine, and observation of its response enables identification of the machine's final state. A transfer sequence is then applied to take the machine to the desired starting state.

A homing sequence is derived from a *homing tree*. A *tree* consists of a root, branches, and leaves. For example, for a homing tree, the root of the tree is the set of all states, a branch is formed by an input symbol, and a leaf by groups of next-states. The procedure for constructing a homing tree is as follows.

Step 1 Apply each input symbol to the states of the machine, each of which develops a leaf which contains the next-states that are obtained from the state-transition function of the machine.

Step 2 Group all the next-states in one group if the corresponding outputs are identical; otherwise group them in different groups.

Step 3 A kth-level leaf becomes terminal if it is identical to a leaf of a level lower than k. When two or more leaves of the kth level are identical, all except one of them may be terminated (ignored).

Step 4 A homing tree is completed whenever (the first time such a situation occurs) a kth-level leaf contains only singleton groups and groups consisting of one state with duplicates.

The following example illustrates this procedure.

Example 10.1.1

Consider the machine M_1 whose transition table is shown in Fig. 10.1.1(a). Following the procedure described above, the homing tree of this machine is constructed and depicted in Fig. 10.1.1(b). Note that this tree is completed at the second-level leaf $(A)(D)(C)(A)$.

From the homing tree of a machine, a homing sequence can be obtained which is the sequence of input symbols along the path from the root of the tree to the leaf that causes the termination of the tree. For example, with reference to the homing tree of machine M_1 in Fig. 10.1.1(b), the homing sequence of this machine is 10. By applying this sequence to the machine and observing its response (output sequence), the machine is brought to a known state. For example,

Initial state (unknown)	Response to homing sequence 10		Final state (known)
A	01	indicating	D
B	11	indicating	A
C	00	indicating	A
D	10	indicating	C

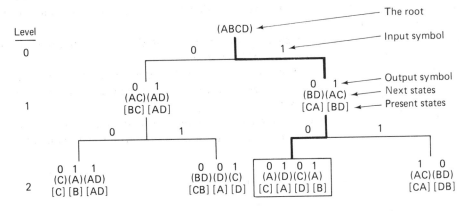

	x	0	1
q			
	A	A, 1	D, 0
	B	A, 0	A, 1
	C	C, 0	B, 0
	D	D, 1	C, 1

(a) Machine M₁

(b) The homing tree of machine M₁

Fig. 10.1.1   Example 10.1.1.

Note that every sequential machine possesses a homing sequence; hence, it can always be brought to a known state indicated by the response to the homing sequence.

The initialization phase can be *preset* if the machine should possess a synchronizing sequence. A synchronizing sequence is obtained from a synchronizing tree, which is constructed as follows:

Step 1   Applying each input symbol to the states of the machine, record *only* their next state in the leaf. Delete all the duplicates of the next states.

Step 2   A kth-level leaf becomes terminal whenever it is identical to a leaf of level lower than k. When two or more identical leaves occur at the kth level, all but one of them may be terminated.

Step 3   A synchronizing tree is completed the first time in such a situation when a leaf becomes a singleton.

This procedure is illustrated by the following example:

Example 10.1.2

Machine M_1 of Fig. 10.1.1(a) possesses a synchronizing tree, which is shown in Fig. 10.1.2. The synchronizing sequence is the sequence of input symbols along the path from the root of the synchronizing tree to the leaf that causes the termination of the tree. For example,

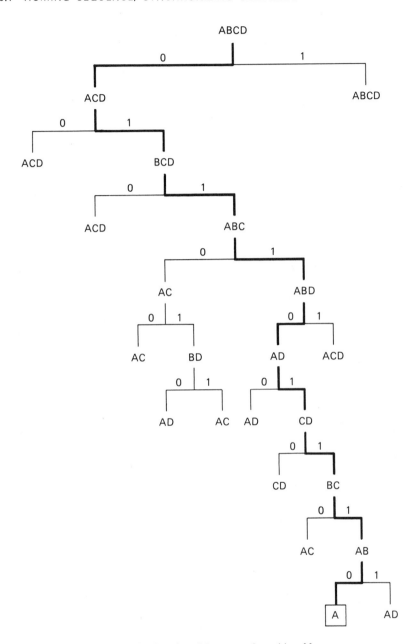

Fig. 10.1.2   Synchronizing tree of machine M_1.

the synchronizing sequence obtained from the synchronizing tree of Fig. 10.1.2 is 011101110. If we apply this sequence to the machine, regardless of its initial state, the machine will land at state A from which we can begin our checking experiment.

Note that not every machine possesses a synchronizing sequence. To prove that this statement is true is a rather simple matter. For example, if we change one of the two A states in the input 0 column of machine M_1 [Fig. 10.1.1(a)] to B, the resulting machine obviously will not possess a synchronizing sequence.

In summary, every machine possesses a homing sequence, but not a synchronizing sequence. A machine can be brought to a known state through use of a homing sequence. In this way, the final (known) state is indicated by the response (output sequence) to the homing sequence. If a machine possesses a synchronizing sequence and it is used to bring the machine to a known state, the experimenter does not need to observe the output sequence to determine the known state.

Exercise 10.1

1. Determine which of the following machines possess a synchronizing sequence and which do not.

x / q	0	1
A	$B, 0$	$A, 1$
B	$C, 1$	$A, 0$
C	$C, 1$	$B, 0$

(a) M_a

x / q	0	1
A	$C, 0$	$D, 0$
B	$D, 0$	$A, 0$
C	$A, 1$	$B, 0$
D	$A, 1$	$C, 0$

(b) M_b

x / q	0	1	z
A	A	C	0
B	A	C	1
C	C	B	0
D	D	A	1

(c) M_c

x / q	0	1
A	$C, 0$	$D, 1$
B	$D, 1$	$C, 0$
C	$B, 1$	$A, 1$
D	$B, 0$	$A, 0$

(d) M_d

x / q	0	1
A	$A, 1$	$C, 1$
B	$C, 0$	$C, 0$
C	$D, 1$	$C, 1$
D	$E, 0$	$D, 1$
E	$B, 1$	$D, 1$

(e) M_e

x / q	0	1	z
A	E	A	1
B	B	B	1
C	E	E	0
D	E	C	0
E	E	D	1

(f) M_f

x / q	a	b	c	d
A	$D, 0$	$A, 1$	$B, 1$	$A, 1$
B	$C, 0$	$B, 1$	$A, 1$	$C, 1$
C	$B, 0$	$C, 1$	$C, 1$	$B, 1$
D	$A, 0$	$D, 0$	$B, 0$	$D, 1$

(g) M_g

Fig. P10.1.1

2. For each of the machines of Fig. P10.1.1,
   (a) Construct a homing tree.
   (b) Find a homing sequence.
   (c) Find the output sequences and the (final) state they indicate.

10.2 Distinguishing Sequence and State Identification

After a machine is brought to a known state by a homing sequence or a synchronizing sequence, we then proceed with the second phase of the fault-detection experiment, the state identification. This is accomplished by repeatedly applying a distinguishing sequence to the machine until all the states of the machine are positively identified.

The construction of the distinguishing tree from which a distinguishing sequence is obtained is similar to that of the homing tree described in the previous section. As a matter of fact, the first three steps are the same as those in constructing a homing tree.

Step 4 A leaf becomes terminal if any group of the leaf contains a state with duplicates.

Step 5 A distinguishing tree is completed whenever (the first time such situation occurs) a kth-level leaf contains only singleton groups.

The reason for having the termination rule described in step 4 is that when the next-states of two states are the same, we cannot trace from that state back, since we cannot determine from which of the two states it comes. Consequently, we cannot use that part of the tree to determine (trace back) the machine's initial state. The reason for having the termination rule described in step 5 should be obvious. When a leaf contains only singleton groups, every initial state can be *uniquely* identified by an output sequence of the machine in response to the input sequence consisting of the input symbols along the path from the root of the tree to the terminal leaf; hence, the construction of the tree is completed.

From the procedures for constructing a distinguishing tree and a homing tree described in this section and the previous one, it is noticed that a homing tree of a machine may also be a distinguishing tree of the machine as the one in Fig. 10.1.1(b). In other words, the homing sequence 10 of this machine, which can be used to identify its final state based on its output sequence, is also the distinguishing sequence of the machine that can be used to identify its initial state, as follows:

Initial state		Response to the distinguishing sequence 10		Final state
A	indicating ←	01	indicating →	D
B	indicating ←	11	indicating →	A
C	indicating ←	00	indicating →	A
D	indicating ←	10	indicating →	C

Of course, not every homing tree is a distinguishing tree, nor is every distinguishing tree a homing tree. For example, consider the machine M_2 in Fig. 10.2.1(a), whose homing tree is given in Fig. 10.2.1(b), which is, however, *not* a distinguishing tree of the machine. The homing sequence obtained from this homing tree is 00. The possible output sequences of this machine in response to this homing sequence and the initial and final states they indicate are as follows:

Initial state	Response to homing sequence 00		Final state
or $\left.\begin{array}{c} A \\ C \end{array}\right\}$	indicating ←	01 indicating →	A
or $\left.\begin{array}{c} B \\ D \end{array}\right\}$	indicating ←	10 indicating →	B

It is seen that the initial state of the machine cannot be identified by the sequence 00. Therefore, it is *not* a distinguishing sequence of this machine; it is only a homing sequence.

A distinguishing tree sometimes contains a homing tree as a subtree of it. For example, consider machine M_3 of Fig. 10.2.2(a), whose distinguishing tree is given in Fig. 10.2.2(b). The distinguishing sequence obtained from this tree is 10. The possible output sequences of this machine in response to this sequence and the initial and final states they indicate are given below.

Initial state	Response to distinguishing Sequence 10		Final state
C	indicating ←	$\left.\begin{array}{c} 00 \\ 10 \end{array}\right\}$ indicating →	A
B	indicating ←		
A	indicating ←	01 indicating →	B

Notice that the initial state of the machine can be *uniquely* determined by the output sequence. At the same time, from the response to this sequence, the final state of the machine can also be determined. Since, in general, a homing sequence is shorter than a distinguishing sequence, the former is used for the initialization of a fault-detection experiment. For example, the homing sequence of machine M_3 is the input sequence 0 [see Fig. 10.2.2(b)], which is shorter than the distinguishing sequence 10 and may

x q	0	1
A	B, 0	D, 0
B	A, 1	C, 0
C	D, 0	A, 0
D	A, 1	B, 0

(a) Machine M_2

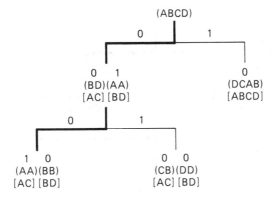

(b) The homing tree of machine M_2 which is not a distinguishing tree

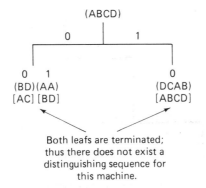

Both leafs are terminated;
thus there does not exist a
distinguishing sequence for
this machine.

(c) Attempt to construct a distinguishing tree from machine M_2

Fig. 10.2.1 Homing tree of machine M_2.

x q	0	1
A	A, 0	C, 0
B	A, 0	B, 1
C	B, 1	A, 0

(a) Machine M_3

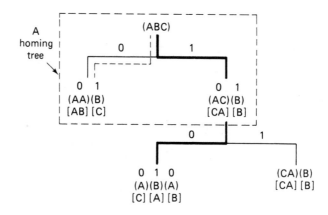

(b) A distinguishing tree of machine M_3 which contains a homing tree of the machine

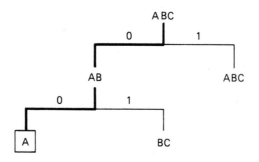

(c) Synchronizing tree of machine M_3

Fig. 10.2.2 Homing, synchronizing, and distinguishing trees of machine M_3.

be used to indicate the final state of the machine by the output sequence as follows:

Response to homing sequence 0	Final state
0 $\xrightarrow{\text{indicating}}$	A
1 $\xrightarrow{\text{indicating}}$	B

Furthermore, it is important to note that *not* every machine possesses a distinguishing sequence. For instance, machine M_2 does not have a distinguishing sequence, since all the leaves of level 1 are terminated before the distinguishing tree can be completed, as seen in Fig. 10.2.1(c).

DEFINITION 10.2.1

A machine is *diagnosable* if it possesses a distinguishing sequence.

DEFINITION 10.2.2

A machine is *strongly connected* if for every pair of states q_i and q_j of a machine M there exists an input sequence which takes M from q_i to q_j.

It is easy to show that machines M_1, M_2, and M_3 are strongly connected and machines M_1 and M_3 are diagnosable. To make machine M_2 diagnosable, we may add an auxiliary output z_a to machine M_2, as shown in Fig. 10.2.3. This modified machine, M_2', now becomes diagnosable.

q \ x	0	1
A	B, 00	D, 0–
B	A, 10	C, 0–
C	D, 01	A, 0--
D	A, 11	B, 0–

Fig. 10.2.3 Machine M_2'.　　　　Added auxiliary output z_a

Unless stated otherwise, it is assumed that every machine under test is strongly connected, minimal, and diagnosable, with no growing states.

Example 10.2.1

To illustrate the second phase of deriving a fault-detection experiment, let us consider a simple example of a three-state machine such as machine M_3. Machine M_3 possesses a homing sequence 0 and a synchronizing sequence 00; we may use either the homing sequence or the synchronizing sequence to bring the machine to a known state. Suppose the synchronizing sequence 00 is used; the machine is supposed to be brought to state A. To verify whether this is indeed the case or not, we need to apply the distinguishing sequence 10 and observe the output. If the output sequence is 01, we have positively identified state A. At this time the machine is supposed to be at state B. To verify this, the distinguishing sequence 10 is applied once again to the machine to see whether the output sequence is 10. If it is, state B has then been positively identified. At this time, the machine is supposed to be at state A, but without testing it, we really do not know. So, we need to apply the distinguishing sequence again to verify whether, at this moment, the machine is indeed in state A or not. If the output sequence of the distinguishing sequence is 01, we identify the state to be A; otherwise, it is not. If the machine does not have any faults, it should now be in state B. Since the transition $A \xrightarrow{\;10\;} B$ has already been verified, it does not need to be reverified.

At this point, we need to identify state C. Before we are able to do this, we need first to bring the machine to state C. Right now the machine is in state B, which unfortunately does not have a direct transition to state C [see Fig. 10.2.2(a)]. What we have to do is first to apply the distinguishing sequence to bring the machine to state A and then apply an input symbol 1 to take the machine to state C. This is shown in Fig. 10.2.4.

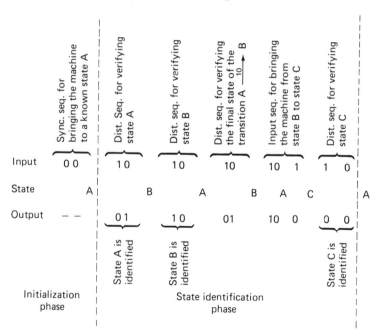

Fig. 10.2.4   Identification of the three states of machine M_3.

After the machine is brought to state C, we apply the distinguishing sequence 10. If the output sequence is 00, state C is identified; otherwise, the machine is faulty. The complete state-identification phase is shown in Fig. 10.2.4.

Exercise 10.2

1. Identify the four states of machine M_1.

2. (a) Determine which of the seven machines of problem 1 of Exercise 10.1 are diagnosable.
    (b) For each of the diagnosable machines found in (a), construct an experiment to identify its states.

10.3 Design of Complete Fault-Detection Experiment

As mentioned before, a fault-detection experiment for a sequential machine consists of three phases: the initialization phase, the state-identification phase, and the transition-verification phase. In the previous two sections we discussed the initialization and state-identification phases. In this section we proceed to the third phase, the transition verification phase.

Example 10.3.1

Let us construct the transition-verification phase of the fault-detection experiment for machine M_3, whose first two phases were presented in Example 10.2.1.

Each transition of a sequential machine, represented by

$$\text{Present state} \xrightarrow{\text{input symbol/output symbol}} \text{Next-state}$$

is verified by the following two steps:

Step 1 Apply the input symbol to the machine and observe its output.
Step 2 Verify the next state using a distinguishing sequence.

For example, see the last three input symbols of the partial experiment for checking machine M_3 in Fig. 10.2.4, which are repeated in Fig. 10.3.1. It is seen that since this input symbol 1 is applied to state A and followed by the distinguishing sequence 10, in addition to identifying state C, it also verifies the transition $A \xrightarrow{1/0} C$ of this machine.

With reference to Fig. 10.3.2, after the state-identification phase is completed and before the transition-verification phase begins, it is needed to verify that the final state at which the machine lands is indeed state A. This can be easily accomplished by applying the distinguishing sequence 10 to the machine and observing its output sequence, as indicated in Fig. 10.3.2. Since the experiment

Present state	Response to 10	Final state
A $\xleftarrow{\text{indicating}}$	01	$\xrightarrow{\text{indicating}}$ B

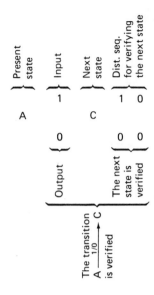

Fig. 10.3.1 An example of verifying a transition of a sequential machine after its states are identified.

which may simply be represented by

$$A \xrightarrow{10/01} B$$

has been verified in the state-identification phase, the machine is now in state B, from which we start our transition verifications. It is seen that after having verified the transitions

$$B \xrightarrow{0/0} A$$

$$B \xrightarrow{1/1} B$$

$$A \xrightarrow{0/0} A$$

we land at state B. At this time, all the transitions starting with states A and B have already been verified; we need to take the machine from state B to state C. This is done by applying the input sequence 101 as follows:

$$B \quad \overset{10}{\underset{10}{}} \quad A \quad \overset{1}{\underset{0}{}} \quad C$$

Note that it is not necessary to verify states A and C in the sequence above as final states in response to input sequences 10 and 1, respectively. This is because, prior to this point, the experiment

$$B \xrightarrow{10/10} A$$

and the transition

$$A \xrightarrow{1/0} C$$

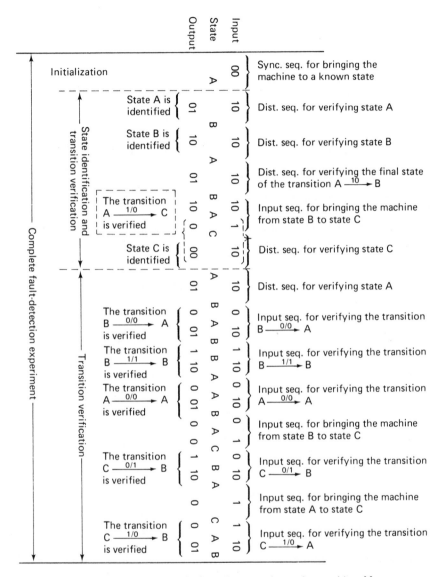

Fig. 10.3.2   Complete fault-detection experiment for machine M_3.

have been verified. After the machine is brought to state C, we start to verify the remaining unverified two transitions, the two transitions starting with state C. The verification experiment for checking the transitions

$$C \xrightarrow{0/1} B$$

and

$$C \xrightarrow{1/0} A$$

is shown in the last part of the experiment. The complete fault-detection experiment for machine M_3 is shown in Fig. 10.3.2, and the complete input and expected output sequences of this experiment are

start
↓

Testing input sequence: 001010101011010010110010010101110

Expected output sequence: --011001100000100111 0001001100001

which is of a length of 33 input symbols.

It should be noted that identifying states of a machine in different order and/or verifying its transitions in different order will result in fault-detection experiments with different lengths. For example, in the experiment above, suppose that we verify the transition $C \xrightarrow{1/0} A$ first and then the transition $C \xrightarrow{0/1} B$, as shown below:

The transition
$C \xrightarrow{1/0} A$
is verified

$\left\{ \begin{array}{cc} & C \\ 0 & 1 \\ & A \\ 01 & 10 \\ & B \end{array} \right\}$ Input seq.
for verifying
$C \xrightarrow{1/0} A$

$\left\{ \begin{array}{cc} & B \\ 0 & 0 \\ & A \\ 0 & 1 \\ & C \end{array} \right\}$ Input seq.
for bringing
the machine
to state C

The transition
$C \xrightarrow{0/1} B$
is verified

$\left\{ \begin{array}{cc} & C \\ 1 & 0 \\ & B \\ 10 & 10 \\ & A \end{array} \right\}$ Input seq.
for verifying
$C \xrightarrow{0/1} B$

Comparing the length of this sequence with the length of the subsequence presented in Fig. 10.3.2, it is found that the former is one input symbol longer than the latter.

Although the three phases must be considered as logically distinct by the experimenter, it is important to point out that, in actual practice, it is not necessary at all to separate them; we should try to mix up and overlap the subsequences used for state identification and transition verification whenever possible to shorten the sequence. This is best illustrated by use of an example.

Example 10.3.2

Now we reconstruct a fault-detection experiment for machine M_3 and try to shorten the experiment by mixing up and overlapping the subsequences used for state identification and transition verification. This is shown in Fig. 10.3.3. It is seen that only the verifications of the transitions

$$B \xrightarrow{1/1} B$$

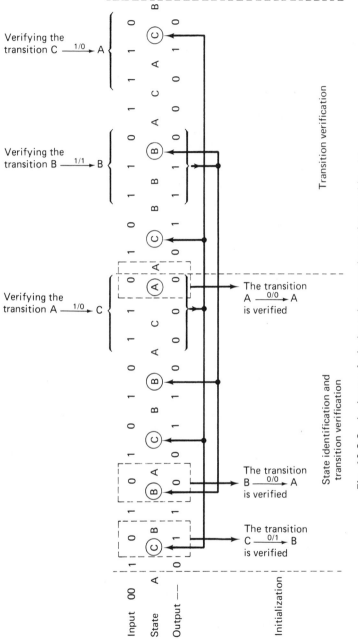

Fig. 10.3.3 A shorter fault-detection experiment for machine M_3.

and

$$C \xrightarrow{1/0} A$$

need to be added to the sequence used for experiment initialization and state identification given in Fig. 10.2.4.

Figure 10.3.3 shows an example of designing a fault-detection experiment for machine M_3 by inserting the states between the input symbols 1 and 0 of the distinguishing sequence, as they become known. For example, right after the transition of $A \xrightarrow{1/0} C$ is verified, we can insert state C (we shall circle the inserted states for differentiating them from the others) whenever we see the pattern in the sequence

For example, we can insert state C between input symbols 1 and 0 of the distinguishing sequence applied to state A as

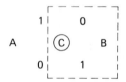

Transition $C \xrightarrow{0/1} B$ is verified.

By so doing, the transition $C \xrightarrow{0/1} B$ is verified without making any extra effort. Similarly, after the transition $B \xrightarrow{1/1} B$ is verified, we can insert state B between the input symbols 1 and 0 of the distinguishing sequence applied to state B, whenever such a pattern appears in the sequence. For example,

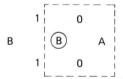

Transition $B \xrightarrow{0/0} A$ is verified.

Notice that, at the same time, the transition $B \xrightarrow{0/0} A$ is verified. Finally, after the transition $C \xrightarrow{1/0} A$ is verified, we can insert state A between input symbols 1 and 0 of the distinguishing sequence applied to state C, whenever such a pattern appears in the sequence. From Fig. 10.3.3, it is seen that we can insert state A as follows:

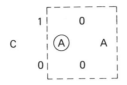

Transition A $\xrightarrow{0/0}$ A is verified.

With the insertion of state A, the transition $A \xrightarrow{0/0} A$ is thus verified. The complete input sequence, state, and output sequence of this fault-detection experiment design is shown in Fig. 10.3.3. Therefore, the testing input sequence and expected output sequence are

start
↓

Testing input sequence: 0010101010110101101110

Expected output sequence: —0110011000011100010

The length of this sequence is 22, which is two thirds of the length of the sequence presented in Example 10.3.1.

Exercise 10.3

1. Design a complete fault-detection experiment for machine M_1.

2. Find a complete fault-detection experiment for each of the diagnosable machines found in problem 2 of Exercise 10.2.

Bibliographical Remarks

The transition checking approach to the fault-detection problem was first introduced by Hennie [1]. This subject is extensively treated in references 2–4. The problem of better organizing a checking experiment for the class of minimized, strongly connected sequential machines which had distinguishing sequences was attacked by Kime [5], Gönenc [6], and Kohavi and Kohavi [7]. For the class of machines which do not have a distinguishing sequence, Kime [8] showed how the addition of a single input column to a machine will enable that machine to be strongly connected and have a distinguishing sequence. Kohavi and Lavallee [9] employed the technique of adding output terminals to construct a definitely diagnosable machine. The problem of designing checking experiments for machines which do not have distinguishing sequences without the necessity of modifying the given machine to make it possess one was attacked by Hsieh [10, 11].

References

1. HENNIE, F. C., "Fault Detecting Experiments for Sequential Circuits," *Proc. Fifth Ann. Symp. Switching Circuit Theory and Logical Design*, 1964, pp. 95–110.

2. CHANG, H. Y., R. C. DORR, and R. A. ELIOTT, "Logic Simulation and Fault Analysis

of a Self-Checking Switching Processor," in *Proc. 1972 IEEE-ACM Design Automation Workshop*, June 1972, pp. 128–135.

3. GILL, A., "State Identification Experiments in Finite Automata," *Information and Control*, Vol. 4, 1961, pp. 132–154.

4. GINSBURG, S., "On the Length of the Smallest Uniform Experiment Which Distinguishes the Terminal States of a Machine," *J. ACM*, Vol. 5, 1958, pp. 266–280.

5. KIME, C. R., "A Failure Detection Method for Sequential Circuits," Department of Electrical Engineering, University of Iowa, Iowa City, Iowa, *Technical Report 66–13*, 1966.

6. GÖNENC, G., "A Method for the Design of Fault Detection Experiments," *IEEE Trans. Computers*, Vol. C-19, June 1970, pp. 551–558.

7. KOHAVI, I. and Z. KOHAVI, "Variable-Length Distinguishing Sequences and Their Application to the Design of Fault-Detection Experiments," *IEEE Trans. Computers*, Vol. C-17, pp. 792–795, August 1968.

8   KIME, C. R., "An Organization for Checking Experiments on Sequential Circuits," *IEEE Trans. Electronic Computers*, Vol. EC-15, pp. 113–115, Feb. 1966.

9. KOHAVI, A., and P. LAVALLEE, "Design of Sequential Machines with Fault Detection Capabilities," *IEEE Trans. Electronic Computers*, Vol. EC-16, pp. 473–484, August 1967.

10. HSIEH, E. P., "Optimal Checking Experiments for Sequential Machines," Ph.D. Dissertation, Department of Electrical Engineering, Columbia University, New York, 1969 (available from University Microfilms).

11. HSIEH, E. P., "Checking Experiments for Sequential Machines," *IEEE Trans. Computers*, Vol. C-20, October 1971, pp. 1152–1166.

11

Digital Design Using
Integrated Circuits

In this chapter, four digital design methods using integrated circuits are presented. Depending on the number of gates and their equivalent contained in an integrated circuit chip, digital integrated circuits are classified into three classes; small-scale integration (SSI), medium-scale integration (MSI), and large-scale integration (LSI). Some commonly used available digital integrated-circuit (IC) chips are introduced. Various recent MOS/LSI semiconductor memory devices, including read only memory (ROM) and programmable logic array (PLA), are discussed. Finally, digital (circuit and systems) design using TTL/SSI, TTL/MSI, and MOS/LSI IC chips are presented and illustrated by examples.

Digital design includes both logic design and circuit design. Logic design is concerned only with the logical aspect of the design problem, whereas circuit design concerns the electronic-circuit realization of the logic design after it is obtained. Linking the logic design with the circuit design is accomplished by a mapping from logic values 0 and 1 of a logic circuit to two voltage intervals, V_L (low-voltage interval) and V_H (high-voltage interval), of an integrated circuit, which may be SSI, MSI, or LSI, as shown below.

$$\text{logic } 0 \longleftrightarrow V(0) = V_L \text{ or } V_H$$
$$\text{logic } 1 \longleftrightarrow V(1) = V_H \text{ or } V_L$$

Two types of logic, positive and negative, are most frequently used:

Positive logic	Negative logic
logic 0 $\longleftrightarrow$ $V(0) = V_L$	logic 0 $\longleftrightarrow$ $V(0) = V_H$
logic 1 $\longleftrightarrow$ $V(1) = V_H$	logic 1 $\longleftrightarrow$ $V(1) = V_L$

Between the two, positive logic is used more frequently than negative logic. Unless otherwise stated, positive logic will be assumed throughout this chapter.

There are five distinctly different digital design approaches. They are:

1. The gate-flip-flop approach, using SSI IC's.
2. The functional block diagram approach, using MSI/LSI IC's.
3. The RAM approach, using MSI/LSI IC's.
4. The PLA approach, using LSI IC's.
5. The microcomputer simulation approach, using LSI IC's.

The first four approaches will be presented in Sections 11.2, 11.3, and 11.4, after digital integrated circuits are introduced in Section 11.1. The fifth approach will be presented in Chapter 12.

11.1 Digital Integrated Circuits

In the past decade digital (circuit and system) design has undergone dramatic change. Today digital designers rarely build any components or devices which are available in integrated-circuit form. This is because digital integrated circuits are not only convenient and easy to use but also cost less. In this chapter, some commonly used integrated circuits will be introduced, and their important characteristics will be discussed.

Integrated-circuit chips are classified into small-scale integration (SSI), medium-scale integration (MSI), and large-scale integration (LSI) according to their complexity.

DEFINITION 11.1.1

A *gate equivalent circuit* is a basic unit of measure of relative digital-circuit complexity. The number of gate equivalent circuits is that number of individual logic gates that would have to be interconnected to perform the same function.

DEFINITION 11.1.2

LSI is a concept whereby a complete major subsystem function is fabricated as a single microcircuit. In this context a major subsystem or system, whether logical or linear, is considered to be one that contains 100 or more equivalent gates or circuitry of similar complexity.

DEFINITION 11.1.3

MSI is a concept whereby a complete subsystem or system function is fabricated as a single microcircuit. The subsystem or system is smaller than for LSI but, whether digital or linear, is considered to be one that contains 12 or more equivalent gates or circuitry of similar complexity.

DEFINITION 11.1.4

SSI circuits are integrated circuits of less complexity than MSI.

Among various different types of modern integrated-circuit logic families— resistor–transistor logic (RTL), direct-coupled transistor logic (DCTL), resistor-capacitor transistor logic (RCTL), diode–transistor logic (DTL), transistor–transistor logic (TTL), complementary-transistor logic (CTL), emitter-coupled logic (ECL), metal-oxide semiconductor (MOS), complementary metal-oxide semiconductor (CMOS), and integrated injection logic (I²L)—TTL and MOS are most commonly used at the present time, CMOS is becoming popular, and I²L is a promising new-comer.

Because of its high speed and low cost, TTL logic offers by far the largest number of standard SSI gates and MSI complex devices. The most commonly used TTL/SSI and TTL/MSI integrated-circuit series is the SN54/SN74 series (SN stands for semiconductor network), or simply the 54/74 series. The 54 series was introduced in 1964 primarily for the military market, where size, power consumption, and reliability requirements were paramount. The operating characteristics of the 54 series were guaranteed in the temperature range $-55°C$ to $125°C$. Soon the low-cost industrial versions of the 54 series, called the 74 series, were introduced, which were primarily for commercial use, with guaranteed operating characteristics over a $0°C$ to $70°C$ range. Not only are individual members of the series 54/74 TTL families compatible, but they also have the following typical characteristics in common:

1. Supply voltage $V_{CC} = +5$ V.
2. The high voltage level $V_H = +5$ V, which denotes logic 1.
3. The low voltage level $V_L = 0$ V (ground), which denotes logic 0.

Within the TTL logic families, there are several variations:

1. Basic type (5400/7400 series gates).
2. Low-power-dissipation version (54L00/74L00 series gates).
3. High-speed version (54S00/74S00 series gates).
4. Low-power-dissipation and very high-speed version (54LS00/74LS00 series gates).
5. Super Schottky version (54SS00/74SS00 series gates).

Some available 54/74 series TTL/SSI gates and flip-flops are given in Table 11.1.1.

Available TTL/MSI devices include:
Data Selectors/Multiplexers
Encoders
Code Converters
Decoders/Demultiplexers
Arithmetic Devices
Comparators
Counters
Shift Registers
Memories

Data Selectors/Multiplexers Data selectors/multiplexers are combinational devices controlled by a selector address that routes one of many input signals to the output. They can be considered semiconductor equivalents to multiposition switches or stepping devices. Data selector/multiplexers may be used for (1) data selection, (2) time-division multiplexing, and (3) function generation.

Encoders   Encoders are circuits with many inputs that generate the address of the active input. If a system design guarantees only one encoder input active, the encoder logic is very simple and can be implemented with gates. If several inputs can be active at one time, a simple encoder would generate the logic OR of their addresses, which is probably undesirable (i.e., inputs 2 and 4 active would generate address 6). A priority encoder generates the address of the active input with the highest priority. The priority is preassigned according to the position at the inputs.

TABLE 11.1.1   Some Available 54/74 Series TTL/SSI
Integrated Circuits

Gate	Function	Device
NAND/NOR	Quad 2-input NAND	54/7400, 54/74L00, 54H/74H00
	Quad 2-input NAND, open collector	54/7401, 54H/74H01
	Quad 2-input NOR	54/7402
	Quad 2-input NAND, open collector	54/7403
	Hex inverters	54/7404, 54/74L04, 54H/74H04
	Hex inverter, open collector	54H/74H05
	Hex inverter buffer/driver, open collector	54/7406
	Hex buffer/driver, open collector	54/7407
	Quad 2-input AND	54/7408
	Quad 2-input AND, open collector	54/7409
	Triple 3-input NAND	54/7410, 54/74L10, 54H/74H10
	Triple 3-input AND	54H/74H11
	Hex inverter buffer/driver, open collector	54/7416
	Hex buffer/driver, open collector	54/7417
	Dual 4-input NAND	54/7420, 54/74L20, 54H/74H20
	Dual 4-input AND	54H/74H21
	Dual 4-input NAND, open collector	54H/74H22
	Quad 2-input NAND, open collector	54/7426
	8-input NAND	54/7430, 54/74L30, 54H/74H30
	Dual 4-input NAND buffer	54/7440 54H/74H40
AND–OR–INVERT	Expandable dual 2-wide 2-input AND–OR–INVERT	54/7450, 54H/74H50
	Dual 2-wide 2-input AND–OR–INVERT	54/7451, 54/74L51, 54H/74H51
	Expandable 4-wide 2-2-2-3-input AND–OR	54/7452
	Expandable 4-wide 2-2-2-3-input AND–OR–INVERT	54/7453, 54H/74H53
	2-2-2-3-input AND–OR–INVERT	54/7454, 54/74L54, 54H/74H54
	Expandable 2-wide 4-input AND–OR–INVERT	54/74L55, 54H/74H55
Expander	Dual 4-input	54/7460, 54H/74H60
	Triple 3-input	54H/74H61
	3-2-2-3-input AND–OR	54H/74H62

TABLE 11.1.1   (Continued)

Gate	Function	Device
	Edge-triggered J-K	54/7470
	J-K master–slave	54/7472
	Dual J-K master–slave	54/7473
	Dual D-type edge triggered	54/7474
	Dual J-K master–slave, preset and clear	54/7476
	J-K master–slave	54/74104
	J-K master–slave	54/74105
	Dual J-K master–slave	54/74107
	Monostable nonretriggerable	54/74121
	R-S master–slave	54/74L71
	J-K master–slave	54/74L72
	Dual J-K master–slave	54/74L73
Flip-flop	Dual D-type edge-triggered	54/74L74
	Dual J-K master–slave	54/74L78
	J-K master–slave	54H/74H71
	J-K master–slave	54H/74H72
	Dual J-K master–slave	54H/74H73
	Dual D-type edge-triggered	54H/74H74
	Dual J-K master–slave with preset and clear	54H/74H76
	Dual J-K master–slave	54H/74H78
	J-K, negative edge-triggered	54H/74H101
	J-K, negative edge-triggered	54H/74H102
	Dual J-K, negative edge-triggered	54H/74H103
	Dual J-K, negative edge-triggered	54H/74H106
	Dual J-K, negative edge-triggered	54H/74H108

Code Converters   Code converters are combinational devices which convert information from one code into another. Commonly used code converters include binary-to-BCD (binary-coded decimal) converters and BCD-to-binary converters, among others.

Decoders/Demultiplexers   There are two categories of decoders/demultiplexers, logic decoders/demultiplexers, and display decoders/drivers. Logic decoders/demultiplexers select and activate a particular output as specified by the address. They can act as minterm generators in random and control logic. Display decoders/drivers generate numeric codes such as seven-segment and then provide the codes to a driver or drive the displays directly.

Arithmetic Devices   Arithmetic devices available in TTL/MSI form are full adders, carry look-ahead adders, and arithmetic logic units.

Comparators   Comparator systems fall into two classes: identity comparators, which detect whether or not two words are identical, and magnitude comparators, which also detect which of the two words is larger. Magnitude comparators are more complex and tend to be slower.

Counters   Counters are found in almost every kind of digital system. They are used not only for counting but also for system operation sequencing, frequency division, and mathematical manipulation as well. Counters are made up of flip-flops, thus are memory systems in the sense that they "remember" how many clock pulses have been applied to the input.

There are two types of counters: the ripple (asynchronous) counter and the synchronous counter. The ripple counter is simple in logic and therefore easy to design. The ripple counter is limited in its speed of operation. Since the flip-flops in the ripple counter are not under the command of a single clock pulse, it is an asynchronous counter. The synchronous counter, on the other hand, is in general more difficult to design compared with the ripple counter. All the flip-flops of a synchronous counter are under the control of the same clock pulse; therefore, the synchronous counters eliminate the cumulative flip-flop delay of the ripple counter. Repetition rate is limited only by the delay of any one flip-flop plus delays introduced by control gating. The synchronous and asynchronous sequential circuit analysis and design procedures presented in Chapter 9 can be applied to the analysis and design of synchronous and ripple counters, respectively.

Shift Registers   A shift register is a group of cascaded flip-flops. Each flip-flop output is connected to the input of the following flip-flop, and a common clock pulse is applied to all flip-flops, clocking them synchronously. Shift registers are used in digital systems for temporary information storage and data manipulation and transferring. In addition, they can be used in many counting circuits, including simple counters, available modulo counters, up/down counters, and increment counters. The input and output of a shift register may be of serial and parallel types; accordingly, the shift register may be classified into five classes:

1.   Serial-in/serial-out shift registers.
2.   Parallel-in/serial-out shift registers.
3.   Serial-in/parallel-out shift registers.
4.   Parallel-in/parallel-out shift registers.
5.   Parallel-in/parallel-out bidirectional shift registers.

All of them are available in integrated circuits.

Memories   Semiconductor memories are also made up of flip-flops. They fall into two basic categories: random access (read/write) memory (RAM) and read-only memory (ROM). A RAM is capable of read and write operations. It has nondestructible readout, but *data in the memory are lost when the power is shut off.* In a ROM, the data content is fixed, readout is nondestructible, and the *data are retained indefinitely even when the power is shut off.*

Some available 54/74 series TTL/MSI combinational and sequential integrated circuits are given in Tables 11.1.2 and 11.1.3, respectively. The descriptions of these integrated cricuits can be found in Texas Instruments, Inc.'s *The TTL Data Book*.

The advent of large-scale integrated arrays has aroused interest in the old and basic idea of using complete arrays to execute combinational logic functions. Previously, the arrays had been considered highly inefficient solutions, even though the

**TABLE 11.1.2 Some Available 54/74 Series
Combinational TTL/MSI Integrated Circuits**

MSI	Function	Device
Data selector/ multiplexer	16-bit	54/74150
	8-bit, with strobe	54/74151
	8-bit	54/74152
	Dual 4-line-to-1-line	54/74153
	Quad 2-input	54/74157
Encoder	Full BCD priority encoder	54/74147
	Cascadable octal priority encoder	54/74148
Code converter	6-line-BCD to 6-line	54/74184
	Binary, or 4-line to 4-line	
	BCD 9's/BCD 10's converters	
	6-bit-binary to 6-bit-BCD converter	54/74185A
Decoder/ demultiplexer	BCD-to-decimal (4 to 10 lines)	54/7442
	Excess-3-to-decimal (4 to 10 lines)	54/7443
	Excess-3-Gray-to-decimal (4 to 10 lines)	54/7444
	BCD-to-decimal decoder/driver (4 to 10 lines)	54/7445
	BCD-to-7-segment decoder/driver	54/7446
	BCD-to-7-segment decoder/driver	54/7447
	BCD-to-7-segment decoder/driver	54/7448
	BCD-to-7-segment decoder/driver	54/7449
	BCD-to-decimal decoder/driver	74141
	BCD-to-decimal decoder/driver	54/74145
	4-line-to-16-line decoder/demultiplexer	54/74154
	Dual 2-line-to-4-line decoder/demultiplexer	54/74155
	Dual 2-line-to-4-line decoder/demultiplexer, open collector	54/74156
Arithmetic device	Gated full-adder	54/7480
	2-bit full-adder	54/7482
	4-bit full-adder	54/7483
	Quad 2-input exclusive-OR gate	54/7486
	4-bit arithmetic logic unit	54/74181
	Look-ahead carry generator	54/74182
Comparator	4-bit magnitude comparator	54/7485, 54/74L85, 54/74S85

design was simple, primarily because of the great waste of circuits. However, as the cost per function and the size have decreased, the ease of design has again become appealing. The arrays, most often called ROM pairs and programmable logic arrays (PLA), have become almost a panacea for digital design problems. ROM pairs and PLA integrated circuits are MOS/LSI.

ROM pairs The circuit schematic of a typical MOS ROM pair is shown in Fig. 11.1.1(a). Each "cell" of the array provides the NAND function, and each path from

TABLE 11.1.3   Some Available 54/74 Series Sequential
TTL/MSI Integrated Circuits

Function	Device
	Ripple (asynchronous) counters
Decade	54/74L90, 54/74LS196, 54/7490, 54/7490A, 54/74176, 54/74196
4-bit binary	54/74L93, 54/74LS197, 54/74293, 54/7493A, 54/74177, 54/74197
Divide-by-12	54/7492A
	Synchronous counters
Decade	54/74160, 54/74162
Decade up/down	54/74L192, 54/74190, 54/74LS190, 54/74192, 54/74LS192
4-bit binary	54/74L193, 54/74191, 54/74S191, 54/74193, 54/74LS193,
	54/74161, 54/74163
	Shift registers
8-bit serial-in, serial-out	54/7491A, 54/74L91
4-bit parallel-in, serial-out	54/7494
8-bit parallel-in, serial-out	54/74165, 54/74166
8-bit serial-in, parallel-out	54/74164, 54/74L164
4-bit parallel-in,	54/74L95, 54/74L99, 54/74LS295, 54/74LS95A, 54/74LS195
parallel-out	54/74178, 54/74179, 54/7495A, 54/74195, 54/74S195
5-bit parallel-in,	54/7496, 54/74L96
parallel-out	
8-bit parallel-in,	54/74199
parallel-out	
4-bit parallel-in	
parallel-out (bidirectional)	54/74194, 54/74S194, 54/74LS194
8-bit parallel-in	
parallel-out (bidirectional)	54/74198
	Random-access memory (RAM)
16-bit register file	54/74170
16-bit multiple-port	74172
register file	
16-bit read/write memory	54/7481A, 54/7484A
64-bit read/write memory	7489
256-bit read/write	
memory	74200, 74S200, 74S206
	Read-only memory (ROM)
256-bit	54/7488A
256-bit programmable	74188A
512-bit programmable	54/74186
1024-bit	54/74187

an input to an output passes through two such cells. The standard-logic schematic diagram of a ROM pair is depicted in Fig. 11.1.1(b). Note that the outputs of the first (upper) ROM are capable of realizing any minterms of a switching function of n variables, where n is the number of inputs of the ROM pair. For example, for $n = 2$ the upper ROM can realize the following (all possible) minterms of two variables:

A	B	Minterm
0	0	$m_0 = \bar{A}\bar{B}$
0	1	$m_1 = \bar{A}B$
1	0	$m_2 = A\bar{B}$
1	1	$m_3 = AB$

The second (lower) ROM provides the OR function; thus, it can produce switching functions of n variables. For instance, the four functions of two variables described below can be realized by the ROM pair in Fig. 11.1.1(b).

Inputs		Outputs			
A	B	F_1	F_2	F_3	F_4
0	0	1	1	0	0
0	1	1	0	1	0
1	0	0	1	0	1
1	1	0	0	1	1

The coupled matrix arrays consisting of two separate ROM's in Fig. 11.1.1(a) may also be presented in the array-logic schematic of Fig. 11.1.1(c). In Fig. 11.1.1(c) the node structure for the coupled ROM arrays is clearly shown. The dark circles within each ROM serve to indicate the presence of a gate on the ROM chip. A block-diagram representation of this ROM pair is shown in Fig. 11.1.1(d).

A ROM pair may contain from 1,000 to 20,000 summing nodes; it is usually programmed by a machine controlled by a computer. The desired input/output function table (truth table) is described on a sequenced deck of standard 80-column computer cards according to a specific format. The number of cards and the data-card format depend upon the type of the chip and the manufacturer. They are described in the manufacturers' products catalogs.

PLA ROM pairs can only realize combinational logic. In order for them to be able to implement sequential logic, memory elements must be added, which becomes a PLA. The simplest form of PLA contains the AND–OR ROM pair with clocked flip-flops in the feedback path, as shown in Fig. 11.1.2(a). Comparing this diagram with the one in Fig. 9.3.1, it is seen that a PLA can realize any sequential machine in

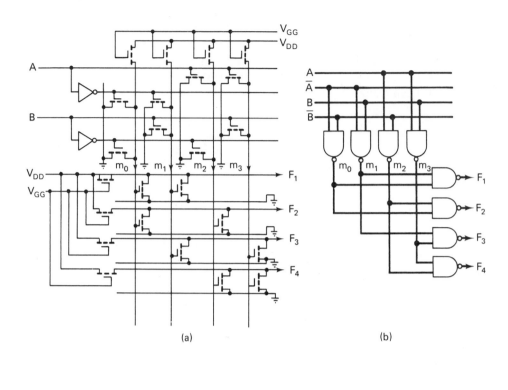

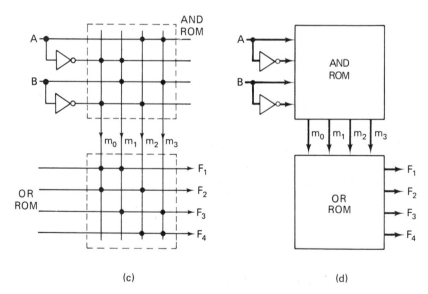

Fig. 11.1.1 (a) Circuit schematic of a ROM-pair; (b) standard-logic schematic; (c) array-logic schematic; (d) block diagram.

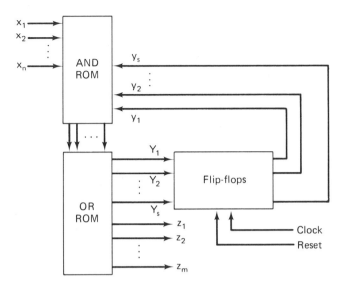

$$Y_i = Y_i\,(x_1,\,x_2,\,\ldots,\,x_n;\,y_1,\,y_2,\,\ldots,\,y_s),\qquad i = 1,\,2,\,\ldots,\,s.$$
$$z_j = z_j\,(x_1,\,x_2,\,\ldots,\,x_n;\,y_1,\,y_2,\,\ldots,\,y_s),\qquad j = 1,\,2,\,\ldots,\,m.$$

Fig. 11.1.2 Block diagram of PLA.

a straightforward manner. Major advantages of the PLA result from the fact that the ROM-pair provides a matrix for design and production automation and that a wide variety of sequential logic functions are economically obtained by designing for a single modification of the gate mask during circuit fabrication. A detailed block diagram of a PLA is shown in Fig. 11.1.3. It should be noted that any commercial PLA can be used as a ROM pair (with all the flip-flops of the PLA unconnected). The block diagram of a typical PLA, TMS 2000, is shown in Fig. 11.1.4.

Some available MOS/LSI RAM's, ROM's, and PLA's are given in Table 11.1.4. The descriptions of these integrated circuits can be found in Texas Instruments, Inc.'s *Memory Data Book*.

Digital designs using TTL/SSI, TTL/MSI, and MOS/LSI integrated circuits are presented in the following three sections.

11.2 Digital Design Using TTL/SSI Integrated Circuits

In this section, we present digital design using ready-made TTL/SSI integrated circuits. For TTL, the logic 0 and 1 are represented by voltage intervals $0 \sim 0.8$ V and $3.6 \sim 5$ V, respectively. The dc power supply (V_{cc}) for TTL is $+5$ V. The design is best illustrated by use of examples.

The first example shows that any gate can be realized by using NAND gates (NOR-gates), since the nand operation (nor operation) is functionally complete.

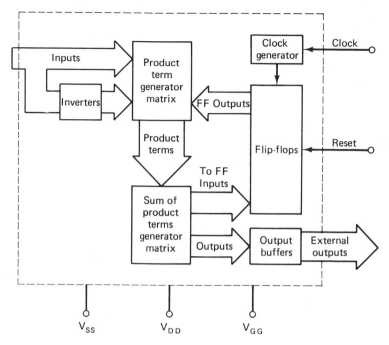

Fig. 11.1.3 Detailed block diagram of PLA.

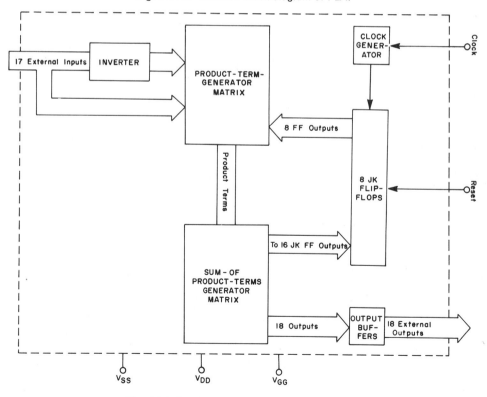

Fig. 11.1.4 Block diagram of PLA TMS 2000.

TABLE 11.1.4 Some Available MOS/LSI Integrated Circuits

Function	Device
Random-access memory (RAM)	
256-bit	TMS 1101 JC, NC
Fully decoded 1024-bit	TMS 1103 NC
High-speed content-addressable	TMS 4000 JC, NC
256-bit	TMS 4003 JR, NC
2048-bit dynamic	TMS 4020 NC
1024-bit	TMS 4023 NC
2048 dynamic	TMS 4025 NC
Read-only memory (ROM)	
2560-bit dynamic	TMS 2300 JC, NC
Row output character generator	TMS 2400 JC, NC
2560-bit static	TMS 2500 JC, NC
2048-bit static	TMS 2600 JC, NC
USASCII-to selectric/	
selectric-to-USASCII code converter	TMS 2602 JC, NC
EBCDIC-to-USASCII code converter	TMS 2603 JC, NC
USASCII-to-EBCDIC/	
selectric-to-EBCDIC code generator	TMS 2604 JC, NC
USASCII, BAUDOT, SELECTRIC	
EBCDIC code generator	TMS 2605 JC, NC
3071-bit static	TMS 2700 JC, JM, NC
1024-bit static	TMS 2800 JC, NC
1280-bit static	TMS 2900 JC, NC
Series character generator	TMS 4100 JC, NC
4096-bit static	TMS 4400 JC, NC
Programmable logic array (PLA)	
17-input, 18-output, 8-JKFF PLA	TMS 2000
13-input, 10-output, 10-JKFF PLA	TMS 2200

Example 11.2.1

Construct an AND-gate, an OR-gate, a NOR-gate, and XOR (EXCLUSIVE-OR)-gate using NAND-gates only.

The AND, OR, NOR, XOR operations can be expressed in terms of NAND operations as

$$A \cdot B = \overline{\overline{A \cdot B}} = \overline{A \uparrow B} \tag{11.2.1}$$

$$A + B = \overline{\overline{A + B}} = \bar{A} \uparrow \bar{B} \tag{11.2.2}$$

$$A \downarrow B = \overline{A + B} = \bar{A} \cdot \bar{B} = \overline{\overline{\bar{A} \cdot \bar{B}}} = \overline{\bar{A} \uparrow \bar{B}} \tag{11.2.3}$$

$$A \oplus B = A\bar{B} + \bar{A}B = (\bar{A} + \bar{B})A + (\bar{A} + \bar{B})B$$

$$= \overline{AB} \cdot A + \overline{AB} \cdot B = (\overline{\overline{AB} + A}) + (\overline{\overline{AB} + B})$$
$$= (\overline{\overline{\overline{A} + \overline{B}} + A}) + (\overline{\overline{\overline{A} + \overline{B}}} + \overline{B})$$
$$= (\overline{(\overline{A} + \overline{B}) \cdot A}) \cdot (\overline{(\overline{A} + \overline{B}) \cdot B})$$
$$= (\overline{A \cdot B \cdot A}) \cdot (\overline{A \cdot B \cdot B})$$
$$= ((A \uparrow B) \uparrow A) \uparrow ((A \uparrow B) \uparrow B) \qquad (11.2.4)$$

The logic-circuit realizations of these gates using exclusively NAND-gates are shown in Fig. 11.2.1(a). Notice that all the NAND-gates in these realizations are two-input NAND-gates and each realization has four NAND-gates or less. The selection of SN 7400 quadruple two-input NAND-gates for the implementations of these logic circuits appears to be most suitable. They are shown in Fig. 11.2.1(b).

Example 11.2.2

Consider the realization of the switching function

$$F(A, B, C, D) = AB + AC + CD + BD \qquad (11.2.5)$$

It can be easily shown that this function is minimal, and the logic-circuit realization of it is shown in Fig. 11.2.2(a). A suitable TTL/SSI integrated-circuit implementation of the logic circuit of Fig. 11.2.2(a) is shown in Fig. 11.2.2(b), in which a single SN 74H52 chip is used.

The following two examples illustrate the TTL/SSI integrated-circuit realization of sequential logic.

Example 11.2.3

Design a Moore-type binary serial adder using TTL/SSI integrated circuits.

The binary serial adder has two inputs x_1 and x_2 and one output z. Each present state of the adder corresponds to a carry from the previous addition which is also considered in determining the present output. Define four states, A_0, A_1, B_0, and B_1, of this machine as follows:

State A_0: no carry with state output 0
State A_1: no carry with state output 1
State B_0: carry with state output 0
State B_1: carry with state output 1

The transition table (Moore-machine description) of the binary adder is shown in Fig. 11.2.3(a). Suppose that the four states are coded as follows:

State	$y_1 y_2$
A_0	00
A_1	01
B_0	11
B_1	10

Then the binary-coded transition table obtained from this state assignment is shown in Fig. 11.2.3(b). If J-K flip-flops are chosen to use in the realization of this machine, the minimized excitation functions of the J-K flip-flops are found to be [see Fig. 11.2.3(c)]

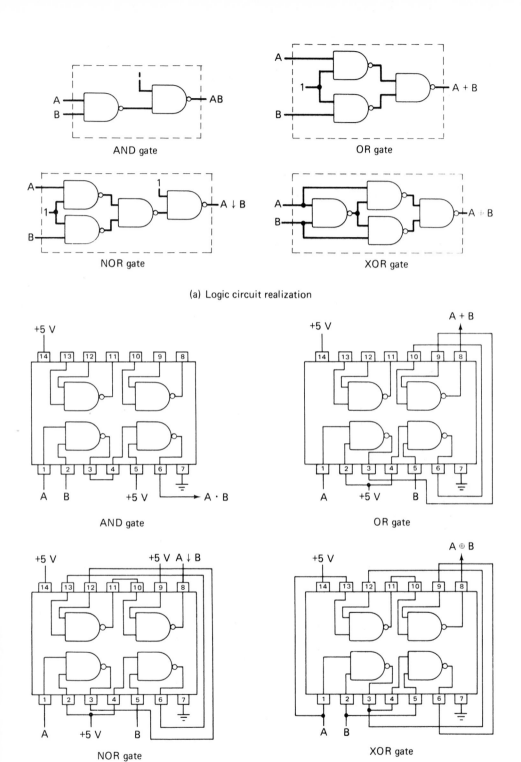

(a) Logic circuit realization

(b) SN7400 realization

Fig. 11.2.1 Example 11.2.1.

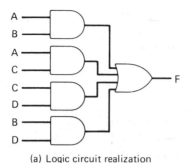

(a) Logic circuit realization

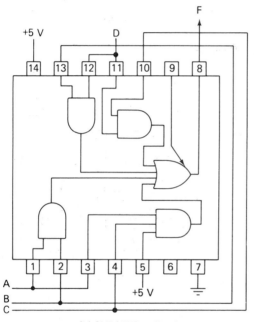

(b) SN74H52 realization   **Fig. 11.2.2** Example 11.2.2.

$$J_1 = x_1 x_2 \qquad\qquad\qquad\qquad\qquad (11.2.6)$$

$$K_1 = \bar{x}_1 \bar{x}_2 = \overline{x_1 + x_2} \qquad\qquad\qquad (11.2.7)$$

$$J_2 = \bar{x}_1 x_2 + x_1 \bar{x}_2 + x_1 \bar{y}_1 + \bar{x}_1 y_1 = (x_1 \oplus x_2) + (x_1 \oplus y_1) \qquad (11.2.8)$$

$$K_2 = \bar{x}_1 \bar{x}_2 \bar{y}_1 + x_1 x_2 y_1 = \overline{x_1 + x_2 + y_1} + x_1 x_2 y_1 \qquad (11.2.9)$$

An integrated-circuit realization using four SN 7400, two SN 7411, one SN 7432, and one SN 7473 is shown in Fig. 11.2.3(d).

Example 11.2.4

A particular sequence of numbers is provided as inputs to a lock, which "remembers" each number and opens at the termination of the correct sequence. Design a sequential system (sequential lock) that has two inputs (x_1, x_2), one output (z), and four stable states (R, E, B, C). The proper input sequence for a logic "1" output (lock opens) should be 00–01–11

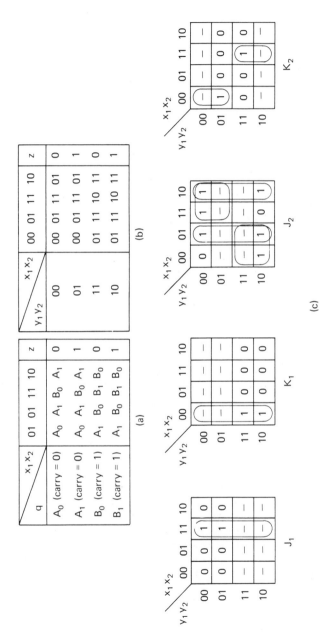

Fig. 11.2.3 (a) Transition table; (b) binary coded transition table; (c) excitation functions; (d) TTL/SSI realization.

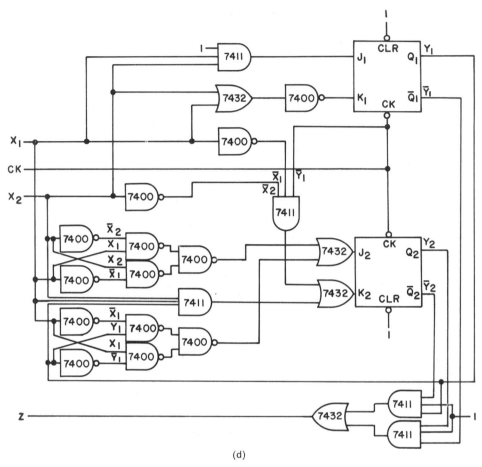

(d)

Fig. 11.2.3 (Continued)

(R–B–C); that is, the output of all states is zero with the exception of state "C," which has an output of 1. The system should return to state "R" (reset) whenever the input is 00. The system should go to state "E" (error) whenever the input sequence is incorrect. Additionally, the system should remain in state "E" (error) until the input is 00, which returns it to the reset state. Assume that TTL/SSI J-K flip-flops and NAND-gates are to be used.

The transition table of this lock is shown in Fig. 11.2.4(a). With the state assignment indicated, the binary-coded transition table is shown in Fig. 11.2.4(b). The Karnaugh maps of the excitation functions and their minimization are shown in Fig. 11.2.4(c). The minimized excitation functions are

$$J_1 = x_1 + y_2 x_2 = \bar{x}_1 \uparrow (y_2 \uparrow x_2) \tag{11.2.10}$$

$$K_1 = \bar{x}_1 \bar{x}_2 = (\bar{x}_1 \uparrow \bar{x}_2) \uparrow (\bar{x}_1 \uparrow \bar{x}_2) \tag{11.2.11}$$

$$J_1 = \bar{x}_1 \bar{y}_1 x_2 = (\bar{x}_1 \uparrow \bar{y}_1 \uparrow x_2) \uparrow (\bar{x}_1 \uparrow \bar{y}_1 \uparrow x_2) \tag{11.2.12}$$

$$K_2 = y_1 + \bar{x}_1 + \bar{x}_2 = \bar{y}_1 \uparrow x_1 \uparrow x_2 \tag{11.2.13}$$

Finally, an integrated-circuit realization using TTL/SSI is given in Fig. 11.2.4(d).

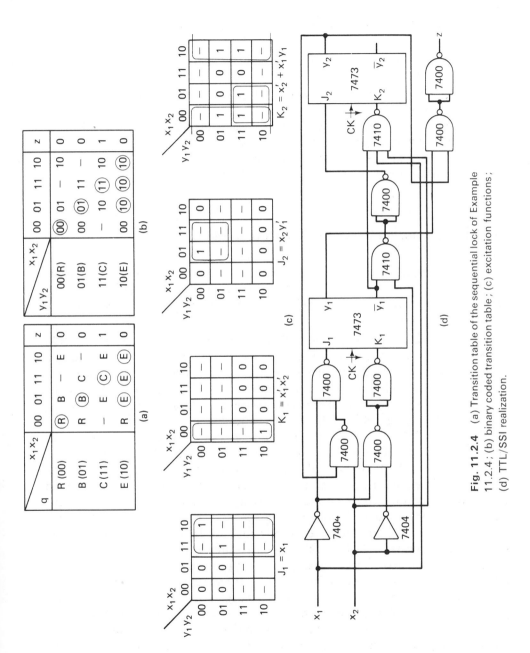

Fig. 11.2.4 (a) Transition table of the sequential lock of Example 11.2.4; (b) binary coded transition table; (c) excitation functions; (d) TTL/SSI realization.

423

Exercise 11.2

1. Implement the logic-circuit diagram of the binary serial adder in Fig. 9.2.1 using TTL/SSI chips.

2. Implement the logic-circuit diagram of Fig. 9.3.2(f) using any 74 series TTL/SSI integrated circuits.

3. Find an economical circuit realization for the logic circuit of Fig. 9.4.2(e) using 74 series-integrated circuits.

4. Convert the D flip-flops in the asynchronous circuit of Fig. 9.5.10 into J-K flip-flops and then realize it using 74 series flip-flops and gates.

5. (a) Construct a Mealy-type transition table of the combination lock of Example 11.2.4.
    (b) Realize the transition table of part (a) using 74 series flip-flops and gates.

11.3   Digital Design Using TTL/MSI Integrated Circuits

As the microelectronics technology continues to advance, integrated circuits are made more and more complex and the circuit cost per gate or function becomes cheaper and cheaper. In general, complex MSI/LSI devices offer the following advantages over SSI gates:

1.   Simplified mechanical construction. Since the packing density of MSI/LSI is greatly increased (compared to that of SSI), more functions can be constructed on a printed circuit board, which simplifies the mechanical construction of the system.

2.   Reduced interconnections. As a direct result of 1, the numbers of solder joints, backplane wiring, and connectors in a system using MSI/LSI devices as building blocks would be in general much less compared to that using SSI gates.

3.   Reduced power consumption and total heat generation. However, the increased density possible with MSI can result in higher power and heat densities.

4.   Low cost. MSI/SSI and LSI/SSI cost comparisons must consider true total cost, not only the purchase price of the integrated circuits. An unavoidable overhead cost is associated with each IC, attributed to testing, handling, insertion, and soldering plus the appropriate share of connectors, PC boards, power supplies, cabinets, and so on. When this cost is added to semiconductor cost, MSI offers more economical solutions, even in cases when the MSI components are more expensive.

5.   Decreased design, debugging, and servicing costs and time. Because MSI/LSI devices are functional subsystems, it is much easier and faster to design a system with them. Functional partitioning also simplifies debugging and service.

6.   Improved system reliability. Mean time between failure of an MSI is roughly the same as that of an SSI device, so a reduction in package count increases system mean time between failure. Moreover, the reduced interconnections improve reliability.

Because of these obvious advantages, MSI/LSI is generally accepted in the regular and repetitive portions of digital designs.

In this section we present digital design using TTL/MSI integrated circuits, digital design using MOS/LSI integrated circuits will be discussed in the next section.

Digital design using TTL/MSI integrated circuits is best illustrated by examples.

Example 11.3.1

As mentioned in Section 11.1, the logic decoder/demultiplexer can act as minterm generators in random logic. For example, SN 75154 is a 24-pin TTL/MSI 4-line-to-16-line

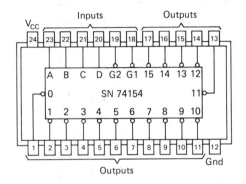

(a)

Inputs						Outputs															
G1	G2	D	C	B	A	0	1	2	3	4	5	6	7	8	9	10	11	12	13	14	15
L	L	L	L	L	L	L	H	H	H	H	H	H	H	H	H	H	H	H	H	H	H
L	L	L	L	L	H	H	L	H	H	H	H	H	H	H	H	H	H	H	H	H	H
L	L	L	L	H	L	H	H	L	H	H	H	H	H	H	H	H	H	H	H	H	H
L	L	L	L	H	H	H	H	H	L	H	H	H	H	H	H	H	H	H	H	H	H
L	L	L	H	L	L	H	H	H	H	L	H	H	H	H	H	H	H	H	H	H	H
L	L	L	H	L	H	H	H	H	H	H	L	H	H	H	H	H	H	H	H	H	H
L	L	L	H	H	L	H	H	H	H	H	H	L	H	H	H	H	H	H	H	H	H
L	L	L	H	H	H	H	H	H	H	H	H	H	L	H	H	H	H	H	H	H	H
L	L	H	L	L	L	H	H	H	H	H	H	H	H	L	H	H	H	H	H	H	H
L	L	H	L	L	H	H	H	H	H	H	H	H	H	H	L	H	H	H	H	H	H
L	L	H	L	H	L	H	H	H	H	H	H	H	H	H	H	L	H	H	H	H	H
L	L	H	L	H	H	H	H	H	H	H	H	H	H	H	H	H	L	H	H	H	H
L	L	H	H	L	L	H	H	H	H	H	H	H	H	H	H	H	H	L	H	H	H
L	L	H	H	L	H	H	H	H	H	H	H	H	H	H	H	H	H	H	L	H	H
L	L	H	H	H	L	H	H	H	H	H	H	H	H	H	H	H	H	H	H	L	H
L	L	H	H	H	H	H	H	H	H	H	H	H	H	H	H	H	H	H	H	H	L
L	H	X	X	X	X	H	H	H	H	H	H	H	H	H	H	H	H	H	H	H	H
H	L	X	X	X	X	H	H	H	H	H	H	H	H	H	H	H	H	H	H	H	H
H	H	X	X	X	X	H	H	H	H	H	H	H	H	H	H	H	H	H	H	H	H

H = High level, L = Low level, X = Irrelevant

(b)

Fig. 11.3.1 (a) Pin description of SN 74154; (b) function table; (c) functional block diagram and schematics of inputs and outputs; (d) realization of $F(A, B, C, D) = \Sigma\,(3, 5, 7, 10, 11, 12, 13, 14, 15)$ using an SN 74154.

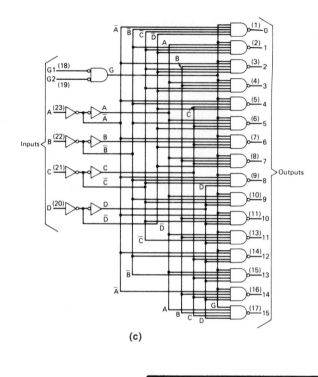

(c)

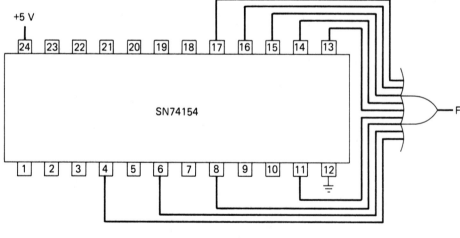

(d)

Fig. 11.3.1 (Continued)

decoder/demultiplexer, which can be used to generate any of the $2^{16} = 65{,}536$ possible switching functions of four variables. The pin description, function table, and functional block diagram and schematics of inputs and outputs of SN 74154 are shown in Figs. 11.3.1(a), (b), and (c), respectively. From the functional block diagram of Fig. 11.3.1(c), it is seen that the

16 outputs of this chip produce exactly the 16 minterms of four-variable switching functions. For example, the sum-of-products form of the function of Eq. (11:2.5) is found to be

$$F(A, B, C, D) = \Sigma\,(3, 5, 7, 10, 11, 12, 13, 14, 15) \qquad (11.3.1)$$

which can be realized by ORing the 3, 5, 7, 10, 11, 12, 13, 14, and 15 outputs, which correspond to the minterms m_3, m_5, m_7, m_{10}, m_{11}, m_{12}, m_{13}, m_{14}, and m_{15}, respectively, as shown in Fig. 11.3.1(d).

A standard synchronous up/down-counter, if clocked a sufficient number of times, will reset when the maximum counting sequence has been exceeded; that is, a 4-bit counter will count to 15 (BCD) and reset. The counting sequence may be modified, however, such that the counter will count to some number n and reset ($n <$ maximum count). One method that may be used to construct an n-counter is shown below.

Example 11.3.2

The counter (SN 74193) increments on each clock pulse and the output of the counter (SN 74193) is compared (SN 7485) to some number n [see Fig. 11.3.2(a)]. When the output of the counter is equal to n, the counter is reset. Design an up counter that counts to 10_{10} and resets. The counter should be preset so that the counting sequence begins at 0000.

The pin descriptions of the chips SN 74193 and SN 7485 are shown in Fig. 11.3.2(b), and the counter that counts up to 10_{10} is shown in Fig. 11.3.2(c).

Example 11.3.3

Suppose that it is desired to design a modulo-12 counter with seven-segment display. The required counting sequence is as follows:

Decimal	Q_D	Q_C	Q_B	Q_A	
1	0	0	0	1	
2	0	0	1	0	
3	0	0	1	1	
4	0	1	0	0	
5	0	1	0	1	
6	0	1	1	0	
7	0	1	1	1	
8	1	0	0	0	
9	1	0	0	1	
10	1	0	1	0	
11	1	0	1	1	
12	1	1	0	0	immediately
{13}	{1	1	0	1	$\longrightarrow$ 0 0 0 1}
1	0	0	0	1	
2	0	0	1	0	etc.

Design this counter using TTL/MSI integrated circuits.

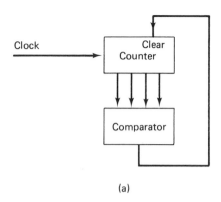

(a)

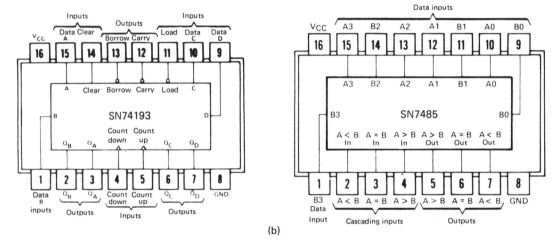

(b)

Fig. 11.3.2 (a) A scheme for designing modulo-*n* counter;
(b) pin descriptions of SN 74193 and SN 7485; (c) a modulo-10
counter.

A design of this counter is shown in Fig. 11.3.3. In this design, one SN 74193 4-bit up/down-counter, one SN 74185A binary-to-BCD converter, two SN 7446 BCD-to-seven-segment decoders/drivers, two DL 707 seven-segment displays, two SN 7404 inverters, and one SN 7410 NAND-gates are used.

Up to this point, two methods for realizing sequential machines have been presented. One is using TTL/SSI flip-flops and gates. The logic synthesis procedure and circuit design of this method were discussed in Sections 9.3 and 11.2, respectively. The other is using the functional block-diagram approach, in which each functional block is filled in with an MSI chip, as shown in the two examples above.

A third approach described below is using an MSI or LSI RAM plus a register and a selector device, as shown in the block diagram of Fig. 11.3.4. This scheme can

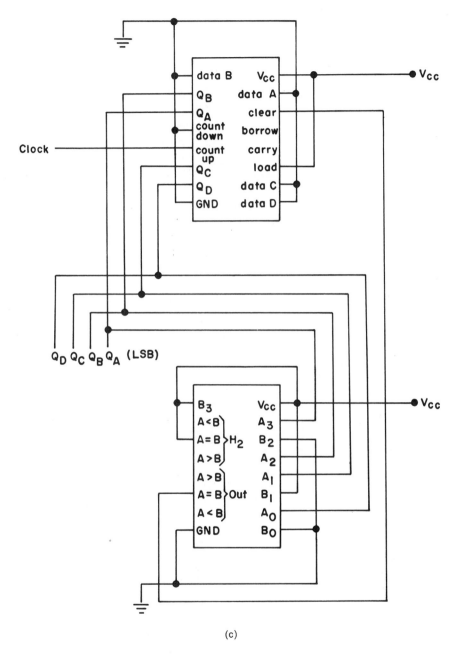

(c)

Fig. 11.3.2   (Continued)

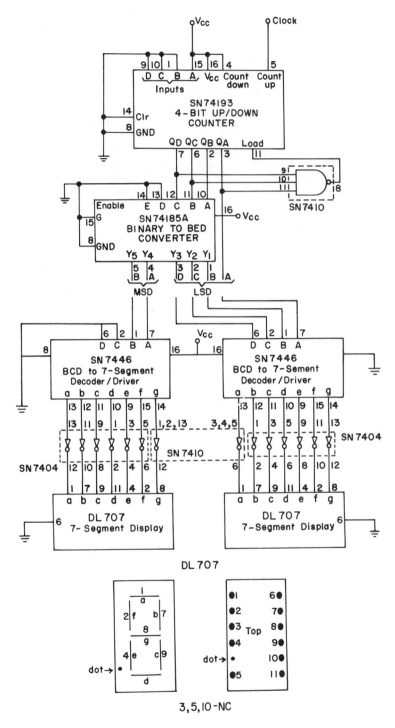

Fig. 11.3.3 A modulo-12 counter with seven-segment displays.

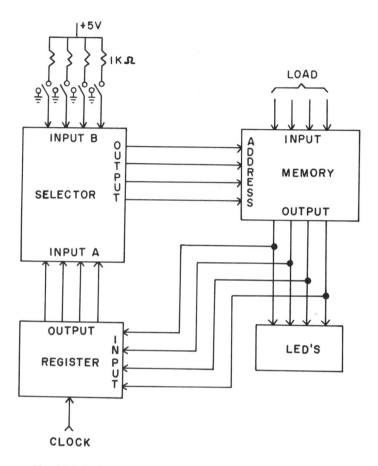

Fig. 11.3.4   Block diagram of the system for realizing a Moore machine.

realize any Moore-type sequential machine with ease. The procedure for converting a Moore machine into a Mealy machine is straightforward. For example, consider the Moore-machine description of the binary serial adder in Fig. 11.2.3(a), which, for convenience, is repeated in Table 11.3.1(a). The conversion of this Moore machine into a Mealy machine calls for the deletion of the state-output z column and the addition of an output to each transition of the machine. The resulting transition table, which is of Mealy type, is shown in Table 11.3.1(b). It is observed that states A_0 and A_1 are equivalent and so are states B_0 and B_1. Combining the equivalent states A_0 and A_1 (B_0 and B_1) into one state and denoting it $A(B)$, we obtain the minimized Mealy machine in Table 11.3.1(c).

To obtain a Moore machine from a Mealy machine, it is first necessary to split every state of the Mealy machine if different output values are associated with the transitions into that state. For example, state A of Table 11.3.1(c) can be reached

TABLE 11.3.1 Example for Illustrating the Moore Machine/Mealy Machine Conversion

Converting Moore machine into Mealy machine ⟶

(a)

q \\ x_1x_2	00	01	11	10
A_0 (carry = 0)	A_0	A_1	B_0	A_1
A_1 (carry = 0)	A_0	A_1	B_0	A_1
B_0 (carry = 1)	A_1	B_0	B_1	B_0
B_1 (carry = 1)	A_1	B_0	B_1	B_0

(b)

q \\ x_1x_2	00	01	11	10	z
A_0	$A_0, 0$	$A_1, 1$	$B_0, 0$	$A_1, 1$	0
A_1	$A_0, 0$	$A_1, 1$	$B_0, 0$	$A_1, 1$	1
B_0	$A_1, 1$	$B_0, 0$	$B_1, 1$	$B_0, 0$	0
B_1	$A_1, 1$	$B_0, 0$	$B_1, 1$	$B_0, 0$	1

A_0, A_1 are equivalent
B_0, B_1 are equivalent

(c)

q \\ x_1x_2	00	01	11	10
A (carry = 0)	$A, 0$	$A, 1$	$B, 0$	$A, 1$
B (carry = 1)	$A, 1$	$B, 0$	$B, 1$	$B, 0$

Converting Mealy machine into Moore machine ⟶

from either state A or state B. But since different outputs are associated with these transitions, state A must be replaced by two equivalent states, A_0 with an output 0 and A_1 with an output 1, as shown in Table 11.3.1(b). Every transition to A with a 0 output is directed to A_0, and every transition to A with a 1 output, to A_1. Applying the same procedure to state B yields the transition table of Table 11.3.1(b), which can be transformed to the Moore machine of Table 11.3.1(a).

The following examples illustrate the realization of Moore machines using a RAM. In these examples, the memory device used is a 16-word 4-bit RAM (SN 7489). The RAM output is connected to a LED, which displays the output. After appropriate output data have been determined, it should be loaded (written) into the RAM. The purpose of the selector (SN 74157) is to select four of eight inputs (A's or B's), which are used as address locations. The selector simply connects inputs A or B directly to the output. The selector output may be reset (all zero) by enabling terminal 15. The shift registor (SN 7459A) is used in a parallel-in/parallel-out mode. The output of the RAM is tied to the input of the register. When the register is clocked, the data pass to the output and become the address of the RAM memory location; thus, the previous RAM output becomes the next address.

The address, data, and control inputs and data outputs of SN 7489, SN 74157, and SN 7495A are described below.

SN 7489 64-bit read/write memory:
> A, B, C, and D: four address inputs
> D_1, D_2, D_3, and D_4:   four data inputs
> S_1, S_2, S_3, and S_4:   four data outputs
> ME: memory enable
> WE: write enable

SN 74157 quad 2-to-1-line data selector/multiplexer:
> $1A, 2A, 3A, 4A$, and $1B, 2B, 3B, 4B$:   two sets of four data inputs
> $1Y, 2Y, 3Y$, and $4Y$:   four outputs
> G: strobe
> S: select

SN 7495A 4-bit parallel-access shift register:
> A, B, C, D: four inputs
> Q_A, Q_B, Q_C, Q_D: four outputs
>   Serial input (SI)
>   Mode control
>   Clock 1:   Right shift
>   Clock 2:   Left shift

The block diagram of the system for realizing Moore machines using these components is depicted in Fig. 11.3.5.

Example 11.3.4

Design a counter that will count consecutively from zero to nine and reset the tenth clock pulse. The counter should increment on each clock pulse (the only input to the system) and display the present count on a LED.

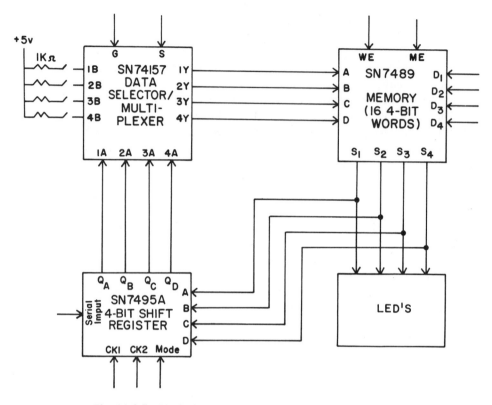

Fig. 11.3.5 Block diagram of the system for realizing Moore machines using SN 7489, SN 74157, SN 7495A, and LED's.

The circuit realization of the decade counter is shown in Fig. 11.3.6. The programming of the RAM in this design involves the following steps:

Step 1   Switch the Program/Run Switch to Program.

Step 2   Each address is called for, one at a time, by Program Address Select.

Step 3   Each address is programmed by Program Source Switches.

Step 4   After the addresses are set, the Program/Run Switch is set to Run.

Step 5   With address (0000) as a starting state, it is loaded into the shift register on the rise of the incoming clock.

Step 6   When the clock goes low, the contents of the shift register are fed into the RAM as the next address.

Step 7   When the clock goes high again, the contents of this new memory location are loaded into the shift register.

Step 8   The clock goes to the low state, causing the shift register to feed the new address to the RAM.

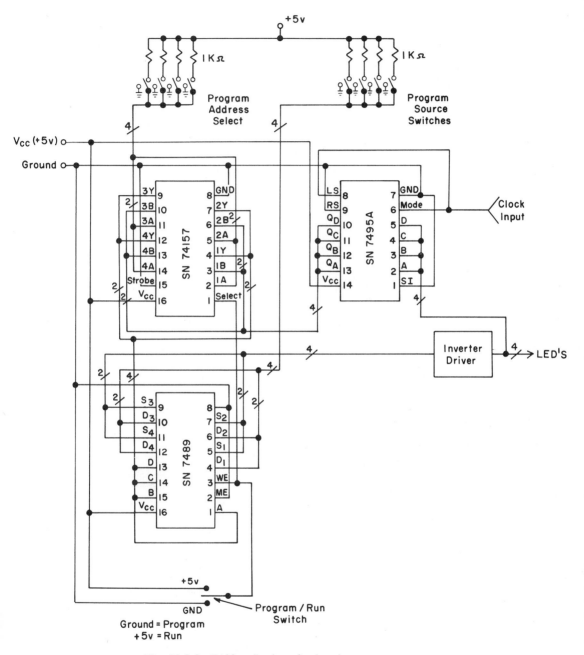

Fig. 11.3.6   RAM realization of a decade counter.

Step 9   Repeat steps 7 and 8 until the state of 9 is reached.

Step 10   Fill the contents of memory locations of addresses 1001 through 1111 with 0000 (the starting state). The memory contents of the SN 7489 RAM looks like this:

Program Stored in the RAM

Memory address	Stored data
D C B A	$S_4 S_3 S_2 S_1$
0 0 0 0	0 0 0 1
0 0 0 1	0 0 1 0
0 0 1 0	0 0 1 1
0 0 1 1	0 1 0 0
0 1 0 0	0 1 0 1
0 1 0 1	0 1 1 0
0 1 1 0	0 1 1 1
0 1 1 1	1 0 0 0
1 0 0 0	1 0 0 1
1 0 0 1	0 0 0 0
1 0 1 0	0 0 0 0
1 0 1 1	0 0 0 0
1 1 0 0	0 0 0 0
1 1 0 1	0 0 0 0
1 1 1 0	0 0 0 0
1 1 1 1	0 0 0 0

Erroneous states {1 0 1 0 … 1 1 1 1} — Reset to zero

(Present state)   (Next-state)

Any "accident" that will give an incorrect counting state exceeding 9 will, on the next pulse, give the reset state.

Example 11.3.5

Design a system that will operate an elevator in a building that has four floors. The system should have two inputs (x_1 and x_2) and four possible states. Each state should have an output such that a LED may be utilized to display the selected floor in BCD form. For purposes of simplification, the system may receive only one set of inputs (x_1 and x_2) at a time. The system should be designed such that its next-state is determined both by the inputs and its present state; that is, the RAM address will consist of the two inputs plus two outputs from the previous state of the RAM. The system will be similar to the block diagram shown in Fig. 11.3.4, with the exception of two shift register inputs which will be disconnected from the RAM output and will become x_1 and x_2, as shown in Fig. 11.3.7.

Denote the four floors by G, 1, 2, and 3. The transition diagram is shown in Fig. 11.3.8(a). If we code the four states as

State q	State variables $y_1 y_2$
G	00
1	01
2	10
3	11

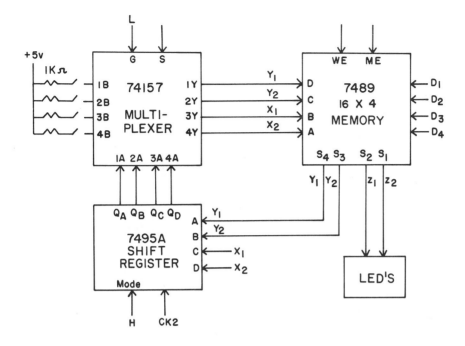

NOTE: I. L and H denote V_L and V_H, respectively.

2.	WE	ME	OPERATION
	L	L	WRITE
	H	L	READ

3.	S (74157)	OUTPUT
	L	Y = A
	H	Y = B

Fig. 11.3.7 Block diagram of the system for realizing the elevator controller of Example 11.3.5.

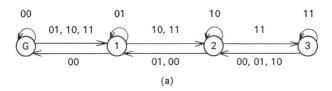

(a)

Fig. 11.3.8 (a) Transition diagram of the elevator controller; (b) coded transition table; (c) circuit realization.

y_1y_2 \\ x_1x_2	00	01	11	10	z_1z_2
00	00	01	01	01	00
01	00	01	10	10	01
11	10	10	11	10	11
10	01	01	11	10	10

Y_1Y_2

(b)

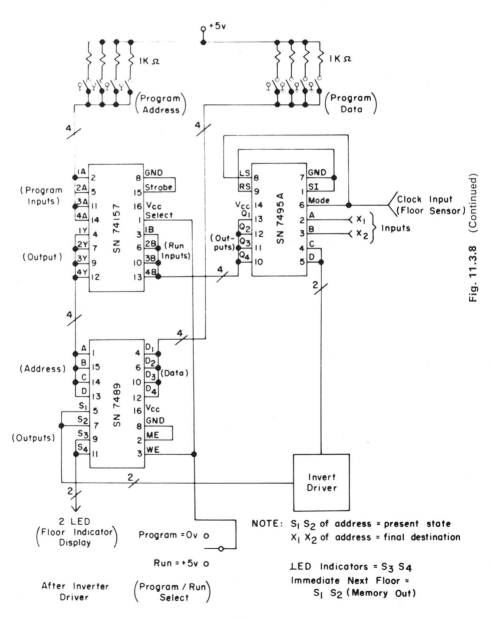

NOTE: $S_1 S_2$ of address = present state
$X_1 X_2$ of address = final destination

LED Indicators = $S_3 S_4$
Immediate Next Floor =
$S_1 S_2$ (Memory Out)

Fig. 11.3.8 (Continued)

(c)

the coded transition table is given in Fig. 11.3.8(b). The circuit realization is shown in Fig. 11.3.8(c). The elevator floor wanted is placed into the inputs (x_1, x_2). On a clock pulse the present state of the elevator (S_1, S_2) is fed along with the request (x_1, x_2) into memory through the shift register used as intermediate storage. The next-state will give the immediate next floor to go to, until the request has finally been reached. The program that implements the coded transition table of Fig. 11.3.8(b) is shown below.

Program Stored in the RAM

State	Memory address $S_1 S_2 x_1 x_2$	Stored data $S_1 S_2 S_3 S_4$
G	0 0 0 0	0 0 0 0
	0 0 0 1	0 1 0 1
	0 0 1 0	0 1 1 0
	0 0 1 1	0 1 1 1
1	0 1 0 0	0 0 0 0
	0 1 0 1	0 1 0 1
	0 1 1 0	1 0 1 0
	0 1 1 1	1 0 1 1
2	1 0 0 0	0 1 0 0
	1 0 0 1	0 1 0 1
	1 0 1 0	1 0 1 0
	1 0 1 1	1 1 1 1
3	1 1 0 0	1 0 0 0
	1 1 0 1	1 0 0 1
	1 1 1 0	1 0 1 0
	1 1 1 1	1 1 1 1

$$\underbrace{y_1 y_2}_{\text{Present state}} \underbrace{x_1 x_2}_{\text{Present input}} \qquad \underbrace{y_1 y_2}_{\text{Next-state}} \underbrace{z_1 z_2}_{\text{Present output}}$$

From the two examples above it is seen that this approach is very systematic and easy to apply. The biggest advantage of this approach is, however, its generality and flexibility. For example, in Example 11.3.4, by changing the program in RAM, we can have any modulo-n counter, for $2 \leq n \leq 16$. Modulo-n counters with $n > 16$ may also be obtained by this system by simply replacing the SN 7489 chip with a larger RAM chip.

As another example, the system of Fig. 11.3.7 used in Example 11.3.5 for realizing the elevator controller may also be used to realize other entirely different sequential systems by, again, simply writing a different program and storing it in the RAM of the system. For instance, the binary serial adder and the sequential lock, whose TTL/ SSI designs were presented in Examples 11.2.3 and 11.2.4, respectively, may also be realized by using the system of Fig. 11.3.7 with an appropriate program stored in the RAM. This is shown in the following examples.

Example 11.3.6

Design the binary serial adder described in Example 11.2.3 using the system of Fig. 11.3.7.

The Moore-type transition table of the binary serial adder was given in Fig. 11.2.3(a). The circuit diagram for realizing this machine is the same as the one in Fig. 11.3.8(c), except that in this design the S_3 output of the SN 7489 chip will be used as the sum output and the S_4 output will not be used. The program that implements the next-state function is as follows:

Program Stored in the RAM

State	Memory address $S_1 S_2 x_1 x_2$	Stored data $S_1 S_2 S_3 S_4$
A_0	0 0 0 0	0 0 0 ×
	0 0 0 1	0 1 0 ×
	0 0 1 0	1 0 0 ×
	0 0 1 1	1 1 0 ×
A_1	0 1 0 0	0 1 1 ×
	0 1 0 1	1 0 1 ×
	0 1 1 0	1 1 1 ×
	0 1 1 1	0 0 1 ×
B_0	1 0 0 0	1 0 1 ×
	1 0 0 1	1 1 1 ×
	1 0 1 0	0 0 1 ×
	1 0 1 1	0 1 1 ×
B_1	1 1 0 0	1 1 0 ×
	1 1 0 1	0 0 0 ×
	1 1 1 0	0 1 0 ×
	1 1 1 1	1 0 0 ×

$\qquad\qquad\nearrow\quad\nwarrow\qquad\qquad\nearrow\quad\nwarrow$

Present   Present   Next-   Present
state   input   state   output

Example 11.3.7

Design the sequential lock described in Example 11.2.4 using the system of Fig. 11.3.7.

The Moore-type transition table of the sequential lock was given in Fig. 11.2.4(a). The circuit diagram for realizing this machine is again the same as the one in Fig. 11.3.8(c), except that the S_3 output of the SN 7489 chip will be connected to the lock and the S_4 output will be connected to an alarm system. The lock will open whenever $S_3 = 1$. The program that implements this machine is as follows:

Program Stored in the RAM

State	Memory address $S_1S_2x_1x_2$	Stored data $S_1S_2S_3S_4$
R	0 0 0 0	0 0 0 0
	0 0 0 1	0 1 0 0
	0 0 1 0	1 0 0 1
	0 0 1 1	1 0 0 1
B	0 1 0 0	0 0 0 0
	0 1 0 1	1 0 0 1
	0 1 1 0	1 0 0 1
	0 1 1 1	1 1 1 0
E	1 0 0 0	0 0 0 0
	1 0 0 1	1 0 0 1
	1 0 1 0	1 0 0 1
	1 0 1 1	1 0 0 1
C	1 1 0 0	0 0 0 0
	1 1 0 1	1 0 0 1
	1 1 1 0	1 0 0 1
	1 1 1 1	1 0 0 1

Present state   Present input   Next-state   Present output

Exercise 11.3

1. SN 74195, a parallel-in/parallel-out shift register is another very commonly used TTL/MSI chip. One of its applications is to construct data-conversion devices. For example, two SN 74195's can construct a serial-to-parallel data converter and a parallel-to-serial data converter, as shown in Fig. P11.3.1. Verify their operations by examining the circuit of Fig. P11.3.1(a) for a serial data input 0011011 and the circuit of Fig. P11.3.1(b) for a parallel data input 0011011.

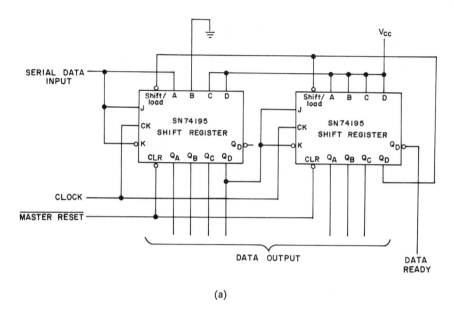

(a)

(b)

Fig. P11.3.1

2. Design a complete 8-bit binary serial adder as shown in Fig. P11.3.2. TTL/MSI are to be used. Give a detailed diagram, including chips used and all the interconnections.

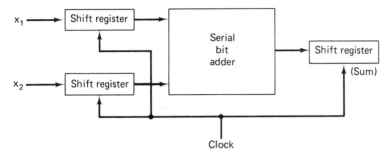

Fig. P11.3.2

3. Design a modulo-60 up-counter using the scheme of Fig. 11.3.2(a).

4. Draw a block diagram of a 12-hour digital clock using two modulo-60 counters and one modulo-12 counter (such as the one obtained in Example 11.3.3) and having second, minute, and hour seven-segment displays.

5. Convert each of the following Mealy machines into Moore machines.

q \ x	0	1
A	C, 0	B, 0
B	A, 0	D, 0
C	B, 1	A, 1
D	D, 1	C, 0

(a) M_a

q \ x	0	1
A	A, 0	C, 1
B	B, 1	A, 1
C	B, 0	C, 0
D	D, 1	D, 0

(b) M_b

q \ $x_1 x_2$	00	01	11	10
A	B, 0	A, 1	C, 0	B, 0
B	C, 1	A, 1	B, 0	A, 1
C	B, 0	B, 0	A, 1	B, 0

(c) M_c

Fig. P11.3.5

6. Realize each of the Moore machines obtained in problem 5 using the system of Fig. 11.3.4.

7. (a) Determine the minimum length and the minimum size of the RAM that will be required to realize each of the machines in Fig. P11.3.7 using the system of Fig. 11.3.4.

q \ σ	a	b	c	z
A	C	A	D	0
B	B	C	E	1
C	A	E	A	1
D	B	D	C	0
E	E	A	B	1

(a) M_d

q \ σ	a	b	c
A	C, 1	A, 0	D, 0
B	B, 0	C, 0	E, 1
C	A, 1	E, 1	A, 0
D	B, 0	D, 0	C, 0
E	E, 1	A, 1	B, 0

(b) M_e

Fig. P11.3.7

q \ x	0	1	z_1z_2
A	E	F	10
B	D	F	01
C	F	D	11
D	A	C	00
E	C	A	10
F	B	B	01

(c) M_f

q \ x	0	1
A	B, 0	C, 1
B	D, 0	C, 1
C	A, 0	E, 0
D	E, 1	F, 1
E	G, 0	F, 0
F	B, 0	D, 1
G	D, 0	E, 0

(d) M_g

Fig. P11.3.7 (Continued)

(b) Use a suitable TTL/MSI RAM plus other necessary MSI chips to implement each of the machines of part (a).

8. Repeat problem 7 for each of the incompletely specified machines in Fig. P11.3.8.

q \ σ	a	b	c
A	C, 0	E, 1	—
B	C, 0	E, –	—
C	B, –	C, 0	A, –
D	B, 0	C, –	E, –
E	—	E, 0	A, –

(a) M_h

q \ x	0	1
A	—	F, 0
B	B, 0	C, 0
C	E, 0	A, 1
D	B, 0	D, 0
E	F, 1	D, 0
F	A, 0	—

(b) M_i

q \ x_1x_2	00	01	11	10
A	B, 00	—	—	A, 11
B	B, 00	F, 01	—	B, 01
C	—	D, 10	C, 00	E, 11
D	B, 00	D, 10	—	—
E	B, 00	—	—	E, 11
F	—	F, 01	G, 00	—
G	—	A, 10	G, 00	A, 11

(c) M_j

q \ σ	a	b	c	d	e	f	g
A	—	E, –	–, 0	C, –	—	F, –	—
B	A, 0	B, –	E, –	D, –	—	—	—
C	–, 1	—	—	—	D, –	E, –	F, –
D	D, –	F, –	—	—	A, 0	C, –	—
E	—	—	A, –	F, –	–, 1	—	B, –
F	B, –	—	B, 1	—	C, –	—	D, –

(d) M_k

Fig. P11.3.8

9. Find a RAM realization of the sequential machine described in Example 9.3.2. A detailed circuit block diagram is required.

10. Information is transferred in a digital system using a 4-bit code. Bits 1 through 3 are information bits and bit 4 is a parity bit. The value of the parity bit is chosen so that the

number of 1's (including the parity bit) in the 4-bit representation of the information is even. Design a sequential circuit that will indicate with a 1 output whenever an error is present in the received information.

(a) TTL/SSI flip-flops and gates are to be used.

(b) TTL/MSI RAM's, multiplexers, and shift registers are to be used.

11. Suppose that it is desired to design the control unit of an automatic toll collector which is to be used on a toll highway to speed up traffic flow. This unit is to count the amount of change placed into the collector. Suppose the toll of the highway is 35 cents. If 35 cents is deposited, the go light is flashed on and a change collect signal is sent out to collect the coins; otherwise, the stop light is to remain on. Design a gate-flip-flop integrated-circuit realization and an MSI integrated-circuit realization of this control unit.

11.4 Digital Design Using MOS/LSI Integrated Circuits

In this section, we present the method of designing digital systems using PLA's, the fourth digital design approach. This approach, again, is best illustrated by examples. The first two examples will be combinational circuit design using ROM pairs (PLA's with flip-flops unconnected), and the last two will be sequential circuit design using PLA's.

Example 11.4.1

Any two-level AND–OR combinational circuit can be implemented by a ROM pair. The grid connections of the AND and OR ROM's are done by a computer-controlled machine. For example, the function $f(A, B, C, D) = AB + AC + CD + BD$ of Example 11.2.2 can be realized by the ROM pair shown in Fig. 11.4.1.

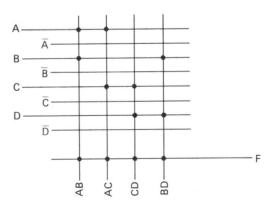

Fig. 11.4.1 ROM-pair realization of $F(A, B, C, D) = AB + AC + CD + BD$.

Example 11.4.2

Another useful application of ROM pair is in the design of code conversion. For example, the most frequently used decimal-to-binary conversion can be implemented by a ROM pair. The truth table of the decimal-to-binary conversion is

9	8	7	6	5	4	3	2	1	0	B_4	B_3	B_2	B_1
0	0	0	0	0	0	0	0	0	1	0	0	0	0
0	0	0	0	0	0	0	0	1	0	0	0	0	1
0	0	0	0	0	0	0	1	0	0	0	0	1	0
0	0	0	0	0	0	1	0	0	0	0	0	1	1
0	0	0	0	0	1	0	0	0	0	0	1	0	0
0	0	0	0	1	0	0	0	0	0	0	1	0	1
0	0	0	1	0	0	0	0	0	0	0	1	1	0
0	0	1	0	0	0	0	0	0	0	0	1	1	1
0	1	0	0	0	0	0	0	0	0	1	0	0	0
1	0	0	0	0	0	0	0	0	0	1	0	0	1

and the ROM-pair realization is shown in Fig. 11.4.2.

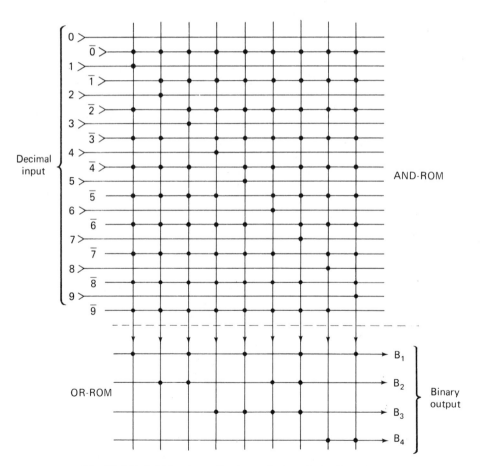

Fig. 11.4.2 ROM-pair realization of the decimal-to-binary converter.

Example 11.4.3

Design the binary serial adder and the sequential lock of Examples 11.2.3 and 11.2.4 using a **PLA** (such as TMS 2000).

The designs are shown in Fig. 11.4.3.

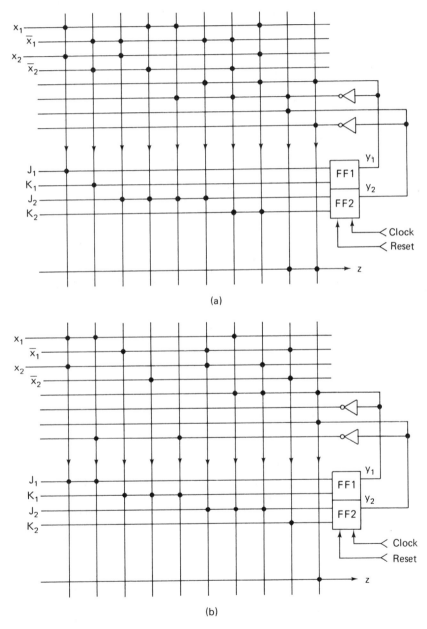

(a)

(b)

Fig. 11.4.3   (a) PLA design of the serial adder; (b) PLA design of the sequential lock.

Example 11.4.4

Design a modulo-12 counter with seven-segment display of Example 11.3.2 using a PLA.

The transition table of the counter with seven-segment decimal display output is shown in Fig. 11.4.4(a). Applying the application tables of the J-K flip-flop to the next-state function of this machine yields the excitation functions shown in Fig. 11.4.4(b), from which the following minimized functions are obtained:

$$J_1 = y_2 y_3 y_4 \qquad J_3 = \bar{y}_1 y_4 + \bar{y}_2 y_4$$
$$K_1 = y_2 + y_3 y_4 \qquad K_3 = y_1 y_2 + y_4$$
$$J_2 = \bar{y}_1 y_3 y_4 \qquad J_4 = \bar{y}_1 + \bar{y}_2$$
$$K_2 = y_1 + y_3 y_4 \qquad K_4 = 1$$

These functions are realized in the upper right corner of the PLA in Fig. 11.4.4(c). The decoded outputs for the two seven-segment decimal display can be obtained from the binary output by the ROM pair of the PLA. The complete design of the modulo-12 counter with seven-segment decimal displays is shown in Fig. 11.4.4(c).

Exercise 11.4

1. Use a ROM pair (PLA with flip-flops unconnected) to realize each of the following switching functions.
   (a) $f(A, B, C, D) = \Sigma(0, 3, 4, 5, 8, 12, 14, 15)$
   (b) $f(A, B, C, D, E) = \Sigma(1, 2, 5, 7, 9, 13, 16, 17, 18, 19, 25, 26, 28, 30)$
   (c) $f(A, B, C, D, E, F) = \Sigma(0, 2, 4, 6, 7, 9, 11, 15, 18, 20, 26, 32, 36, 42, 43, 44, 50, 55, 58, 61, 63)$

2. Use a ROM pair to implement each of the following code conversions.
   (a) Binary-to-Gray
   (b) Binary-to-6,3,1,−1 BCD
   (c) Binary-to-XS3
   (d) Binary-to-2421
   (e) Binary-to-two-out-of-five

TABLE P11.4.2

Decimal value	Binary	Gray	6,3,1,−1	XS3	2421	Two-out-of-five
0	0000	0000	0000	0011	0000	00011
1	0001	0001	0010	0100	0001	00101
2	0010	0011	0101	0101	0010	01001
3	0011	0010	0100	0110	0011	10001
4	0100	0110	0110	0111	0100	00110
5	0101	0111	1001	1000	1011	01010
6	0110	0101	1011	1001	1100	10010
7	0111	0100	1010	1010	1101	01100
8	1000	1100	1101	1011	1110	10100
9	1001	1101	1111	1100	1111	11000

Fig. 11.4.4 — (a) Transition table of the modulo-12 counter.

Y₁ Y₂ Y₃ x/Y₄	Clock	z₁ z₂ z₃ z₄	a b c d e f g	a b c d e f g	Display
0 0 0 0	0001	0 0 0 0	1 1 1 1 1 1 0	1 1 1 1 1 1 0	0
0 0 0 1	0010	0 0 0 1	1 1 1 0 0 0 0	1 1 1 0 0 0 0	1
0 0 1 0	0011	0 0 1 0	0 1 1 0 1 0 0	0 1 1 0 1 0 0	2
0 0 1 1	0100	0 0 1 1	1 1 0 1 1 0 1	1 1 0 1 1 0 1	3
0 1 0 0	0101	0 1 0 0	1 1 1 1 0 0 1	1 1 1 1 0 0 1	4
0 1 0 1	0110	0 1 0 1	0 1 1 0 0 1 1	0 1 1 0 0 1 1	5
0 1 1 0	0111	0 1 1 0	1 0 1 1 0 1 1	1 0 1 1 0 1 1	6
0 1 1 1	1000	0 1 1 1	1 0 1 1 1 1 1	1 0 1 1 1 1 1	7
1 0 0 0	1001	1 0 0 0	1 1 1 0 0 0 0	1 1 1 0 0 0 0	8
1 0 0 1	1010	1 0 0 1	1 1 1 1 1 1 1	1 1 1 1 1 1 1	9
1 0 1 0	1011	1 0 1 0	1 1 1 1 0 1 1	1 1 1 1 0 1 1	—
1 0 1 1	0000	1 0 1 1	0 1 1 1 0 1 0	0 1 1 1 0 1 0	—
1 1 0 0	0000	— — — —	0 1 1 — — 0 0	0 1 1 — — 0 0	—
1 1 0 1	0000	— — — —	0 1 1 — — 0 0	0 1 1 — — 0 0	—
1 1 1 0	0000	— — — —	0 1 1 — — 0 0	0 1 1 — — 0 0	—
1 1 1 1	0000	— — — —	0 1 1 — — 0 0	0 1 1 — — 0 0	—
	Y₁ Y₂ Y₃	Binary output	Decoded outputs for seven-segment decimal display		Display

seven-segment layout:

```
   a
 f   b
   g
 e   c
   d
```

(a)

Fig. 11.4.4 (a) Transition table of the modulo-12 counter; (b) excitation functions and their minimization; (c) PLA realization of modulo-12 counter with seven-segment display.

449

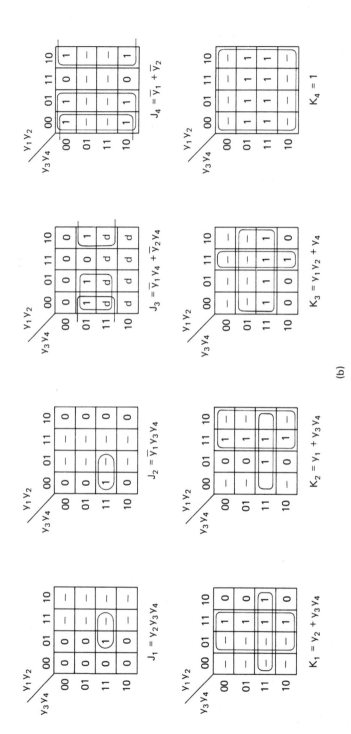

Fig. 11.4.4 (Continued)

450

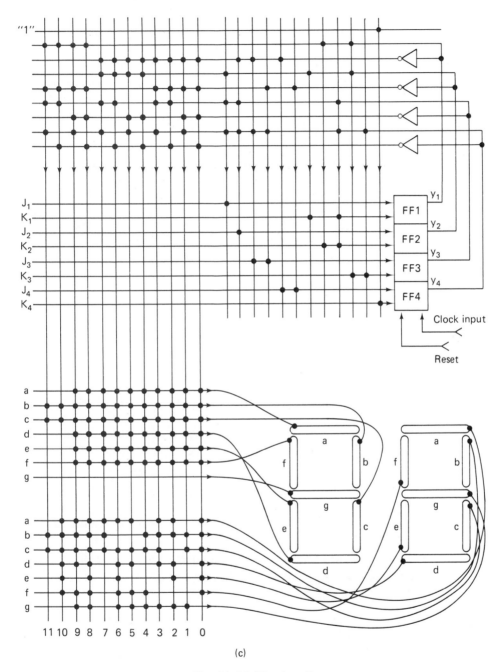

(c)

Fig. 11.4.4 (Continued)

451

3. Design the decade counter (with seven-segment display) described in Example 11.3.4 using a PLA.

4. Design the elevator controller described in Example 11.3.5 using a PLA.

5. Realize the machines M_a–M_k whose transition tables were given in Figs. P11.3.5, P11.3.7, and P11.3.8 using PLA's.

Bibliographical Remarks

References 1–3 provide detailed information about several TTL integrated circuits manufactured by three major manufacturers. References 4–6 discuss 54/74 integrated circuits and their applications, and reference 7 introduces Fairchild semiconductor's series TTL/MSI and applications. References 8 and 9 provide design techniques and applications of semiconductor memory in general and MOS/LSI in particular, respectively. Available MOS/LSI integrated circuits can be found in references 10–12. Finally, references 13–16 are recent articles on IIL technology.

References

1. *The TTL Data Book*, Texas Instruments, Inc.

2. *TTL Data Book*, Fairchild Semiconductor.

3. *Signetics Digital 54/7400 Data Book*, Signetics Corporation.

4. HIBBERD, R. G., *Integrated Circuits*, McGraw-Hill, New York, 1969.

5. MORRIS, R. L., and J. R. MILLER, eds., *Designing with TTL Integrated Circuits*, McGraw-Hill, New York, 1971.

6. LEE, S. C., *Digital Circuits and Logic Design*, Prentice-Hall, Englewood Cliffs, N.J., 1976.

7. *The TTL Applications Handbook*, Fairchild Semiconductor, August 1973.

8. CARR, W. N., and J. P. MIZE, *MOS/LSI Design and Application*, McGraw-Hill, New York, 1972.

9. LUECKE, G., J. P. MIZE, and W. N. CARR, *Semiconductor Memory Design and Application*, McGraw-Hill, New York, 1973.

10. *The Semiconductor Memory Data Book*, Texas Instruments, Inc.

11. *MOS/LSI Standard Products Catalog*, Texas Instruments, Inc.

12. *McMOS Handbook*, Motorola Semiconductor Products, Inc.

13. HORTON, R. L., J. ENGLADE, and G. McGEE, "I²L Takes Bipolar Integration a Significant Step Forward," *Electronics*, January 23, 1975, pp. 83–89.

14. ALTSTEIN, J., "I²L: Today's Versatile Vehicle for Tomorrow's Custom LSI," *EDN*, February 20, 1975, pp. 34–38.

15. HAFFNER, W. D., "A New Transistor Logic Family," *Popular Electronics*, January 1976, pp. 56–57.

12

Digital Design Using Microprocessors for Simulation

In Chapter 11, four approaches to digital design—the gate-flip-flop approach, the functional block-diagram approach, the RAM approach, and the PLA approach—were presented. In this chapter, we shall present a fifth method, the microcomputer simulation approach, which is the software simulation of digital logic using microcomputers. The microprocessor replaces hardwired logic by storing program sequences in the ROM rather than implementing these sequences with gates, flip-flops, counters, etc. There are at least five advantages of using this new method:

1. Manufacturing costs of products can be significantly reduced.
2. Products can get to the market faster, providing the user with the opportunity to increase product sales and market share.
3. Product capability is enhanced, allowing manufacturers to provide customers with better products, which can frequently command a higher price in the marketplace.
4. Development costs and time are reduced.
5. Product reliability is increased, which leads to a corresponding reduction in both service and warranty costs.

Although there are many microprocessors and microcomputers available, the method of designing digital systems using any of them is essentially the same. In this chapter, we shall use the Intel 8080A microprocessor and the Intel MDS-800 microcomputer development system to illustrate the method, since they are very popularly used by digital designers.

12.1 Typical 8080 Microcomputer System

A *microprocessor* is an arithmetic and logical unit plus control unit realized on a small number of LSI chips. A monolithic microprocessor provides the basic arithmetic and the central processing unit (CPU) of the computer. A *microcomputer* is then simply a computer whose CPU is a microprocessor. A functional block diagram of a typical 8080 microcomputer system is shown in Fig. 12.1.1(a). Like any other computer, an 8080 microcomputer has the following three basic components:

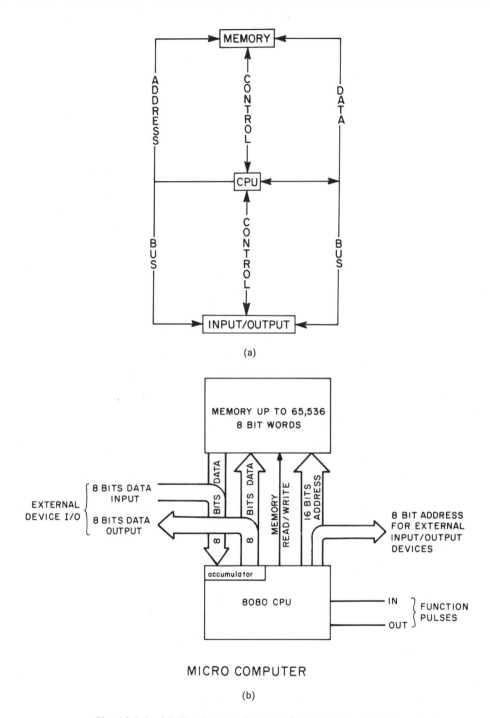

MICRO COMPUTER

(b)

Fig. 12.1.1 (a) The three basic parts of a computer; (b) sizes and flow directions of data bus and address bus of an 8080 microcomputer.

1. A central processor unit (CPU).
2. A memory.
3. Input/output (I/O) ports.

The memory serves as a place to store *instructions*, the coded pieces of information that direct the activities of the CPU, and *data*, the coded pieces of information that are processed by the CPU. A group of logically related instructions stored in memory is referred to as a *program*. The CPU "reads" each instruction from memory in a logically determined sequence, and uses it to initiate processing actions. If the program sequence is coherent and logical, processing the program will produce intelligible and useful results.

The memory is also used to store the data to be manipulated as well as the instructions which direct the manipulation. The program must be organized such that the CPU does not read a noninstruction word when it expects to see an instruction. The CPU can rapidly access any data stored in memory; but often the memory is not large enough to store the entire data bank required for a particular application. The problem can be resolved by providing the computer with one or more *input ports*. The CPU can address these ports and input the data contained there. The addition of input ports enables the computer to receive information from external equipment (such as a paper tape reader or floppy disk) at high rates of speed and in large volumes.

A computer also requires one or more *output ports* that permit the CPU to communicate the result of its processing to the outside world. The output may go to a display, for use by a human operator, to a peripheral device that produces "hard copy," such as a line printer, to a peripheral storage device, such as a floppy disk unit, or the output may constitute process control signals that direct the operations of another system, such as an automated assembly line. Like input ports, output ports are addressable. The input and output ports together permit the processor to communicate with the outside world.

The CPU unifies the system. It controls the functions performed by the other components. The CPU must be able to fetch instructions from memory, decode their binary contents, and execute them. It must also be able to reference memory and I/O ports as necessary in the extension of instructions. In addition, the CPU should be able to recognize and respond to certain external control signals, such as INTERRUPT and WAIT requests.

There are three types of buses linking various parts of an 8080 microcomputer: address bus, data bus, and control bus. The sizes of these buses are 16 bits, 8 bits, and 6 bits, respectively. Since the size of the address bus is 16 bits, the maximum number of directly addressable memory locations of an 8080 microcomputer is 65,536. Therefore, the maximum size of the memory is bounded by this number. The memory can be all RAM, or all ROM, or a combination of the two as long as the total memory size does not exceed 65,536 words, which is in general sufficient for most applications. Each word of an 8080 microcomputer is 8 bits; thus, all the data buses are 8 bits, and so are all the data registers in the microcomputer. Data (8 bits of information) coming into the memory from an I/O port and going out of the microcomputer from the memory to an I/O port must go through a special data register (8-bit) called the *accumulator* or *A-register*, shown in Fig. 12.1.1(b).

The block diagram of a typical 8080 microcomputer is shown in Fig. 12.1.2. It consists of a CPU, a clock generator and driver, a system controller and bus driver, ROM's, RAM's, I/O programmable communication interface, and I/O programmable peripheral interface. They are described below.

A. CPU (8080A)

The 8080A microprocessor is a complete 8-bit parallel central processor unit (CPU) for use in general-purpose digital computer systems. It is fabricated on a single LSI chip (see Fig. 12.1.3). The 8080A transfers data and internal state information by means of an 8-bit, bidirectional three-state data bus (D_0–D_7). Memory and peripheral device addresses are transmitted over a separate 16-bit three-state address bus (A_0–A_{15}). Six timing and control outputs (SYNC, DBIN, WAIT, WR, HLDA, and INTE) emanate from the 8080, while four control inputs (READY, HOLD, INT, and RESET), four power inputs ($+12$ V, $+5$ V, -5 V, and GND), and two clock inputs ($\phi 1$ and $\phi 2$) are accepted by the 8080A.

The 8080A CPU consists of the following functional units:

1. Register array and address logic.
2. Arithmetic and logic unit (ALU).
3. Instruction register and control section.
4. Bidirectional, three-state data bus buffer.

Figure 12.1.4 illustrates the functional blocks within the 8080A CPU.

Registers The register section consists of a static RAM array organized into six 16-bit registers:

1. Program counter (PC). The program counter maintains the memory address of the current program instruction and is incremented automatically during every instruction fetch. (See Fig. 12.1.5.)

2. Stack pointer (SP). The stack pointer maintains the address of the next available stack location in memory. The stack pointer can be initialized to use any portion of read/write memory as a stack. The stack pointer is decremented when data are "pushed" onto the stack and incremented when data are "popped" off the stack (i.e., the stack grows "downward"). This is depicted in Fig. 12.1.6.

3. Six 8-bit general-purpose registers arranged in pairs, referred to as B, C; D, E; and H, L. The six general-purpose registers can be used either as single registers (8-bit) or as register pairs (16-bit).

4. A temporary register pair called W, Z. This temporary register pair is not program-addressable and is only used for the internal execution of instructions.

Eight-bit data bytes can be transferred between the internal bus and the register array by means of the register-select multiplexer. Sixteen-bit transfers can proceed

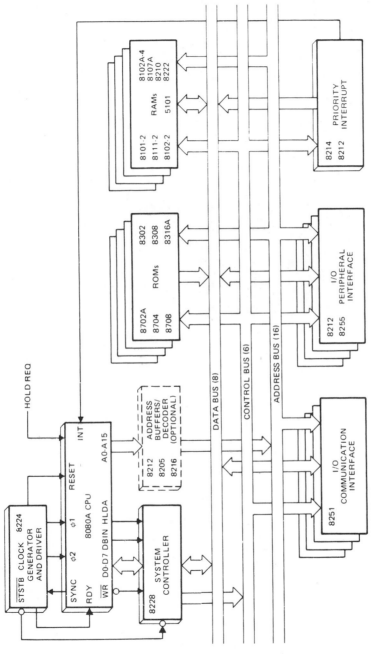

Fig. 12.1.2 Block diagram of a typical 8080 microcomputer.

457

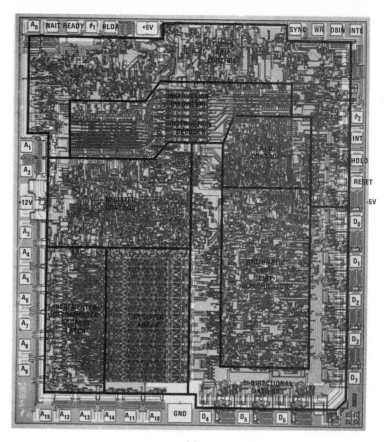

(a)

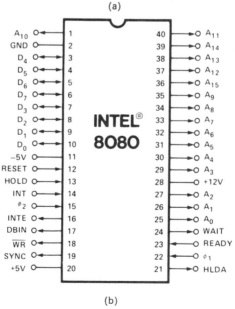

(b)

Fig. 12.1.3 (a) 8080A photomicrograph; (b) 8080A pin configuration.

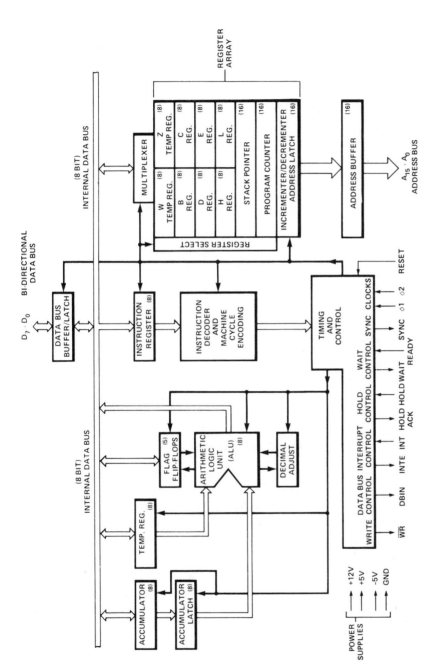

Fig. 12.1.4 8080A CPU functional block diagram.

459

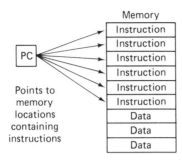

Fig. 12.1.5 Program counter.

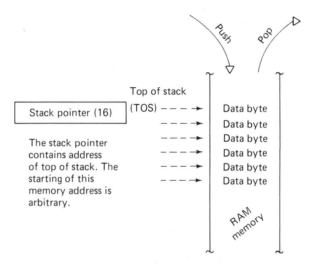

Fig. 12.1.6 Stack pointer.

between the register array and the address latch or the incrementer/decrementer circuit. The address latch receives data from any of the three register pairs and drives the 16 address output buffers (A_0–A_{15}), as well as the incrementer/decrementer circuit. The incrementer/decrementer circuit receives data from the address latch, and sends them to the register array. The 16-bit data can be incremented or decremented or simply transferred between registers.

Arithmetic and Logic Unit (ALU)   The ALU contains the following registers:

1. An 8-bit accumulator (ACC).
2. An 8-bit temporary accumulator (ACT).
3. A 5-bit flag register: zero, carry, sign, parity, and auxiliary carry.
4. An 8-bit temporary register (TMP).

Arithmetic, logical, and rotate operations are performed in the ALU. The ALU is fed by the temporary register and the temporary accumulator and carry flip-flop. The result of the operation can be transferred to the internal bus or to the accumulator;

the ALU also feeds the flag register. The temporary register receives information from the internal bus and can send all or portions of it to the ALU, the flag register, and the internal bus. The accumulator can be loaded from the ALU and the internal bus and can transfer data to the temporary accumulator and the internal bus.

Instruction Register and Control   During an instruction fetch, the first byte of an instruction (containing the OP code) is transferred from the internal bus to the 8-bit instruction register. The contents of the instruction register are, in turn, available to the instruction decoder. The output of the decoder, combined with various timing signals, provides the control signals for the register array, ALU, and data buffer blocks. In addition, the outputs from the instruction decoder and external control signals feed the timing and state control section, which generate the state and cycle timing signals.

Data Bus Buffer   This 8-bit bidirectional three-state buffer is used to isolate the CPU's internal bus from the external data bus (D_0–D_7). In the output mode, the internal bus contents are loaded into an 8-bit latch, which, in turn, drives the data bus output buffers. The output buffers are switched off during input or nontransfer operations. During the input mode, data from the external data bus are transferred to the internal bus.

B.   Clock Generator and Driver (8224)

These comprise a single-chip clock generator/driver for the 8080A CPU and are controlled by a crystal, selected by the designer, to meet a variety of system speed requirements. Also included are circuits to provide power-up reset, advance status strobe, and synchronization of ready.

C.   System Controller and Bus Driver (8228)

These are a single-chip system controller and bus driver for a 8080 microcomputer system (MCS-80) and generate all signals required to directly interface MCS-80 family RAM, ROM, and I/O components.

D.   ROM's (8702A, 8704, 8708, 8302, 8308, 8316A)

All the MCS-80 family ROM's are erasable and electrically programmable.

E.   RAM's (8101-2, 8111-2, 8102-2, 8102A-4,
     8107B-4, 8210, 8222)

The 8101-2, 8111-2, 8102-2, and 8102A-4 are static MOS RAM's. The 8107B-4 is a 4096-word-by-1-bit dynamic MOS RAM designed for memory applications where very low cost and large bit storage are important design objectives. The 8210 and 8222 are Schottky bipolar RAM's.

F. I/O Programmable Communication Interface (8251)

This is a universal synchronous/asynchronous receiver/transmitter (USART) chip designed for data communications in microcomputer systems.

G. I/O Programmable Peripheral Interface
(8212, 8255)

The 8212 I/O port consists of an 8-bit latch with three-state output buffers along with control and device selection logic. Also included is a service request flip-flop for the generation and control of interrupts to the microprocessor.

A complete design and debugging tool which allows the integration of both microcomputer hardware and software development is often referred to as a *microcomputer development system*. For example, the Intellec Microcomputer Development System (MDS) is such a system. It operates under control of an 8080 microcomputer, which supervises all system resources such as main memory, I/O peripheral devices, Intellec bus facilities, and optional system facilities such as DMA (Direct Memory Access) and ICE (In-Circuit Emulator). The Intellec MDS is a self-contained modular microcomputer development system, which consists of the following:

1. The 8080 CPU module. This module contains an Intellec 8080 CPU, an 8-bit microprocessor. The CPU provides 78 instructions, unlimited subroutine nesting, vectored interrupt, DMA capabilities, and a 16-line address bus which is associated with a bidirectional eight-line data bus.

2. 16K of RAM memory. This can be expanded up to a maximum of 12K of PROM and 64K of RAM in increments of 6K (PROM) or 16K (RAM) 8-bit bytes.

3. The Monitor module. This module contains the Intellec MDS Monitor and all MDS peripheral interface hardware.

4. Front panel control module. This module contains circuits for controlling the front panel operations.

The basic Intellec MDS capabilities may be significantly enhanced by the addition of the following optional modules.

1. The In-Circuit Emulator (ICE) module. This module permits a user to build his system without auxiliary hardware or software test equipment. Instead, he uses the Intellec MDS to control and monitor the excution of his system. Thus, his prototype system also becomes his production system.

To the user's system, the In-Circuit Emulator looks like a replacement for his CPU chip. Specifically, the ICE-80 module looks like a replacement for a user's 8080 CPU chip. It plugs into the user's system in place of his CPU and performs all the functions of that CPU. The other end of the ICE module connects to the Intellec MDS, where it interacts with the MDS software.

ICE-80 allows the user to test his system even though it is not completely built. Some or all of the storage of the user's system can be contained in Intellec MDS memory during system development. Similarly, some or all of the user's peripherals can be the standard peripherals attached to the MDS system. Whenever the emulator system makes a memory or I/O access, it first consults an address map to determine the physical location of a logical memory address or I/O port.

ICE-80 can also be used as a diagnostic tool with the MDS even when the user plans to build no hardware. For example, it can be used to debug software residing in the MDS.

2. Additional I/O modules. The user may install additional I/O modules, up to 256 I/O ports.

3. The Direct Memory Access (DMA) module (8257). The primary function of a DMA is to generate, upon a peripheral request, a sequential memory address which will allow the peripheral to access or deposit data directly from or to memory.

The following software is provided for the user of the Microcomputer Development System Diskette Operating System (MDS–DOS):

1. Intel System Implementation Supervisor (ISIS). ISIS requires an Intellec MDS with a minimum of 32K RAM memory.

2. ISIS Text Editor.

3. ISIS 8080 Macro Assembler.

4. Monitor.

5. In-Circuit Emulator/80 (ICE-80).

12.2 8080 Instruction Set and Assembly Language Programming

When a computer is designed, the engineers provide the central processing unit (CPU) with the ability to perform a particular set of operations. The CPU is designed such that a specific operation is performed when the CPU control logic decodes a particular instruction. Consequently, the operations that can be performed by a CPU define the computer's *instruction set.*

Each computer instruction allows the programmer to initiate the performance of a specific operation. All computers implement certain arithmetic operations in their instruction set, such as an instruction to add the contents of two registers. Often logical operations (e.g., OR the contents of two registers) and register operate instructions (e.g., increment a register) are included in the instruction set. A computer's instruction set will have instructions that move data between registers, between a register and memory, and between a register and an I/O device. Most instruction sets also provide *conditional instructions.* A conditional instruction specifies an operation to be performed only if certain conditions have been met; for example, jump to a particular instruction if the result of the last operation was zero. Conditional instructions provide a program with a decision-making capability.

By logically organizing a sequence of instructions into a coherent program, the programmer can "tell" the computer to perform a very specific and useful function.

The computer, however, can only execute programs whose instructions are in a binary-coded form (i.e., a series of 1's and 0's) called *machine code.* Because it would be extremely cumbersome to program in machine code, programming languages have been developed. There are programs available which convert the programming language instruction into a machine code that can be interpreted by the processor.

One type of programming language is *assembly language*. A unique assembly language mnemonic is assigned to each of the computer's instructions. The programmer can write a program (called the *source program*) using these mnemonics and certain operands; the source program is then converted into machine instructions (called the *object code*). Each assembly language instruction is converted into one machine code instruction (one or more bytes) by an *assembler* program.

For convenience, machine code instructions of microcomputers are usually represented in hexadecimal numbers [0 through *F*, see Fig. 12.2.1(a)]. To show the program size reduction and the programming ease of the assembly language versus the machine language, a program for "selecting the larger of two numbers stored in memory and storing it in a third location" in machine code and assembly language is given in Fig. 12.2.1(b).

Hexa-decimal	Binary	Decimal	Machine code	Assembly	
0	0000	0			
1	0001	1	3A		
2	0010	2	FF		
3	0011	3	01		
4	0100	4	21	LDA	Y
5	0101	5	D3	LXI	H,X
6	0110	6	00	CMP	M
7	0111	7	BE	JC	GO
8	1000	8	DA	MOV	A,M
9	1001	9	43	GO: STA	Z
A	1010	10	01		
B	1011	11	7E		
C	1100	12	32		
D	1101	13	88		
E	1110	14	01		
F	1111	15			
(a)				(b)	

Fig. 12.2.1 (a) Hexadecimal numbers and their corresponding binary and decimal numbers. (b) A program for "selecting the larger of two numbers stored in memory and storing it in a third location" in machine code and assembly language.

Figure 12.2.2 shows a programming model of a CPU; the accumulator (A register); registers B, C, D, E, H, and L; stack pointer; and program counter were introduced in the previous section. In addition, there are five condition flags associated with the execution of instructions on the 8080A. They are Zero, Sign, Parity, Carry, and

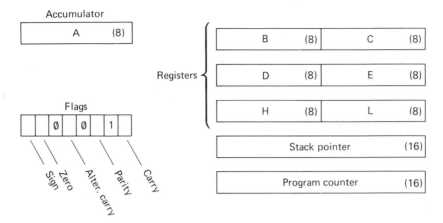

Fig. 12.2.2   A programming model of CPU.

Auxiliary Carry and are each represented by a 1-bit register in the CPU. A flag is "set" by forcing the bit to 1; "reset" by forcing the bit to 0. Unless indicated otherwise, when an instruction affects a flag, it affects it in the following manner:

   1.   Zero: If the result of an instruction has the value 0, this flag is set; otherwise, it is reset.

   2.   Sign: If the most significant bit of the result of the operation has the value 1, this flag is set; otherwise, it is reset.

   3.   Parity: If the modulo-2 sum of the bits of the result of the operation is 0 (i.e., if the result has even parity), this flag is set; otherwise, it is reset (i.e., if the result has odd parity).

   4.   Carry: If the instruction resulted in a carry (from addition), or a borrow (from subtraction or a comparison) out of the high-order bit, this flag is set; otherwise, it is reset.

   5.   Auxiliary Carry: If the instruction caused a carry out of bit 3 and into bit 4 of the resulting value, the auxiliary carry is set; otherwise, it is reset. This flag is affected by single precision additions, subtractions, increments, decrements, comparisons, and logical operations, but is principally used with additions and increments preceding a DAA (Decimal Adjust Accumulator) instruction.

   Memory for the 8080A is organized into 8-bit quantities called *bytes*. Each byte has a unique 16-bit binary address corresponding to its sequential position in memory. The 8080A can directly address up to 65,536 bytes of memory, which may consist of both read-only memory (ROM) elements and random-access memory (RAM) elements (read/write memory). A typical memory layout is shown in Fig. 12.2.3 on page 466.

   Data in the 8080A are stored in the form of 8-bit binary integers:

Data word

D_7	D_6	D_5	D_4	D_3	D_2	D_1	D_0

MSB   LSB

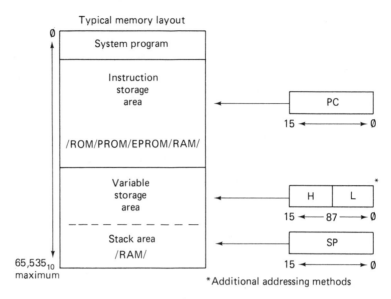

Fig. 12.2.3 A typical memory layout.

When a register or data word contains a binary number, it is necessary to establish the order in which the bits of the number are written. In the Intel 8080A, bit 0 is referred to as the *least significant bit* (LSB), and bit 7 (of an 8 bit number) is referred to as the *most significant bit* (MSB).

The 8080A program instructions may be one, two, or three bytes in length. Multiple byte instructions must be stored in successive memory locations; the address of the first byte is always used as the address of the instructions. The exact instruction format will depend on the particular operation to be executed.

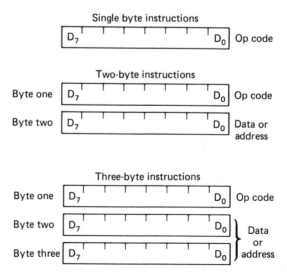

Often the data that are to be operated on are stored in memory. When multibyte numeric data are used, the data, like instructions, are stored in successive memory locations, with the least significant byte first, followed by increasingly significant bytes. The 8080A has four different modes for addressing data stored in memory or in registers:

1. Direct—Bytes 2 and 3 of the instruction contain the exact memory address of the data item (the low-order bits of the address are in byte 2, the high-order bits in byte 3).

2. Register—The instruction specifies the register or register pair in which the data are located.

3. Register indirect—The instruction specifies a register pair which contains the memory address where the data are located (the high-order bits of the address are in the first register of the pair, the low-order bits in the second).

4. Immediate—The instruction contains the data itself. This is either an 8-bit quantity or a 16-bit quantity (least significant byte first, most significant byte second).

Unless directed by an interrupt or branch instruction, the execution of instructions proceeds through consecutively increasing memory locations. A branch instruction can specify the address of the next instruction to be executed in one of two ways:

1. Direct—The branch instruction contains the address of the next instruction to be executed. (Except for the "RST" instruction, byte 2 contains the low-order address and byte 3 the high-order address.)

2. Register indirect—The branch instruction indicates a register pair which contains the address of the next instruction to be executed. (The high-order bits of the address are in the first register of the pair, the low-order bits in the second.)

The RST instruction is a special 1-byte call instruction (usually used during interrupt sequences). RST includes a 3-bit field; program control is transferred to the instruction whose address is eight times the contents of this 3-bit field.

Before describing the 8080 instructions, the following symbols and abbreviations are first presented on page 468.

The 8080 instruction set includes five different types of instructions:

1. Data-transfer group—move data between registers or between memory and registers.

2. Arithmetic group—add, subtract, increment, or decrement data in registers or in memory.

3. Logical group—AND, OR, EXCLUSIVE-OR, compare, rotate, or complement data in registers or in memory.

4. Branch group—conditional and unconditional jump instructions, subroutine call instructions, and return instructions.

5. Stack, I/O, and machine control group—includes I/O instructions, as well as instructions for maintaining the stack and internal control flags.

Symbols	Meaning	Symbols	Meaning
accumulator	Register A	rh	The first (high-order) register of a designated register pair.
addr	16-bit address quantity		
data	8-bit data quantity	rl	The second (low-order) register of a designated register pair.
data 16	16-bit data quantity		
byte 2	The second byte of the instruction	PC	16-bit program counter register (PCH and PCL are used to refer to the high-order and low-order 8 bits, respectively).
byte 3	The third byte of the instruction		
port	8-bit address of an I/O device		
r,r1,r2	One of the registers A,B,C,D,E,H,L		
DDD,SSS	The bit pattern designating one of the registers A,B,C,D,E,H,L (DDD= destination, SSS=source) :	SP	16-bit stack pointer register (SPH and SPL are used to refer to the high-order and low-order 8 bits, respectively).

DDD or SSS	Register name
111	A
000	B
001	C
010	D
011	E
100	H
101	L

Symbols	Meaning
r_m	Bit m of the register ·r (bits are number 7 through 0 from left to right).
Z,S,P,CY,AC	The condition flags: Zero, Sign, Parity, Carry, and Auxiliary Carry, respectively.
( )	The contents of the memory location or registers enclosed in the parentheses.
←	"Is transferred to"
∧	Logical AND
∀	EXCLUSIVE-OR
∨	Inclusive-OR
+	Addition
−	Two's-complement subtraction
*	Multiplication
←→	"Is exchanged with"
—	The one's complement [e.g., $(\bar{A})$]
n	The restart number 0 through 7
NNN	The binary representation 000 through 111 for restart number 0 through 7, respectively.

rp	One of the register pairs : B represents the B,C pair with B as the high-order register and C as the low-order register; D represents the D,E pair with D as the high-order register and E as the low-order register; H represents the H,L pair with H as the high-order register and L as the low-order register; SP represents the 16-bit stack pointer register.
RP	The bit pattern designating one of the register pairs B,D,H,SP :

RP	Register pair
00	B-C
01	D-E
10	H-L
11	SP

The 8080 instructions of these five groups are given in Table 12.2.1. The same instruction set presented in alphabetical order and the machine codes are shown in Table 12.2.2.

TABLE 12.2.1 The 8080 Instruction Set

Group	Mnemonic	Verbal description	Symbolic description
Data transfer group	MOV r1,r2	Move register	(r1) ⟵ (r2)
	MOV r, M	Move from memory	(r) ⟵ ((H)(L))
	MOV M, r	Move to memory	((H)(L)) ⟵ (r)
	MVI r, data	Move immediate	(r) ⟵ (byte 2)
	MVI M, data	Move to memory immediate	((H)(L)) ⟵ (byte 2)
	LXI rp, data 16	Load register pair immediate	(rh) ⟵ (byte 3), (rl) ⟵ (byte 2)
	LDA addr	Load accumulator direct	(A) ⟵ ((byte 3) (byte 2))
	STA addr	Store accumulator direct	(byte 3) (byte 2) ⟵ (A)
	LHLD addr	Load H and L direct	(L) ⟵ ((byte 3) (byte 2)), (H) ⟵ ((byte 3) (byte 2) + 1)
	SHLD addr	Store H and L direct	((byte 3) (byte 2)) ⟵ (L), ((byte 3) (byte 2) + 1) ⟵ (H)
	LDAX rp	Load accumulator indirect	(A) ⟵ ((rp))
	STAX rp	Store accumulator indirect	((rp)) ⟵ (A)
	XCHG	Exchange H and L with D and E	(H) ⟷ (D), (L) ⟷ (E)
Arithmetic group	ADD r	Add register	(A) ⟵ (A) + (r)
	ADD M	Add memory	(A) ⟵ (A) + ((H)(L))
	ADI data	Add immediate	(A) ⟵ (A) + (byte 2)
	ADC r	Add register with carry	(A) ⟵ (A) + (r) + (CY)
	ADC M	Add memory with carry	(A) ⟵ (A) + ((H)(L)) + (CY)
	ACI data	Add immediate with carry	(A) ⟵ (A) + (byte 2) + (CY)
	SUB r	Subtract register	(A) ⟵ (A) − (r)
	SUB M	Subtract memory	(A) ⟵ (A) − ((H)(L))
	SUI data	Subtract immediate	(A) ⟵ (A) − (byte 2)
	SBB r	Subtract register with borrow	(A) ⟵ (A) − (r) − (CY)
	SBB M	Subtract memory with borrow	(A) ⟵ (A) − ((H)(L)) − (CY)
	SBI data	Subtract immediate with borrow	(A) ⟵ (A) − (byte 2) − (CY)
	INR r	Increment register	(r) ⟵ (r) + 1
	INR M	Increment memory	((H)(L)) ⟵ ((H)(L)) + 1
	DCR r	Decrement register	(r) ⟵ (r) − 1
	DCR M	Decrement memory	((H)(L)) ⟵ ((H)(L)) − 1
	INX rp	Increment register pair	(rh)(rl) ⟵ (rh)(rl) + 1
	DCX rp	Decrement register pair	(rh)(rl) ⟵ (rh)(rl) − 1
	DAD rp	Add register pair to H and L	(H)(L) ⟵ (H)(L) + (rh)(rl)
	DAA	Decimal adjust accumulator	

469

TABLE 12.2.1 (Continued)

Group	Mnemonic	Verbal description	Symbolic description
	ANA r	AND register	$(A) \leftarrow (A) \wedge (r)$
	ANA M	AND memory	$(A) \leftarrow (A) \wedge ((H)(L))$
	ANI data	AND immediate	$(A) \leftarrow (A) \wedge (\text{byte } 2)$
	XRA r	EXCLUSIVE-OR register	$(A) \leftarrow (A) \veebar (r)$
	XRA M	EXCLUSIVE-OR memory	$(A) \leftarrow (A) \veebar ((H)(L))$
	XRI data	EXCLUSIVE-OR immediate	$(A) \leftarrow (A) \veebar (\text{byte } 2)$
	ORA r	OR register	$(A) \leftarrow (A) \vee (r)$
	ORA M	OR memory	$(A) \leftarrow (A) \vee ((H)(L))$
	ORI data	OR immediate	$(A) \leftarrow (A) \vee (\text{byte } 2)$
Logical group	CMP r	Compare register	$(A) - (r)$
	CMP M	Compare memory	$(A) - ((H)(L))$
	CPI data	Compare immediate	$(A) - (\text{byte } 2)$
	RLC	Rotate left	$(A_{n+1}) \leftarrow (A_n),\ (A_0) \leftarrow (A_7),\ (CY) \leftarrow (A_7)$
	RRC	Rotate right	$(A_n) \leftarrow (A_{n+1}),\ (A_7) \leftarrow (A_0),\ (CY) \leftarrow (A_0)$
	RAL	Rotate left through carry	$(A_{n+1}) \leftarrow (A_n),\ (CY) \leftarrow (A_7),\ (A_0) \leftarrow (CY)$
	RAR	Rotate right through carry	$(A_n) \leftarrow (A_{n+1}),\ (CY) \leftarrow (A_0),\ (A_7) \leftarrow (CY)$
	CMA	Complement accumulator	$(A) \leftarrow (\bar{A})$
	CMC	Complement carry	$(CY) \leftarrow (\overline{CY})$
	STC	Set carry	$(CY) \leftarrow 1$
Branch group	JMP addr.	Jump	$(PC) \leftarrow (\text{byte } 3)\ (\text{byte } 2)$
	J condition addr	Conditional jump	If (CCC), $(PC) \leftarrow (\text{byte } 3)\ (\text{byte } 2)$

Condition	CCC
NZ—not zero (Z = 0)	000
Z—zero (Z = 1)	001
NC—no carry (CY = 0)	010
C—carry (CY = 1)	011
PO—parity odd (P = 0)	100
PE—parity even (P = 1)	101
P—plus (S = 0)	110
M—minus (S = 1)	111

TABLE 12.2.1 (Continued)

Group	Mnemonic	Verbal description	Symbolic description
Branch group	CALL addr	Call	$((SP) - 1) \leftarrow (PCH), ((SP) - 2) \leftarrow (PCL),$ $(SP) \leftarrow (SP) - 2, (PC) \leftarrow (byte\ 3)\ (byte\ 2)$
	Condition addr	Condition	If (CCC) (see above). $((SP) - 1) \leftarrow (PCH),$ $((SP) - 2) \leftarrow (PCL), (SP) \leftarrow (SP) - 2,$ $(PC) \leftarrow (byte\ 3)\ (byte\ 2)$
	RET	Return	$(PCL) \leftarrow ((SP)), (PCH) \leftarrow ((SP) + 1), (SP) \leftarrow (SP) + 2$
	R condition	Conditional return	If (CCC) (see above), $(PCL) \leftarrow ((SP)),$ $(PCH) \leftarrow ((SP) + 1), (SP) \leftarrow (SP) + 2$
	RST n	Restart	$((SP) - 1) \leftarrow (PCH), ((SP) - 2) \leftarrow (PCL),$ $(SP) \leftarrow (SP) - 2, (PC) \leftarrow 8 * (NNN)$
	PCHL	Jump H and L indirect-move H and L to PC	$(PCH) \leftarrow (H)$ $(PCL) \leftarrow (L)$
Stack, I/O, and machine control group	PUSH rp	Push	$((SP) - 1) \leftarrow (rh), ((SP) - 2) \leftarrow (rl), (SP) \leftarrow (SP) - 2$
	PUSH PSW	Push processor status word	$((SP) - 1) \leftarrow (A), ((SP) - 2)_0 \leftarrow (CY), ((SP) - 2)_1 \leftarrow 1,$ $((SP) - 2)_2 \leftarrow (P), ((SP) - 2)_3 \leftarrow 0, ((SP) - 2)_4 \leftarrow (AC),$ $((SP) - 2)_5 \leftarrow 0, ((SP) - 2)_6 \leftarrow (Z), ((SP) - 2)_7 \leftarrow (S)$ $(SP) \leftarrow (SP) - 2$
	POP rp	Pop	$(rl) \leftarrow ((SP)), (rh) \leftarrow ((SP) + 1), (SP) \leftarrow (SP) + 2$
	POP PSW	Pop processor status word	$(CY) \leftarrow ((SP))_0, (P) \leftarrow ((SP))_2, (AC) \leftarrow ((SP))_4$ $(Z) \leftarrow ((SP))_6, (S) \leftarrow ((SP))_7, (A) \leftarrow ((SP) + 1)$ $(SP) \leftarrow (SP) + 2$
	XTHL	Exchange stack top with H and L	$(L) \leftrightarrow ((SP)), (H) \leftrightarrow ((SP) + 1)$
	SPHL	Move HL to SP	$(SP) \leftarrow (H)(L)$
	IN port	Input	$(A) \leftarrow (data)$
	OUT port	Output	$(data) \leftarrow (A)$
	EI	Enable interrupts	
	DI	Disable interrupts	
	HLT	Halt	
	NOP	No operation	

TABLE 12.2.2   8080 Instructions and Their Instruction
Codes

Mnemonic	Description	D_7	D_6	D_5	D_4	D_3	D_2	D_1	D_0
ACI	Add immediate to A with carry	1	1	0	0	1	1	1	0
ADC M	Add memory to A with carry	1	0	0	0	1	1	1	0
ADC r	Add register to A with carry	1	0	0	0	1	S	S	S
ADD M	Add memory to A	1	0	0	0	0	1	0	1
ADD r	Add register to A	1	0	0	0	0	S	S	S
ADI	Add immediate to A	1	1	0	0	0	1	1	0
ANA M	And memory with A	1	0	1	0	0	1	1	0
ANA r	And register with A	1	0	1	0	0	S	S	S
ANI	And immediate with A	1	1	1	0	0	1	1	0
CALL	Call unconditional	1	1	0	0	1	1	0	1
CC	Call on carry	1	1	0	1	1	1	0	0
CM	Call on minus	1	1	1	1	1	1	0	0
CMA	Complement A	0	0	1	0	1	1	1	1
CMC	Complement carry	0	0	1	1	1	1	1	1
CMPM	Compare memory with A	1	0	1	1	1	1	1	0
CMP r	Compare register with A	1	0	1	1	1	S	S	S
CNC	Call on no carry	1	1	0	1	0	1	0	0
CNZ	Call on no zero	1	1	0	0	0	1	0	0
CP	Call on positive	1	1	1	1	0	1	0	0
CPE	Call on parity even	1	1	1	0	1	1	0	0
CPI	Compare immediate with A	1	1	1	1	1	1	1	0
CPO	Call on parity odd	1	1	1	0	0	1	0	0
CZ	Call on zero	1	1	0	0	1	1	0	0
DAA	Decimal adjust A	0	0	1	0	0	1	1	1
DAD B	Add B & C to H & L	0	0	0	0	1	0	0	1
DAD D	Add D & E to H & L	0	0	0	1	1	0	0	1
DAD H	Add H & L to H & L	0	0	1	0	1	0	0	1
DAD SP	Add stack pointer to H & L	0	0	1	1	1	0	0	1
DCR M	Decrement memory	0	0	1	1	0	1	0	1
DCR r	Decrement register	0	0	D	D	D	1	0	1
DCX B	Decrement B & C	0	0	0	0	1	0	1	1
DCX D	Decrement D & E	0	0	0	1	1	0	1	1
DCX H	Decrement H & L	0	0	1	0	1	0	1	1
DCX SP	Decrement stack pointer	0	0	1	1	1	0	1	1
DI	Disable Interrupt	1	1	1	1	0	0	1	1
EI	Enable Interrupts	1	1	1	1	1	0	1	1
HLT	Halt	0	1	1	1	0	1	1	0
IN	Input	1	1	0	1	1	0	1	1
INR M	Increment memory	0	0	1	1	0	1	0	0
INR r	Increment register	0	0	D	D	D	1	0	0
INX B	Increment B & C registers	0	0	0	0	0	0	1	1
INX D	Increment D & E registers	0	0	0	1	0	0	1	1
INX H	Increment H & L registers	0	0	1	0	0	0	1	1
INX SP	Increment stack pointer	0	0	1	1	0	0	1	1
JC	Jump on carry	1	1	0	1	1	0	1	0
JM	Jump on minus	1	1	1	1	1	0	1	0

TABLE 12.2.2 (Continued)

Mnemonic	Description	Instruction code							
		D_7	D_6	D_5	D_4	D_3	D_2	D_1	D_0
JMP	Jump unconditional	1	1	0	0	0	0	1	1
JNC	Jump on no carry	1	1	0	1	0	0	1	0
JNZ	Jump on no zero	1	1	0	0	0	0	1	0
JP	Jump on positive	1	1	1	1	0	0	1	0
JPE	Jump on parity even	1	1	1	0	1	0	1	0
JPO	Jump on parity odd	1	1	1	0	0	0	1	0
JZ	Jump on zero	1	1	0	0	1	0	1	0
LDA	Load A direct	0	0	1	1	1	0	1	0
LDAX B	Load A indirect	0	0	0	0	1	0	1	0
LDAX D	Load A indirect	0	0	0	1	1	0	1	0
LHLD	Load H & L direct	0	0	1	0	1	0	1	0
LXI B	Load immediate register Pair B & C	0	0	0	0	0	0	0	1
LXI D	Load immediate register Pair D & E	0	0	0	1	0	0	0	1
LXI H	Load immediate register Pair H & L	0	0	1	0	0	0	0	1
LXI SP	Load immediate stack pointer	0	0	1	1	0	0	0	1
MVI M	Move immediate memory	0	0	1	1	0	1	1	0
MVI r	Move immediate register	0	0	D	D	D	1	1	0
MOV M, r	Move register to memory	0	1	1	1	0	S	S	S
MOV r, M	Move memory to register	0	1	D	D	D	1	1	0
MOV r1, r2	Move register to register	0	1	D	D	D	S	S	S
NOP	No operation	0	0	0	0	0	0	0	0
ORA M	Or memory with A	1	0	1	1	0	1	1	0
ORA r	Or register with A	1	0	1	1	0	S	S	S
ORI	Or immediate with A	1	1	1	1	0	1	1	0
OUT	Output	1	1	0	1	0	0	1	1
PCHL	H & L to program counter	1	1	1	0	1	0	0	1
POP B	Pop register pair B & C off stack	1	1	0	0	0	0	0	1
POP D	Pop register pair D & E off stack	1	1	0	1	0	0	0	1
POP H	Pop register pair H & L off stack	1	1	1	0	0	0	0	1
POP PSW	Pop A and Flags off stack	1	1	1	1	0	0	0	1
PUSH B	Push register Pair B & C on stack	1	1	0	0	0	1	0	1
PUSH D	Push register Pair D & E on stack	1	1	0	1	0	1	0	1
PUSH H	Push register Pair H & L on stack	1	1	1	0	0	1	0	1
PUSH PSW	Push A and Flags on stack	1	1	1	1	0	1	0	1
RAL	Rotate A left through carry	0	0	0	1	0	1	1	1
RAR	Rotate A right through carry	0	0	0	1	1	1	1	1
RC	Return on carry	1	1	0	1	1	0	0	0
RET	Return	1	1	0	0	1	0	0	1
RLC	Rotate A left	0	0	0	0	0	1	1	1
RM	Return on minus	1	1	1	1	1	0	0	0
RNC	Return on no carry	1	1	0	1	0	0	0	0
RNZ	Return on no zero	1	1	0	0	0	0	0	0
RP	Return on positive	1	1	1	1	0	0	0	0
RPE	Return on parity even	1	1	1	0	1	0	0	0
RPO	Return on parity odd	1	1	1	0	0	0	0	0
RRC	Rotate A right	0	0	0	0	1	1	1	1

TABLE 12.2.2   (Continued)

Mnemonic	Description	D_7	D_6	D_5	D_4	D_3	D_2	D_1	D_0
		Instruction code							
RST	Restart	1	1	A	A	A	1	1	1
RZ	Return on zero	1	1	0	0	1	0	0	0
SBB M	Subtract memory from A with borrow	1	0	0	1	1	1	1	0
SBB r	Subtract register from A with borrow	1	0	0	1	1	S	S	S
SBI	Subtract immediate from A with borrow	1	1	0	1	1	1	1	0
SHLD	Store H & L direct	0	0	1	0	0	0	1	0
SPHL	H & L to stack pointer	1	1	1	1	1	0	0	1
STA	Store A direct	0	0	1	1	0	0	1	0
STAX B	Store A indirect	0	0	0	0	0	0	1	0
STAX D	Store A indirect	0	0	0	1	0	0	1	0
STC	Set carry	0	0	1	1	0	1	1	1
SUB M	Subtract memory from A	1	0	0	1	0	1	1	0
SUB r	Subtract register from A	1	0	0	1	0	S	S	S
SUI	Subtract immediate from A	1	1	0	1	0	1	1	0
XCHG	Exchange D & E, H & L Registers	1	1	1	0	1	0	1	1
XRA M	Exclusive Or memory with A	1	0	1	0	1	1	1	0
XRA r	Exclusive Or register with A	1	0	1	0	1	S	S	S
XRI	Exclusive Or immediate with A	1	1	1	0	1	1	1	0
XTHL	Exchange top of stack, H & L	1	1	1	0	0	0	1	1

Notes: DDD or SSS—000 B—001 C—010 D—011 E—100 H—101 L—110 Memory—111 A.

In addition to the 8080 machine instructions, there are also pseudo-instructions recognized by the assembler. A pseudo-instruction is written in the same fashion as the machine instructions but does not cause any object code to be generated. It acts merely to provide the assembler with information to be used subsequently while generating object code. The general assembler language format of a pseudo-instruction is

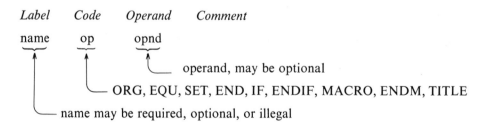

Label   *Code*   *Operand*   *Comment*

name   op   opnd

operand, may be optional

ORG, EQU, SET, END, IF, ENDIF, MACRO, ENDM, TITLE

name may be required, optional, or illegal

Names on pseudo-instructions are not followed by a colon, as are labels. *Names* are required in the label field of MACRO, EQU, and SET pseudo-instructions. The label fields of the remaining pseudo-instructions may contain optional labels, exactly like the labels on machine instructions. In this case, the label refers to the memory location immediately following the last previously assembled machine instruction. If present, names may be 1–5 characters long. The 8080 pseudo-instructions are shown in Table 12.2.3.

TABLE 12.2.3   8080 Pseudo-Instructions

Label	Code	Operand	Description
oplab:	ORG	exp A 16-bit address	The assembler's location counter is set to the value of exp, which must be a valid 16-bit memory address. The next machine intruction or data byte(s) generated will be assembled at address exp, exp + 1, etc.
name Required name	EQU	exp An expression	The symbol "name" is assigned the value of exp by the assembler. Whenever the symbol "name" is encountered subsequently in the assembly, this value will be used.
name Required name	SET	exp An expression	The symbol "name" is assigned the value of exp by the assembler. Whenever the symbol "name" is encountered subsequently in the assembly, this value will be used unless changed by another SET instruction.
oplab:	END	exp An expression	The END statement signifies to the assembler that the physical end of the program has been reached, and that generation of the object program and (possibly) listing of the source program should now begin.
oplab:	IF	exp An expression	The assembler evaluates exp. If exp evaluates to zero, the statements between IF and ENDIF are ignored. Otherwise, the intervening statements are assembled as if the IF and ENDIF were not present. IF-ENDIF pseudo-instructions can be nested to eight levels.
		Statements	
oplab:	ENDIF		
name Required name	MACRO	list A list of expressions	The assembler accepts the statements between MACRO and ENDM as the definition of the macro named "name." Upon encountering "name" in the code field of an instruction, the assembler substitutes the parameters specified in the operand field of the instruction for the occurrences of "list" in the macro definition, and assembles the statements.
		Statements	
oplab:	ENDM		
oplab	TITLE	string String of ASCII characters enclosed in quotation marks	The string of up to 66 characters specified in the TITLE pseudo-instruction is printed beneath the page header on all pages following the specification of the title until a new title is specified.

The following two examples are helpful for understanding the three types of instructions, 1-byte, 2-byte, and 3-byte; how to use Table 12.2.2 to find their instruction code in hexadecimal; and the execution sequence of a program.

Example 12.2.1

MOV B, A, ADI 7EH, JMP A69D are examples of one-, two, and three-byte instructions, respectively.

Mnemonic	Contents of memory in mnemonic	Contents of memory in machine code	Contents of memory in hexadecimal numbers
MOV B,A	MOV B,A	0 1 D D D S S S	
(One-byte instruction)		B = 000 ⇓ ⇓ A = 111	
		0 1 0 0 0 1 1 1	47
Data			
ADI 7EH	ADI	1 1 0 0 0 1 1 0	C6
(Two-byte instruction)	7E	0 1 1 1 1 1 1 0	7E
Address			
JMP A69DH	JMP	1 1 0 0 0 0 1 1	C3
(Three-byte instruction)	9D	1 0 0 1 1 1 0 1	9D
	A6	1 0 1 0 0 1 1 0	A6

Note that both address and data can be described in hexadecimal, binary, or decimal. They are indicated by a letter H, B, and blank, respectively. For example, the 7EH in the above instruction can be expressed by 01111110B (in binary) or 126 (in decimal).

Example 12.2.2

Each instruction is executed in a certain sequence. Since a program is made up of instructions, it must be executed in a sequence specified by the execution sequences of the individual instructions and the order of the instructions in the program. For example, consider the execution sequence of the following program:

Mnemonic	Comment
IN 1	Input a data from port 1 to the accumulator.
ADI 10	Add 10 (decimal 10) to the number in the accumulator.
OUT 2	Output the number in the accumulator to port 2.

The execution sequence of each individual instruction is

Instruction	Step number
IN	1A, 1B, 1C, 1D
ADI	2A, 2B, 2C
OUT	3A, 3B, 3C, 3D

which are shown in Fig. 12.2.4. Therefore, the execution sequence of this program is 1A, 1B, ..., 3D.

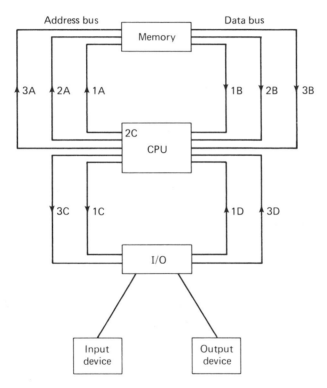

Fig. 12.2.4 An example of illustrating the execution sequence of a program.

The following are several examples of 8080 assembly language programs.

Example 12.2.3

Write a program to do the following simple task.

"Get a data from, say, port 1 and store it in memory, and then get a second data from port 1 and store it in memory, and so on. Repeat this process 10 times and then stop."

A flow chart describing this problem and an 8080 assembly language program which implements it are shown below.

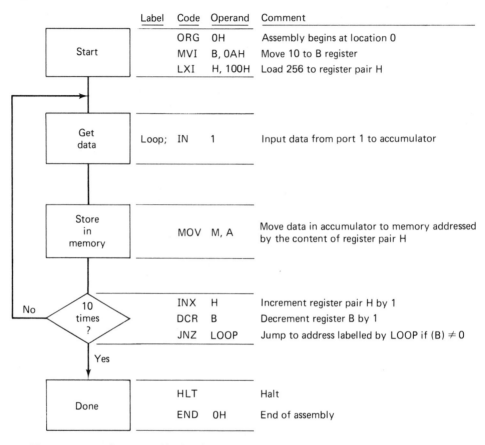

Label	Code	Operand	Comment
	ORG	0H	Assembly begins at location 0
	MVI	B, 0AH	Move 10 to B register
	LXI	H, 100H	Load 256 to register pair H
Loop;	IN	1	Input data from port 1 to accumulator
	MOV	M, A	Move data in accumulator to memory addressed by the content of register pair H
	INX	H	Increment register pair H by 1
	DCR	B	Decrement register B by 1
	JNZ	LOOP	Jump to address labelled by LOOP if (B) ≠ 0
	HLT		Halt
	END	0H	End of assembly

The program, after assembly by the 8080 assembler, in hexadecimal looks like this:

Program in

ADRS	HEX	MNEMONIC
00	06	MVI
	0A	
02	21	LXI
	00	
	01	
05	DB	IN
	03	
07	77	MOV
08	23	INX
09	05	DCR
0A	C2	JNZ
	05	
	00	
0D	76	HLT

Memory   Label = 0050

Example 12.2.4

Suppose that the problem of Example 12.2.3 is modified slightly to the one described by the flow chart below. The program and its assembled data are shown on the right.

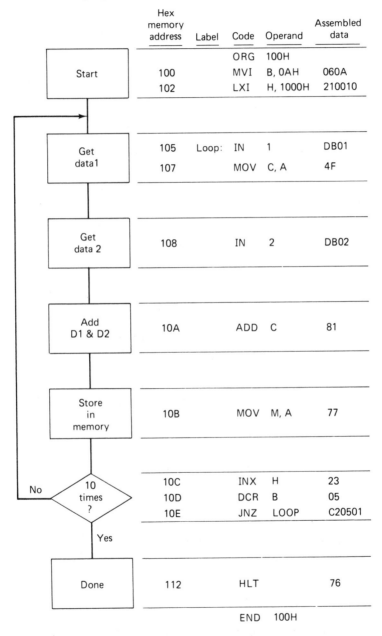

Hex memory address	Label	Code	Operand	Assembled data
		ORG	100H	
100		MVI	B, 0AH	060A
102		LXI	H, 1000H	210010
105	Loop:	IN	1	DB01
107		MOV	C, A	4F
108		IN	2	DB02
10A		ADD	C	81
10B		MOV	M, A	77
10C		INX	H	23
10D		DCR	B	05
10E		JNZ	LOOP	C20501
112		HLT		76
		END	100H	

A third example is a further extension of the problem in Example 12.2.4, in that the program will involve two loops.

Example 12.2.5

In the problem of Example 12.2.4, suppose that after having received data D1 and D2 from input ports 1 and 2, we want to compare them first. Add D1 and D2 if they are not equal. When they are equal, get another D2 from input port 2 and compare it with D1 again, and so on. The remaining part is unchanged.

A flow chart describing the modified version of the problem of Example 12.2.4 and a program that implements it are given on page 481.

Exercise 12.2

Write an 8080A assembly language program for performing each of the following operations.

1. Transfer a block of 10 pieces of data from one memory location to another.

2. Sort 10 integers in ascending order.

3. Add 10 binary integers stored in 20 consecutive memory locations and store the result in a memory location.

4. Multiply two unsigned binary integers and store the result in a memory location.

5. Add two signed numbers in 2's-complement form and store the result in a memory location.

12.3 Microcomputer Simulation of Digital Design

In this section, the simulation of a digital design using an 8080 microcomputer is presented. A digital design simulation program is usually written in the 8080 assembly language. After it is assembled, the machine codes generated are programmed in an erasable programmable read-only memory (EPROM) by a PROM programmer. This microprogrammed EPROM is then plugged into a PROM socket of a microcomputer system design kit (SDK) such as Intel SDK-80 (see Fig. 12.3.1). Data input from I/O devices and generated from the program reside in the RAM part of the memory.

A. Simulation of General System Designs

The technique described above may be applied to any system design problems. For example, the system design of Example 12.2.3 may be accomplished by using an SDK-80 system with a PROM containing the machine language program presented in Example 12.2.3. When we decide to change or modify the design, such as to those as described in Examples 12.2.4 and 12.2.5, *all we need to do is to change or modify the program.*

B. Simulation of Digital Circuits

One of the most widely used applications of the microcomputer system design kit is the replacement of hardwired logic. Since the 8080 machine instruction set includes

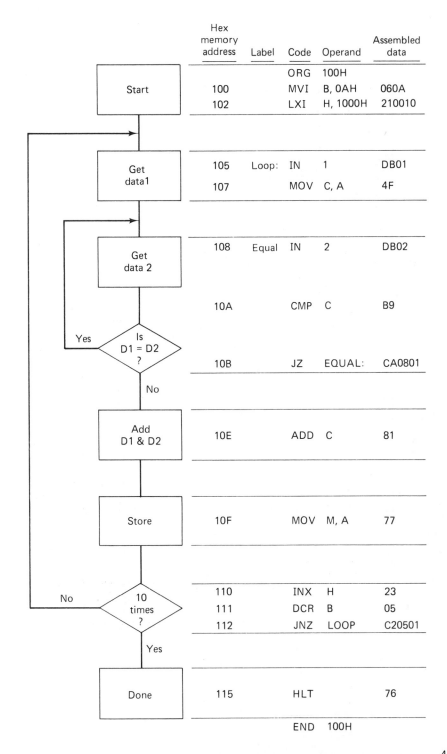

Hex memory address	Label	Code	Operand	Assembled data
		ORG	100H	
100		MVI	B, 0AH	060A
102		LXI	H, 1000H	210010
105	Loop:	IN	1	DB01
107		MOV	C, A	4F
108	Equal	IN	2	DB02
10A		CMP	C	B9
10B		JZ	EQUAL:	CA0801
10E		ADD	C	81
10F		MOV	M, A	77
110		INX	H	23
111		DCR	B	05
112		JNZ	LOOP	C20501
115		HLT		76
		END	100H	

Flowchart labels: Start, Get data1, Get data 2, Is D1 = D2 ? (Yes / No), Add D1 & D2, Store, 10 times ? (No / Yes), Done

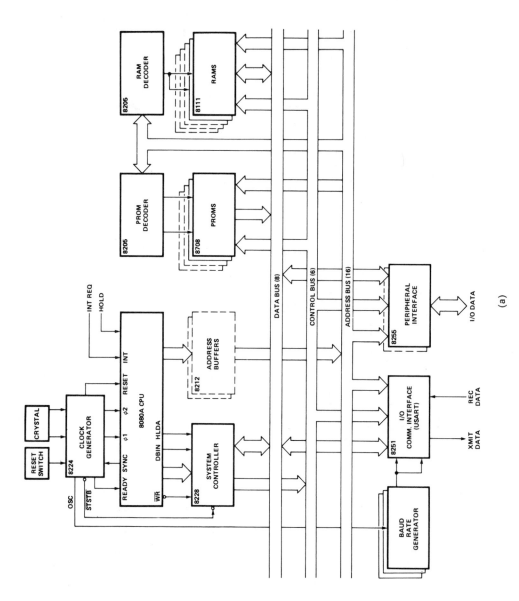

Fig. 12.3.1 (a) SDK-80 functional block diagram; (b) assembled board.

Fig. 12.3.1 (Continued)

483

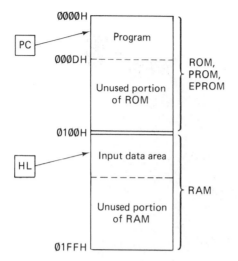

Fig. 12.3.2 Memory map of the program of Example 12.2.3.

all the basic logical operations (AND, OR, NAND, NOR, and XOR), an 8080 micro-computer can simulate any logical gate. For example, an AND-gate can be simulated by a program as follows:

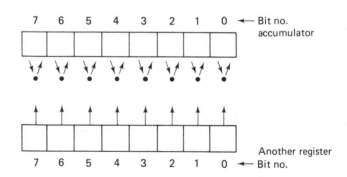

```
LDA   SRCA    :Load first set of inputs from SCRA
MOV   BA      :Save in the B register
LDA   SRCB    :Load second set of inputs from SRCB
ANA   B       :And B with A
STA   DST     :Save result in DST
```

Others can be simulated in a similar way.

As described in Section 9.2, a flip-flop can be described by a characteristic function. For example, the characteristic function of the J-K flip-flop, which completely describes the operation (the transition table) of the flip-flop, is

$$Y = Jy' + K'y$$

This function can be simulated by ANDing J and y' (obtaining it from complementing y), and K' and y, then ORing them. The delay between y and Y may be provided in several ways. One way is to use the NOP instruction in a loop. The timing of the delay is set by the time required for executing an NOP operation multiplied by the number of times that the NOP operation has been executed.

The following are several examples of digital design using the microcomputer simulation technique. An Intel 8080 microcomputer system (such as MDS-800) is assumed to be used.

Example 12.3.1

Software implementation of the combinational circuit of Fig. 12.3.3.

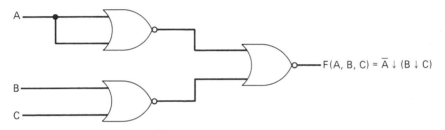

$$F(A, B, C) = \bar{A} \downarrow (B \downarrow C)$$

Fig. 12.3.3

There are several ways to simulate it. For example, it can be simulated by writing an assembly language program to implement (1) the function $F(A, B, C) = \bar{A} \downarrow (B \downarrow C)$, or (2) its simplified form $F(A, B, C) = AB + AC$, or (3) the truth table of the function. Among them, the program of implementing $AB + AC$ appears to be the simplest which is presented below.

Computer Program

Address in hex	Hex code	Label	Assembly	Program	Comment
3000			ORG	3000H	;THIS IS A PROGRAM TO SIMULATE THE CIRCUIT ;OF FIG 12.3.3. ;SEND INPUTS THROUGH THE CONSOLE.THE ;OUTPUT WOULD BE DISPLAYED.
F803		CI	EQU	0F803H	
F809		CO	EQU	0F809H	
3000	CD03F8		CALL	CI	;GET A. ASCII CODES OF 0 AND 1 ARE 30H AND ;31H
3003	D630		SUI	30H	;30H−30H=0 AND 31H−30H=1
3005	320131		STA	3101H	;STORE A IN MEMORY LOCATION 3101H
3008	CD03F8		CALL	CI	;GET B
300B	D630		SUI	30H	;30H−30H=0 AND 31H−30H=1
300D	320231		STA	3102H	;STORE B IN MEMORY LOCATION 3102H
3010	CD03F8		CALL	CI	;GET C

Address in hex	Hex code	Label	Assembly	Program	Comment
3013	D630		SUI	30H	;30H−30H=0 AND 31H−30H=1
3015	320331		STA	3103H	;STORE C IN MEMORY LOCATION 3103H
3018	3A0131		LDA	3101H	;LOAD A TO ACC. A
301B	47		MOV	B,A	;REG. B HAS VALUE OF A
301C	3A0231		LDA	3102H	;LOAD B TO ACC. A
301F	4F		MOV	C,A	;REG. C HAS VALUE OF B
3020	3A0331		LDA	3103H	;LOAD C TO ACC. A
3023	5F		MOV	E,A	;REG. E HAS VALUE OF C
3024	1600		MVI	D,0	;SET REG. D=0
3026	78		MOV	A,B	
3027	A1		ANA	C	
3028	4F		MOV	C,A	;REG. C HAS THE VALUE A·B
3029	78		MOV	A,B	
302A	A3		ANA	E	;ACC. A HAS THE VALUE A·C
302B	B1		ORA	C	;ACC. A HAS THE VALUE A·B+A·C
302C	C630		ADI	30H	
302E	4F		MOV	C,A	
302F	CD09F8		CALL	CO	;OUTPUT THE RESULT
3000			END	3000H	

Note: The instructions SUI 30H in this program may be omitted.

Example 12.3.2

Software implementation of the binary serial adder.

Again, there are several ways to implement this device. One way is to write a program to implement its transition table which was given in Table 9.3.2(a). The implementation of this device through the simulation of the sequential circuit of Fig. 9.2.1 is left to the reader as an exercise (problem 5).

Computer Program

Address in hex	Hex code	Label	Assembly	Program	Comment
3000			ORG	3000H	;THIS PROGRAM IS A SOFTWARE SIMULATION ;OF BINARY SERIAL ADDER.
F803		CI	EQU	0F803H	
F809		CO	EQU	0F809H	
3000	CD03F8		CALL	CI	;GET X1
3003	E67F		ANI	7FH	
3005	D630		SUI	30H	
3007	320135		STA	3501H	;STORE X1
300A	CD03F8		CALL	CI	;GET X2
300D	E67F		ANI	7FH	
300F	D630		SUI	30H	

Address in hex	Hex code	Label	Assembly	Program	Comment
3011	320235		STA	3502H	;STORE X2
3014	CD03F8		CALL	CI	;GET Y
3017	E67F		ANI	7FH	
3019	D630		SUI	30H	
301B	320335		STA	3503H	;STORE Y
					;CALCULATION OF SUM
301E	210135		LXI	H,3501H	
3021	46		MOV	B,M	;GET X1
3022	23		INX	H	
3023	7E		MOV	A,M	;GET X2 IN TO ACC.
3024	A8		XRA	B	
3025	23		INX	H	
3026	46		MOV	B,M	
3027	A8		XRA	B	;GET RESULT IN A ACC
3028	320435		STA	3504H	;STORE SUM
302B	210135		LXI	H,3501H	;LOAD H WITH 3501
302E	7E		MOV	A,M	;GET X1
302F	A0		ANA	B	:COMPUTE X1,Y,RESULT IN ACC
3030	320A35		STA	350AH	
3033	23		INX	H	
3034	7E		MOV	A,M	;GET X2 IN ACC
3035	A0		ANA	B	;COMPUTE X2,Y
3036	47		MOV	B,A	;B HAS X2,Y
3037	3A0A35		LDA	350AH	;A HAS X1,Y
303A	B0		ORA	B	;A HAS X1,Y OR X2,Y
303B	47		MOV	B,A	;B HAS X1,Y OR X2,Y
303C	7E		MOV	A,M	;GET X2
303D	2B		DCX	H	
303E	4E		MOV	C,M	;GET X1,H=3501
303F	A1		ANA	C	;AND X1,X2
3040	B0		ORA	B	;A HAS X1,X2 OR X1,Y OR X2,Y
3041	320535		STA	3505H	
3044	0630		MVI	B,30H	
3046	79		MOV	A,C	
3047	80		ADD	B	;ASCII EQU OF X1 IS 4
3048	320135		STA	3501H	
304B	23		INX	H	
304C	7E		MOV	A,M	;GET X2
304D	80		ADD	B	
304E	320235		STA	3502H	;STORE X2 IN 3502
3051	23		INX	H	
3052	7E		MOV	A,M	;GET Y
3053	80		ADD	B	
3054	320335		STA	3503H	
3057	23		INX	H	
3058	7E		MOV	A,M	
3059	80		ADD	B	
305A	320435		STA	3504H	

Address in hex	Hex code	Label	Assembly	Program	Comment
305D	23		INX	H	
305E	7E		MOV	A,M	
305F	80		ADD	B	
3060	320535		STA	3505H	
3063	210135		LXI	H,3501H	;PUT X1 IN H
3066	4E		MOV	C,M	;PUT X1 IN C
3067	CD09F8		CALL	CO	;PRINT X1
306A	23		INX	H	
306B	4E		MOV	C,M	;GET X2
306C	CD09F8		CALL	CO	;PRINT X2
306F	23		INX	H	
3070	4E		MOV	C,M	;GET Y
3071	CD09F8		CALL	CO	;PRINT Y
3074	23		INX	H	
3075	4E		MOV	C,M	;GET SUM
3076	CD09F8		CALL	CO	;PRINT SUM
3079	23		INX	H	
307A	4E		MOV	C,M	;GET CARRY
307B	CD09F8		CALL	CO	;PRINT CARRY
3000			END	3000H	

Example 12.3.3

Software implementation of the eight-bit odd-parity checker.

The block diagram of the eight-bit odd-parity checker is shown in Fig. 12.3.4(a) and the odd-parity eight-bit BCD codes are given in Fig. 12.3.4(b).

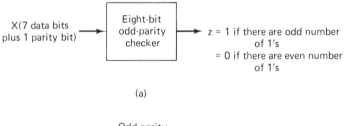

X(7 data bits plus 1 parity bit) → Eight-bit odd-parity checker → $z = 1$ if there are odd number of 1's
= 0 if there are even number of 1's

(a)

	Odd parity	
Parity bit	BCD	Final representation
1	0000000	10000000
0	0000001	00000001
0	0000010	00000010
1	0000011	10000011
⋮	⋮	⋮
0	1111111	01111111

(b)

Fig. 12.3.4 Example 12.3.3.

Computer Program

Address in hex	Hex code	Label	Assembly	Program	Comment
3000			ORG	3000H	;8-BIT ODD-PARITY CHECKER PROGRAM.
F803		CI	EQU	0F803H	
F809		CO	EQU	0F809H	
3000	210835		LXI	H,3508H	
3003	CD03F8	LP1 :	CALL	CI	
3006	E67F		ANI	7FH	
3008	D630		SUI	30H	
300A	77		MOV	M,A	
300B	28		DCX	H	
300C	7D		MOV	A,L	
300D	D600		SUI	00H	
300F	CA1530		JZ	LP2	
3012	C30330		JMP	LP1	
3015	210835	LP2 :	LXI	H,3508H	
3018	7E		MOV	A,M	
3019	28	LP3 :	DCX	H	
301A	46		MOV	B,M	
301B	A8		XRA	B	
301C	47		MOV	B,A	
301D	7D		MOV	A,L	
301E	D601		SUI	01H	
3020	CA2730		JZ	LP4	
3023	78		MOV	A,8	
3024	C31930		JMP	LP3	
3027	78	LP4 :	MOV	A,B	
3028	320035		STA	3500H	
302B	210835		LXI	H,3508H	
302E	7E	LP5 :	MOV	A,M	
302F	C630		ADI	30H	
3031	4F		MOV	C,A	
3032	CD09F8		CALL	CO	
3035	7D		MOV	A,L	
3036	D601		SUI	01H	
3038	CA3F30		JZ	LP6	
303B	2B		DCX	H	
303C	C32E30		JMP	LP5	
303F	0E20	LP6 :	MVI	C,20H	
3041	CD09F8		CALL	CO	
3044	CD09F8		CALL	CO	
3047	CD09F8		CALL	CO	
304A	3A0035		LDA	3500H	
304D	C630		ADI	30H	
304F	4F		MOV	C,A	
3050	CD09F8		CALL	CO	
3053	0E0D		MVI	C,0DH	
3055	CD09F8		CALL	CO	
3058	0E0A		MVI	C,0AH	
305A	CD09F8		CALL	CO	
305D	76		HLT		
3000			END	3000H	

Example 12.3.4

Software implementation of SN 74H87 true-complement zero-one circuit whose input/output relation is described in Fig. 12.3.5.

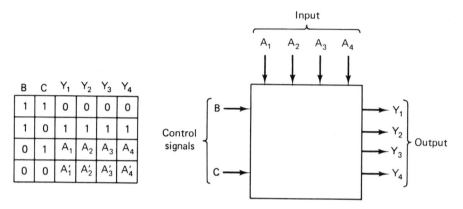

B	C	Y_1	Y_2	Y_3	Y_4
1	1	0	0	0	0
1	0	1	1	1	1
0	1	A_1	A_2	A_3	A_4
0	0	A_1'	A_2'	A_3'	A_4'

Fig. 12.3.5 Example 12.3.4.

Computer Program

Address in hex	Hex code	Label	Assembly	Program	Comment
3000			ORG	3000H	;THIS PROGRAM IS A SOFTWARE
					;IMPLEMENTATION OF SN 74H87 OF
					;FIG. 12.3.5.
					;DATA INPUTS MUST BE PUNCHED FIRST AND
					;FOLLOWED BY CONTROL INPUTS.
F803		CI	EQU	0F803H	
F809		CO	EQU	0F809H	
3000	CD03F8		CALL	CI	;GET A1
3003	E67F		ANI	7FH	;STRIP PARITY BIT OFF
3005	320035		STA	3500H	
3008	CD03F8		CALL	CI	;GET A2
300B	E67F		ANI	7FH	
300D	320135		STA	3501H	
3010	CD03F8		CALL	CI	;GET A3
3013	E67F		ANI	7FH	
3015	320235		STA	3502H	
3018	CD03F8		CALL	CI	;GET A4
301B	E67F		ANI	7FH	
301D	320335		STA	3503H	
3020	CD03F8		CALL	CI	;GET B
3023	E67F		ANI	7FH	
3025	320435		STA	3504H	

Address in hex	Hex code	Label	Assembly	Program	Comment
3028	CD03F8		CALL	CI	;GET C
302B	E67F		ANI	7FH	
302D	320535		STA	3505H	
3030	210435		LXI	H,3504H	
3033	7E		MOV	A,M	;GET B
3034	FE30		CPI	30H	;IS IT 0 ?
3036	CA4630		JZ	LP1	;YES
3039	23		INX	H	;NO
303A	7E		MOV	A,M	;GET C
303B	FE30		CPI	30H	;IS IT 0 ?
303D	CA4330		JZ	LP2	;YES B=1,C=0
3040	C3A630		JMP	LP7	
3043	C39530	LP2:	JMP	LP6	
3046	23	LP1:	INX	H	
3047	7E		MOV	A,M	;B=0
3048	FE30		CPI	30H	;IS C=0 ?
304A	CA5030		JZ	LP3	;YES
304D	C38F30		JMP	LP5	;NO
3050	C35330	LP3:	JMP	LP4	
3053	210035	LP4:	LXI	H,3500H	;OUTPUT IS THE COMPLEMENT OF THE INPUT.
3056	7E		MOV	A,M	
3057	FE30		CPI	30H	
3059	CA6030		JZ	LP8	
305C	3D		DCR	A	
305D	C36130		JMP	LP12	
3060	3C	LP8:	INR	A	
3061	77	LP12:	MOV	M,A	
3062	23		INX	H	
3063	7E		MOV	A,M	
3064	FE30		CPI	30H	
3066	CA6D30		JZ	LP9	
3069	3D		DCR	A	
306A	C36E30		JMP	LP13	
306D	3C	LP9:	INR	A	
306E	77	LP13:	MOV	M,A	
306F	23		INX	H	
3070	7E		MOV	A,M	
3071	FE30		CPI	30H	
3073	CA7A30		JZ	LP10	
3076	3D		DCR	A	
3077	C37B30		JMP	LP14	
307A	3C	LP10:	INR	A	
307B	77	LP14:	MOV	M,A	
307C	23		INX	H	
307D	7E		MOV	A,M	
307E	FE30		CPI	30H	
3080	CA8730		JZ	LP11	
3083	3D		DCR	A	

Address in hex	Hex code	Label	Assembly	Program	Comment
3084	C38830		JMP	LP15	
3087	3C	LP11:	INR	A	
3088	77	LP15:	MOV	M,A	
3089	CDB730		CALL	PRT	
308C	C3CE30		JMP	DONE	
308F	CDB730	LP5:	CALL	PRT	;OUTPUT IS EQUAL TO THE INPUT
3092	C3CE30		JMP	DONE	
3095	0E31	LP6:	MVI	C,31H	;OUTPUT 1
3097	CD09F8		CALL	CO	
309A	CD09F8		CALL	CO	
309D	CD09F8		CALL	CO	
30A0	CD09F8		CALL	CO	
30A3	C3CE30		JMP	DONE	
30A6	0E30	LP7:	MVI	C,30H	;OUTPUT 0
30A8	CD09F8		CALL	CO	
30AB	CD09F8		CALL	CO	
30AE	CD09F8		CALL	CO	
30B1	CD09F8		CALL	CO	
30B4	C3CE30		JMP	DONE	
30B7	210035	PRT:	LXI	H,3500H	
30BA	4E		MOV	C,M	
30BB	CD09F8		CALL	CO	
30BE	23		INX	H	
30BF	4E		MOV	C,M	
30C0	CD09F8		CALL	CO	
30C3	23		INX	H	
30C4	4E		MOV	C,M	
30C5	CD09F8		CALL	CO	
30C8	23		INX	H	
30C9	4E		MOV	C,M	
30CA	CD09F8		CALL	CO	
30CD	C9		RET		
30CE	0E0A	DONE:	MVI	C,0AH	
30D0	CD09F8		CALL	CO	
30D3	0E0D		MVI	C,0DH	
30D5	CD09F8		CALL	CO	
3000			END	3000H	

From the discussion above we conclude that any digital circuits, combinational as well as sequential, can be conveniently obtained by a microcomputer system design kit (such as SDK-80) and a microprogrammed PROM with a simulation program.

Exercise 12.3

Use the microcomputer simulation technique described in this section to implement each of the following digital designs. An Intel 8080 microcomputer system is assumed to be used.

1. Simulate the following gates:
   (a) OR
   (b) NAND
   (c) NOR
   (d) XOR (EXCLUSIVE-OR)

2. Simulate the following flip-flops:
   (a) D flip-flop
   (b) T flip-flop
   (c) S-R flip-flop
   (d) J-K flip-flop

3. Simulate the following TTL/SSI IC chips.
   (a) SN 7404 hex inverters.
   (b) SN 7408 quad, two-input AND.
   (c) SN 7411 triple, three-input AND.
   (d) SN 7474 dual, D-type positive edge-triggered flip-flops with preset and clear.
   (e) SN 7476 master–slave J-K flip-flops.

4. Simulate the following TTL/MSI IC chips.
   (a) SN 74185A 6-bit-binary to 6-bit-BCD converter.
   (b) SN 74154 4-line-to-16-line decoder/demultiplexer.
   (c) SN 7483 4-bit full adder.
   (d) SN 74193 up-down binary counter.
   (e) SN 7495A 4-bit parallel-in/parallel-out shift register.
   (f) SN 7485 4-bit magnitude comparator.

5. Simulate the binary serial adder circuit of Fig. 9.2.1.

6. Design the counter of Example 11.3.4.

7. Design the parity checker described in problem 10 of Exercise 11.3.

8. Design the automatic toll collector of problem 11 of Exercise 11.3.

9. Design a binary multiplier.

10. Design a digital clock.

Bibliographical Remarks

The material presented in this chapter is new and can be found in the following references.

References

1. CUSHMAN, R. H., "Intel 8080: The First of the Second-Generation Microprocessors," *EDN*, Vol. 19, No. 9, 1974, pp. 30–36.

2. *Intel 8080 Microcomputer System User's Manual*, Intel Corporation, 1975.

3. *Intel 8080 Assembly Language Programming Manual*, Intel Corporation, 1975.

4. *Intel MDS-800 Operator's Manual*, Intel Corporation, 1975.

5. *Intel MDS-DOS Operator's Manual*, Intel Corporation, 1975.

6. *Intel MCS-80 System Design Kit User's Guide*, Intel Corporation, 1975.

7. *The Bugbook*, I, II, III, IV, and V, E & L Instruments, Inc.

8. OSBORNE, A., *An Introduction to Microcomputers*, Vol. I, Basic Concepts, Adam Osborne and Associates, 1976.

9. OSBORNE, A., *8080 Programming for Logic Design*, Adam Osborne and Associates, 1976.

Index

A

Activity vector, 222
Adaptive experiment, 387
Antisymmetric function, 136
a-number, 124
Application table, 353, 354
Arithmetic device, 409
Assembly language, 464
Asynchronous sequential circuits, 365
Atoms, 26, 66
Automaton, 236

B

Boolean:
 algebra, 24
 derivative, 74
 difference, 74
 differential calculus, 73
 functions, 35
 lattice, 19
 multiple partial derivative, 76
 partial derivative, 74
 product, 49
 sum, 49

C

Canonical forms, 35, 52, 62, 63, 162
 product-of-sums, 41, 52, 162
 sum-of-products, 41, 52, 162
Characteristic function, 345
Closed fault, 196

Comparator, 409
Compatible input sequence, 255
Compatible states, 255
Complementary dual, 142
Completely monotonic function, 121
Complete test set, 195
Constant function, 35
Convergence point, 196
Counters, 410

D

Dagger function, 141
DALG-II, 226, 229
D-algorithm, 222
Data selectors, 407
D-calculus, 210
D-chain, 226
D-cube, 222
D-drive, 222
Decoders, 409
Demultiplexers, 409
Derivative, 92
 multiple total, 92
 total, 92
Differential, 89
 multiple partial, 90
 partial, 89
 total, 91
 total, *p*-, 97
 total, *q*-, 97
D-intersection, 219
Distinguishability equivalence, 251

Distinguishing sequence, 391
Distinguishing tree, 391
Distributive inequality, 13
Dominant faults, 196
Dual of a function, 142

E

Empty sequence, 300
Encoders, 408
Equivalent sequential machines, 251
Equivalent states, 251
Equivalent switching functions, 36
Excitation functions, 349
Excitation table, 349
Expansion theorem, 50

F

Fault detection experiment, 386
Fault set graph, 197
Flip-flops, 341
 D, 341
 J-K, 346
 S-R, 344
 T, 346
Functionally complete functions, 137
Fuzzy:
 algebra, 159
 function, 160
 logic, 157

G

Generalized:
 complement, 58
 DeMorgan's theorem, 61
 expansion theorem, 64
 Shannon's theorem, 62
Generators, 249
 basic, 249
 unassigned, 249

H

Hazards, 373
 dynamic, 377
 essential, 375
 static, 374
Homing sequence, 386
Homing tree, 387

I

In-circuit emulator (ICE), 462
Inclusion diagram, 4
Input-restrictive sequential machine, 256
Integrated circuits (IC's):
 CMOS, 407
 CTL, 407
 DCTL, 406
 DTL, 407
 ECL, 407
 I^2L, 407
 LSI, 405
 MOS, 407
 MSI, 405
 RCTL, 406
 RTL, 406
 SSI, 405
 TTL, 407

J

J-K flip-flop, 346

K

Karnaugh maps, 97, 113, 131
k-monotonic function, 121

L

Lattice, 1, 11, 247, 275, 281
 Boolean, 17
 complemented, 20
 distributive, 19
 modular, 17
Literal, 40
Literal proposition, 206

M

Machines:
 finite-state, 236
 Mealy, 238, 313, 431
 Moore, 237, 313, 431
MacLaurin series, 85, 86
m-class logic, 157
m-class logic function, 162
Memories, 410
Microcomputers, 453
Microprocessors, 453

Mixed-symmetric function, 152
Modular inequality, 13
Monotonically decreasing function, 105, 185
Monotonically increasing function, 105, 185
Monotonic function, 104
Monotonic two-level circuits, 184
Multiple faults, 199
Multiplexers, 407
Multivalued logic, 147

N

n-monotonic functions, 121
Nonregular sets, 300

O

One-bias function, 185
Output consistent:
  S.P. lattice, 253
  S.P. partition, 252

P

Parallel decomposition, 289
Partition:
  block of, 242
  O.C.S.P., 252
  S.P., 242
Pierce's arrow, 141
Preset experiment, 388
Prime singular cube, 213
Primitive D-cube of failure, 213
Primitive flow table, 378
Programmable logic array (PLA), 405
Projection function, 35
Propagation D-cube, 214
Pseudo-instruction, 475

Q

Quine-McCluskey method, 113

R

Rabin-Scott nondeterministic machine, 313
Races:
  critical, 368
  noncritical, 368
Random access memory (RAM), 410, 428

Read only memory (ROM), 405
Reduction factor, 155
Regular expression, 299, 301
  reverse of, 303
Regular sets, 299
Rotation operation, 58

S

Self-dual, 144
Sensitivity function, 92
  total, 93
Sensitizing fault, 196
Sensitizing line value, 195
Sequential circuits, 339
  asynchronous, 339
  clocked, 339
  fundamental-mode, 365
  pulse-mode, 360
  synchronous, 339
  unclocked, 339
Sequential machines:
  completely specified, 252
  diagnosable, 395
  incompletely specified, 254
  strongly-connected, 395
Serial decomposition, 284
Sets:
  ordered, 1
  partially ordered, 2
  power, 3
  totally ordered, 2
  well-ordered, 8
Shannon's theorem, 49
Sheffer's stroke, 141
Shift register, 410
Singular cover, 212
Singular cube, 213
Spanning fault, 195
Special switching functions:
  monotonic, 108
  threshold, 116
  symmetric, 123
  unate, 110
S.P. lattice, 281
S.P. partition, 241, 242, 281
State assignment, 262, 369
State minimization, 251
Sublattice, 22
Subset construction, 316

Symmetric function, 123, 149
  a-number, 123
  characteristic set *A*, 124
  mixed, 152
  partially, 123
  totally, 123
Synchronizing sequence, 388
Synchronizing tree, 388

T

Tape:
  recognized by a machine, 299
  reverse, 303
Test cube, 222
TEST-DETECT, 226, 229
Threshold function, 116
Total derivative, 92
Total differential, 91
Total sensitivity function, 93
Total variation, 93

Truth table, 36
T set, 195

U

Unate function, 110, 118, 119
Undetectable fault set, 197

V

Vector switching algebra, 59
Vector switching function, 61

W

Weight of a function, 138
Weights, 116

Z

Zero-bias function, 185